An ATLAS of the SMALLER MAPS in ORIENTABLE and NONORIENTABLE SURFACES

The CRC Press Series on

DISCRETE MATHEMATICS AND ITS APPLICATIONS

Series Editor

Kenneth H. Rosen, Ph.D.

AT&T Bell Laboratories

The CRC Handbook of Combinatorial Designs, *Charles J. Colbourn and Jeffrey H. Dinitz*

Frames and Resolvable Designs: Uses, Constructions, and Existence, *Steven Furino, Ying Miao, and Jianxing Yin*

Handbook of Discrete and Computational Geometry, *Jacob E. Goodman and Joseph O'Rourke*

Design Theory, *Charles C. Lindner and Christopher A. Rodgers*

Network Reliability: Experiments with a Symbolic Algebra Environment, *Daryl D. Harms, Miroslav Kraetzl, Charles J. Colbourn, and John S. Devitt*

Handbook of Applied Cryptography, *Alfred J. Menezes, Paul C. van Oorschot, and Scott A. Vanstone*

Quadratics, *Richard A. Mollin,*

Fundamental Number Theory with Applications, *Richard A. Mollin*

Cryptography: Theory and Practice, *Douglas R. Stinson*

Graph Theory and Its Applications, *Jonathan Gross and Jay Yellen*

Combinatorial Algorithms:Generation Enumeration, and Search, *Donald L. Kreher and Douglas R. Stinson*

Introduction to Information Theory and Data Compression, *Darrel R. Hankerson Greg A. Harris, and Peter D. Johnson*

Handbook of Discrete and Combinatorial Mathematics, *Kenneth H. Rosen*

Abstract Algebra Applications with Maple, *Richard Klima, Ernest Stitzinger, and Neil Sigmon*

An Atlas of The Smaller Maps in Orientable and Nonorientable Surfaces, *David Jackson and Terry Visentin*

Algebraic Number Theory, *Richard A. Mollin*

An ATLAS of the SMALLER MAPS in ORIENTABLE and NONORIENTABLE SURFACES

DAVID M. JACKSON

Department of Combinatorics & Optimization
University of Waterloo
Ontario, Canada

TERRY I. VISENTIN

Department of Mathematics
University of Winnepeg
Winnepeg, Canada

CHAPMAN & HALL/CRC

Boca Raton London New York Washington, D.C.

Library of Congress Cataloging-in-Publication Data

Jackson, D.M.
An atlas of the smaller maps in orientable and nonorientable surfaces / David M. Jackson and Terry I. Visentin.
p. cm. -- (CRC Press series on discrete mathematics and its applications)
Includes bibliographical references and index.
ISBN 1-58488-207-7 (alk. paper)
1. Mappings (Mathematics) 2. Surfaces. I. Visentin, Terry I. II. Title. III. Series

QA360 .J33 2000
511.3′3--dc21 00-059007

Direct all inquiries to CRC Press LLC, 2000 N.W. Corporate Blvd., Boca Raton, Florida 33431.

International Standard Book Number 1-58488-207-7
Library of Congress Card Number 00-059007
Printed in the United States of America 1 2 3 4 5 6 7 8 9 0
Printed on acid-free paper

Preface

The enumerative theory of maps impinges on several areas of mathematics and mathematical physics and is the subject of extensive research. For example, the generating series for monopoles (maps with exactly one vertex) in orientable surfaces can be used to obtain the virtual Euler characteristic for the moduli spaces of complex algebraic curves. Monopoles in locally orientable surfaces are used for the case of real algebraic curves. The partition function for the ϕ^4-model of two dimensional quantum gravity is, before the double scaling limit is taken, the generating series for quadrangulations (maps whose faces are each bounded by four edges). The number of pairs of vertex partitions and face partitions for which there is a prescribed number of maps arises in the theory of "dessins d'enfants," involving the action of the absolute Galois group (the automorphism group of the field of algebraic numbers) on maps. Moreover, certain combinatorial and algebraic aspects of the theory are used to determine the number of ramified coverings of the Riemann sphere by a surface of given genus with prescribed ramification points of given ramification orders (a classical problem in complex analysis); to determine the degree of the Lyashko-Looijenga map (a question in singularity theory); to determine the number of characteristic classes of certain types of bundles over corresponding moduli spaces (a question in enumerative algebraic geometry); to determine symmetric functions whose connexion coefficients are the same as those for the class algebra of the symmetric group (a question in symmetric function theory).

The most familiar maps are probably the triangulations (maps whose faces are bounded by three edges) of the sphere. Historically, considerable attention has been devoted to the enumeration of classes of rooted maps in the sphere (or, equivalently, the plane), originally as part of an enumerative attack on what was, at the time, the Four Colour Conjecture.

Two conjectures were crucial to the inception of the Atlas. The first, the *Quadrangulation Conjecture*, involves the construction of a natural bijection between rooted maps and rooted quadrangulations. The second, the *b-Conjecture*, involves the determination of a new invariant of rooted maps. There is evidence that the invariant is associated with nonorientability. The first of these conjectures caused us to realize that such a compendium would be useful, and the second spurred us to bring it to the present state. Accounts of these conjectures are given in Section 4.3 and Section 4.4, partly as a working example of the use of the Atlas and partly because the conjectures are of considerable interest in their own right.

We hope that readers will find the Atlas of value in their research, and that there will be some who are engaged by the two perplexing conjectures.

We benefitted in several ways from a number of people. Daniel Brown provided us with drawings of monopoles in locally orientable surfaces on up to 4 edges. These provided additional checks on our work and helped us to produce drawings of monopoles in polygonal representations of nonorientable surfaces of genus 3 and 4. John Stembridge made available to us his accelerated Maple procedure for computing Jack symmetric functions and his collection of Jack symmetric functions for partitions of up to 16. We thank Gilberto Bini and John Harer for many discussions about the b-Conjecture and its significance to the moduli spaces of real and complex curves, and Claude Itzykson and Malcolm Perry for several long conversations about the Quadrangulation Conjecture and maps in general, and for explaining to us their significance in mathematical physics. Aleksander Zvonkin gave us valuable comments about the Absolute Galois group and its relationship to certain maps. This prompted us to include the material on nonrealizable partitions. The value of having an exhaustive collection of the smaller maps became clear during work on the enumerative theory of maps, much of it with Ian Goulden, and DMJ would like to thank him for continuing productive research collaboration.

Waterloo and Winnipeg
September 2000

D.M. Jackson and T.I. Visentin

Contents

Part I

Maps

Chapter 1

Introduction

This Atlas gives a complete listing of maps and hypermaps with a small number of edges for both orientable and nonorientable surfaces, and the numbers of their rootings. It is intended as an aid to research both in the study of maps and in their application to other areas of mathematics, where there appears to be a marked need for an exhaustive list of maps of manageable size.

We decided to include all maps on up to 5 edges in orientable surfaces and 4 edges in nonorientable surfaces. These limits on the number of edges were selected for three reasons. The first reason was a practical one: maps with more edges are too numerous to include in a manageable way. The second reason was a perceptual one: maps with more edges are difficult to visualize in polygonal representations of surfaces of type higher than two, and their inclusion, even if feasible given their number, does not promote the purpose behind the compilation of this Atlas. The third reason is a mathematical one: maps satisfying the above conditions are sufficiently complex combinatorially to exhibit a good range of combinatorial properties for testing purposes; in other words, a conjecture that holds for each of the maps in the Atlas stands a fair chance of being true in general.

The material is organized into three parts. Part I contains an introduction to maps and the use of the Atlas and an account of two substantial conjectures about maps; Part II, the Atlas itself, consists of drawings of maps in polygonal representations of surfaces and the set of all embeddings of their associated graphs; Part III contains tables of numbers of maps and tables of two important polynomials, namely, the genus distribution and the b-polynomial.

The purpose of the introductory material of Part I is to explain the construction of the Atlas, to give an introduction to maps and their axiomatization, and to provide some combinatorial information about two substantial conjectures about maps, as well as to draw attention to some of the areas in which maps arise in an essential way.

Some familiarity with maps is assumed, and a number of technical terms have been used for expository purposes in the opening sections. Definitions of these terms, however, have been included for completeness, but appear where they are first used in a technical way.

1.1 Organization of the Atlas

The Atlas consists of collections of maps, map numbers and map polynomials. The format of each of these is described in the introduction to the appropriate collection so that the information is readily available when the collection is used.

1.1.1 Summary of the content of the Atlas

The type of information that is explicitly provided in the Atlas is given below under the separate headings.

Background for maps (Chapter 2)

The principal definitions for maps are given in Chapter 2, with a few examples illustrating the use of the Atlas.

Axiomatization of maps (Chapter 3)

The axiomatizations of maps in both orientable and nonorientable surfaces are given in Chapter 3 in terms of permutations that encode the local structure at each vertex of a map.

Generating series (Chapter 4)

Chapter 4 gives the background material on symmetric functions that is required for the formulation of the genus series for maps and, implicitly, hypermaps.

Conjectures (Chapter 4)

Detailed accounts of the Quadrangulation Conjecture and the b-Conjecture are given in Section 4.3 and Section 4.4, together with the background on Jack symmetric functions. In addition, examples are given illustrating each of the conjectures.

The main collections of maps (Chapters 5 and 6)

The main collections of maps are given in Chapter 5 and Chapter 6 for maps in orientable and nonorientable surfaces, respectively. In the case of orientable surfaces, all maps on up to 5 edges are given and in the case of nonorientable surfaces, all maps on up to 4 edges are given. The maps are indexed by genus, number of edges and number of vertices. Each entry gives the vertex partition, face partition and number of rootings. The individual roots themselves have not been explicitly specified in the interests of keeping the size of the Atlas within manageable bounds.

Subcollections of maps (Chapter 7)

Triangulations and quadrangulations, which are of independent interest, are given in the same form in Section 7.1 and Section 7.2. Hypermaps are given in Section 7.3 in a similar format that has been adapted to show symmetry.

Associated graphs (Chapter 8)

There are 142 nonisomorphic graphs associated with the maps that appear in the main collection. These comprise all connected graphs (with loops and multiple edges allowed) on at most 5 edges. These graphs are listed in Chapter 8 according to number of edges and number of vertices, and, with each graph, the list of its embeddings is given in increasing order of genus. The genus distribution is given for each graph at the end of the list.

Numbers of maps (Chapters 9 and 10)

Section 9.1 and Section 9.2 give the numbers of rooted maps in orientable and nonorientable surfaces, respectively, for given vertex partitions and face partitions, and are organized by genus and number of edges. This information is summarized in Section 9.3 where the numbers of maps are listed for a given number of edges, and also for a given number of edges and vertices. Among these tables are some that include counts of maps which have not been drawn.

Section 10.1 and Section 10.2 give the numbers of unrooted maps in orientable and nonorientable surfaces, respectively, for given vertex partitions and face partitions, and are organized by genus and number of edges.

Nonrealizable pairs of partitions (Chapter 11)

The pairs of partitions that satisfy the Euler-Poincaré Formula, but for which there is no map with these as their vertex and face partitions, are given in Chapter 11. They are listed separately for orientable and nonorientable surfaces.

Map polynomials (Chapter 12)

The b-polynomials are given in Section 12.1 and are indexed by type and number of edges. The genus distributions for graphs in orientable surfaces are given in Section 12.2. This summarizes the information given in Chapter 8.

1.1.2 Preparation of the tables

The initial terms in the genus series $M(\mathbf{x}, \mathbf{y}, z, b)$ and $M(x, y, z, b)$ for rooted maps in orientable ($b = 0$) surfaces and locally orientable ($b = 1$) surfaces were computed using the symbolic computer algebra system Maple and the symmetric function package [31] (called SF). The tables of coefficients were typeset directly from Maple output, without manual intervention, to eliminate the possibility of transcription errors. The set of drawings of maps were prepared manually from lists of permutations that satisfy the axiomatizations for the surfaces. The numbers of maps within each class were checked against the tables. The genus distribution and the lists of k-realizable pairs of proper partitions were computed directly from the computer representation of the collections of maps, and were then typeset directly without manual intervention.

1.2 Further reading

Although there is presently no monograph on the enumerative theory of maps in arbitrary surfaces and the application of the mathematical background of this theory to other fields, there are sufficiently many research papers where information on this subject may be found. It has not been our intention that the introductory material of the first four chapters should serve more than its deliberately circumscribed role of providing a useful background to the construction of the Atlas and the two conjectures. However, the applications of the theory that already have been mentioned to major areas of mathematics are interesting in themselves, so the larger role has been served in a small way by the inclusion of a few references to these applications in the bibliography, but with no claims to exhaustivity in this respect.

The bibliography contains primary, secondary and tertiary sources. The primary sources contain the proofs of results that are stated in Part I. The expressions for the genus series and the character theoretic material that are needed for this purpose are instances of such results. In the secondary sources the principal emphasis is the use of enumerative arguments to address questions that arise in other major areas of mathematics, and for which techniques associated with maps play a prominent role. These sources have been selected because they provide access for combinatorialists first to the mainly enumerative aspects of these areas and then to the questions in which these aspects arise. The use of polygonal identification with the technical details of the enumerative arguments for certain moduli space calculations is an instance of this. The tertiary sources relate to current work in these major areas where there may be enumerative arguments which, while being important to such areas, are not the main concern. These references are included for the reader who wishes to have instances of enumerative arguments in other areas of mathematics and to understand precisely how it is that questions in these areas are combinatorialized. The combinatorialization of ramified coverings of the sphere is an example of such a question. Conspicuous by its absence from Part I is any mention of the use of integral representations for the genus series for maps and the relationship to matrix models. These representations are important, but they were not used in the construction of the Atlas, and are therefore omitted. There are cases in which such integrals, which are integrals over Hermitian complex and real symmetric matrices with respect to a Gaussian measure, can be explicitly evaluated in an appropriate ring. A few references to matrix models and instances of the evaluation of such matrices have been included as tertiary sources.

No attempt has been made to be exhaustive in any one of these types of sources. However, the references will provide the interested reader with a starting point of access to several areas where substantial use is made of maps and the associated enumerative techniques.

For the axiomatizations for maps in orientable and locally orientable surfaces see Tutte [32]; for the axiomatization of hypermaps see Walsh [37]; for aspects of embedding theory see Gross and Tucker [15]; for the enumeration of several classes of maps in the sphere see Tutte [33, 34, 35, 36]; for enumerative techniques in combinatorial theory see Goulden and Jackson [11]; for a development of the theory of symmetric functions see Macdonald [24]; for an exposition of the properties of Jack symmetric functions also see Stanley [30] and Knop and Sahi [21]; for the connexion of Jack symmetric functions to statistical mechanics see Lesage, Pasquier and Serban [22]; for a proof of the functional relationship between the genus series for rooted quadrangulations and all rooted maps, a factorization of matrix integrals and a brief account of the connexion to 2-dimensional quantum gravity see Jackson, Perry and

Visentin [18]; for a proof of the extension of this to Eulerian maps and hypermaps see Jackson and Visentin [19]; for the development of the b-Conjecture see Goulden and Jackson [9]; for further work on the Quadrangulation Conjecture and the b-Conjecture see Brown [5]; for the use of monopoles in the determination of the virtual Euler characteristic for the moduli spaces of complex curves see Harer and Zagier [16]; for an account of the connexion of the b-Conjecture to moduli spaces see Goulden, Harer and Jackson [10]; for some enumerative questions associated with moduli spaces see [7]; for other enumerative questions in algebraic geometry see Ernstrom and Kennedy [8], Looijenga [23], Khovanskii and Zdravkovska [20] and Caporaso and Harris [6]; for an account of the theory of "dessins d'enfants" see [27]; for an account of drawing curves over number fields see Shabat and Voevodsky [28]; for connexions between plane trees and algebraic numbers see Shabat and Zvonkin [29] and Adrianov and Shabat [1]; for Belyi's Theorem see [3]; for an account of field theoretic approaches to the enumeration of maps see 't Hooft [13], Bessis, Itzykson and Zuber [4] and Itzykson and Drouffe [17]; for techniques for matrix integrals see Mehta [25]; for a combinatorial construction for ramified coverings of Riemann surfaces see Hurwitz [14]; for combinatorial aspects of singularity theory see Arnold [2]; for an account of random surfaces and their applications see [26]; for some combinatorial aspects of random matrices see Hanlon, Stanley and Stembridge [12].

Chapter 2

Surfaces and maps

The technical details for the construction of the Atlas have been kept to a minimum, although it has been necessary to include the axiomatizations for maps in orientable and locally orientable surfaces and, without proof, an explicit form for the generating series for maps in locally orientable surfaces.

The specific expressions for the generating series that are given are included because they enabled us to confirm, by explicit computation, that the correct number of maps have been constructed in each class. They are also required in the description of the two conjectures. Other explicit series, for certain subclasses of maps, are also known, but were not essential to the construction of the Atlas and so have not be given. The axiomatizations of maps in both orientable and locally orientable surfaces have been included since these were essential, first for generating the encodings for maps themselves, and then for testing that the maps constructed in each class were inequivalent.

2.1 Representation of maps and surfaces

2.1.1 Principal definitions

The principal definitions that are needed in the use of this Atlas are collected in this section and are followed in Section 2.2 by examples that give instances of these definitions, together with further commentary.

Surfaces and orientation

Throughout the Atlas a *surface* is assumed to be a 2-manifold without boundary. It is *orientable* if the clockwise and anticlockwise senses around each point can be designated consistently; otherwise the surface is *nonorientable*. A sense may be selected for the orientable surface, without loss of generality, to produce an *oriented* surface (that is, a surface with a specified orientation). Throughout, it is therefore assumed that orientable surfaces are to have the *anticlockwise* sense as orientation. The term "orientable surface" will therefore be used to mean "oriented surface with anticlockwise orientation."

Maps and equivalence of maps

A *map* is a 2-cell embedding of a graph, with loops and multiple edges allowed, in a surface which may be orientable or nonorientable: this is to say, if the edges of a map are deleted, then the surface decomposes into regions homeomorphic to open discs. The open discs correspond to the *faces* of the map. Two maps are *equivalent* if there is a diffeomorphism between the surfaces that maps one map onto the other.

The map given below, for example, is a 2-cell embedding in the sphere since the deletion of its edges decomposes the sphere into three discs. However, if the same map (the cyclic order of incidence of edges with the vertex must therefore be unchanged for it to be the same map) is embedded in the torus, on deletion of the edges the torus decomposes into either two discs and a punctured torus, or one disc and a cylinder, so the embedding is not a 2-cell embedding.

A *map invariant* is a property of a map that is unchanged by any diffeomorphism of the surface. To show that two maps $\mathfrak{m}$ and $\mathfrak{m}'$ are inequivalent, it is therefore necessary to find a map invariant ξ such that $\xi(\mathfrak{m}) \neq \xi(\mathfrak{m}')$ under the action of some diffeomorphism of the surface. On the other hand, to show that $\mathfrak{m}$ and $\mathfrak{m}'$ are equivalent it is necessary to exhibit a diffeomorphism.

Combinatorially, a map is uniquely defined by the cyclic orders in which edges are incident with vertices. This will be explained in detail in the discussion of the combinatorial axiomatization of maps in locally orientable surfaces given in Chapter 3.

This axiomatization greatly assists the recognition of the inequivalence and equivalence of maps, since it eliminates the necessity of establishing the nonexistence or existence of a diffeomorphism.

Genus and type

Orientable surfaces are classified by their number of handles, and this number is called the *genus* of the surface. There is a unique orientable surface of each genus. The orientable surfaces in increasing order of genus (in brackets) are: the *sphere* (0), the *torus* (1), the *double torus* (2), ... The nonorientable surfaces are classified by their number of crosscaps and for these surfaces it is this number which is called the genus of the surface. The term genus will therefore only be used in contexts where the orientability of the surface has been specified. There is a unique nonorientable surface of each positive genus. The nonorientable surfaces in increasing order of genus (in brackets) are: the *projective plane* (1), the *Klein bottle* (2), the *crosscapped torus* (3), the *doubly crosscapped torus* (4) ....

For sets which contain maps on both orientable and nonorientable surfaces, it is necessary to use *type* rather than genus. The type of a map $\mathfrak{m}$ is given by the *Euler-Poincaré Formula*

$$\#\text{vertices}(\mathfrak{m}) - \#\text{edges}(\mathfrak{m}) + \#\text{faces}(\mathfrak{m}) = 2 - 2\,\text{type}(\mathfrak{m}).$$

For orientable surfaces, genus and type are the same. However, for nonorientable surfaces, the type is half the genus. Thus the type of the sphere is 0, the type of the projective plane is $1/2$, the torus and the Klein bottle have the same type, which is equal to 1, the type of the crosscapped torus is $3/2$ and the type of the double torus and the doubly crosscapped torus is 2. It follows that if the type is a half-integer, then the surface is nonorientable, but if the type is a positive integer, then the orientability of the surface cannot be determined from type alone. The term *locally orientable* refers collectively to orientable and nonorientable surfaces.

2.1.2 Polygonal representation of orientable surfaces

An orientable surface can be represented by a regular polygon with sides directed in a particular way, under the convention that pairs of identically labelled edges, which are always oppositely directed with reference to a cyclic tour of the polygon (as in the figures below), are identified as directed edges. Under this identification of edges the *corner points* in the regular polygon are identified as one. The polygon is called a *polygonal representation* of the surface. In the commentary it is necessary to distinguish between the edges of the polygonal boundary, the edges of the map and the sides of edges of the map. In view of this, the edges in the polygonal representation are called *lines*.

Although there are other polygonal representations, it is convenient to use one in which it is the pairs of opposite lines that are labelled. The polygonal representations of the orientable surfaces which appear in the Atlas are labelled according to this convention and are given below.

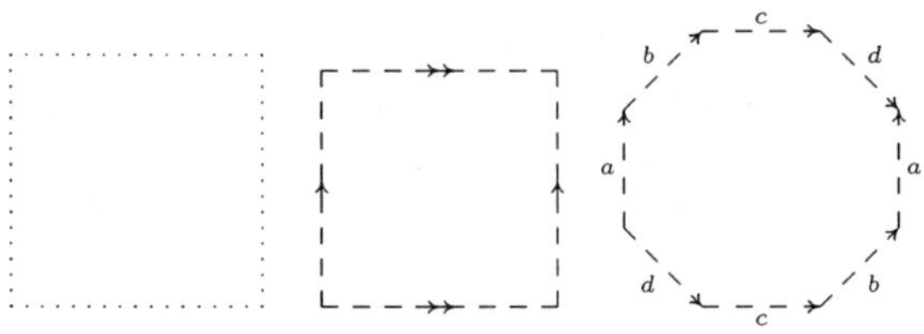

sphere torus double torus

The labels on lines are retained, although technically this convention makes them redundant, as a convenience in the case of the double torus. Because of the conditions placed on the Atlas, these are the only orientable surfaces that arise for the maps that have been drawn.

2.1.3 Polygonal representation of nonorientable surfaces

A nonorientable surface can also be represented by a regular polygon with sides directed in a particular way, under the convention that identically labelled edges are identified as directed edges. In the particular way of directing edges there is to be at least one pair of identically labelled edges that are similarly directed with reference to a cyclic tour of the polygon. The corner points of the regular polygon are identified under this construction. The polygonal representations of the nonorientable surfaces which appear in the Atlas are given below.

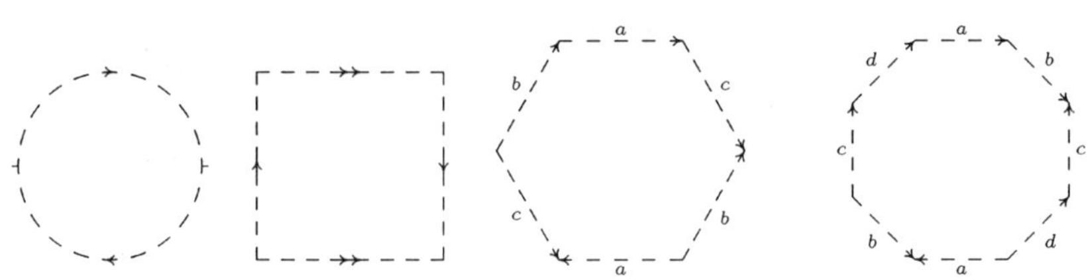

projective plane Klein bottle crosscapped torus doubly crosscapped torus

As in the case of the orientable surfaces, the edges in the polygonal boundary are called *lines*. The convention is again adopted that it is the opposite lines that are identically labelled. Because of the conditions placed on the Atlas, these are the only nonorientable surfaces that that have been drawn. Although this convention makes the labels on the lines redundant, they are retained as a convenience in the case of the crosscapped torus and the doubly crosscapped torus.

In a nonorientable surface there are two cyclic senses at each point and these are distinguishable. However, neither of the senses can be identified specifically as anti-clockwise or clockwise.

Intersections of lines of a polygon with edges of a map

The effect of an edge of a map meeting a (directed) line in the polygonal boundary is determined by considering a dotted path drawn parallel to the edge at distance ε (throughout ε is small) from it to distinguish a side of the edge. The reappearance of the edge and the dotted path at the corresponding points on the opposite (directed) line in the polygonal boundary must be consistent with the identification of these directed lines. There are two cases:

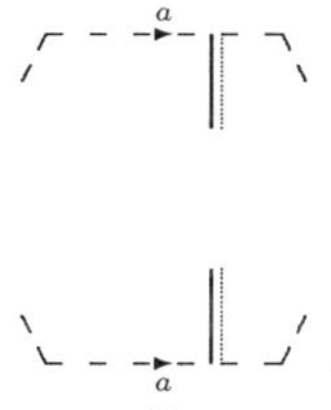

(i)

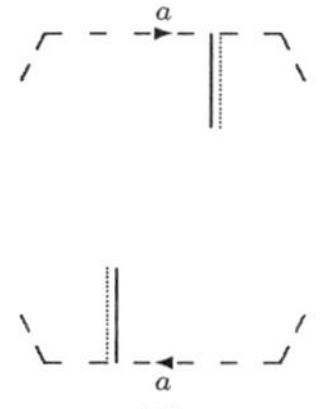

(ii)

For orientable surfaces, of course, only case (i) occurs. In case (ii), there is a "twist" in the edge indicating that the edge has passed through a *crosscap*. This can occur only for nonorientable surfaces.

Occasionally it is convenient to allow an edge to pass through the identified corner points. In the nonorientable case some care must be exercised in checking the sidedness of such edges. The following two diagrams indicate the trajectory of a path at distance ε from one side of such an edge.

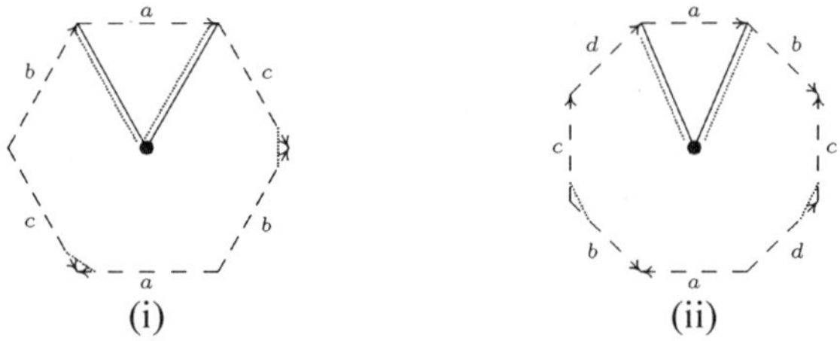

It should be noticed that there are segments of the path which appear in other corners of the polygonal representation (for instance, where the heads of b and c meet in the first diagram). It is clear by tracing such paths that the edge in case (i) is twisted while the edge in case (ii) is not.

2.1.4 Rooting, associated graph and dual

Vertex and face partitions

The *vertex partition* of a map is a partition of $2n$ that lists, in its parts, the degrees of each vertex, where the degree of a vertex is the number of edges incident with it and n is the number of edges. The vertex partition of a map $\mathfrak{m}$ is denoted by $\nu(\mathfrak{m})$. The notation $\theta \vdash m$ indicates that θ is a partition of m. The convention adopted for partitions is that the parts appear in nonincreasing order. The number of parts in θ is denoted by $l(\theta)$ and the sum of the parts is denoted by $|\theta|$. Thus $l(\nu(\mathfrak{m}))$ is the number of vertices of $\mathfrak{m}$ and $|\nu|$ is the sum of the vertex degrees, and therefore twice the number of edges.

Similarly, the *face partition* $\phi(\mathfrak{m})$ of a map $\mathfrak{m}$ is a partition that lists, in its parts, the degrees of each face, where the degree of a face is the number of edges in its boundary.

Both the vertex and face partitions are, of course, map invariants and, by the Euler-Poincaré Formula, they satisfy

$$l(\nu(\mathfrak{m})) - n + l(\phi(\mathfrak{m})) = 2 - 2t,$$

where $|\nu(\mathfrak{m})| = |\phi(\mathfrak{m})| = 2n$ and t is the type of $\mathfrak{m}$.

The root of a map

The enumeration of maps is assisted by removing the symmetries of maps. This is done by *rooting* the maps. A *rooted map* is a map which is obtained by distinguishing an edge, a direction along this edge, and a side of the edge. The distinguished edge is called the *root edge*, the origin of this edge is called the *root vertex*, and the face on the distinguished side of the root edge is called the *root face*. For maps in orientable surfaces, it suffices to select a directed edge as root in orientable surfaces. One side of an edge is identifiable, through the orientation of the surface, as the right-hand side and the other as the left-hand side. The rooting is indicated by placing an arrow on the root edge which directs it away from the root vertex, and throughout this Atlas the face on the right-hand side of this directed edge is selected to be the root face.

For maps in nonorientable surfaces, the sides of an edge cannot be identified specifically as the right-hand side or the left-hand side, so the root is indicated by placing an arrow parallel to the root edge, but at a small distance from it on the distinguished side.

Two rooted maps are *equivalent* if there is a diffeomorphism between the surfaces that sends one map onto the other, and that maps the root of one onto the root of the other. A rooted map therefore has only the trivial automorphism. The number of inequivalent ways of rooting a map will simply be called the *number of rootings* of the map and this number divides twice the number of edges for maps in orientable surfaces, and divides four times the number of edges for maps in nonorientable surfaces.

The associated graph and the genus distribution

A graph may have many embeddings in many surfaces. For each such map, the graph is called its *associated graph*, and it is uniquely defined. The graphs that arise as associated graphs of maps in the Atlas are denoted by $\mathfrak{g}_i$, where i is a serial number that is not related to the map in an essential way.

Without some familiarity with the polygonal representations of the torus and higher surfaces, it may be difficult to visualize a map concretely. The associated graph can be useful in this respect since it immediately provides the incidence structure of the map.

Let $\mathfrak{g}$ be a graph with loops and multiple edges allowed. Let $d_{\mathfrak{g}}(g)$ be the number of maps (not rooted), in an orientable surface of genus g, with $\mathfrak{g}$ as their associated graph. The polynomial

$$\sum_{g\geq 0} d_{\mathfrak{g}}(g)u^g$$

in an indeterminate u is called the *genus distribution* for $\mathfrak{g}$.

The dual of a map

Let $\mathfrak{m}$ be a map in a locally orientable surface. The *dual* of $\mathfrak{m}$ is the unique map $\overline{\mathfrak{m}}$ whose vertex set is the set of faces of $\mathfrak{m}$ and whose edge set is the multiset of vertex pairs $\{u, v\}$ of $\overline{\mathfrak{m}}$, one for each edge common to the boundaries of the faces of $\mathfrak{m}$ that correspond to u and v. This determines a unique cyclic order for the edges incident with each vertex of $\overline{\mathfrak{m}}$.

This construction can be realized geometrically as follows. Put a vertex v_f in the interior of each face f of $\mathfrak{m}$. These vertices form the vertex set of $\overline{\mathfrak{m}}$. For each face f of $\mathfrak{m}$, join the midpoint of the edges in the boundary of f to v_f in such a way that these lines do not meet except at v_f. Each edge of $\overline{\mathfrak{m}}$ is formed as the concatenation $v_\alpha P v_\beta$ of two lines $v_\alpha P$ and $P v_\beta$, where α and β are faces of $\mathfrak{m}$ with a common edge in their boundaries, with midpoint P.

Dualization is an involution on the set of all maps, and a map that is equivalent to its dual is said to be *self-dual.* Dualization can be extended to an involution for the set of all rooted maps by the provision of a convention for the root induced in the dual by the root of the preimage.

Hypermaps and their rooting

A *hypermap* is a map (2-face colourable) whose faces may be coloured with two colours (*black* and *white*, for typographic convenience) such that edges separate only faces of different colours. The convention is imposed that the black faces (shown as shaded in the figures) are called *hyperfaces* and the white faces are called *hyperedges.* The *hyperface partition* lists the degrees of the hyperfaces in nonincreasing order and the *hyperedge partition* lists the degrees of the hyperedges in nonincreasing order. A *symmetric* hypermap is a map that is invariant under interchange of the two colours. If a hypermap $\mathfrak{h}$ is a 2-face-colouring of a map $\mathfrak{m}$, then the number of rootings of $\mathfrak{h}$ equals the number of rootings of $\mathfrak{m}$ if $\mathfrak{h}$ is a symmetric hypermap, and is half the number of rootings of $\mathfrak{m}$ otherwise.

Specialization to maps is achieved by requiring the hyperedges to be digons (faces of degree two), and by then identifying the edges of each digon. The hyperfaces of the hypermaps then become the faces of the map that is constructed in this way, and the hyperedges become its edges. This construction is called (hyperedge) *digon conflation.* A *rooted hypermap* is a 2-face-coloured map with the convention adopted throughout that the root face is to be a hyperface. Under digon conflation, the root hyperface of the hypermap becomes the root face of the corresponding map.

The generating series for hypermaps can be expressed in a very compact form which can be specialized to obtain the generating series for maps on locally orientable surfaces. More details about this are given in Chapter 4. Moreover, hypermaps appear prominently in the examination of one of the conjectures.

Nonrealizable pairs of proper partitions

Suppose that α and β are partitions of $2n$ for which there is some nonnegative integer g such that the Euler-Poincaré condition $l(\alpha) - n + l(\beta) = 2 - 2g$ holds. Such a pair of partitions is called *proper*. A pair (α, β) of partitions for which there is no map with α as vertex partition and β as face partition is called *nonrealizable*, and is called *realizable* if there is such a pair of proper partitions. Clearly, by duality, (α, β) is realizable if and only if (β, α) is realizable.

If (α, β) is a pair of proper partitions that is realizable by exactly k maps then the pair is said to be a *k-realizable* pair of proper partitions. Thus, for example, nonrealizable pairs are 0-realizable.

2.1.5 The entry for a map in the Atlas

The entries that occur in this Atlas are drawings of (unrooted) maps in polygonal representations of surfaces. Since it is occasionally necessary to distinguish between a map and its drawing on a particular choice of polygonal representation of a surface, the entries will be referred to as *map diagrams* (or *diagrams* when the word "map" is understood from the context). Each map is assigned an *index number* of the form $a \cdot b$, in which a is g or $\tilde{g}$, g is the genus and the $\tilde{}$ indicates that the surface is nonorientable: b is a serial number and it is not related to the map in an essential way. The map whose index number in the Atlas is $a \cdot b$ is denoted throughout by $\mathfrak{m}_{a \cdot b}$, and the index number appears in the top left-hand corner of the entry for the map. The vertex partition appears at the bottom left-hand corner of each map, the face partition appears at the bottom right-hand corner, and the number of inequivalent ways in which a map can be rooted is given in the top right-hand corner of the diagram.

Maps that are also hypermaps are listed in Section 7.3 with a similar format. The bottom right-hand corner in this case gives two partitions, namely, the hyperedge partition and hyperface partition, in some order. For a symmetric hypermap these are the same and the number of rootings is given in the top right-hand corner. For hypermaps that are not symmetric, it should be understood that either of the two partitions given in the bottom right-hand corner can be the hyperedge partition (the other being the hyperface partition), and the number of rootings for each possibility is given in the top right-hand corner. These two numbers are necessarily equal. Thus a symmetric hypermap is recognized by the presence of only one number in the top right-hand corner of an entry, whereas the presence of two numbers in the top right-hand corner indicates that the hypermap is not symmetric.

Chapter 8 contains a reindexing of the collection of maps in orientable surfaces by their associated graph, so that the list of all embeddings of this graph can be seen at a glance.

2.2 Examples of the definitions

Various aspects of these definitions are illustrated below, with a number of additional observations.

Polygonal representation of surfaces

Consider the following map diagrams:

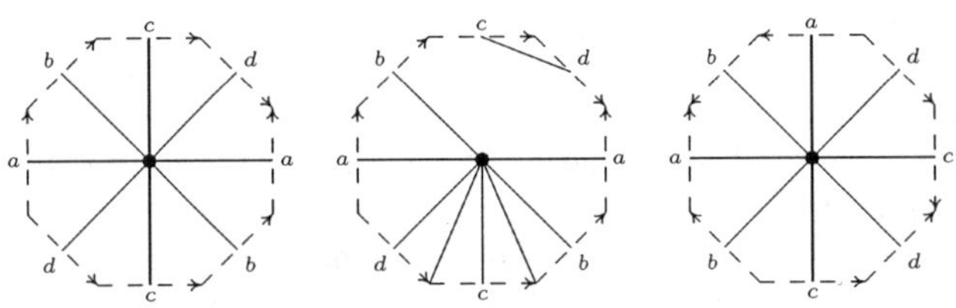

(i) (ii) (iii)

Diagrams (i) and (ii) show maps $\mathfrak{m}_{2\cdot 1}$ and $\mathfrak{m}_{2\cdot 2}$, respectively, as they appear in the Atlas. Diagram (iii) gives a drawing of $\mathfrak{m}_{2\cdot 2}$ in a different polygonal representation of the double torus in which it is no longer the opposite lines that are identified. To confirm that (ii) and (iii) do indeed give the same map, it is necessary to visualize how the half-edges match up at the intersections with lines to form edges of the map, and then to compare the cyclic order of edges around the vertex in each diagram.

While (ii) may appear to be an awkward drawing of $\mathfrak{m}_{2\cdot 2}$ compared with (iii), it is important to use a uniform polygonal representation of the double torus throughout the Atlas to avoid mistaking (i) and (iii), for example, as drawings of the same map. The map $\mathfrak{m}_{2\cdot 1}$ is correspondingly awkward to draw in the polygonal representation used in (iii). In short, no single choice of polygonal representation permits every map to be drawn in the visually most appealing way, and the benefit of adopting a single polygonal representation for a surface is seen in the ease with which maps may be compared. This is true for nonorientable surfaces as well.

Inequivalence of maps

The maps

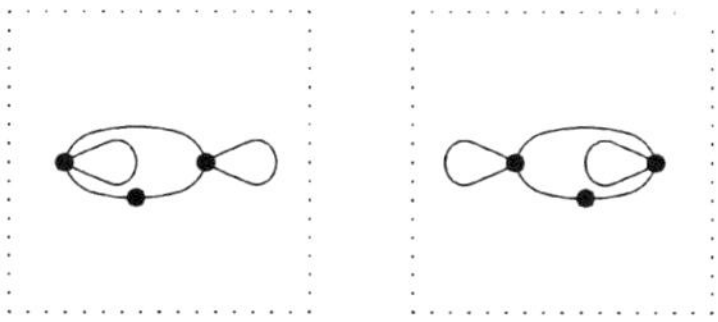

$\mathfrak{a}$ $\mathfrak{b}$

in the sphere, which appear as $\mathfrak{m}_{0\cdot 223}$ and $\mathfrak{m}_{0\cdot 224}$ in the Atlas, are inequivalent. To establish this, it is necessary to specify a map invariant in which the two differ. Since the vertex partition is the same for each map, as is the face partition, neither of these invariants is refined enough to distinguish the maps. A more refined invariant can be found by considering the cyclic sequences of vertices encountered in tours of boundaries of faces. These are called *face sequences*.

Consider one of the faces of degree 4 in $\mathfrak{a}$ and make a tour of this face by tracing a path within distance ε of its boundary in a direction consistent with the orientation of the surface (in this case, the sphere). Let a denote the vertex incident with the loop that is part of the boundary, let b denote the bivalent vertex and let c denote the remaining vertex. The cyclic order of vertices encountered in the tour is abca. Now consider each of the faces of degree 4 in $\mathfrak{b}$ in turn. For each there is only one assignment of the symbols $\mathsf{a}, \mathsf{b}, \mathsf{c}$ to the vertices according to the rule given above. The cyclic order of vertices in the tours of these faces is acba in both cases. The two face sequences abca and acba are different, so since face sequences are invariant under diffeomorphisms of the surface, the maps are not equivalent.

Vertex and face partition, and genus

If a map $\mathfrak{m}$ has 1 vertex of degree 2 and 2 vertices of degree 4, then its vertex partition is the partition $(4, 4, 2)$, which is denoted more compactly as $\nu(\mathfrak{m}) = [4^2\,2]$. Since $\nu(\mathfrak{m})$ is a partition of 10, $\mathfrak{m}$ has 5 edges.

There are three maps in an orientable surface with vertex partition $[4^2\,2]$ and face partition $[9\,1]$: they are $\mathfrak{m}_{1\cdot 442}$, $\mathfrak{m}_{1\cdot 443}$ and $\mathfrak{m}_{1\cdot 444}$ Thus since $\nu(\mathfrak{m}_{1\cdot 442}) = [4^2\,2]$, then $\mathfrak{m}_{1\cdot 442}$ has 3 vertices. Since $\phi(\mathfrak{m}_{1\cdot 442}) = [9\,1]$, then $\mathfrak{m}_{1\cdot 442}$ has 2 faces. Moreover,

$[4^2\,2] \vdash 10$ so $\mathfrak{m}_{1\cdot 442}$ has 5 edges. Then, by the Euler-Poincaré Formula, the type g of $\mathfrak{m}_{1\cdot 442}$ is given by $0 = 2 - 2g$, so $g = 1$, which is consistent with the surface being a torus. In addition, $\mathfrak{m}_{1.443}$ and $\mathfrak{m}_{1.444}$ are different embeddings of $\mathfrak{g}_{75}$ in the torus. Each of these three maps has 10 rootings, so there are 30 rooted maps with vertex partition $[4^2\,2]$ and face partition $[9\,1]$.

The associated graph and genus distribution

From Section 12.2, the genus distribution for $\mathfrak{g}_{18}$, the graph with four edges and one vertex, is

$$3 + 11u + 4u^2.$$

Thus this graph has 3 embeddings in the sphere, 11 embeddings in the torus and 4 embeddings in the double torus, each being a monopole.

Additional information about rootings is available from the entry for $\mathfrak{g}_{18}$ in Chapter 8. In the sphere, 1 of the maps has two rootings, 1 map has four rootings and 1 map has eight rootings (14 rooted maps); in the torus 1 map has two rootings, 3 maps have four rootings and 7 maps have eight rootings (70 rooted maps); in the double torus 1 map has one rooting, 1 map has four rootings and 2 maps have eight rootings (21 rooted maps). Then the corresponding polynomial for *rooted* embeddings of the graph is

$$14 + 70u + 21u^2.$$

The dual of a map

The construction of the dual of $\mathfrak{m}_{0\cdot 69}$ is given by

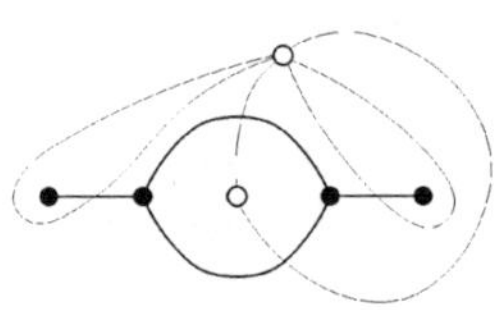

in which the vertices and edges of $\mathfrak{m}_{0\cdot 69}$ are drawn following the usual conventions. The vertices of the dual are hollow dots and the comparatively thinly drawn edges are the edges of the dual. The dual can be redrawn and identified as $\mathfrak{m}_{0\cdot 30}$.

The above construction sets up a bijection between the edges of a map and the edges of its dual since an edge of the map crosses a unique edge of the dual. The root edge of the dual is therefore taken to be the corresponding edge in the dual under this bijection. The root vertex and the root face of the dual correspond under this bijection to the root face and root vertex, respectively, of the map.

Hypermaps

The map $\mathfrak{m}_{0\cdot31}$ is 2-face-colourable and is therefore a hypermap. It is not symmetric since its hyperedge and hyperface partitions are not equal, a condition which is necessary, but not sufficient for symmetry. The two colourings of $\mathfrak{m}_{0\cdot31}$ are given below as $\mathfrak{h}$ and $\mathfrak{h}'$.

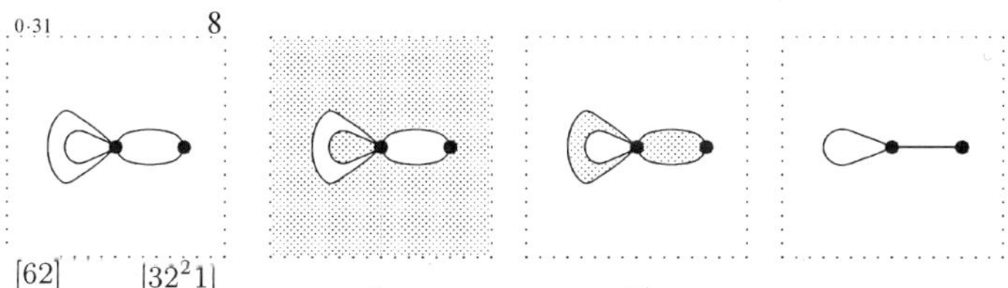

$\mathfrak{h}$ $\mathfrak{h}'$ $\mathfrak{m}$

Of these, $\mathfrak{h}$ has two hyperedges (the two white faces), so under digon conflation, this hypermap corresponds to $\mathfrak{m}$, which is identified as $\mathfrak{m}_{0\cdot4}$. On the other hand, $\mathfrak{h}'$ has no hyperedges that are digons, and is therefore unaffected by digon conflation.

The map $\mathfrak{m}_{0\cdot21}$ has 2 distinct rootings and is 2-face-colourable. The corresponding hypermaps are given below as (i) and (ii).

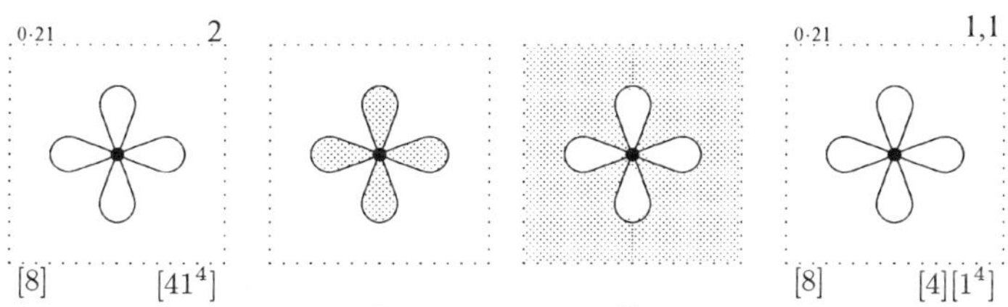

(i) (ii)

Each has one rooting: (i) has hyperedge partition $[4]$ and hyperface partition $[1^4]$; (ii) has hyperedge partition $[1^4]$ and hyperface partition $[4]$. This information is summarized in the rightmost diagram, which is taken from the subcollection of hypermaps in Section 7.3.

The map $\mathfrak{m}_{0\cdot30}$ is 2-face colourable and has 4 distinct rootings. However, it corresponds to the single symmetric hypermap

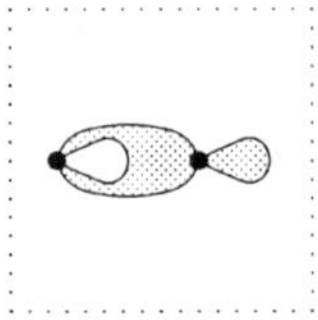

which has 4 rootings. Its hyperedge partition and hyperface partition are both $[3\,1]$.

2.3 Using the Atlas

Obtaining answers to some elementary questions about maps will provide examples of the use of the Atlas and will show how it has been organized.

2.3.1 Example: Do the vertex and face partitions determine the number of rootings?

Find minimum counterexamples, first in orientable surfaces and then in nonorientable surfaces, to the assertion that the number of rootings of a map is determined by the vertex and face partitions alone.

The maps $\mathfrak{m}_{0\cdot 38}$ and $\mathfrak{m}_{0\cdot 39}$ have, respectively, 8 and 4 rootings. Among the maps on orientable surfaces, with the same vertex partition and the same face partition but with different numbers of rootings, that then have the fewest edges, these maps have the fewest vertices. This can be ascertained by inspecting the maps of the Atlas serially from genus 0. The vertex and face partitions are $[6\,1^2]$, and the maps are *self-dual*, and have 4 edges and 3 vertices. From Chapter 8, they are embeddings of the same graph $\mathfrak{g}_{25}$ in the sphere.

The first such example for nonorientable surfaces occurs for 3 edges and 1 vertex. The maps are $\mathfrak{m}_{\tilde{1}\cdot 6}$ and $\mathfrak{m}_{\tilde{1}\cdot 7}$, with vertex and face partitions $[6]$ and $[4\,1^2]$, respectively, and with 6 and 3 rootings. From Chapter 8, they are embeddings of the same graph $\mathfrak{g}_7$ in the projective plane.

The original assertion is therefore false for both orientable and nonorientable surfaces.

2.3.2 Example: Maps with the same vertex and face partitions but different associated graphs

Are there maps with the same vertex and face partitions, but with different associated graphs?

There are certainly examples of such maps, and these can be found by searching the Atlas serially from genus 0. An instance of a pair of such maps is $\mathfrak{m}_{1\cdot428}$ and $\mathfrak{m}_{1\cdot429}$, both of which have vertex partition $[5\,3\,2]$ and face partition $[9\,1]$. From Chapter 8, the associated graph of these are $\mathfrak{g}_{70}$ and $\mathfrak{g}_{73}$, respectively. However, these associated graphs are not isomorphic since one has three loops while the other has one.

The duals $\mathfrak{m}_{1\cdot160}$ and $\mathfrak{m}_{1\cdot161}$ of $\mathfrak{m}_{1\cdot428}$ and $\mathfrak{m}_{1\cdot429}$, respectively, are, however, embeddings of the same graph, namely, $\mathfrak{g}_{49}$.

Smallest examples on each surface can be found by searching, from the beginning, the portion of the Atlas for the specified genus. The smallest example in the sphere is found in this way to be $\mathfrak{m}_{0\cdot45}$ and $\mathfrak{m}_{0\cdot46}$, both of which are self-dual. Their associated graphs are $\mathfrak{g}_{26}$ and $\mathfrak{g}_{27}$, respectively. The smallest example in the torus can be found in the same way to be $\mathfrak{m}_{1\cdot35}$ and $\mathfrak{m}_{1\cdot36}$. Their associated graphs are $\mathfrak{g}_{22}$ and $\mathfrak{g}_{21}$, respectively. The duals of these maps are $\mathfrak{m}_{1\cdot26}$ and $\mathfrak{m}_{1\cdot27}$, respectively, and are embeddings of the same graph, namely, $\mathfrak{g}_{19}$.

2.3.3 Examples of nonrealizability

Nonrealizable pairs of proper partitions

For a given surface, are there nonrealizable pairs of proper partitions?

From Chapter 11, there are no nonrealizable pairs of proper partitions for maps of genus greater than 1 that have no more than 8 edges in locally orientable surfaces. Note that this includes only one of the two surfaces of type 1. That is to say, there are pairs of proper partitions which are not realizable in the torus, yet there are none which cannot be realized in the Klein bottle, at least for all maps on at most 8 edges.

k-realizable pairs of partitions

Are there 6-realizable pairs of proper partitions that are realized by maps of genus 1 with 5 edges, and which are the maps that realize these pairs of partitions?

The restriction to maps of genus one with five edges is imposed to make the example of manageable size. From Section 10.1, the 6-realizable pairs of proper partitions for maps with five edges in the torus are

1. $([8\,1^2],[8\,2])$;
2. $([8\,1^2],[6\,4])$;
3. $([6\,3\,1],[8\,2])$;
4. $([5\,3\,2],[9\,1])$;
5. $([5\,2^2\,1],[10])$.

The left member of each pair is a vertex partition and the right member is a face partition. These are found in the Subsection entitled **Genus 1: 5 edges**, where there are two tables, with subtitles $[3{:}\,2]$ (3 vertices and 2 faces) and $[4{:}\,1]$ (4 vertices and 1 face).

The maps themselves, with these as vertex and face partitions, can be found in Section 5.2 of the Atlas by searching, in reverse lexicographic order, the set of maps with five edges, first on the vertex partition and then, among these, searching on the face partition. The maps found in this way are

1. $\mathfrak{m}_{1\cdot295}$, $\mathfrak{m}_{1\cdot296}$, $\mathfrak{m}_{1\cdot297}$, $\mathfrak{m}_{1\cdot298}$, $\mathfrak{m}_{1\cdot299}$, $\mathfrak{m}_{1\cdot300}$;
2. $\mathfrak{m}_{1\cdot308}$, $\mathfrak{m}_{1\cdot309}$, $\mathfrak{m}_{1\cdot310}$, $\mathfrak{m}_{1\cdot311}$, $\mathfrak{m}_{1\cdot312}$, $\mathfrak{m}_{1\cdot313}$;
3. $\mathfrak{m}_{1\cdot370}$, $\mathfrak{m}_{1\cdot371}$, $\mathfrak{m}_{1\cdot372}$, $\mathfrak{m}_{1\cdot373}$, $\mathfrak{m}_{1\cdot374}$, $\mathfrak{m}_{1\cdot375}$;
4. $\mathfrak{m}_{1\cdot426}$, $\mathfrak{m}_{1\cdot427}$, $\mathfrak{m}_{1\cdot428}$, $\mathfrak{m}_{1\cdot429}$, $\mathfrak{m}_{1\cdot430}$, $\mathfrak{m}_{1\cdot431}$;
5. $\mathfrak{m}_{1\cdot483}$, $\mathfrak{m}_{1\cdot484}$, $\mathfrak{m}_{1\cdot485}$, $\mathfrak{m}_{1\cdot486}$, $\mathfrak{m}_{1\cdot487}$, $\mathfrak{m}_{1\cdot488}$.

This completes the listing of the maps that satisfy the criteria.

The index of the associated graph for each of these maps can be found by first obtaining the associated graph, and by then searching through Section 12.2, first by the number of edges and then by the number of vertices. The corresponding associated graphs are, respectively,

1. $\mathfrak{g}_{58}$ for each of the 6 maps; $[1^6]$
2. $\mathfrak{g}_{58}$ for each of the 6 maps; $[1^6]$
3. $\mathfrak{g}_{63}$, $\mathfrak{g}_{63}$, $\mathfrak{g}_{63}$, $\mathfrak{g}_{63}$; $\mathfrak{g}_{63}$, $\mathfrak{g}_{63}$; $[4\,2]$
4. $\mathfrak{g}_{72}$, $\mathfrak{g}_{72}$; $\mathfrak{g}_{70}$; $\mathfrak{g}_{73}$, $\mathfrak{g}_{73}$, $\mathfrak{g}_{73}$; $[3\,2\,1]$
5. $\mathfrak{g}_{90}$; $\mathfrak{g}_{91}$, $\mathfrak{g}_{91}$; $\mathfrak{g}_{92}$; $\mathfrak{g}_{92}$; $\mathfrak{g}_{93}$; $[2^2\,1^2]$.

In this listing the partition $[2^2\,1^2]$, for example, indicates that there are four graphs, two with 2 embeddings in the torus and two with 1 embedding in the torus such that the vertex partition is $[5\,2^2\,1]$ and the face partition is $[10]$. The five partitions in the list are partitions of 6.

2.4 An application of k-realizable partitions

There is a relationship between k-realizable pairs of partitions, and therefore maps, and the study of the absolute Galois group, and a brief discussion of this is included here as an example of the use of the Tables. The reader is directed to the references cited in the Introduction for further details.

2.4.1 The absolute Galois group

Some preliminaries are required. Let $f \in \mathbb{Q}[x]$ be an irreducible polynomial, and let $\mathbb{E}$, where $\mathbb{Q} \subseteq \mathbb{E}$, be the splitting field extension of $f(x)$. Let $\sigma: \mathbb{E} \longrightarrow \mathbb{E}$ be an automorphism of $\mathbb{E}$ such that $\sigma(a) = a$ for all $a \in \mathbb{Q}$. The group of all such automorphisms is denoted by $\mathrm{Gal}(\mathbb{E}/\mathbb{Q})$, and is finite if the extension is algebraic. Let $\mathbb{K}$ be a field such that $\mathbb{Q} \subseteq \mathbb{K} \subseteq \mathbb{E}$. Then, by the main theorem of Galois Theory, there exists a subgroup H of $\mathrm{Gal}(\mathbb{E}/\mathbb{Q})$ such that $\mathbb{K} = \mathbb{E}^H$, where $\mathbb{E}^H = \{e \in \mathbb{E}: h(e) = e, \text{for all } h \in H\}$, so subfields are in correspondence with subgroups. Finally, recall that if $\mathrm{Gal}(\mathbb{E}/\mathbb{Q}) = \{1, h_1, \ldots, h_s\}$ is a finite group and if $f(\alpha) = 0$, then $h_i(\alpha)$, for $i = 1, \ldots, s$, are called the *conjugates* of α. In general, if $e \in \mathbb{E}$, then $h_i(e)$ is a conjugate of e.

The algebraic numbers are the elements $a \in \mathbb{C}$ that generate a finite extension $\mathbb{Q}(a)$ of $\mathbb{Q}$. They form a field $\bar{\mathbb{Q}}$, the algebraic closure of $\mathbb{Q}$ in $\mathbb{C}$. Then $\Gamma = \mathrm{Gal}(\bar{\mathbb{Q}}/\mathbb{Q})$ is called the *absolute Galois group*, and it is one of the most interesting but intractable groups in mathematics.

2.4.2 Belyi functions

Let $\mathbb{X}$ be a smooth compact connected Riemann surface of genus $g \geq 0$ and let $f: \mathbb{X} \longrightarrow \bar{\mathbb{C}}$, where $\bar{\mathbb{C}} = \mathbb{C} \cup \{\infty\}$, be a non-constant meromorphic function on $\mathbb{X}$. Let n be the degree of f. Then, for $w \in \bar{\mathbb{C}}$, $f^{-1}(w)$ consists of n distinct points, for all but a finite number of w, called *critical values*, for which the equation $f(z) = w$

has multiple roots. The multiplicities of the roots of $f(z) = w$ form a partition of n. A function f that has at most three critical values, namely, $\{0, 1, \infty\}$, is called a *Belyi function*. A Belyi function f for which the points of $f^{-1}(1)$ have multiplicity exactly two (so $n = 2m$) is called a *pure* Belyi function. Thus, if f is an arbitrary Belyi function, then $4f(1 - f)$ is a pure Belyi function. The pair $(\mathbb{X}, f)$ is called a *Belyi pair*.

Let f be a pure Belyi function. Then the preimage of the segment $[0, 1]$ is a map $\mathfrak{m}$. The vertices of $\mathfrak{m}$ are the points of $f^{-1}(0)$, and the degree of each vertex is the multiplicity of the corresponding root. Informally, the preimages of $[0, 1]$ under f may be regarded as half-edges, each incident with exactly one vertex, and the preimage pairs of 1 indicate how half-edges are paired to form edges. Each face contains exactly one element of $f^{-1}(\infty)$, which is therefore a pole of f, and the multiplicity of this pole is the degree of the corresponding face. Thus the partitions ν of $f^{-1}(0)$ and ϕ of $f^{-1}(\infty)$ are, respectively, the vertex and face partitions of $\mathfrak{m}$. Conversely, by a particular case of Riemann's Existence Theorem, for any map $\mathfrak{m}$ there exists a corresponding Belyi pair $(\mathbb{X}, f)$ that is unique up to an automorphism of $\mathbb{X}$.

The Riemann surfaces $\mathbb{X}$ for which Belyi functions exist are very specific and are given by Belyi's Theorem. This states that a meromorphic function $f: \mathbb{X} \longrightarrow \bar{\mathbb{C}}$ with at most three critical values exists if and only if the Riemann surface $\mathbb{X}$ is defined over the field $\bar{\mathbb{Q}}$ of algebraic numbers.

A naive, but correct, way to understand "defined over the field $\bar{\mathbb{Q}}$" is to note that $\mathbb{X}$ may be represented by a curve $F(x, y) = 0$, where $F \in \bar{\mathbb{Q}}[x, y]$. This is a polynomial with coefficients that are algebraic numbers. Then the Belyi function f is a rational function in x, y whose coefficients are algebraic numbers, so $f \in \bar{\mathbb{Q}}(x, y)$. The absolute Galois group acts on the coefficients of F and f, replacing them with conjugates, to give another Belyi pair that corresponds to another map. Moreover, for any non-identity element of Γ there exists a map that is not fixed under the action of this element, so the action is faithful. Vertex and face partitions are combinatorial invariants of this action and, since the number of maps with given vertex and face partitions is finite, every Galois orbit is finite. In fact, the set of maps having the same vertex and face partitions is either an orbit or a union of orbits.

Let $G < \Gamma$ be a subgroup of Γ that fixes a given map $\mathfrak{m}$. The orbit of $\mathfrak{m}$ under the action of Γ is finite, so G has finite index in Γ. By the main theorem of Galois Theory, the subgroups of Γ of finite index are in one-to-one correspondence with field extensions of $\mathbb{Q}$, and so to each group G there corresponds the field $\mathbb{F}$ of all algebraic numbers fixed by G. The minimal Galois extension of $\mathbb{Q}$ containing $\mathbb{F}$ corresponds to the maximal normal subgroup of Γ that is contained in G. This is the *moduli field* of the map $\mathfrak{m}$, and is common to the whole orbit. Its degree is equal to the number of maps in the orbit and is therefore bounded by the number of maps that have the same vertex and face partitions.

This explains the interest in finding all of the k-realizable pairs of partitions that correspond to maps of genus g. When the maps are restricted to be trees, then for any k there is a finite number of explicitly described infinite series of k-realizable partitions, together with a finite number of "sporadic" k-realizable partitions with an explicit bound on their weight. However, even for maps in the plane, very little is known. When $k = 1$, the maps have $\mathbb{Q}$ as their moduli field, and the maps of the five Platonic solids are obviously among the sporadic maps for this case.

Chapter 3

The axiomatization and the encoding of maps

Each map can be associated with a unique element in a particular group and a finite set, where the action of the group on the set satisfies certain conditions. This association is given as an axiomatization, one for maps in orientable surfaces and another for maps in locally orientable surfaces, and in each case a permutation representation is given. The listing of maps is then reduced to the listing of permissible permutations. By specifying a particular action of these permutations on a given set that satisfies the conditions and by adopting certain conventions, these permutations can be put in one-to-one correspondence with maps with labels attached to edges. Such maps are called *labelled maps*. A rooted map is recovered from a labelled map by erasing the attached labels and assigning a root in a canonical way.

Although the axiomatization for maps in locally orientable surfaces gives an alternative encoding for maps in orientable surfaces, it requires twice the number of labels, which is prohibitive, and is therefore not used for this purpose. The additional labels are required for conveying information about twists in edges, and edges have no twists in orientable surfaces. From the enumerative point of view, however, the axiomatizations are equally important. They permit explicit expressions to be obtained for the genus series for maps in orientable and locally orientable surfaces. The genus series for maps in nonorientable surfaces is then obtained as the difference of these.

3.1 Orientable surfaces

3.1.1 Axiomatization for maps in orientable surfaces

A *connected premap* in an orientable surface is an ordered pair (ρ, τ) of permutations acting on a set S of $2n$ elements such that

Axiom 1. $\rho^2 = \iota$, where ι is the identity,

Axiom 2. for any $a \in \mathcal{S}$, then $a, \rho a$ are distinct,

Axiom 3. the group $\langle \tau, \rho \rangle$ acts transitively on $\mathcal{S}$.

The disjoint cycles of the permutation

$$\varphi = \tau\rho$$

correspond to the faces of the map and a cycle of length k under this correspondence is associated with a face of degree k. Thus φ is called the face permutation. The partition that gives the cycle structure of φ is the face partition of the map.

Each labelled map corresponds to a connected premap, and each connected premap corresponds to a unique labelled map. The correspondence can be made bijective by making ρ an arbitrary, but fixed, fixed-point free involution on $\mathcal{S}$.

Actions

For the purposes of drawing the maps, it is necessary to supply an action of ρ and τ on the set $\mathcal{S}$. If $\mathfrak{m}$ is a rooted map with n edges, in an orientable surface, let $\mathcal{S} = \{1, \ldots, 2n\}$. Each edge has two ends, which are called *end positions*. The elements of $\mathcal{S}$ are assigned to the $2n$ end positions so that $2k+1, 2k+2$ are the end position labels of an edge for $k = 0, \ldots, n-1$, and so that the directed edge $\overrightarrow{(1,2)}$ is the root edge.

In this realization,

$$\rho$$

is the involution whose m cycles are pairs of end symbols at different ends of an edge, and is called the *edge permutation*. Let v be a vertex of the map and let $(a_1, \ldots, a_k)$ be the cyclic list of end symbols encountered in a tour of v from an arbitrarily chosen end symbol a_1 in the anticlockwise sense. Then τ is the permutation whose disjoint cycles are associated in pairs with each vertex v and have the form $(a_1, \ldots, a_k)$. The degree of v is k. The permutation τ is called the *vertex permutation* or the *rotation system* for the map in an orientable surface.

It is convenient to regard the two ends of an edge as belonging, notionally, to two different halves of the same edges. These two halves are called *half-edges*.

3.1.2 Encoding a map as a permutation

The example

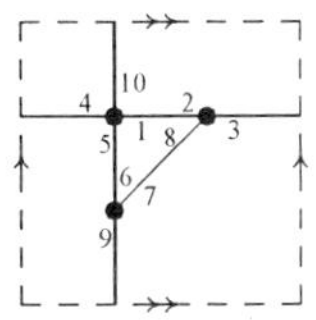

illustrates how the faces and face partition are encoded by these actions in the specific case of the labelling of $\mathfrak{m}_{1\cdot 457}$. The vertex permutation, the fixed edge permutation and the face permutation are, respectively,

$$\begin{aligned} \tau &= (1\,10\,4\,5)(2\,8\,3)(6\,9\,7), \\ \rho &= (1\,2)(3\,4)(5\,6)(7\,8)(9\,10), \\ \varphi = \tau\rho &= (1\,8\,6)(2\,10\,7\,3\,5\,9\,4). \end{aligned}$$

The labels $\{1, \dots, 2n\}$ (where $n = 5$) have been assigned to the $2n$ end positions of a map with n edges so that an edge has its ends labelled $2k + 1$, $2k + 2$, for some $k \geq 0$. The convention is adopted that the labels on the edge $\{i, j\}$ are positioned so that i is on the right-hand side of the directed edge $\overrightarrow{(i, j)}$ and that j is on the right-hand side of the directed edge $\overrightarrow{(j, i)}$. This is possible only when the surface is orientable. By convention, 1 is selected to fix the root. This specifies a mutually incident vertex, edge and face triple. The vertex is the unique vertex incident with the half-edge labelled 1, the edge is $\{1, 2\}$ and the face is the (unique) face on the right of the directed edge $\overrightarrow{(1, 2)}$.

Note that τ has cycle type $[4\,3^2]$, so there is one vertex of degree 4 and two vertices of degree 3; and that ϕ has cycle type $[7\,3]$, so there is one face of degree 7 and one face of degree 3. Each cycle of the face permutation specifies a face. This face has the property that the labels appearing on it, by their assignment to edge positions, are listed by the cycle in the order in which they are encountered in a clockwise tour inside the boundary of the face.

If the edges are deleted the surface decomposes into two discs, corresponding to the two faces, so this is indeed a 2-cell embedding in the torus.

3.1.3 Construction of the set of all rooted maps

The set of all rooted maps in orientable surfaces is obtained by first determining the set of all rotation systems. It suffices to show the process in a particular case. As an example, the axiomatization is used to construct the set of all rooted maps for the specific case of two edges ($n = 2$). These are given in a preliminary form in Figure 3.1 on page 34 and in final form in Figure 3.2 on page 35.

The construction involves working with the set of all permutations of $1, \ldots, 2n$, where n is the number of edges, and selecting each permutation in turn to be τ. The strategy has two steps. First, for each of the 24 ($= (2n)!$) such permutations, a labelled map or a labelled union of maps is constructed. Second, drawings are discarded which are not connected since, for these, the group $\langle \tau, \rho \rangle$ does not act transitively on $1, \ldots, 4$, and then recognize equivalence of maps through the axiomatization. Finally, one map is selected at random from each equivalence class. These then correspond to the rooted maps, and are given in Figure 3.2.

The canonical fixed-point free involution is $\rho = (1\,2)(3\,4)$. If the premap is connected, the genus can be computed directly from the Euler-Poincaré Formula, and this then uniquely determines the surface on which the map is to be drawn.

Step 1: Having selected one of the permutations to be τ, it is a straightforward matter to draw the corresponding object in the selected surface since τ gives the cyclic order of end labels of half-edges incident with each vertex, and ρ indicates how these labels are to be connected. These drawings are given in Figure 3.1 and are indexed by the choice of τ. The columns in Figure 3.1 are labelled by the vertex partition of the labelled maps appearing in the same column. Of course, it may be necessary for one or more edges to cross a line of the polygonal representation of the surface. However, since the correct surface has been selected, it is known that there is a correct drawing of the map. If the premap is connected the object is a labelled map. If it is disconnected, it is a labelled union of maps and is to be discarded. All but four of the permutations correspond to labelled maps. Labelled unions of maps have been represented in a compact way in Figure 3.1 with the understanding that

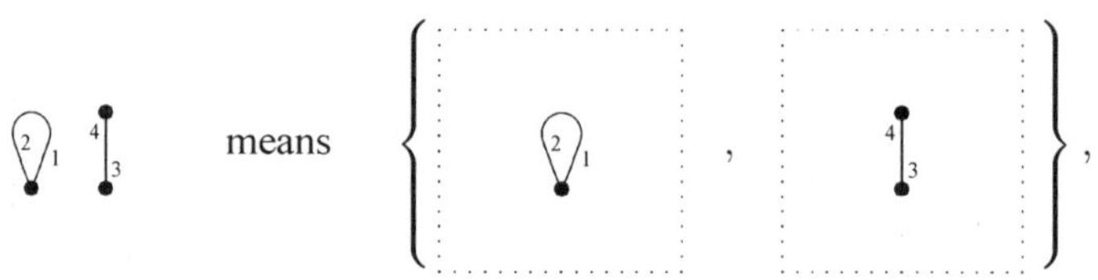

an unordered set of labelled maps. By the Euler-Poincaré Formula applied to each connected component, each map is in the sphere. In general, there will be an arbitrary number of components and the corresponding surfaces are independent of each other.

Although there are twenty labelled maps remaining, some of them correspond to the same rooted map. This can be recognized as follows. The convention is to be that the root edge is to be the directed edge $\overrightarrow{(1,2)}$. For a rooted map with n edges in an orientable surface, there are $2^{n-1}(n-1)!$ ways to assign the labels $3, \ldots, 2n$ to the end positions of the nonroot edges. The corresponding $2^{n-1}(n-1)!$ rooted labelled maps correspond to a single rooted map. When $n = 2$ this multiplicity is equal to 2, and Figure 3.1 indicates with braces the labelled maps that are mutually indistinguishable as rooted maps under the convention that $\overrightarrow{(1,2)}$ is the root edge.

Step 2: Figure 3.2 gives a representative member from each of the equivalence classes of rooted labelled maps, and presents these representatives as rooted maps whose label sets partition $\{1,\ldots,2n\}$. This is done by replacing the edge $\{1,2\}$ by a directed edge, and by then removing all of the labels. There are therefore ten rooted maps with two edges: nine are in the sphere and one is in the torus. Each map is identified with its entry from the Atlas, and its index number is supplied. At the head of each column is an entry from the Atlas. This gives a map, and an integer representing the number of distinct rootings of the map. The corresponding distinct rooted maps are listed beneath it in the same column.

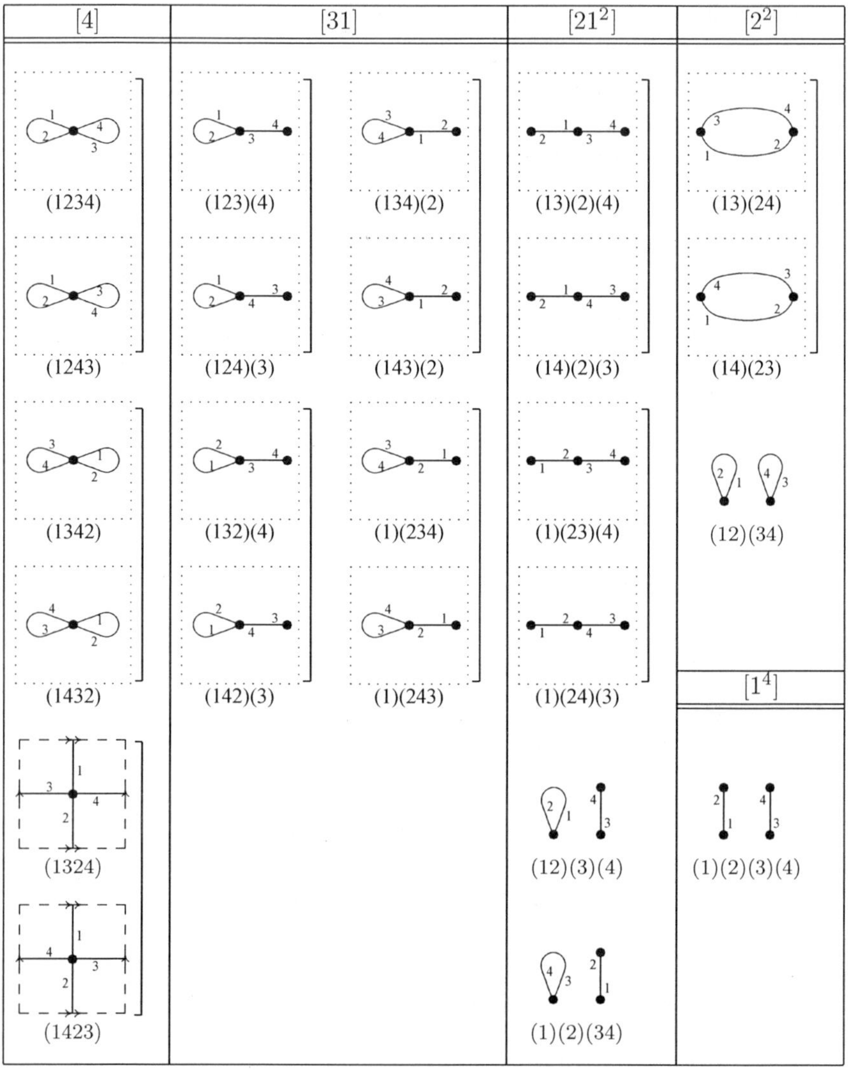

FIGURE 3.1
All labelled maps with 2 edges in orientable surfaces, arranged by vertex partition

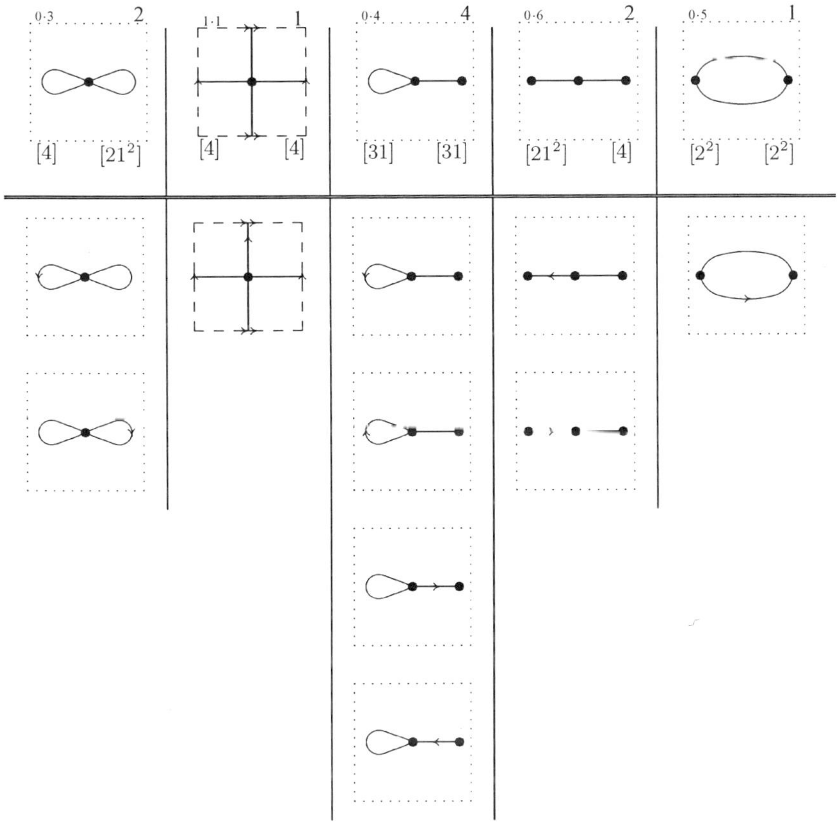

FIGURE 3.2
All rooted maps with 2 edges in orientable surfaces

Different drawings of the same map

Two maps that appear to be different may in fact be equivalent, and care must be taken with this point. The situation is resolved by using the axiomatization. Consider, for example, the "figure of eight" map, denoted by $\mathfrak{a}$, and the "pair of nested loops" map, denoted by $\mathfrak{b}$, given below. The map $\mathfrak{a}$ is assigned labels so that $\rho = (1\,2)(3\,4)$. Then, for this assignment of labels, $\tau = (1\,2\,3\,4)$. The map $\mathfrak{b}$ is assigned labels so that ρ is as before. For the particular labelling that is given, τ is the same as before so, according to the axiomatization, these two maps are equivalent.

name	map	labelled map	τ
$\mathfrak{a}$		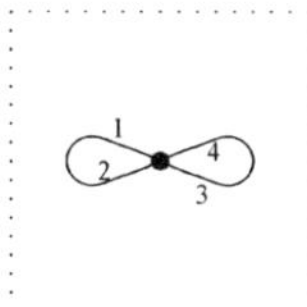	$(1\,2\,3\,4)$
$\mathfrak{b}$	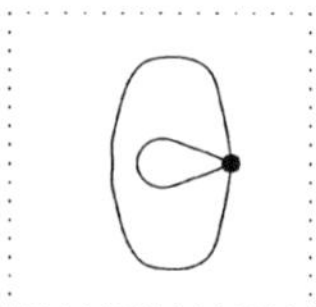	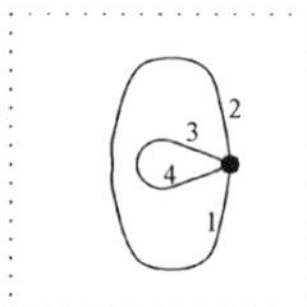	$(1\,2\,3\,4)$

To see why this is so, geometrically, imagine that the larger loop of $\mathfrak{b}$ is smoothly expanded to pass over the sphere without crossing the smaller loop, and is then smoothly contracted. The resulting diagram is precisely $\mathfrak{a}$. This is a diffeomorphism from the sphere to the sphere and maps are, by definition, invariant under these operations. Thus $\mathfrak{a} = \mathfrak{b} = \mathfrak{m}_{0\cdot 3}$.

For maps in surfaces other than the sphere, attention must be paid to edges which are incident with corner points of the polygonal representation. Since the corner points in the polygonal representation of a surface are identified, either none of the corner points is incident with an edge of the map, or exactly two are. If there are two, there may be several ways in which they can be selected, each giving a different, but equivalent drawing. Consider the following embeddings in the double torus.

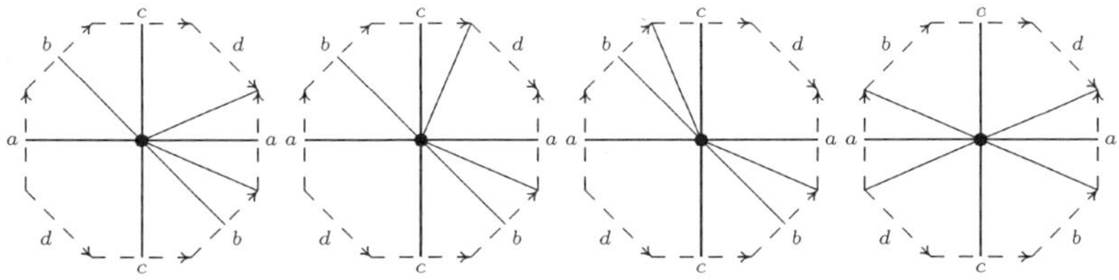

$\mathfrak{a}$ $\mathfrak{b}$ $\mathfrak{c}$ $\mathfrak{d}$

Maps $\mathfrak{a}$ and $\mathfrak{b}$ are both equivalent to $\mathfrak{m}_{2\cdot 3}$. On the other hand, $\mathfrak{c}$ is not a 2-cell embedding of the double torus because one "face" is not homeomorphic to an open disc. Using the axiomatization, one would find that the surface should be a torus in this case. It is also observed that $\mathfrak{d}$ is not a 2-cell embedding in the double torus because the two loops which are incident with corner points of the polygonal representation actually cross.

3.1.4 Determining the number of rootings

The determination of the distinct rootings of a particular map with three edges, namely, $\mathfrak{m}_{1\cdot 6}$, by Step 1 and Step 2 is summarized in Figure 3.3. Notice that there is a question mark at the top right-hand corner, where the number of rootings should appear.

It is known from the Atlas that $\mathfrak{m}_{1.6}$ has three rootings, and the way in which this number is determined is explained as follows. For a map with n edges there are $2n$ possible ways to root it since there are n ways of selecting an edge to be the root edge and 2 ways of assigning a direction to it. The 6 ways of rooting this particular map are listed in the second row of Figure 3.3. By convention the root edge is $\overrightarrow{(1,2)}$. The first pair of rooted labelled maps in the third row are equivalent since the edge-end labels can be assigned to give the same vertex permutation. Moreover, there is no assignment of edge-end labels to the other four rooted maps that will make them equivalent to the first pair. To verify this, it is enough to show that there is no bijection on $\{1,\ldots,6\}$ for which 1 and 2 are fixed points, and that preserves adjacency. The corresponding rooted map is given beneath the underbrace. The second pair of rooted labelled maps are equivalent to each other; and the same holds for the third pair also. Three inequivalent rooted maps are obtained and are given in the fourth row. This verifies that the map $\mathfrak{m}_{1\cdot 6}$ has 3 rootings and resolves the question mark in the top diagram of Figure 3.3.

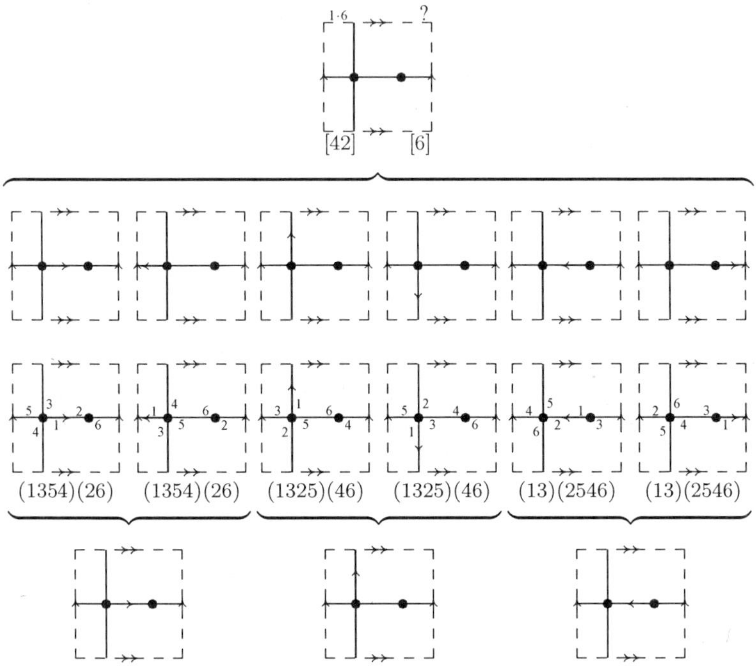

FIGURE 3.3
The number of rootings of $\mathfrak{m}_{1\cdot 6}$

3.2 Locally orientable surfaces

Although maps in locally orientable surfaces are treated differently from maps in orientable surfaces, the general approach is very similar. Again, a map is represented by a single permutation, and each map corresponds to a particular number of labelled maps that can be determined. Since the orientable case has been treated in detail, the discussion of the locally orientable case can be usefully compressed.

3.2.1 Axiomatization for maps in locally orientable surfaces

A *connected premap* in a locally orientable surface is an ordered triple (σ, ρ, τ) of permutations acting on a set S of $4n$ elements such that

Axiom 1. $\rho^2 = \sigma^2 = \iota$, and $\sigma\rho = \rho\sigma$,

Axiom 2. for any $a \in \mathcal{S}$, then $a, \sigma a, \rho a, \sigma\rho a$ are distinct,

Axiom 3. $\tau\sigma = \sigma\tau^{-1}$,

Axiom 4. for each $a \in \mathcal{S}$ the orbits of τ through a and σa are distinct,

Axiom 5. the group $\langle\sigma, \rho, \tau\rangle$ acts transitively on $\mathcal{S}$.

The disjoint cycles of $\varphi = \tau\sigma$ correspond, in pairs of equal length, to the faces of the map, so φ is called the *face permutation.*

Each labelled map corresponds to a connected premap, and each connected premap corresponds to a unique labelled map. The correspondence can be made bijective by making σ and ρ arbitrary, but fixed, fixed-point free involutions on $\mathcal{S}$ satisfying Axioms 1 and 2 above.

Actions

To draw the map it is necessary to supply an action of σ, ρ and τ on $\mathcal{S}$. If m is a map with n edges, let $\mathcal{S} = 1, \ldots, 4n$. Each edge has two ends, and each end has two sides. Thus each edge has four *side-end* positions. Assign the elements of $\mathcal{S}$ to the $4n$ side-end positions so that $4k+1, 4k+2, 4k+3, 4k+4$ are the side-end labels for an edge, for $k = 0, \ldots, n-1$, and that $4k+1, 4k+2$ are at the same end of the edge and $4k+1, 4k+3$ are on the same side of the edge. Then the *edge-side permutation*

$$\sigma$$

is the involution whose $2n$ cycles are pairs of side-end symbols at the same end but different sides of an edge. Similarly, the *edge-end permutation*

$$\rho$$

is the involution whose $2n$ cycles are pairs of side-end symbols at the same side but different ends of an edge. Let v be a vertex of the map and let $(a_1, \ldots, a_{2k})$ be the cyclic list of side-end symbols encountered in a tour of v from an arbitrarily chosen side-end symbol a_1 in the unique local direction $a_2 = \sigma a_1$. Then the *vertex permutation* τ is the permutation whose disjoint cycles are associated in pairs with each vertex v, and have the form $(a_{2k}, a_{2k-2}, \ldots, a_2)$ and $(a_1, a_3, \ldots, a_{2k-1})$. The degree of v is k.

This is the algebraic form of the geometrical statement that a clockwise sense around each vertex cannot be recognized in a nonorientable surface. This is because both cycles are present, but are in opposite senses.

3.2.2 Encoding a map as a permutation

Two examples of the use of the axiomatization for maps in locally orientable surfaces are now given to illustrate how the faces and face partition are encoded by these actions.

The first example is for an encoding of a map in a nonorientable surface (the Klein bottle), specifically the labelling of $\mathfrak{m}_{\tilde{2}\cdot 18}$ given below.

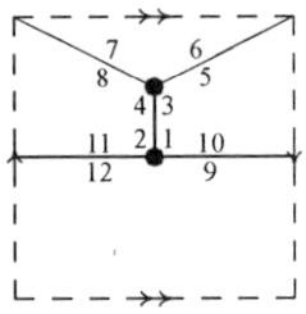

The labels $\{1,\ldots,4n\}$ (where $n=3$) are assigned to the $4n$ side-end positions so that $4k+1$, $4k+2$, $4k+3$, $4k+4$ are the side-end labels for an edge, for some k. Moreover, $4k+1$ and $4k+2$ are to be at the same end of the edge, and $4k+1$ and $4k+3$ are to be on the same side of the edge. The permutations associated with each vertex are $(1\,11\,9)(2\,10\,12)$ and $(4\,5\,7)(3\,8\,6)$. The cycle $(1\,11\,9)$ lists labels in the canonically defined sense $(1,\sigma(1))$. Similarly, the cycle $(2\,10\,12)$ lists labels in the canonically defined sense $(2,\sigma(2))$. A similar statement holds for the other vertex. The vertex, edge-side, edge-end and face permutations are, respectively,

$$\begin{aligned}\tau &= (1\,11\,9)(2\,10\,12)(4\,5\,7)(3\,8\,6),\\ \sigma &= (1\,2)(3\,4)(5\,6)(7\,8)(9\,10)(11\,12),\\ \rho &= (1\,3)(2\,4)(5\,7)(6\,8)(9\,11)(10\,12),\\ \varphi = \tau\sigma\rho &= (1\,5\,6\,4\,11\,12)(2\,8\,7\,3\,10\,9).\end{aligned}$$

Now τ has cycle type $[3^4]$, and dividing the multiplicities by 2 (the cycles appear in pairs) gives $[3^2]$, so there are two vertices of degree 3. Similarly the cycle type of φ is $[6^2]$, and dividing the multiplicities by 2 gives $[6]$, so there is only one face and this has degree 6. In staying on the same side of the boundary in a tour of a face $(1\,5\,6\,4\,11\,12)$ is encountered as the list of labels on the tails of edges, and $(2\,8\,7\,3\,10\,9)$ as the heads of edges. These two cycles have the canonically defined senses $(1,\rho(1))$ and $(2,\rho(2))$, respectively, and these are opposite senses.

The second example illustrates the encoding of a map in an orientable surface (the torus) by means of the axiomatization for maps in locally orientable surfaces. Consider

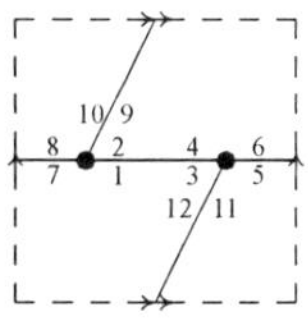

which is a labelling of $\mathfrak{m}_{1\cdot 7}$. Using the same conventions for direction as in the previous example, the vertex, edge-side, edge-end and face permutations are, respectively,

$$\begin{aligned} \tau &= (1\,9\,8)(2\,7\,10)(4\,12\,5)(3\,6\,11), \\ \sigma &= (1\,2)(3\,4)(5\,6)(7\,8)(9\,10)(11\,12), \\ \rho &= (1\,3)(2\,4)(5\,7)(6\,8)(9\,11)(10\,12), \\ \varphi = \tau\sigma\rho &= (1\,12\,8\,4\,9\,5)(2\,6\,10\,3\,7\,11). \end{aligned}$$

Now τ has cycle type $[3^4]$, so there are two vertices of degree 3 (after dividing the multiplicities by 2). Similarly the cycle type of ϕ is $[6^2]$, so there is only face and this has degree 6. In staying on the same side of the boundary in a tour of a face $(1\,12\,8\,4\,9\,5)$ is encountered as the list of labels on the tails of edges, and $(2\,6\,10\,3\,7\,11)$ as the heads of edges. As before, these two cycles have opposite senses.

3.2.3 Constructing the set of all rooted maps

As in the case of maps in orientable surfaces, the set of all rooted maps in locally orientable surfaces is obtained from the set of all rotation systems for locally orientable surfaces. The use of the axiomatization is illustrated in the construction of the set of all rooted maps with one vertex of degree 4. This example has been selected since it is sufficiently small while using surfaces of genus zero (sphere), one (torus, projective plane) and two (Klein bottle). These are given in Figures 3.4, 3.5 and 3.6 on pages 44, 45 and 46.

The construction involves working with the set of all permutations of $1,\ldots,4n$ whose cycle lengths occur an even number of times. Let $\lambda = (\lambda_1, \lambda_2, \ldots)$ be a partition of $2n$ and let λ^2 denote the partition $(\lambda_1, \lambda_1, \lambda_2, \lambda_2, \ldots)$ of $4n$. Let $\mathcal{K}_\lambda$ be the set of all permutations τ of $1,\ldots,4n$ with cycle type λ^2 which satisfy Axiom 3. Then $|\mathcal{K}_\lambda| = |\mathcal{C}_\lambda| 2^{2n-l(\lambda)}$, where $\mathcal{C}_\lambda$ is the conjugacy class of the symmetric group on $2n$ symbols indexed by λ. Note that $\mathcal{K}_\lambda$ is a larger set of permutations than $\mathcal{C}_\lambda$ that is used in the orientable case, in the analogous context.

Fix σ to be $(1\,2)(3\,4)(5\,6)(7\,8)$ and ρ to be $(1\,3)(2\,4)(5\,7)(6\,8)$. In addition, the vertex partition is fixed to be $[4]$, so the set of permutations from which τ is selected is $\mathcal{K}_{[4]}$. There are therefore 48 permutations to consider as instances of τ. The root of a map can be specified uniquely by selecting a single side-end position. The convention is adopted that the side-end position which is assigned the symbol 1 is to be the root.

Step 1: Having selected one of the permutations in $\mathcal{K}_{[4]}$ to be τ, it is a straightforward matter to draw the corresponding object in the correct surface that has been selected by the Euler-Poincaré Formula, although care has to be exercised in respecting directions of lines if an edge crosses a line of the polygonal representation of the surface. If the object is disconnected it is discarded. The remaining are labelled maps.

Although there are 48 labelled maps, some of them correspond to the same rooted map, under the convention that 1 indicates the root. For each map, there are $4^{n-1}(n-1)!$ associated labelled maps. This multiplicity is equal to 4 when $n = 2$, and in Figures 3.4, 3.5 and 3.6 these groups of four are indicated by braces, as in the orientable case.

Step 2: The rooted map that corresponds to a braced group of four labelled maps is given below the group and in the same column. Below this, perhaps spanning several columns, the full entry in the Atlas is given that corresponds to these rooted maps. In this entry, indication of the root has been removed but the number of distinct rootings is retained.

Equivalent maps in a nonorientable surface

As a final example the axiomatization is used to establish whether or not two maps are equivalent. Consider the following two maps.

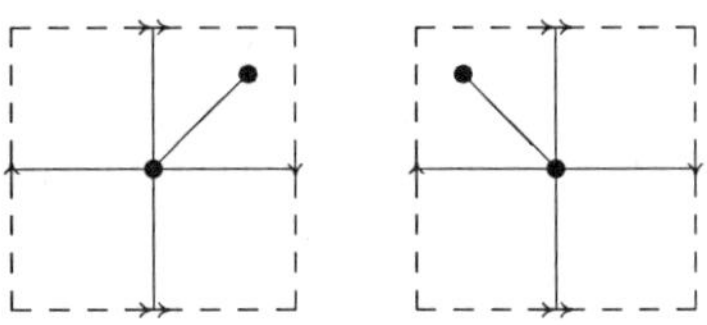

The first is readily identified as $\mathfrak{m}_{\bar{2}\cdot 11}$ and is in the Klein bottle. The second map, which is denoted by $\mathfrak{a}$, is to be identified.

The maps may be labelled as

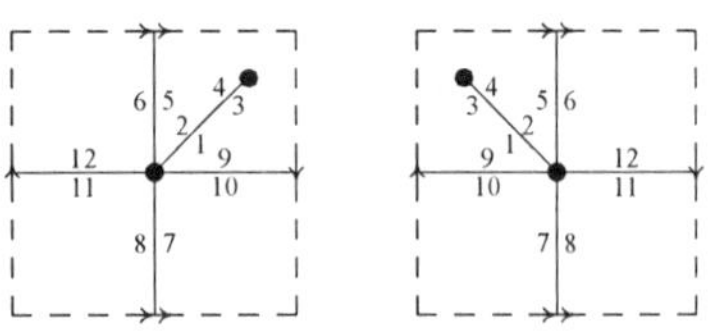

In both cases, the labelling has been assigned so that the edge-side and edge-end permutations are, respectively,

$$\sigma = (1\,2)(3\,4)(5\,6)(7\,8)(9\,10)(11\,12),$$
$$\rho = (1\,3)(2\,4)(5\,7)(6\,8)(9\,11)(10\,12).$$

Moreover, in both cases the vertex permutation is

$$\tau = (1\,5\,12\,8\,10)(2\,9\,7\,11\,6)(3)(4).$$

Then both $\mathfrak{m}_{\tilde{2}\cdot 11}$ and $\mathfrak{a}$ correspond to $\langle\sigma,\rho,\tau\rangle$ so, according to the axiomatization, $\mathfrak{a}$ is identified as $\mathfrak{m}_{\tilde{2}\cdot 11}$.

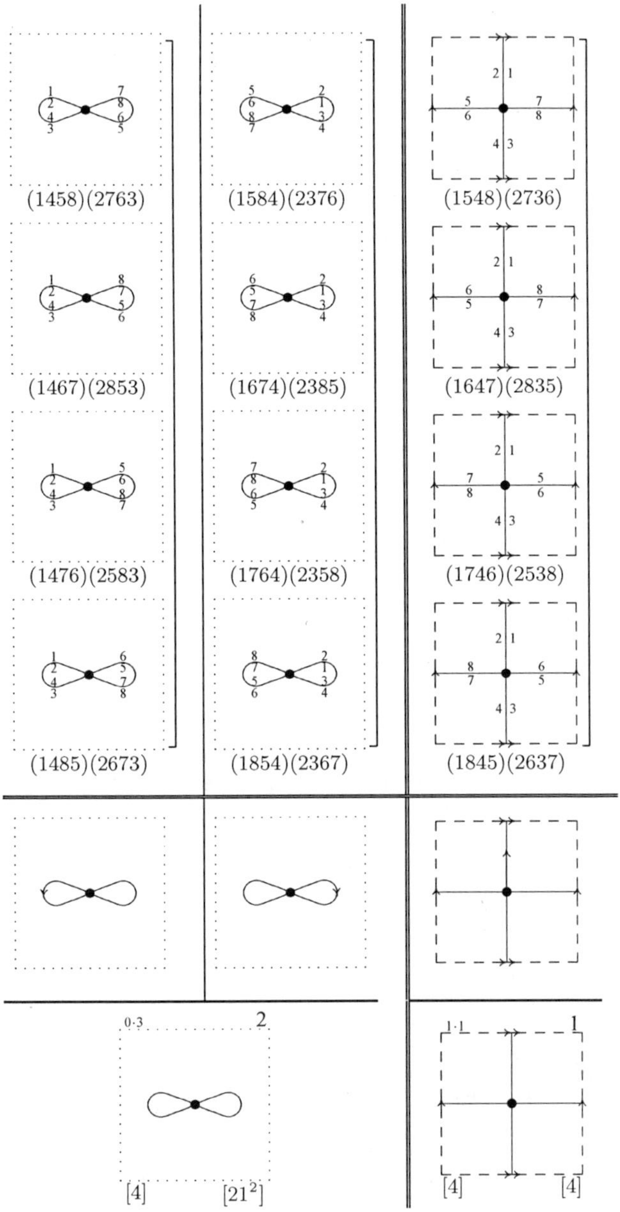

FIGURE 3.4
All labelled maps with vertex partition [4] in the sphere and torus

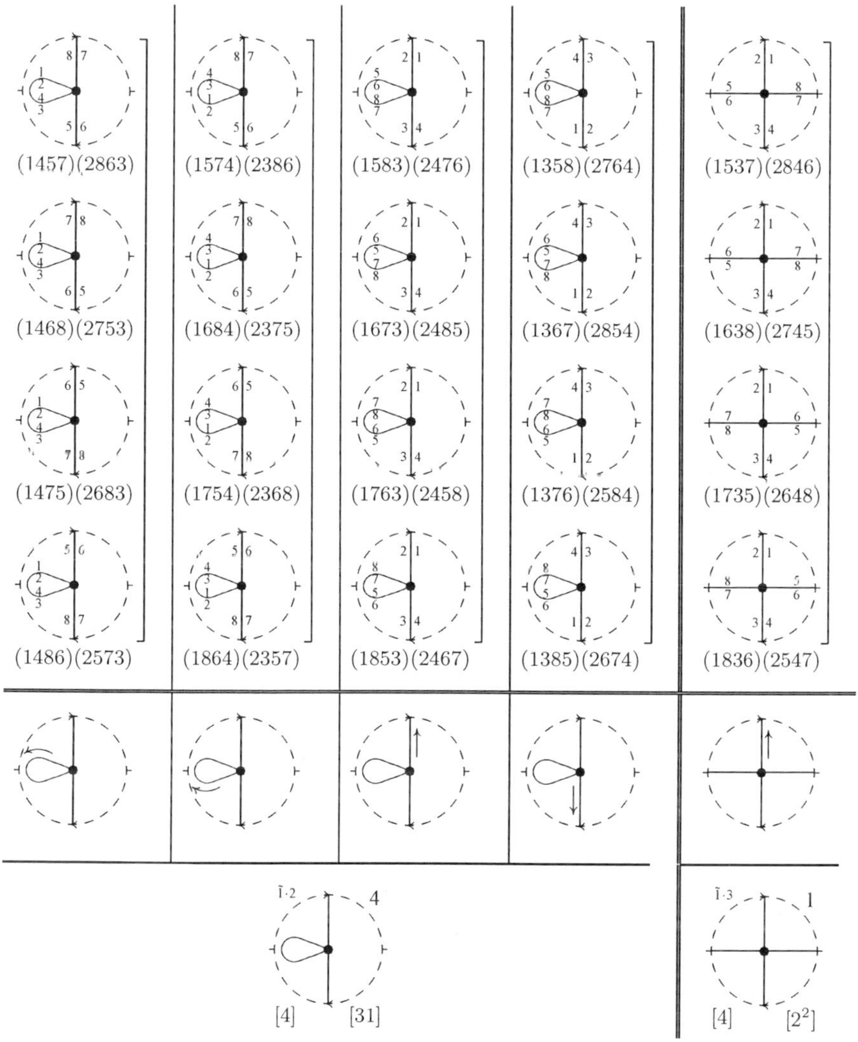

FIGURE 3.5
All labelled maps with vertex partition [4] in the projective plane

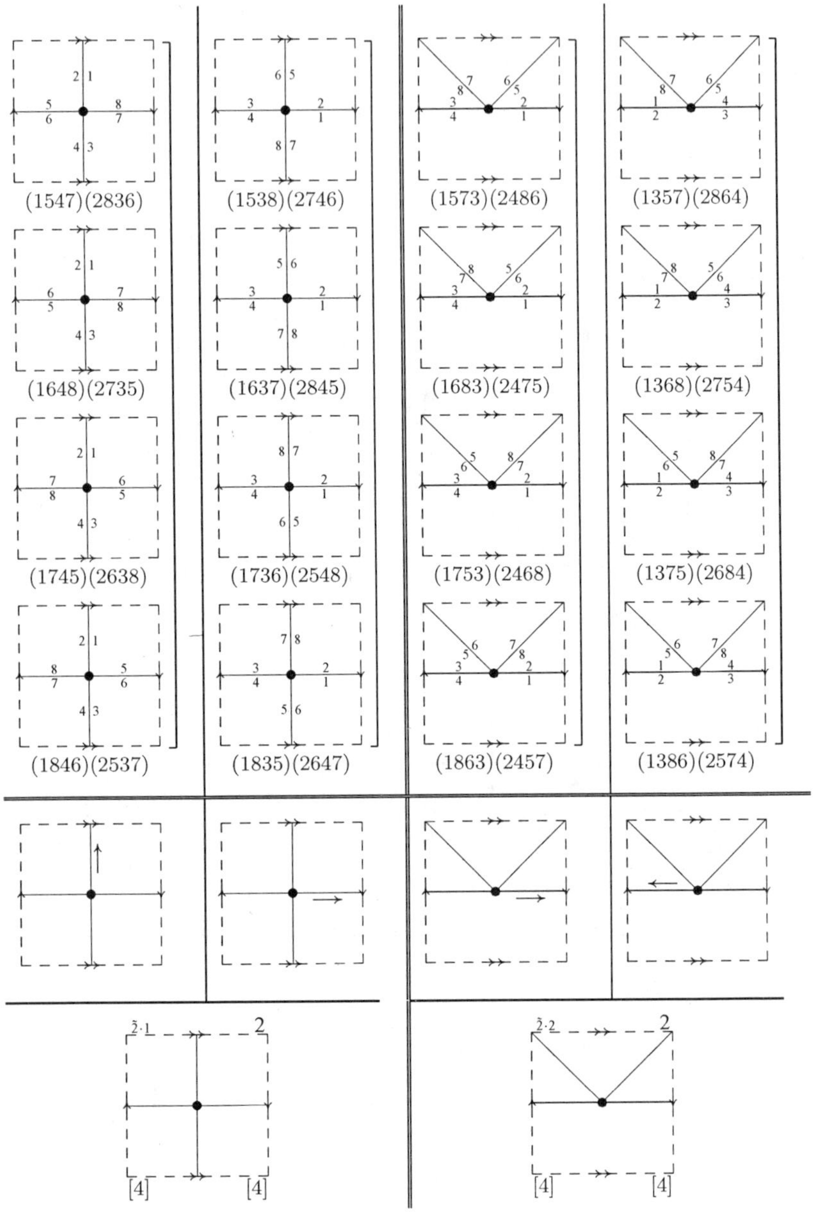

FIGURE 3.6
All labelled maps with vertex partition [4] in the Klein bottle

Chapter 4

Generating series and conjectures

The maps drawn with given vertex and face partition have been checked for pairwise distinguishability by means of the axiomatization. Confirmation that the number of maps that have been drawn is correct is received by determining the appropriate terms of the generating series for maps in both orientable and locally orientable surfaces. Explicit expressions for these generating series are given in this section.

Generating series are known for several classes of rooted maps. These include trees, triangulations in the sphere, quadrangulations in the sphere and all maps in the sphere, monopoles in orientable surfaces, dipoles (maps with exactly two vertices that also have the same degree) in orientable surfaces and monopoles in nonorientable surfaces. However, these were not needed since the enumerative information is available in the series that are given below.

4.1 Generating series for hypermaps

4.1.1 Schur functions and zonal polynomials

Let $x_1, x_2, \ldots$ be indeterminates and let $\mathbf{x} = (x_1, x_2, \ldots)$. Let $p_k(\mathbf{x})$ denote the *power sum symmetric function* $x_1^k + x_2^k + \cdots$ of degree k in $\mathbf{x}$, and let $p_\lambda = p_{\lambda_1} \cdots p_{\lambda_r}$, where $\lambda = (\lambda_1, \ldots, \lambda_r)$ is a partition. Let m_λ denote the sum of $x_1^{\lambda_1} \cdots x_r^{\lambda_r}$ over all distinct permutations of $(\lambda_1, \ldots, \lambda_r)$. This is called a *monomial symmetric function*. The set of all symmetric functions of bounded degree form a ring Λ, and both $\{p_\lambda\}$ and $\{m_\lambda\}$ are bases for this ring. It is convenient to regard it as a ring over rational functions of an indeterminate α over the rationals. The role of α will become apparent in the discussion of the b-Conjecture.

Two particular types of symmetric functions are important in the formulation of the genus series for rooted maps in orientable and locally orientable surfaces. These are the *Schur function* s_θ and the *zonal polynomial* Z_θ, where $\theta \vdash n$. Explicit expressions for them can be given as follows. Let χ^θ be the character of the (ordinary) irreducible representation of the symmetric group indexed by θ and let χ^θ_λ be the evaluation of this character on the conjugacy class indexed by the partition λ.

Let $\mathcal{F}_\theta$ denote the *Ferrers graph*

$$\mathcal{F}_\theta = \{(i,j) \colon 1 \le i \le l(\theta), 1 \le j \le \theta_i\}$$

associated with the partition θ. This is a combinatorial representation of the partition θ. Then the Schur functions and the zonal polynomials can be expressed with respect to the power sum symmetric function basis of Λ with coefficients involving only the evaluation of characters of irreducible representations of the symmetric group and some combinatorial numbers. The expressions are

$$s_\theta = \frac{1}{n!} \sum_{\mu \vdash n} |\mathcal{C}_\mu| \chi^\theta_\mu p_\mu$$

and

$$Z_\theta = \frac{1}{2^n n!} \sum_{\mu \vdash n} \phi^\theta(\mu) p_\mu$$

where $\phi^\theta(\mu) = \sum_{\pi \in \mathcal{K}_\mu} \chi^{2\theta}(\pi)$, $\chi^{2\theta}(\pi)$ is the evaluation of the character $\chi^{2\theta}$ at π and $2\theta = (2\theta_1, 2\theta_2, \ldots)$.

4.1.2 Genus series for rooted hypermaps in orientable and locally orientable surfaces

The generating series that are given are for rooted hypermaps in orientable and locally orientable surfaces, for they can be specialized to give the generating series for maps in orientable and locally orientable surfaces. There is an axiomatization for hypermaps, but it is not needed in its explicit form for the construction of the Atlas, so it is excluded from the discussion.

Let $h(\nu, \phi, \eta; 0)$ and $h(\nu, \phi, \eta; 1)$ be the numbers of hypermaps in orientable and locally orientable surfaces, respectively, with vertex partition ν, hyperface partition ϕ and hyperedge partition η. Then $h(\nu, \phi, \eta; 1) - h(\nu, \phi, \eta; 0)$ is the number of hypermaps with these vertex and hyperface and hyperedge partitions in nonorientable surfaces. Let $\mathbf{x}_\theta$ denote $x_{\theta_1} x_{\theta_2} \cdots$. Let $H(\mathbf{x}, \mathbf{y}, \mathbf{z}, 0)$ and $H(\mathbf{x}, \mathbf{y}, \mathbf{z}, 1)$ be the generating series

$$H(\mathbf{x}, \mathbf{y}, \mathbf{z}, 0) = \sum_{\nu, \phi, \eta \in \mathcal{P}} h(\nu, \phi, \eta; 0) \mathbf{x}_\nu \mathbf{y}_\phi \mathbf{z}_\eta$$

and

$$H(\mathbf{x},\mathbf{y},\mathbf{z},1) = \sum_{\nu,\phi,\eta\in\mathcal{P}} h(\nu,\phi,\eta;1)\mathbf{x}_\nu\mathbf{y}_\phi\mathbf{z}_\eta$$

for these numbers, where $\mathcal{P}$ is the set of all partitions with the empty partition adjoined.

By the Euler-Poincaré Formula, the type of a map is determined by the number of vertices, edges and faces, and this information is available in $H(\mathbf{x},\mathbf{y},\mathbf{z},0)$ and $H(\mathbf{x},\mathbf{y},\mathbf{z},1)$, so they are called the *genus series* for rooted hypermaps in *orientable* and *locally orientable* surfaces, respectively.

Let $\mathbf{p}(\mathbf{x}) = (p_1(\mathbf{x}), p_2(\mathbf{x}), \ldots)$. Then the genus series for rooted hypermaps in orientable surfaces is

$$H(\mathbf{p}(\mathbf{x}),\mathbf{p}(\mathbf{y}),\mathbf{p}(\mathbf{z}),0) = t\frac{\partial}{\partial t}\log\sum_{\theta\in\mathcal{P}} t^{|\theta|}H_\theta s_\theta(\mathbf{x})s_\theta(\mathbf{y})s_\theta(\mathbf{z})\Bigg|_{t=1},$$

and the genus series for rooted hypermaps in locally orientable surfaces is

$$H(\mathbf{p}(\mathbf{x}),\mathbf{p}(\mathbf{y}),\mathbf{p}(\mathbf{z}),1) = 2t\frac{\partial}{\partial t}\log\sum_{\theta\in\mathcal{P}} t^{|\theta|}\frac{1}{H_{2\theta}}Z_\theta(\mathbf{x})Z_\theta(\mathbf{y})Z_\theta(\mathbf{z})\Bigg|_{t=1},$$

where, for $\theta = (\theta_1, \theta_2, \ldots)$,

$$H_\theta = \prod_{(i,j)\in\mathcal{F}_\theta}\left(\theta_i + \tilde{\theta}_j - i - j + 1\right).$$

The partition $\tilde{\theta} = (\tilde{\theta}_1, \tilde{\theta}_2, \ldots)$ is the *conjugate* of θ and is defined by $\mathcal{F}_{\tilde{\theta}} = \mathcal{F}_\theta^t$, where the transpose $\mathcal{F}_\theta^t$ of a Ferrers graph $\mathcal{F}_\theta$ is obtained by reflecting $\mathcal{F}_\theta$ in the diagonal passing through the top left corner.

Although explicit expressions for

$$H(\mathbf{p}(\mathbf{x}),\mathbf{p}(\mathbf{y}),\mathbf{p}(\mathbf{z}),0) \quad\text{and}\quad H(\mathbf{p}(\mathbf{x}),\mathbf{p}(\mathbf{y}),\mathbf{p}(\mathbf{z}),1)$$

have been given in terms of Schur functions and zonal polynomials, in order to obtain $H(\mathbf{x},\mathbf{y},\mathbf{z},0)$ and $H(\mathbf{x},\mathbf{y},\mathbf{z},1)$ from these it is important to observe that the right hand sides are to be resolved with respect to the power sum symmetric functions $p_k(\mathbf{x}), p_k(\mathbf{y})$ and $p_k(\mathbf{z})$. Since the $p_i(\mathbf{x})$ are algebraically independent, $p_k(\mathbf{x})$ can then be replaced by x_k for all $k \geq 1$; and similarly for $p_k(\mathbf{y})$ and $p_k(\mathbf{z})$.

4.1.3 Two algebras

The appearance of Schur functions and zonal polynomials is made less surprising by two observations about the axiomatizations. The axiomatizations provide an encoding of rooted maps as a permutation such that the face permutation is obtained as a product of two permutations. For maps in orientable surfaces it is therefore natural to consider the class algebra of the symmetric group, whose conjugacy classes $\mathcal{C}_\lambda$ are indexed by partitions of n as a natural index. These partitions then correspond to vertex and face partitions. Orthogonal idempotents can be constructed that span the class algebra. These are linear combinations of the C_λ's, where C_λ is a formal sum of the elements of $\mathcal{C}_\lambda$. The scalars are derived from the coefficients of Schur functions with respect to the power sum symmetric function basis of Λ. If $\alpha, \beta, \gamma \vdash n$, then the coefficient of C_γ in $\mathsf{C}_\alpha \mathsf{C}_\beta$ in the class algebra (the *connexion coefficient* for the class algebra) can be determined first by expressing C_α and C_β in terms of the orthogonal idempotents, by multiplying these expressions (this accounts for two Schur functions) to obtain a linear combination of orthogonal idempotents, and by then expressing this in terms of the C_λ. The final step involves a third Schur function.

The product of two of the symmetric functions, in effect, multiplies the vertex permutation and the edge permutation. Multiplication by the third symmetric function, in effect, captures the face permutation, and therefore the face partition. The logarithm appears because the series to which it is applied counts unions of rooted labelled hypermaps, and is exponential in an indeterminate t that marks the number of half-edges. It therefore restricts the enumeration to rooted labelled hypermaps. This is a familiar enumerative device (and was known to Hurwitz at the turn of the century). The differential operator $t\partial/\partial t$, together with the convention that the series is to be regarded as an ordinary series in t, erases the effect of the original labelling which weights each rooted map by an additional multiplicity depending only on the number of edges. This then gives an enumeration of rooted maps.

A similar process can be used in the case of locally orientable surfaces, for, in this case, it is natural to consider the Hecke algebra of the hyperoctahedral group embedded in the symmetric group $\mathfrak{S}_{4n}$ as the stabilizer of a particular fixed point free involution. This algebra is therefore the double coset algebra and its elements have partitions as a natural index. For the presentation that has been used to construct the set of all rooted maps the elements of this algebra are precisely the formal sums, K_λ, of the elements of the double coset $\mathcal{K}_\lambda$ for all $\lambda \vdash n$. There is again a construction for the orthogonal idempotents, which leads to the *connexion coefficients* for this algebra.

4.2 Specialization to maps

Specialization from hypermaps to maps is achieved by digon conflation. Algebraically this corresponds to setting $z_1 = z_3 = z_4 = z_5 = \cdots = 0$, and $z_2 = \sqrt{z}$. Under this substitution for $\mathbf{z}$ the series $H(\mathbf{x}, \mathbf{y}, \mathbf{z}, 0)$ is transformed into $M(\mathbf{x}, \mathbf{y}, z, 0)$, the genus series for rooted maps in orientable surfaces, and $H(\mathbf{x}, \mathbf{y}, \mathbf{z}, 1)$ is transformed into $M(\mathbf{x}, \mathbf{y}, z, 1)$, rooted maps in locally orientable surfaces.

Further specializations are, of course, possible by setting the indeterminates to zero or one. A *smooth* map is a map with no vertices of degree one; a *homeomorphically irreducible* map is a map with no vertices of degree two. Restriction to these classes of maps is achieved algebraically by setting $x_1 = 0$ and $x_2 = 0$, respectively. Face degrees may be restricted as well. For example, *triangulations* are obtained by setting $y_1 = y_2 = y_4 = y_5 = \cdots = 0$, and $y_3 = y$.

Care must be taken that such specializations involve properties of the map and not just the associated graph. For example, the presence of a loop in a map is a graph theoretic property which cannot always be determined by the degrees of vertices and faces in the map. Although a face of degree 1 is always bounded by a loop, the converse is not necessarily true. A simple example is given by $\mathfrak{m}_{0\cdot 15}$ where the loop separates faces of degree 3. Similarly, $\mathfrak{m}_{1\cdot 1}$ has two edges, both of which are loops, but its one face has degree 4. Thus restriction to maps without loops cannot be achieved simply by setting $y_1 = 0$. A similar comment applies for multiple edges. Thus restriction to *simple* maps (maps whose associated graph has no loops or multiple edges) cannot be achieved simply by setting indeterminates to zero or one.

4.2.1 The genus series for maps in orientable surfaces

$M(\mathbf{x}, \mathbf{y}, z, 0)$ is the genus series for rooted maps in orientable surfaces. There are two useful specializations to consider. For this purpose, let f^θ be the evaluation of χ^θ at the identity (the degree of the representation). Let x, y, z be indeterminates marking the number of vertices, faces and edges, respectively.

The first retains vertex partition, face partition and number of edges.

$$M(\mathbf{p}(\mathbf{x}), \mathbf{p}(\mathbf{y}), z, 0) = t\frac{\partial}{\partial t} \log \left(1 + \sum_{\substack{n\geq 1 \\ \theta \vdash 2n}} \frac{(2n)!}{f^\theta} \chi^\theta_{[2^n]} s_\theta(\mathbf{x}) s_\theta(\mathbf{y}) \frac{z^n}{2^n n!} t^{2n} \right) \Bigg|_{t=1} .$$

The second, which is less refined, retains only the number of vertices, edges and faces.

$$M(x, y, z, 0) = t\frac{\partial}{\partial t} \log \left(1 + \sum_{\substack{n\geq 1 \\ \theta \vdash 2n}} \frac{(2n)!}{f^\theta} \chi^\theta_{[2^n]} H_\theta(x) H_\theta(y) \frac{z^n}{2^n n!} t^{2n} \right) \Bigg|_{t=1}$$

where

$$H_\theta(x) = \prod_{(i,j)\in\mathcal{F}_\theta} (x - i + j).$$

The degree of irreducible representations is given explicitly by

$$f^\theta = \frac{\prod_{1\leq i<j\leq m} (\theta_i - i - \theta_j + j)}{\prod_{1\leq i\leq m} (\theta_i - i + j)!},$$

where $m = l(\theta)$. This is known as the *degree formula*.

4.2.2 The genus series for maps in locally orientable surfaces

$M(\mathbf{x}, \mathbf{y}, z, 1)$ is the genus series for rooted maps in locally orientable surfaces. There are two specializations to consider.

The first retains vertex partition, face partition and number of edges.

$$M(\mathbf{p}(\mathbf{x}), \mathbf{p}(\mathbf{y}), z, 1) = 2t\frac{\partial}{\partial t} \log \left(1 + \sum_{\substack{n\geq 1 \\ \theta \vdash 2n}} \frac{f^{2\theta}}{(4n)!} \left([p_2^n]\, Z_\theta \right) Z_\theta(\mathbf{x}) Z_\theta(\mathbf{y}) z^n t^{2n} \right) \Bigg|_{t=1}.$$

The second, again less refined, retains only the number of vertices, edges and faces.

$$M(x, y, z, 1) = 2t\frac{\partial}{\partial t} \log \left(1 + \sum_{\substack{n\geq 1 \\ \theta \vdash 2n}} \frac{f^{2\theta}}{(4n)!} \left([p_2^n]\, Z_\theta \right) H^\star_\theta(x) H^\star_\theta(y) z^n t^{2n} \right) \Bigg|_{t=1},$$

where

$$H^\star_\theta(x) = \prod_{(i,j)\in\mathcal{F}_\theta} (x + 2j - i - 1).$$

4.3 The Quadrangulation Conjecture

We conclude this introductory material with an account of the two conjectures which lay behind the inception of the Atlas.

4.3.1 An informal principle of enumerative combinatorics

The common source of the two conjectures is an "informal principle" in enumerative combinatorics that an algebraic relationship between the generating series for two sets of combinatorial objects implies the existence of a natural bijection between the sets.

In the present context, this suggests that algebraic properties of generating series for classes of maps should be induced directly by combinatorial properties of maps themselves or by invariants associated with the maps. This Introduction to the Atlas is therefore concluded with a discussion of these conjectures and the role served by the Atlas in their study. Although the conjectures are of interest from a purely combinatorial point of view, they also appear to have an impact on substantial questions that arise outside the field of combinatorics. Brief comments are made on the nature of these interconnexions.

The algebraic property that lies behind the Quadrangulation Conjecture is a simple linear functional relationship between the genus series for rooted quadrangulations and all rooted maps. This implies a relationship between the number of rooted quadrangulations and a weighted sum of rooted maps. The conjecture posits the existence of a natural bijection between rooted quadrangulations on the one hand and, on the other hand, rooted maps with certain subconfigurations combinatorially distinguished, or *decorated*, to account for the presence of the weights. The Atlas can be used to provide instances of paired sets of rooted quadrangulations and rooted decorated maps whose elements are in bijective correspondence with each other. A natural bijection that accounts for the functional relationship must therefore be consistent with these setwise actions. What is absent is a description of the elementwise action of a natural bijection that accounts for these setwise actions, and that can be shown to hold in general. Since conjectures similar to the Quadrangulation Conjecture can be formulated for other regular maps (such as triangulations) and, as will be seen, for Eulerian maps, it appears that the Quadrangulation Conjecture is an instance of a deeper and more general combinatorial phenomenon.

4.3.2 Background

The functional relationship between the genus series for rooted quadrangulations and all rooted maps induces a relationship between these classes of maps with monovalent and bivalent vertices removed. This in turn implies a relationship between the partition functions of the ϕ^4-model and the Penner model in 2-dimensional quantum gravity, after an appropriate limit is taken.

The same formalism arises in the approach of 't Hooft to quantum chromodynamics. The partition function can be shown to be given by counting rooted maps in orientable surfaces. For this purpose, it is convenient to make cuts within distance ε parallel to each edge, thereby excising regions homeomorphic to discs. The excised regions correspond to the faces of the map. What is left is a collection of "ribbons," of width 2ε, that meet at junctions corresponding to the original vertices. The two edges of the ribbon are regarded as a quark-antiquark pair that are permanently attached to each other and that sweep out a 2-dimensional surface in space-time. This surface is imagined to be the consequence of the interaction of gluons and can be regarded as a meson. In this model interactions of mesons occur where these surfaces meet, namely, at the junctions corresponding to the vertices of the map.

4.3.3 The conjecture

Let $A(u, x, y, z)$ be the generating series for all rooted maps in orientable surfaces, with respect to genus (marked by the indeterminate u), and the numbers of vertices, faces and edges. Let $Q(u, x, y, z)$ be the corresponding generating series for rooted quadrangulations in orientable surfaces. Then these generating series are known to be related by

$$Q(u^2, x, y, z) = \tfrac{1}{2}\left(A(4u^2, x, x+u, z^2y) + A(4u^2, x, x-u, z^2y)\right).$$

This is a linear functional relationship between Q and A that involves simple transformations of the arguments of $A(u, x, y, z)$. According to the "informal principle" in enumerative combinatorics, the simple transformations of arguments $y \mapsto x + u$ and $z \mapsto z^2y$ suggest that there may be a direct proof that involves elementary operations on the surfaces. Indeed, the occurrence of $x + u$ as the face argument of A suggests that a face of a rooted map is to be replaced by a vertex or a handle, together with the cutting and gluing of faces that is suggested by the occurrence of z^2y as the edge argument of A.

While suggestive, these ideas remain illusory. The only known proof requires quite sophisticated algebraic ideas, including the group algebra of the symmetric group, properties of Schur functions, the factorization of characters evaluated on certain conjugacy classes into the product of a pair of characters evaluated on two particular conjugacy classes, and a bijective mapping that extends this factorization, in effect, to a Cartesian product of characters on these two conjugacy classes. Moreover, there is very little about the proof that can be combinatorialized and, in particular, it does not provide, even indirectly, a basis for a more direct and geometrical explanation of the functional relationship.

Neverthless, it is possible, at least in principle, to accumulate further combinatorial information about the bijection that is sought by examining pairs of sets that correspond under its action and, from this information, to conjecture a natural elementwise action. The rest of this section is concerned with the construction of such pairs of sets.

The functional relationship between genus series indicates that, for orientable surfaces, there is a correspondence between rooted quadrangulations of given genus and rooted maps of not greater genus. To specify this in greater detail, let $q(g, e)$ be the number of rooted quadrangulations of genus g with e edges and let $a(g, f, e)$ be the number of rooted maps of genus g with e edges and f faces. Then the relationship between Q and A is equivalent to the weighted sum

$$q(g, 2e) = \sum_{\gamma=0}^{g} \sum_{f} 2^{2\gamma} \binom{f}{2g - 2\gamma} a(\gamma, f, e),$$

relating the combinatorial numbers $q(g, e)$ and $a(g, f, e)$ and this, in turn, implies a correspondence between the two classes of maps at the elementwise level. One of these classes is the set of rooted quadrangulations while the other requires an additional construction to be performed on the set of all maps to account for the weights in the sum.

Decoration

The weight associated with $a(g, f, e)$ may be realized combinatorially by attaching markers to the map. The markers are to be indistinguishable from each other. In particular, the factor $\binom{f}{2g-2\gamma}$ can be realized by marking $2g - 2\gamma$ of the faces in all possible ways, while the factor $2^{2\gamma}$ can be realized by first selecting canonically 2γ edges, and by then marking a subset of these edges in all possible ways. The "2" corresponds to "marking" or "not marking" a canonically selected edge.

The result of attaching markers to a map will be referred to as a *decorated* map. Clearly, it is necessary to show that in any given situation, the markers can be attached in a canonical way although not necessarily uniquely. One method may be superior to another, but it suffices for present purposes to provide one method. It is convenient to regard the markers as being attached through the action of a *decorating operator*, which is denoted by Δ.

In this terminology, a natural bijection is sought between rooted quadrangulations and the rooted decorated maps.

Examples: some setwise actions

A small example will demonstrate the setwise action of a bijection between rooted quadrangulations and rooted decorated maps. Figure 4.1 shows the 15 rooted quadrangulations of the torus with 4 edges.

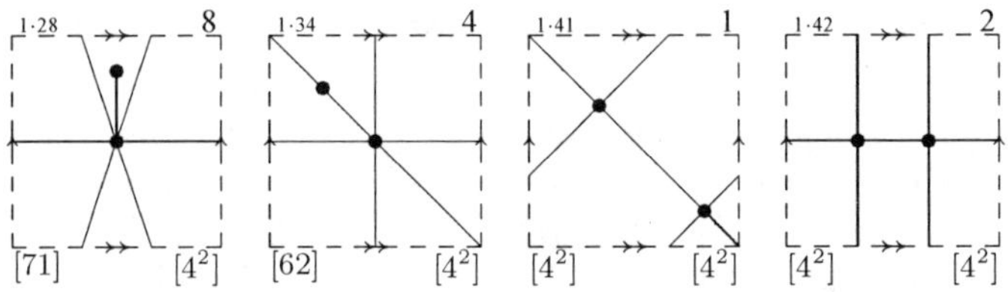

FIGURE 4.1
The 15 rooted quadrangulations of genus 1 with 4 edges

According to the relationship, they can be put into correspondence with the set of decorated maps with 2 edges in the sphere and torus. There are 15 decorated maps in the image set, and these are given in Figure 4.2. They can be accounted for as follows.

Such a decorated map in the sphere must have 2 of its faces marked. From the Atlas it is seen that there are 3 (unrooted) maps of genus 0 with 2 edges and 2 or more faces: $\mathfrak{m}_{0\cdot 3}$, $\mathfrak{m}_{0\cdot 4}$ and $\mathfrak{m}_{0\cdot 5}$. The first row of Figure 4.2 lists the 3 ways of marking precisely 2 of the 3 faces of $\mathfrak{m}_{0\cdot 3}$. There are no canonically selected edges to mark. This produces a set of 6 decorated rooted maps, since $\mathfrak{m}_{0\cdot 3}$ has 2 rootings. The second row of Figure 4.2 contains $\mathfrak{m}_{0\cdot 4}$ and $\mathfrak{m}_{0\cdot 5}$. Each of them has 2 faces, so they both must be marked, and again there are no canonically selected edges to mark. Since these maps have 4 and 1 rootings, respectively, this accounts for 5 of the decorated maps.

Such a decorated map in the torus will have no faces marked, but must have a subset of 2 canonically selected edges marked. There is one rooted map with 2 edges in the torus and it appears in the third row of Figure 4.2. Since it has only 2 edges, both must be in the canonically selected set. The four drawings show the 4 different ways of marking a subset of these. The canonically selected edges are indicated by *broken* lines, *dashed* to indicate a marked edge, *dotted* to indicate an unmarked edge. The third row therefore contains decorated maps. This completes the description of

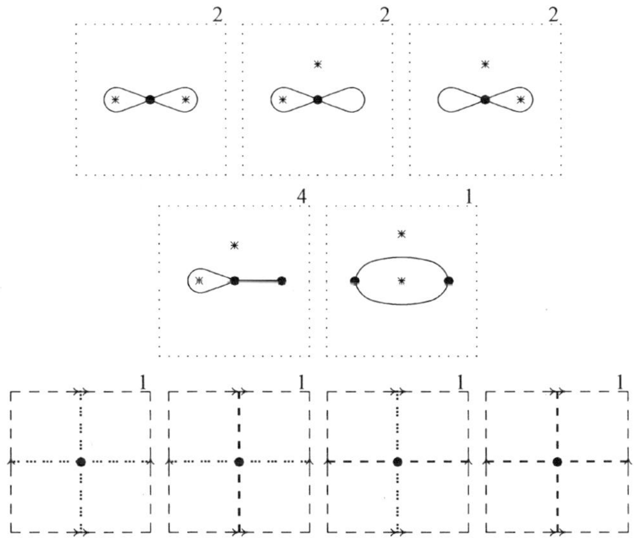

FIGURE 4.2
The 15 decorated maps with 2 edges

the 15 decorated rooted maps corresponding bijectively to the quadrangulations of Figure 4.1.

A bijection for the sphere: the medial construction

For $u = 0$, the surface is the sphere and the relation reduces to

$$Q(0, x, y, z) = A(0, x, x, z^2 y).$$

There is a well known bijection, called the *medial construction*, that accounts for this relationship.

The medial of a map $\mathfrak{m}$ is constructed as follows. Let $\mathcal{V}$ denote the vertex set of $\mathfrak{m}$. A new vertex is put into each face of $\mathfrak{m}$, and let $\mathcal{V}'$ denote this set of vertices. For each face of $\mathfrak{m}$, the vertices encountered in a tour of a face of $\mathfrak{m}$ are joined to the vertex of $\mathcal{V}'$ which is in the face. Let μ denote this map. The *medial* $\mathfrak{m}^*$ of $\mathfrak{m}$ is the map obtained from μ by deleting the edges of $\mathfrak{m}$, and is clearly a quadrangulation. The associated graph of $\mathfrak{m}^*$ has vertex set $\mathcal{V} \cup \mathcal{V}'$. By construction, the edges of $\mathfrak{m}^*$ join vertices in $\mathcal{V}$ to vertices in $\mathcal{V}'$, so $\mathfrak{m}^*$ is bipartite.

If $\mathfrak{m}$ is rooted, let e be its root edge and let v be its root vertex. Let e' be the first edge of μ, that is not an edge of $\mathfrak{m}$, which is encountered in an anticlockwise tour of $\mathfrak{v}$ starting from $\mathfrak{e}$. The root edge of $\mathfrak{m}^*$ is defined to be $\mathfrak{e}'$, and $\mathfrak{v}$ is its tail. This supplies a rooting to $\mathfrak{m}^*$.

The construction is reversed as follows. Since $\mathfrak{m}^\star$ is bipartite it is vertex 2-colourable. The bipartition $\{\mathcal{A},\mathcal{B}\}$ of the rooted quadrangulation $\mathfrak{m}^\star$ can be recovered by defining $\mathcal{A}$ to be the set of all vertices in the same colour class as the root vertex of $\mathfrak{m}^\star$. Then μ is the map obtained by adjoining to the edge set of $\mathfrak{m}^\star$ the diagonals of the faces of $\mathfrak{m}^\star$ that join vertices in $\mathcal{A}$. The map $\mathfrak{m}$ is obtained from μ be deleting the edges of $\mathfrak{m}^\star$, and by then determining the induced rooting.

It follows that the medial construction is a bijection from the set of all rooted maps in the sphere to the set of all rooted bipartite quadrangulations in the sphere.

It can be shown as follows that all quadrangulations in the sphere are bipartite. Let $\mathfrak{q}$ be a quadrangulation and let $\mathfrak{t}$ be one of its spanning trees. Let $\mathcal{V}_\mathfrak{q}$ be the cycle space of $\mathfrak{q}$. This is a vector space over $GF(2)$ consisting of the set of all even spanning subgraphs of $\mathfrak{q}$. Since $\mathfrak{q}$ is a quadrangulation in the sphere, each fundamental cycle of $\mathfrak{q}$ with respect to $\mathfrak{t}$ is a face cycle of $\mathfrak{q}$, and therefore has even length. The fundamental cycles form a basis of $\mathcal{V}_\mathfrak{q}$, so each cycle of $\mathfrak{q}$ is a sum over $GF(2)$ of even length cycles, and so has even length. Thus $\mathfrak{q}$ has no cycles of odd length and is therefore bipartite. It follows that the medial construction is a bijection from the set of all rooted maps in the sphere to the set of all rooted quadrangulations in the sphere.

Because the medial construction requires biparticity, it does not extend to surfaces of higher genera. For example, $\mathfrak{m}_{1\cdot 1}$ is a quadrangulation in the torus and is not bipartite.

4.3.4 Generalization to Eulerian maps and the bijection Ω

More detailed information about the conjectured bijection can be obtained through an enumerative argument that refines the weight function with respect to which the maps are counted and that, in some sense, attains the limit of generalization of the classes of maps for which a conjecture of this type holds in orientable surfaces. The limit is imposed by an algebraic requirement that a particular character evaluation must factorize in a particular way, for without this there appears to be no possibility of a factorization of series that is necessary for the observed additive relation to hold between the logarithms of these series (it will be recalled that the *logarithm* appears in the expression for the genus series). This extension of the Quadrangulation Conjecture concerns maps whose face degrees are even. Such maps are called *dually Eulerian.*

An enumerative argument shows that the genus series $E(u,x,\mathbf{y},z)$ for the set $\mathcal{E}$ of rooted dually Eulerian maps with respect to genus, number of vertices, face partition and number of edges marked, respectively, by $u,x,\mathbf{y},z$ and the genus series

$H(u, x, y^{(1)}, \mathbf{y}^{(2)}, z)$ for rooted hypermaps with respect to the number of hyperfaces and hyperedge partition η marked, respectively, by $y^{(1)}$ and $\mathbf{y}^{(2)}$ are related by

$$E(u^2, x, \mathbf{y}, z) = \frac{1}{2}\left(H(4u^2, x, x+u, \mathbf{y}, z) + H(4u^2, x, x-u, \mathbf{y}, z)\right)\Big|_{y_i \mapsto y_{2i}, i=1,2,\ldots}$$

Let $m_{\mathcal{E}}(g, i, \mathbf{j}, e)$ be the number of rooted dually Eulerian maps of genus g with i vertices, e edges and face partition $\mathbf{j}$, and let $h(\gamma, v, f, \mathbf{j}, e)$ be the number of rooted hypermaps of genus γ with v vertices, f faces, e edges and hyperedge partition $\mathbf{j}$. Then the above relationship between E and H is equivalent to the relationship

$$m_{\mathcal{E}}(g, i, 2\mathbf{j}, e) = \sum_{\substack{\gamma \geq 0; v, f \geq 1 \\ v+f-2(g-\gamma)=i}} 2^{2\gamma} \binom{f}{2g-2\gamma} h(\gamma, v, f, \mathbf{j}, e)$$

between the combinatorial numbers $m_{\mathcal{E}}(g, v, \mathbf{j}, e)$ and $h(g, v, f, \mathbf{j}, e)$. This expresses one class of numbers as a weighted sum of the other class, and from the weight it is possible to characterize a set $\widetilde{\mathcal{H}}$ of decorated rooted hypermaps that are in bijection with the set of rooted Eulerian maps. The conjecture therefore asserts that there is a natural bijection, Ω, which achieves this.

Decoration

To assist in the combinatorial realization of the weighting, it is necessary to introduce the notion of a strut in a map. Let $\mathfrak{s}$ be a submap of a map $\mathfrak{m}$ that contains all of the vertices of $\mathfrak{m}$, that has the same genus as $\mathfrak{m}$ and that has precisely one face. Let $\mathfrak{t}$ be a spanning tree of $\mathfrak{s}$. A *strut* of $\mathfrak{m}$ is an edge of $\mathfrak{s}$ that is not in $\mathfrak{t}$. Let $\mathcal{S}$ denote the set of all struts. If $\mathfrak{m}$ has genus zero, then $\mathfrak{s}$ is a spanning tree and $\mathcal{S}$ is empty. For higher genera, a spanning tree is not a 2-cell embedding in the surface, so the struts serve as a minimal set of edges that can be added to a spanning tree to obtain a 2-cell embedding in the surface. Moreover, it is possible to make a canonical choice of $\mathfrak{s}$ and then $\mathfrak{t}$, so a canonical set of struts can always be selected for any map. In view of this $\mathcal{S}$ is denoted by $\mathcal{S}_{\mathfrak{m}}$.

A natural way of realizing the weight that appears in the sum is the following. Let $\mathfrak{h}$ be a hypermap counted by $h(\gamma, v, f, \mathbf{j}, e)$. The factor $2^{2\gamma}$ is realized by specifying a canonical set $\mathcal{S}_{\mathfrak{h}}$ of 2γ edges of $\mathfrak{h}$ from which subsets can be chosen. This is achieved by observing that a map with $2\gamma + v - 1$ edges has one face, by the Euler-Poincaré formula. Since a spanning tree has $v - 1$ vertices, $\mathcal{S}_{\mathfrak{h}}$ is a set of 2γ edges, so there are $2^{2\gamma}$ ways of selecting a subset of these edges. The factor of $\binom{f}{2g-2\gamma}$ can be realized by selecting an even number of hyperfaces to be distinguished. Thus, for the generalized Quadrangulation Conjecture, a *decorated* map is a hypermap with a distinguished set of struts chosen from a (canonical) set $\mathcal{S}_{\mathfrak{h}}$ and an even subset of distinguished hyperfaces.

4.3.5 Setwise action of the bijection Ω

Let $\mathcal{E}_{g,2\mathbf{j}}$ be the set of all rooted dually Eulerian maps of genus g and face partition $2\mathbf{j}$ (the number of vertices and edges is then determined), and let $\mathcal{H}_{\gamma,f,\mathbf{j}}$ be the set of rooted hypermaps of genus γ, with f hyperfaces and hyperedge partition $\mathbf{j}$ (the number of vertices and edges is then determined). For a map $\mathfrak{m}$, let $\Delta_{p,q}(\mathfrak{m})$ be the set of maps obtained from it by marking p hyperfaces in all possible ways and at most q struts in all possible ways, so $\Delta_{p,q}$ is a mapping that applies decoration to a map. Then the relationship between the combinatorial numbers can be recast as a bijection

$$\Omega\colon \mathcal{E}_{g,2\mathbf{j}} \xrightarrow{\sim} \bigcup_{\gamma=0}^{g} \bigcup_{f=\max(1,2g-2\gamma)}^{2g-2\gamma+\#vertices(\mathfrak{e})-1} \Delta_{2g-2\gamma,2\gamma}(\mathcal{H}_{\gamma,f,\mathbf{j}}) : \mathfrak{e} \longmapsto \mathfrak{h},$$

where the unions are disjoint. Thus, under Ω, a rooted dually Eulerian map $\mathfrak{e}$ with parameters $(g, 2\mathbf{j})$ corresponds to a decorated rooted hypermap $\mathfrak{h}$ with parameters $(\gamma, f, \mathbf{j})$, $2g-2\gamma$ marked hyperfaces and 2γ marked struts.

The enumerative argument does not provide any information about the functional dependence of f and γ from the collection of parameters of $\mathfrak{h}$ on the parameters of the preimage $\mathfrak{e}$. It is possible, however, with the aid of the Atlas, to give instances of pairs of sets between which Ω acts bijectively, and this appears to be the limit of the extent to which enumerative information can assist in the characterization of Ω.

Two examples are now given to illustrate the setwise action of the generalized Quadrangulation Conjecture, with a detailed account of the actual decorations that are used to make the correspondence bijective. In the first example the marking of the hyperfaces is forced since every hyperface is to be marked. In the second example the marking is not forced.

Example 1: The image set of $\mathcal{E}_{1,[8]}$ under Ω

The first example is provided by an examination of $\mathcal{E}_{1,[8]}$. Figure 4.3 shows the 21 rooted maps with one face in the double torus, and these together constitute $\mathcal{E}_{1,[8]}$.

These maps have even face degree and according to the generalized relationship, they can be put into correspondence with the set of decorated hypermaps with hyperedge partition $[4]$ in the sphere, torus and double torus. Now, from the setwise action of Ω,

$$\mathcal{E}_{1,[8]} \xrightarrow{\sim} \Delta_{4,0}\left(\mathcal{H}_{0,4,[4]}\right) \cup \Delta_{2,2}\left(\mathcal{H}_{1,2,[4]}\right).$$

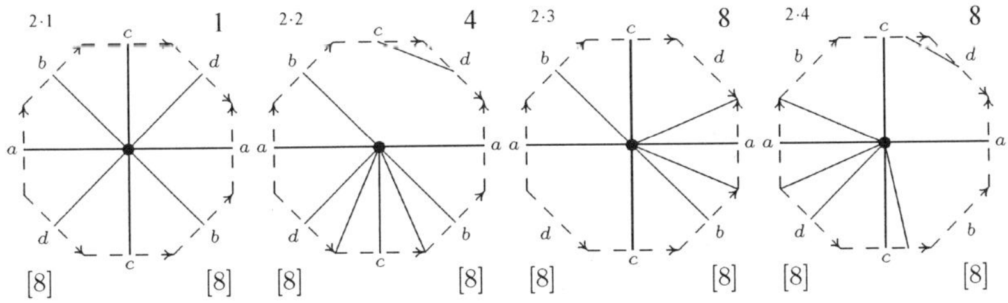

FIGURE 4.3
The 21 rooted maps of genus 2 with face partition [8]

There are 21 maps in the image set, and these are given in Figure 4.4.

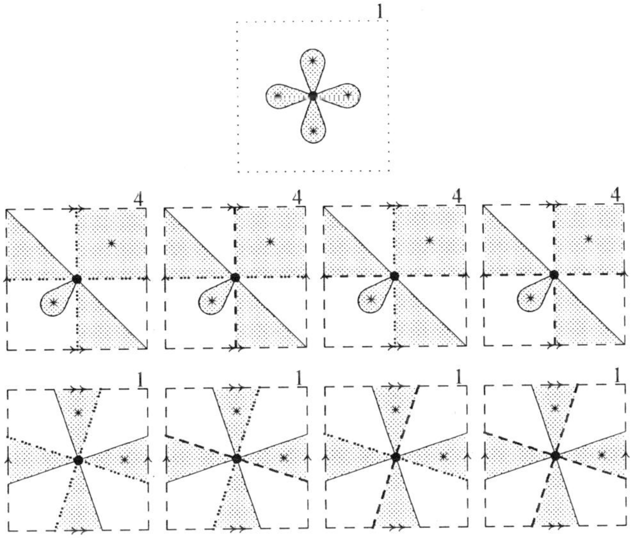

FIGURE 4.4
The 21 decorated hypermaps with hyperedge partition [4]

Such a decorated hypermap in the sphere must have 4 of its hyperfaces marked. There is only one rooted hypermap of genus 0 with hyperedge partition [4] and 4 or more hyperfaces. It is given in the first row of Figure 4.4 and, since it has precisely 4 hyperfaces, under the action of $\Delta_{4,0}$ they are all marked, and no strut is marked. This hypermap appears in the Atlas as $\mathfrak{m}_{0\cdot 21}$. Note that as a rooted map it has 2 rootings, but these correspond to 2 different hypermaps, one with 1 hyperedge and 4 hyperfaces, the other with 4 hyperedges and 1 hyperface.

Such a decorated hypermap in the torus must have 2 of its hyperfaces marked and a subset of its 2 struts marked. The struts can be chosen canonically for each rooted hypermap. However, for brevity, it is convenient to select struts arbitrarily for each hypermap, independent of the rooting. There are 2 different hypermaps of genus 1 with hyperedge partition $[4]$ and 2 or more hyperfaces. The first appears in the second row of Figure 4.4 and since it has precisely 2 hyperfaces, under the action of $\Delta_{2,2}$ they are both marked, and at most two struts are marked. The four drawings show the 4 different ways of marking the struts. The struts are indicated by broken lines, dashed to indicate a marked strut, dotted to indicate an unmarked strut. This hypermap appears in the Atlas as $\mathfrak{m}_{1\cdot 14}$, and as a hypermap it has 4 rootings. So the second row contains 16 decorated hypermaps. The other hypermap of genus 1 with hyperedge partition $[4]$ appears in the third row of Figure 4.4 and since it too has precisely 2 hyperfaces, under the action of $\Delta_{2,2}$ they are both marked and at most 2 struts are marked. Once more, the four drawings in this row show the 4 different ways of marking the struts. This hypermap appears in the Atlas as $\mathfrak{m}_{1\cdot 16}$, and as a hypermap it has 1 rooting.

This completes the description of the 21 decorated hypermaps corresponding to the dually Eulerian maps of Figure 4.3.

Example 2: The image set of $\mathcal{E}_{1,[6\,2]}$ under Ω

The second example is provided by an examination of $\mathcal{E}_{1,[6\,2]}$. Figure 4.5 shows the 40 rooted maps with one face of degree 6 and one face of degree 2 in the torus, and the set of these rooted maps is $\mathcal{E}_{1,[6\,2]}$.

These maps also have even face degree, so they can be put into correspondence with the set of decorated hypermaps with hyperedge partition $[3\,1]$ in the sphere and torus. Now, from the setwise action of Ω,

$$\mathcal{E}_{1,[6\,2]} \xrightarrow{\sim} \Delta_{2,0}(\mathcal{H}_{0,3,[3\,1]}) \cup \Delta_{2,0}(\mathcal{H}_{0,2,[3\,1]}) \cup \Delta_{0,2}(\mathcal{H}_{1,1,[3\,1]})\,.$$

There are 40 decorated rooted hypermaps in the image set, and these are given in Figure 4.6.

Such a decorated hypermap in the sphere must have 2 of its hyperfaces marked. Looking through Section 7.3, it is seen that there are 4 (unrooted) hypermaps of genus 0 with hyperedge partition $[3\,1]$ and 2 or more hyperfaces: $\mathfrak{m}_{0\cdot 22}$, $\mathfrak{m}_{0\cdot 30}$, $\mathfrak{m}_{0\cdot 31}$ and $\mathfrak{m}_{0\cdot 36}$. The set $\mathcal{H}_{0,3,[3\,1]}$ is the set of all rootings of $\mathfrak{m}_{0\cdot 22}$. To this set $\Delta_{2,0}$ is applied. The first row of Figure 4.6 lists the 3 ways of marking precisely 2 of the 3 hyperfaces of $\mathfrak{m}_{0\cdot 22}$. There are no marked struts. This produces a set of decorated rooted hypermaps.

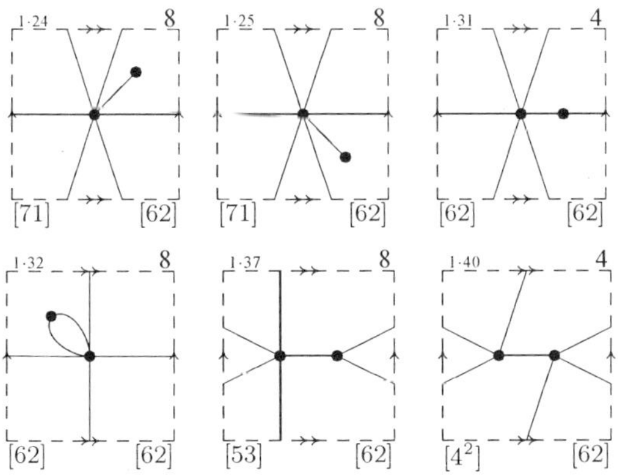

FIGURE 4.5
The 40 rooted maps of genus 1 with face partition $[6\,2]$

The set $\mathcal{H}_{0,2,[3\,1]}$ is the set of all rootings of the other three hypermaps. To this set $\Delta_{2,0}$ is applied, and the resulting maps appear in the second row of Figure 4.6. Each of them has 2 hyperfaces, so they both must be marked, and there are no marked struts. Since there are 4 ways to root each of these hypermaps, this accounts for 24 of the decorated maps.

The set $\mathcal{H}_{1,1,[3\,1]}$ is the set of all rootings of $\mathfrak{m}_{1\cdot 14}$ since there is only hypermap of genus 1 with hyperedge partition $[3\,1]$. As a hypermap it has 4 rootings. To this set $\Delta_{0,2}$ is applied. None of the hyperfaces is to be marked, but a subset of its 2 (canonically chosen) struts are marked in all possible ways. It appears in the third row of Figure 4.6. The four drawings show the 4 different ways of marking the struts. The third row therefore contains 16 decorated hypermaps. This completes the description of the 40 decorated rooted hypermaps corresponding to the dually Eulerian maps of Figure 4.5.

These two examples, and others that may be constructed in a similar way, provide refined setwise actions of Ω. What remains unknown at present, however, is an elementwise description of Ω, and it is hoped that information obtained from such setwise actions will assist the determination of this bijection.

It is not known whether there is an analogous relationship for the case of locally orientable surfaces, since the character theory for the hyperoctahedral group does not appear to support the same type of factorization. Nor is it known whether there is an analogous combinatorial bijection.

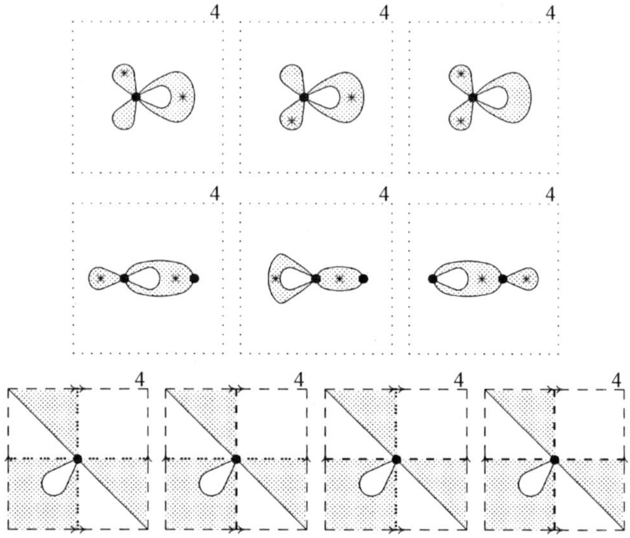

FIGURE 4.6
The 40 decorated hypermaps with hyperedge partition $[3\,1]$

4.4 The b-Conjecture

The algebraic property that lies behind the b-Conjecture is that the genus series for rooted maps in orientable and nonorientable surfaces can be expressed as specializations of a single series $H(\mathbf{x}, \mathbf{y}, \mathbf{z}, b)$ involving Jack symmetric functions in a parameter b, and that, from computational evidence, $H(\mathbf{x}, \mathbf{y}, \mathbf{z}, b)$ appears to have coefficients that are polynomials in the indeterminate b with nonnegative integer coefficients. It is therefore conjectured that b marks a combinatorial invariant of rooted maps. The Atlas can be used to provide the rooted maps that correspond to a given polynomial in b. What is absent, however, is the description of an invariant β of rooted maps such that the generating series for the maps with respect to this invariant is precisely the polynomial in b.

4.4.1 Background

The determination of the virtual Euler characteristic for the moduli spaces of complex algebraic curves can be reduced by an argument from algebraic geometry to a question about the genus series for rooted monopoles in orientable surfaces. For the case of real curves there is an analogous argument that reduces to a question about the genus series for rooted monopoles in locally orientable surfaces. With the aid of $H(\mathbf{x}, \mathbf{y}, \mathbf{z}, b)$ the virtual Euler characteristics for these two cases can be united into a single expression that interpolates between the two. This gives a parametrization of the virtual Euler characteristic for the moduli spaces of real and complex algebraic curves in terms of a parameter b. This parameter takes on the value 0 for complex curves and 1 for real curves. In the b-Conjecture, however, the indeterminate b is conjectured to be associated with an invariant of rooted maps in locally orientable surfaces, so this in turn suggests that the parametrization of the virtual Euler characteristic is a natural one. Indeed, it has been conjectured that this parametrized virtual Euler characteristic is the virtual Euler characteristic of moduli spaces indexed in a natural way by a property in the geometry that is induced by β.

4.4.2 Jack symmetric functions

The strong similarity between $H(\mathbf{x}, \mathbf{y}, \mathbf{z}, 0)$ and $H(\mathbf{x}, \mathbf{y}, \mathbf{z}, 1)$ may be made more apparent by recasting the expressions for them in terms of another symmetric function, the Jack symmetric function.

Let a be a positive real number. An inner product $\langle\ ,\ \rangle_a$ can be defined on the ring Λ of symmetric functions by

$$\langle p_\lambda, p_\mu \rangle_a = \delta_{\lambda\mu} \frac{|\lambda|!}{|C_\lambda|} a^{l(\lambda)}.$$

When $a = 1$, this reduces to the familiar inner product with respect to which the monomial symmetric functions and the complete symmetric functions are dual bases. If $\preceq$ denotes reverse lexicographic order, then the *Jack symmetric functions* $J_\lambda(\mathbf{x}; a)$ are uniquely determined by the following three conditions:

1. $\langle J_\lambda, J_\mu \rangle_a = 0$ for $\lambda \neq \mu$ (orthogonality),
2. $[m_\mu] J_\lambda = 0$ unless $\mu \preceq \lambda$ (triangularity),
3. $[m_{[1^n]}] J_\lambda = n!$ for all $\lambda \vdash n$ (normalization),

where [A]B denotes the coefficient of A in B. In the above instances of this operator the action is well defined in view of the observations that have been made about bases of the space of symmetric functions. The parameter a is called the *Jack parameter.*

For example, the first few Jack symmetric functions are:

$$\begin{aligned} J_{[1]} &= p_1, \\ J_{[1^2]} &= p_1^2 - p_2, \\ J_{[2]} &= p_1^2 + ap_2, \\ J_{[1^3]} &= p_1^3 - 3p_1p_2 + 2p_3, \\ J_{[2\,1]} &= p_1^3 + (a-1)p_1p_2 - ap_3, \\ J_{[3]} &= p_1^3 + 3ap_1p_2 + 2a^2p_3. \end{aligned}$$

The coefficients of the Jack symmetric function, as polynomials in the ground indeterminates $x_1, x_2, \ldots$, are polynomials in a, so a may be regarded as an indeterminate.

The appearance of the Jack symmetric function is attributable to the fact that $s_\theta(\mathbf{x}) = H_\theta J_\theta(\mathbf{x};1)$ is a Schur function, and that $Z_\theta(\mathbf{x}) = J_\theta(\mathbf{x};2)$ is a zonal polynomial.

The norm of J_θ

There is an explicit expression for $\langle J_\theta, J_\theta\rangle_a$, the square of the norm of J_θ. The *arm length* $a_\theta(\mathsf{x})$, *leg length* $l_\theta(\mathsf{x})$ and *hook length* $h_\theta(\mathsf{x})$ of the cell $\mathsf{x} = (s,t)$ of the Ferrers graph $\mathcal{F}_\theta$ are defined by

$$\begin{aligned} a_\theta(\mathsf{x}) &= |\{(s,j) \in \mathcal{F}_\theta : j > t\}|, \\ l_\theta(\mathsf{x}) &= |\{(i,t) \in \mathcal{F}_\theta : i > s\}|, \\ h_\theta(\mathsf{x}) &= a_\theta(\mathsf{x}) + l_\theta(\mathsf{x}) + 1. \end{aligned}$$

$a_\theta(\mathsf{x})$ is the number of cells that are strictly to the right of x and in the same row of $\mathcal{F}_\theta$; $l_\theta(\mathsf{x})$ is the number of cells that are strictly below x and in the same column of $\mathcal{F}_\theta$. For this purpose, the convention is adopted that the rows of $\mathcal{F}_\theta$ decrease weakly in length from top to bottom and are justified to the left. It should be noted from the definitions that $a_\theta(\mathsf{x})$ and $l_\theta(\mathsf{x})$ are numerical invariants associated with the partition θ. Then, in terms of these,

$$\langle J_\theta, J_\theta\rangle_a = \prod_{\mathsf{x}\in\mathcal{F}_\theta} \left(aa_\theta(\mathsf{x}) + l_\theta(\mathsf{x}) + a\right)\left(aa_\theta(\mathsf{x}) + l_\theta(\mathsf{x}) + 1\right).$$

4.4.3 The conjecture

The generating series $H(\mathbf{x}, \mathbf{y}, \mathbf{z}, b)$ is defined as follows. Let

$$H(\mathbf{p}(\mathbf{x}), \mathbf{p}(\mathbf{y}), \mathbf{p}(\mathbf{z}), b) = (1+b)t\frac{\partial}{\partial t} \log \Phi(\mathbf{p}(\mathbf{x}), \mathbf{p}(\mathbf{y}), \mathbf{p}(\mathbf{z}); t, b)\bigg|_{t=1},$$

where

$$\Phi(\mathbf{p}(\mathbf{x}), \mathbf{p}(\mathbf{y}), \mathbf{p}(\mathbf{z}); t, b) = \sum_{\theta \in \mathcal{P}} \frac{t^{|\theta|}}{\langle J_\theta, J_\theta \rangle_{1+b}} J_\theta(\mathbf{x}; 1+b) J_\theta(\mathbf{y}; 1+b) J_\theta(\mathbf{z}; 1+b).$$

This is an explicit expression for $H(\mathbf{p}(\mathbf{x}), \mathbf{p}(\mathbf{y}), \mathbf{p}(\mathbf{z}), b)$ in terms of Jack symmetric functions, but to obtain $H(\mathbf{x}, \mathbf{y}, \mathbf{z}, b)$ from this it is again important to observe that it is necessary to resolve the right hand side of $\Phi(\mathbf{p}(\mathbf{x}), \mathbf{p}(\mathbf{y}), \mathbf{p}(\mathbf{z}); t, b)$ with respect to the power sum symmetric functions $p_k(\mathbf{x}), p_k(\mathbf{y})$ and $p_k(\mathbf{z})$. Since the $p_i(\mathbf{x})$ are algebraically independent, $p_k(\mathbf{x})$ can then be replaced by x_k for all $k \geq 1$; and similarly for $p_k(\mathbf{y})$ and $p_k(\mathbf{z})$.

Implicit in this expression for $H(\mathbf{x}, \mathbf{y}, \mathbf{z}, b)$ is that it does correctly specialize for $b = 0, 1$ to the expressions for $H(\mathbf{x}, \mathbf{y}, \mathbf{z}, 0)$ and $H(\mathbf{x}, \mathbf{y}, \mathbf{z}, 1)$ that have been given in Section 4.2.1 and Section 4.2.2 in terms of Schur functions and zonal polynomials, respectively. Hence, for $b = 0$ and $b = 1$, $H(\mathbf{x}, \mathbf{y}, \mathbf{z}, b)$ is the genus series for rooted hypermaps in orientable and locally orientable surfaces, respectively, with respect to vertex partition (denoted by ν), hyperface partition (denoted by ϕ) and hyperedge partition (denoted by η). Let

$$h(\nu, \phi, \eta; b)$$

be the coefficient of $\mathbf{x}_\nu \mathbf{y}_\phi \mathbf{z}_\eta$ in $H(\mathbf{x}, \mathbf{y}, \mathbf{z}, b)$.

The $h(\nu, \phi, \eta; b)$ are listed for all hypermaps on at most 5 edges in Section 12.1.1. This provides evidence for the conjecture that $h(\nu, \phi, \eta; b)$ is a polynomial in b with coefficients that are nonnegative integers. These polynomials are called *b-polynomials* for hypermaps. The b-Conjecture asserts that the indeterminate b marks an invariant β of rooted hypermaps that is associated with nonorientability (another invocation of the "informal principle"). Moreover, the specializations $b = 0$ and $b = 1$ give the genus series for hypermaps in orientable and locally orientable surfaces, respectively. In this sense, $H(\mathbf{x}, \mathbf{y}, \mathbf{z}, b)$ is a natural refinement of the genus series for rooted hypermaps in locally orientable surfaces given in Section 4.1.2.

Since $h(\nu, \phi, \eta; 0)$ gives the number of rooted hypermaps in orientable surfaces, then $\beta(\mathfrak{h}) = 0$ if $\mathfrak{h}$ is in an orientable surface. However, the action of β on rooted hypermaps in nonorientable surfaces remains perplexing.

Specialization from hypermaps to maps

Let $M(\mathbf{x}, \mathbf{y}, z, b)$ be the series obtained from $H(\mathbf{x}, \mathbf{y}, \mathbf{z}, b)$ by setting $z_1 = z_3 = z_4 = \cdots = 0$, and $z_2 = \sqrt{z}$. Then for $b = 0$ and $b = 1$, $M(\mathbf{x}, \mathbf{y}, z, b)$ is the genus series for rooted maps in orientable and locally orientable surfaces, respectively, with respect to vertex partition, face partition and number of edges. Let

$$m(\nu, \phi, n; b)$$

be the coefficient of $\mathbf{x}_\nu \mathbf{y}_\phi z^n$ in $M(\mathbf{x}, \mathbf{y}, z, b)$.

The $m(\nu, \phi, n; b)$ are listed for all $1 \leq n \leq 5$ in Section 12.1.2. This provides evidence for the conjecture that $m(\nu, \phi, n; b)$ is a polynomial in b with coefficients that are nonnegative integers. These polynomials are called *b-polynomials* for maps. The b-Conjecture asserts that the indeterminate b marks an invariant β of rooted maps that is associated with nonorientability. Moreover, the specializations $b = 0$ and $b = 1$ give the genus series for maps in orientable and locally orientable surfaces, respectively. In this sense, $M(\mathbf{x}, \mathbf{y}, \mathbf{z}, b)$ is a natural refinement of the genus series for rooted maps in locally orientable surfaces given in Section 4.2.

Since $m(\nu, \phi, n; 0)$ gives the number of rooted maps in orientable surfaces, then $\beta(\mathfrak{m}) = 0$ if $\mathfrak{m}$ is in an orientable surface. However, the action of β on rooted maps in nonorientable surfaces remains perplexing.

4.4.4 Examples

It is instructive to look at specific polynomials and the sets of rooted hypermaps or maps for which these are the generating series, for it is from such examples that further information about the putative invariant β may emerge. One example will be given in each case.

b-polynomial for hypermaps

From Section 12.1.1 it is seen that

$$h([6], [3], [3]; b) = 1 + b + 2b^2,$$

the b-polynomial for all rooted hypermaps in locally orientable surfaces with vertex partition $[6]$, hyperface partition $[3]$ and hyperedge partition $[3]$. This set of maps is given below.

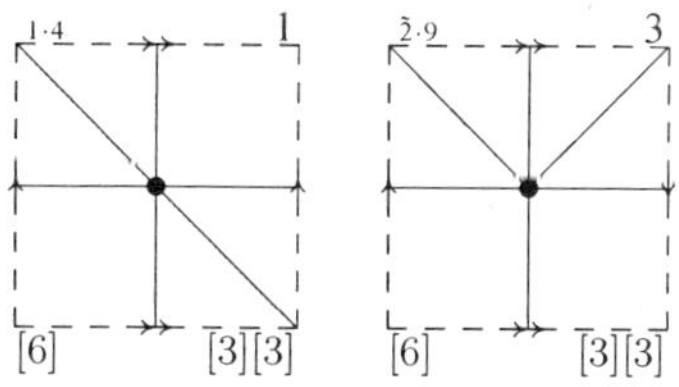

Of course, the constant term 1 in $h([6],[3],[3];b)$ corresponds to the leftmost hypermap, which is embedded in the torus. The other hypermap is embedded in the Klein bottle and has 3 rootings. These 3 rootings must correspond to $b+2b^2$. In other words, one of these rooted hypermaps has $\beta(\mathfrak{h}) = 1$, and for the other 2 rooted hypermaps, $\beta(\mathfrak{h}) = 2$. This example shows that β depends on the rooting and is therefore an invariant of *rooted* hypermaps associated with nonorientability.

b-polynomial for maps

From Section 12.1 it is seen that

$$m([3^2],[6],3;b) = 1 + b + 5b^2,$$

the b-polynomial for all rooted maps in locally orientable surfaces on 3 edges with vertex partition $[3^2]$ and face partition $[6]$. This set of maps is given below.

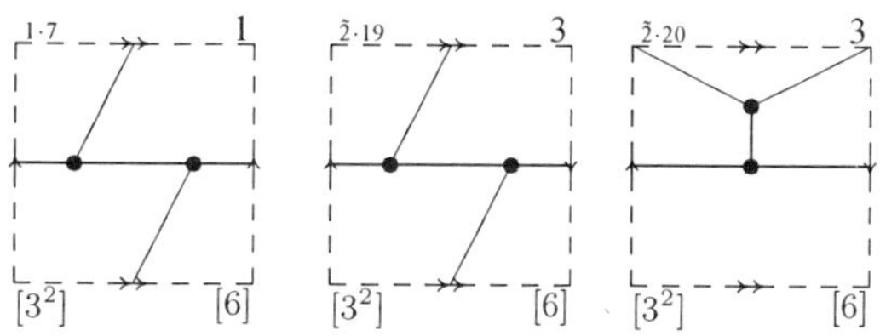

Of course, the constant term 1 in $m([3^2],[6],3;b)$ corresponds to the leftmost map, which is embedded in the torus. The other two maps are embedded in the Klein bottle and each has 3 rootings. Therefore these 6 rootings must correspond to $b + 5b^2$. In other words, one of these maps has $\beta(\mathfrak{m}) = 1$, and for the other 5 maps, $\beta(\mathfrak{m}) = 2$. This example shows that β depends on the rooting and is therefore an invariant of *rooted* maps associated with nonorientability.

Part II

The Atlas

Chapter 5

Maps in orientable surfaces

In this chapter we give the polygonal representation of all maps on up to five edges in orientable surfaces. The maps are listed by *genus, number of edges* and *number of vertices* where these are written in decreasing order of subordination, with the maps arranged in decreasing (lexicographic) order of *vertex partition* and then *edge partition.* When the number of edges or vertices changes, a title indicating this appears at the beginning of a line.

At the foot of each map, the left-hand partition is the vertex partition and the right-hand partition is the face partition. The integer in the top right-hand corner of each map is the number of distinct rootings of the map and the serial number is given in the top left-hand corner.

An annotation is given at the foot of odd pages indicating the genus, number of edges and number of vertices of the first map on the page. The annotation at the foot of the next even page gives the corresponding information for the last map on the page.

5.1 Genus 0 – the sphere

1 edge

1 vertex

0·1 1

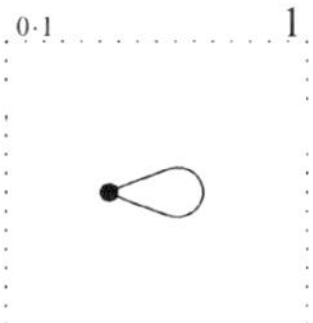

[2] $[1^2]$

1 edge

2 vertices

0·2 1

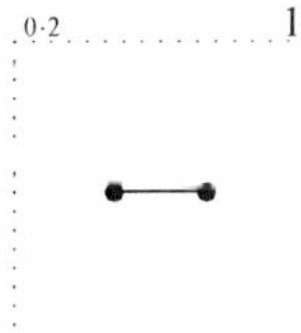

$[1^2]$ [2]

2 edges

1 vertex

0·3 2

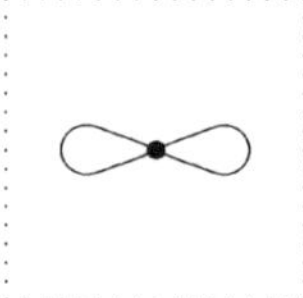

[4] $[21^2]$

2 edges

2 vertices

0·4 4

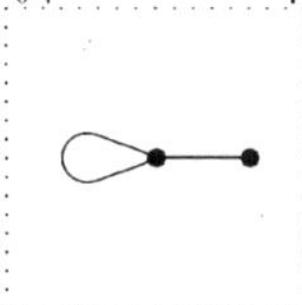

[31] [31]

0·5 1

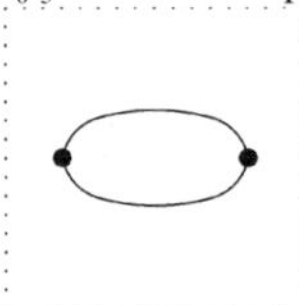

$[2^2]$ $[2^2]$

2 edges

3 vertices

0·6 2

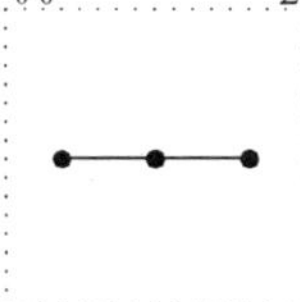

$[21^2]$ [4]

3 edges

1 vertex

0·7 2

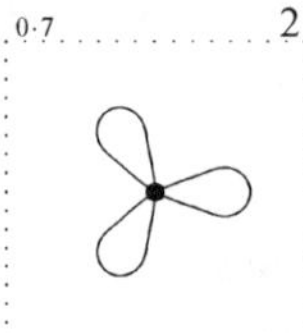

[6] $[31^3]$

0·8 3

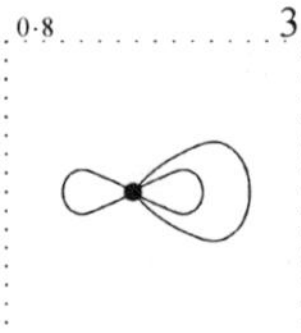

[6] $[2^2 1^2]$

3 edges

2 vertices

0·9 6

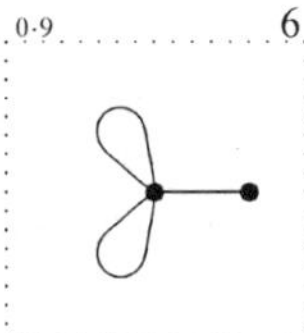

[51] $[41^2]$

0·10 6

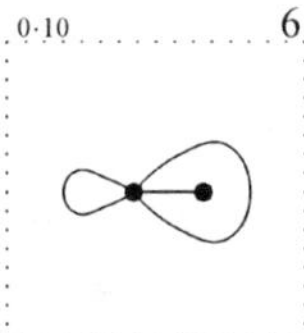

[51] [321]

0·11 6

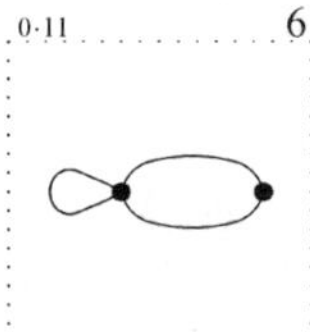

[42] [321]

0·12 3

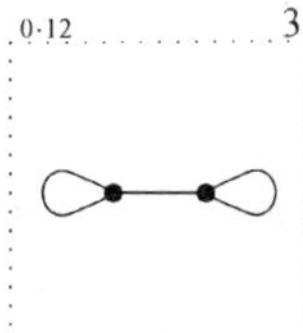

$[3^2]$ $[41^2]$

0·13 1

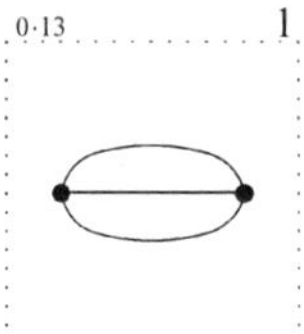

$[3^2]$ $[2^3]$

Genus 0: 3e 3v

3 edges

3 vertices

0·14 6

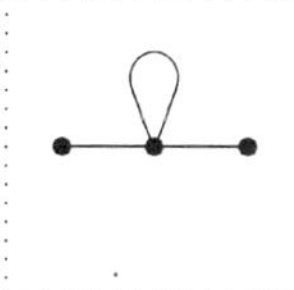

$[41^2]$ $[51]$

0·15 3

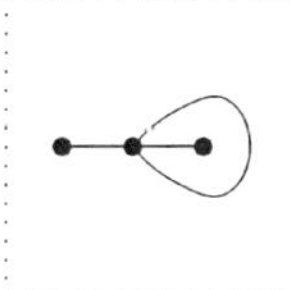

$[41^2]$ $[3^2]$

0·16 6

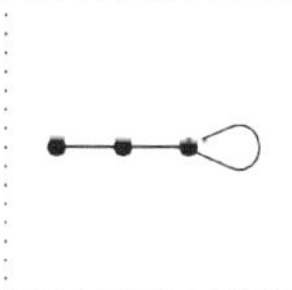

$[321]$ $[51]$

0·17 6

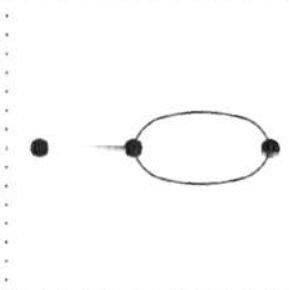

$[321]$ $[42]$

0·18 1

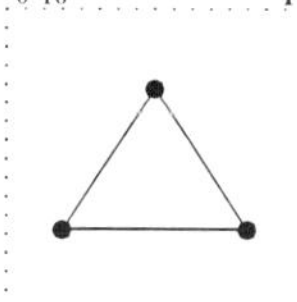

$[2^3]$ $[3^2]$

3 edges

4 vertices

0·19 2

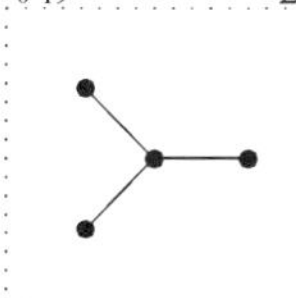

$[31^3]$ $[6]$

0·20 3

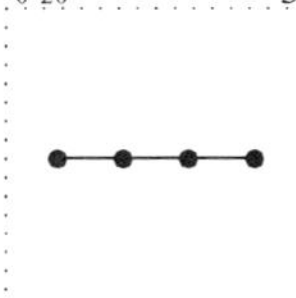

$[2^2 1^2]$ $[6]$

4 edges

1 vertex

0·21 2

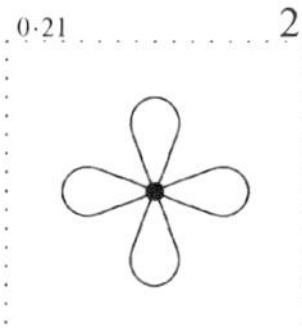

$[8]$ $[41^4]$

0·22 8

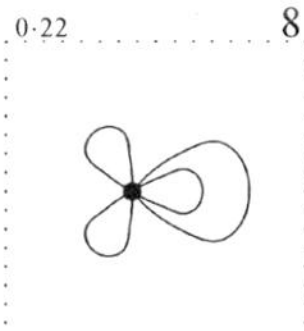

$[8]$ $[321^3]$

0·23 4

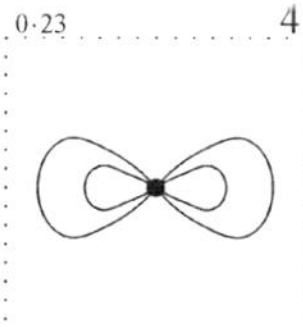

$[8]$ $[2^3 1^2]$

4 edges

2 vertices

0·24 8

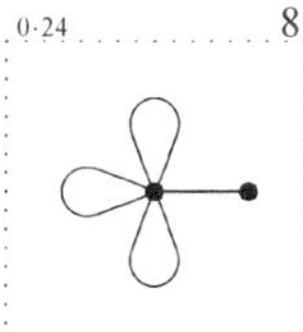

$[71]$ $[51^3]$

0·25 8

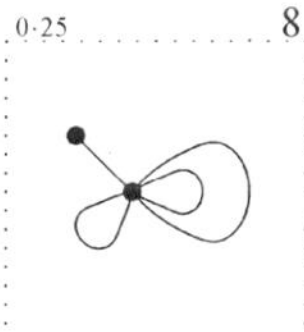

$[71]$ $[421^2]$

0·26 8

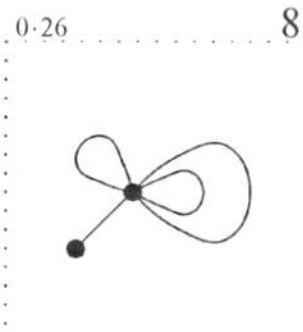

$[71]$ $[421^2]$

0·27 8

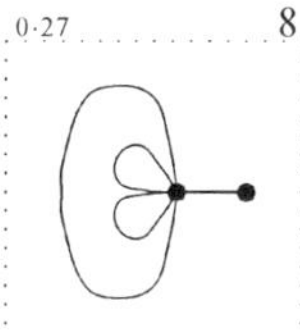

$[71]$ $[3^2 1^2]$

0·28 8

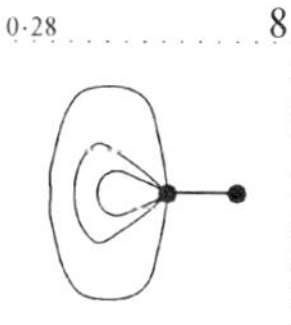

$[71]$ $[32^2 1]$

0·29 8

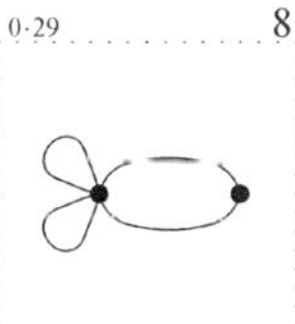

$[62]$ $[421^2]$

0·30 4

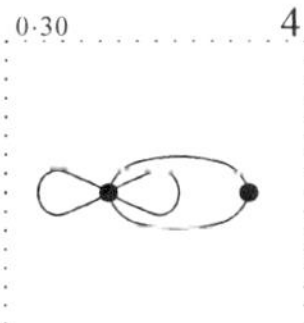

$[62]$ $[3^2 1^2]$

0·31 8

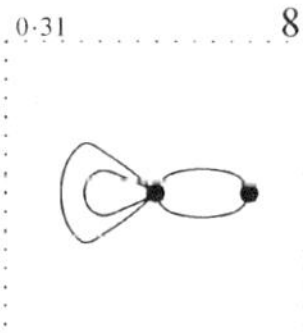

$[62]$ $[32^2 1]$

0·32 8

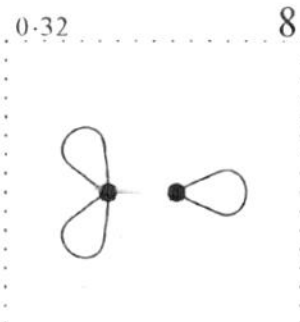

$[53]$ $[51^3]$

0·33 8

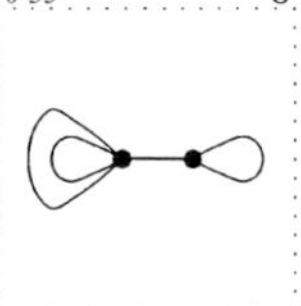

$[53]$ $[421^2]$

0·34 8

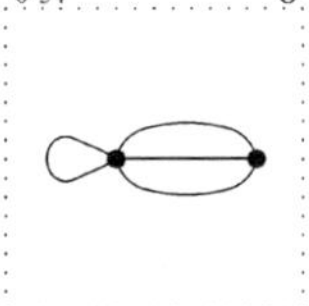

$[53]$ $[32^2 1]$

0·35 4

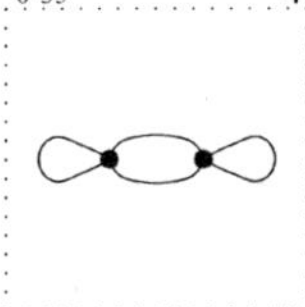

$[4^2]$ $[421^2]$

0·36 4

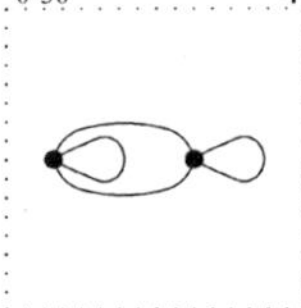

$[4^2]$ $[3^2 1^2]$

0·37 1

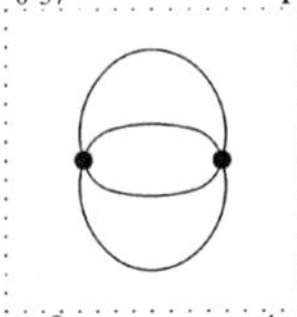

$[4^2]$ $[2^4]$

4 edges

3 vertices

0·38 8

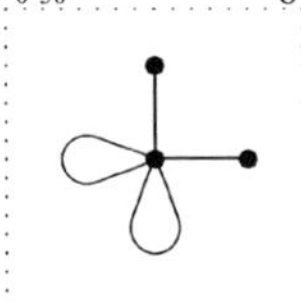

$[61^2]$ $[61^2]$

0·39 4

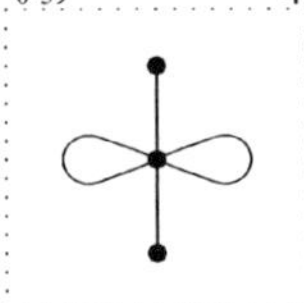

$[61^2]$ $[61^2]$

0·40 8

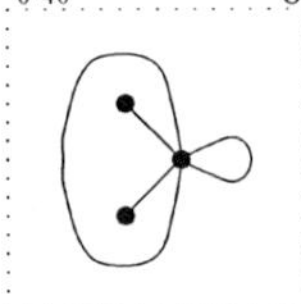

$[61^2]$ $[521]$

0·41 8

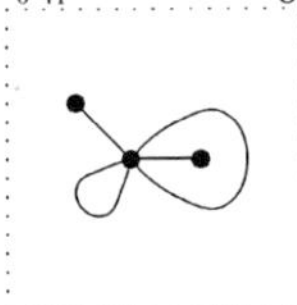

$[61^2]$ $[431]$

0·42 8

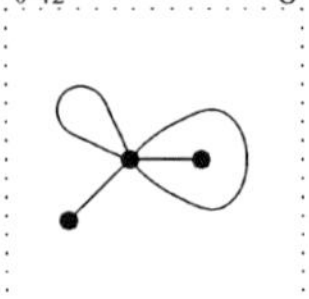

$[61^2]$ $[431]$

0·43 4

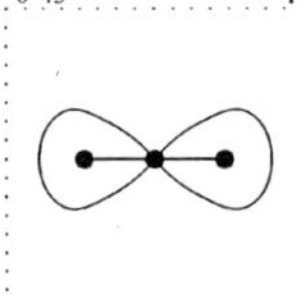

$[61^2]$ $[3^2 2]$

0·44 8

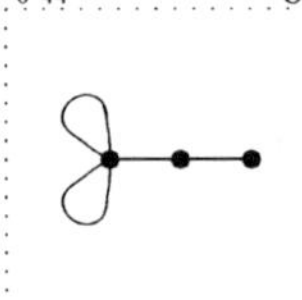

$[521]$ $[61^2]$

0·45 8

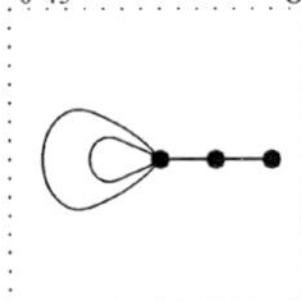

$[521]$ $[521]$

0·46 8

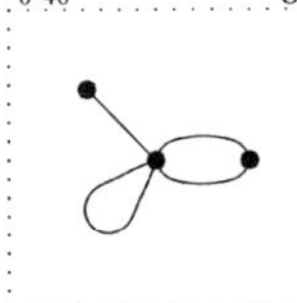

$[521]$ $[521]$

0·47 8

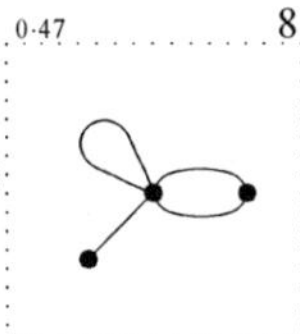

$[521]$ $[521]$

0·48 8

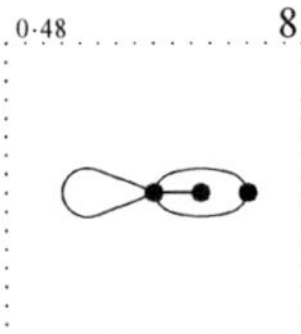

$[521]$ $[431]$

0·49 8

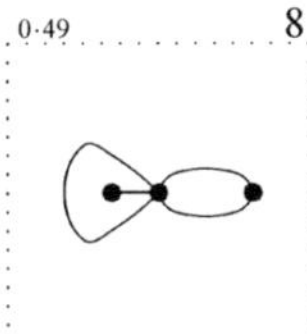

$[521]$ $[3^2 2]$

0·50 8

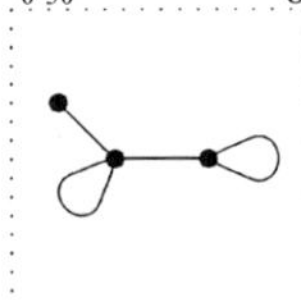

$[431]$ $[61^2]$

0·51 8

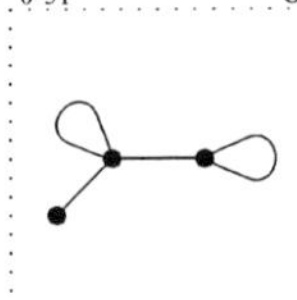

$[431]$ $[61^2]$

0·52 8

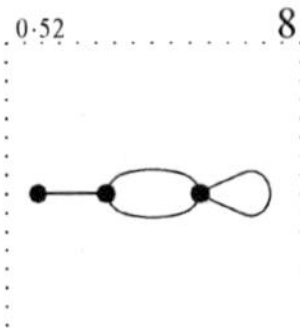

$[431]$ $[521]$

0·53 8

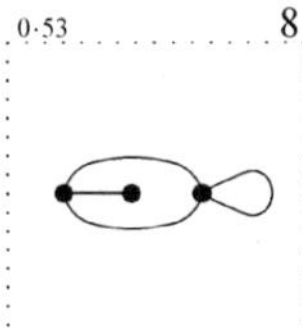

$[431]$ $[431]$

0·54 8

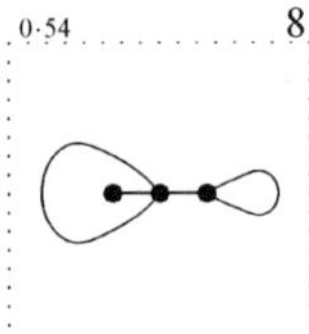

$[431]$ $[431]$

0·55 8

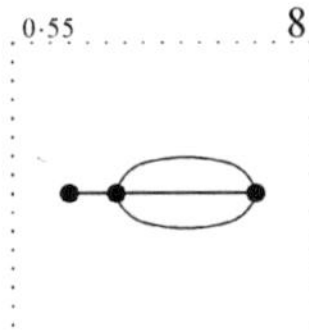

$[431]$ $[42^2]$

0·56 8

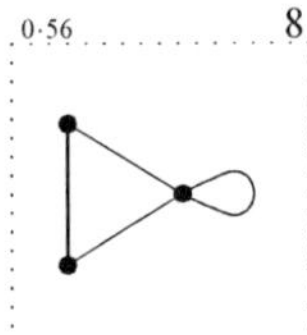

$[42^2]$ $[431]$

0·57 4

$[42^2]$ $[42^2]$

0·58 4

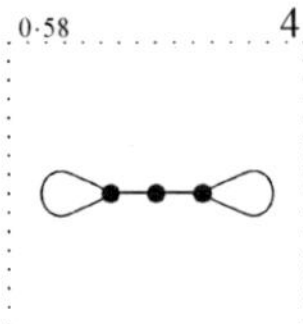

$[3^2 2]$ $[61^2]$

0·59 8

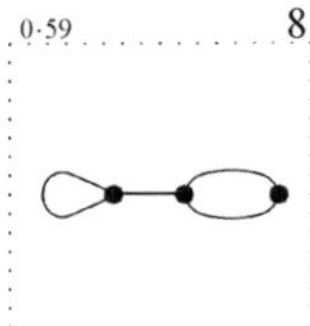

$[3^2 2]$ $[521]$

0·60 4

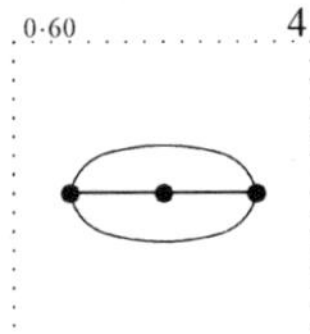

$[3^2 2]$ $[3^2 2]$

Genus 0: 4e 4v

4 edges

4 vertices

0·61 8
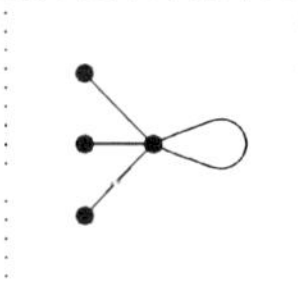
$[51^3]$ $[71]$

0·62 8
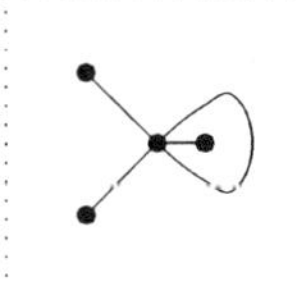
$[51^3]$ $[53]$

0·63 8
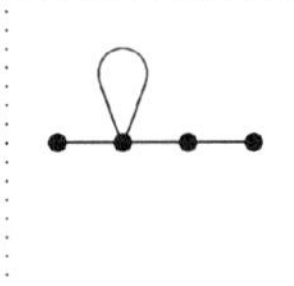
$[421^2]$ $[71]$

0·64 8
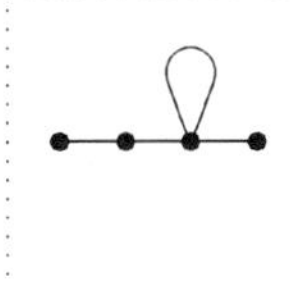
$[421^2]$ $[71]$

0·65 8
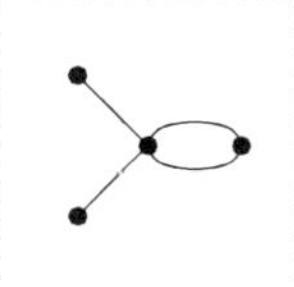
$[421^2]$ $[62]$

0·66 8
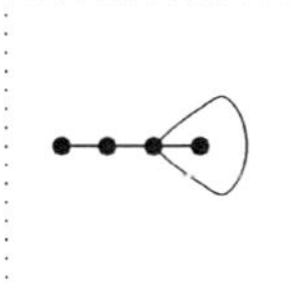
$[421^2]$ $[53]$

0·67 4
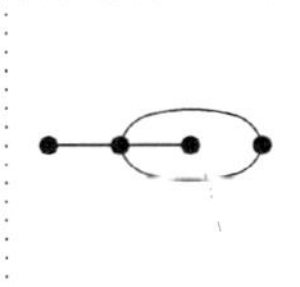
$[421^2]$ $[4^2]$

0·68 8
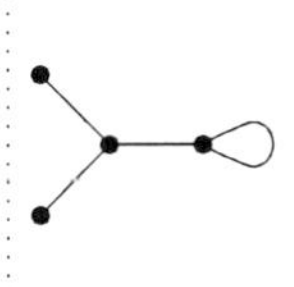
$[3^2 1^2]$ $[71]$

0·69 4
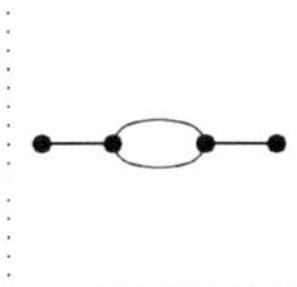
$[3^2 1^2]$ $[62]$

0·70 4
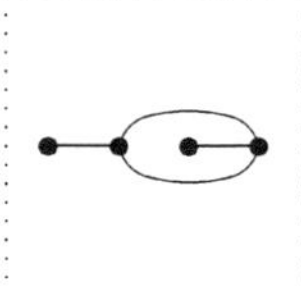
$[3^2 1^2]$ $[4^2]$

0·71 8
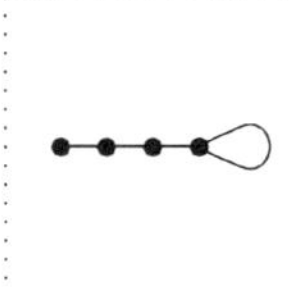
$[32^2 1]$ $[71]$

0·72 8
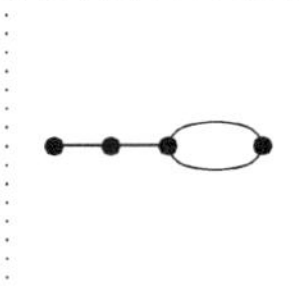
$[32^2 1]$ $[62]$

0·73 8
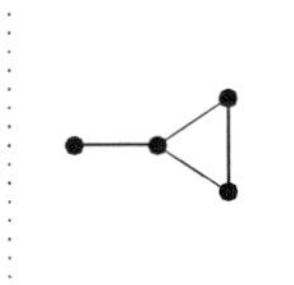
$[32^2 1]$ $[53]$

0·74 1
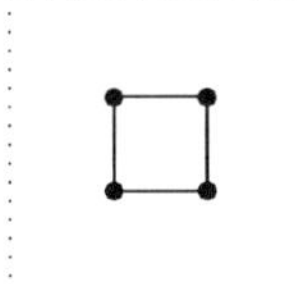
$[2^4]$ $[4^2]$

4 edges

5 vertices

0·75 2
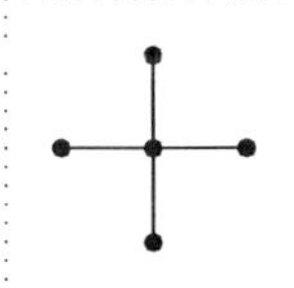
$[41^4]$ $[8]$

0·76 8
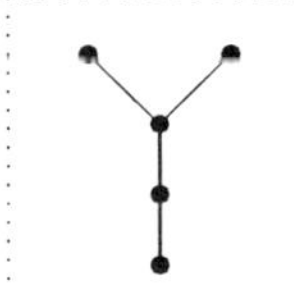
$[321^3]$ $[8]$

0·77 4
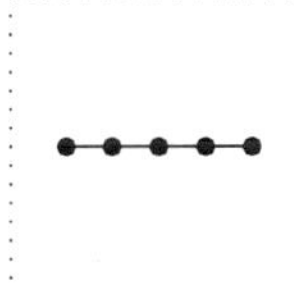
$[2^3 1^2]$ $[8]$

5 edges

1 vertex

0·78 2
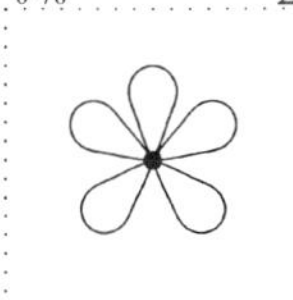
$[10]$ $[51^5]$

0·79 10
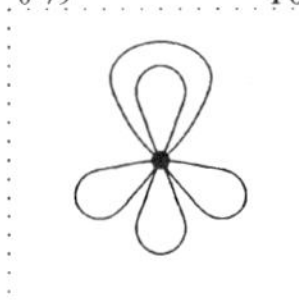
$[10]$ $[421^4]$

0·80 5
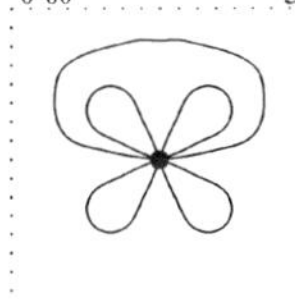
$[10]$ $[3^2 1^4]$

0·81 10
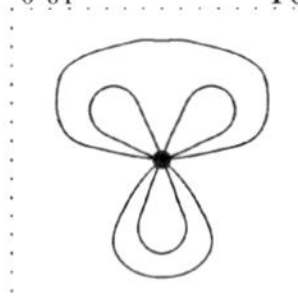
$[10]$ $[32^2 1^3]$

0·82 10

$[10]$ $[32^2 1^3]$

0·83 5
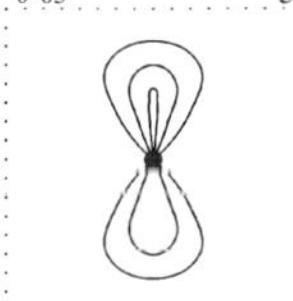
$[10]$ $[2^4 1^2]$

5 edges

2 vertices

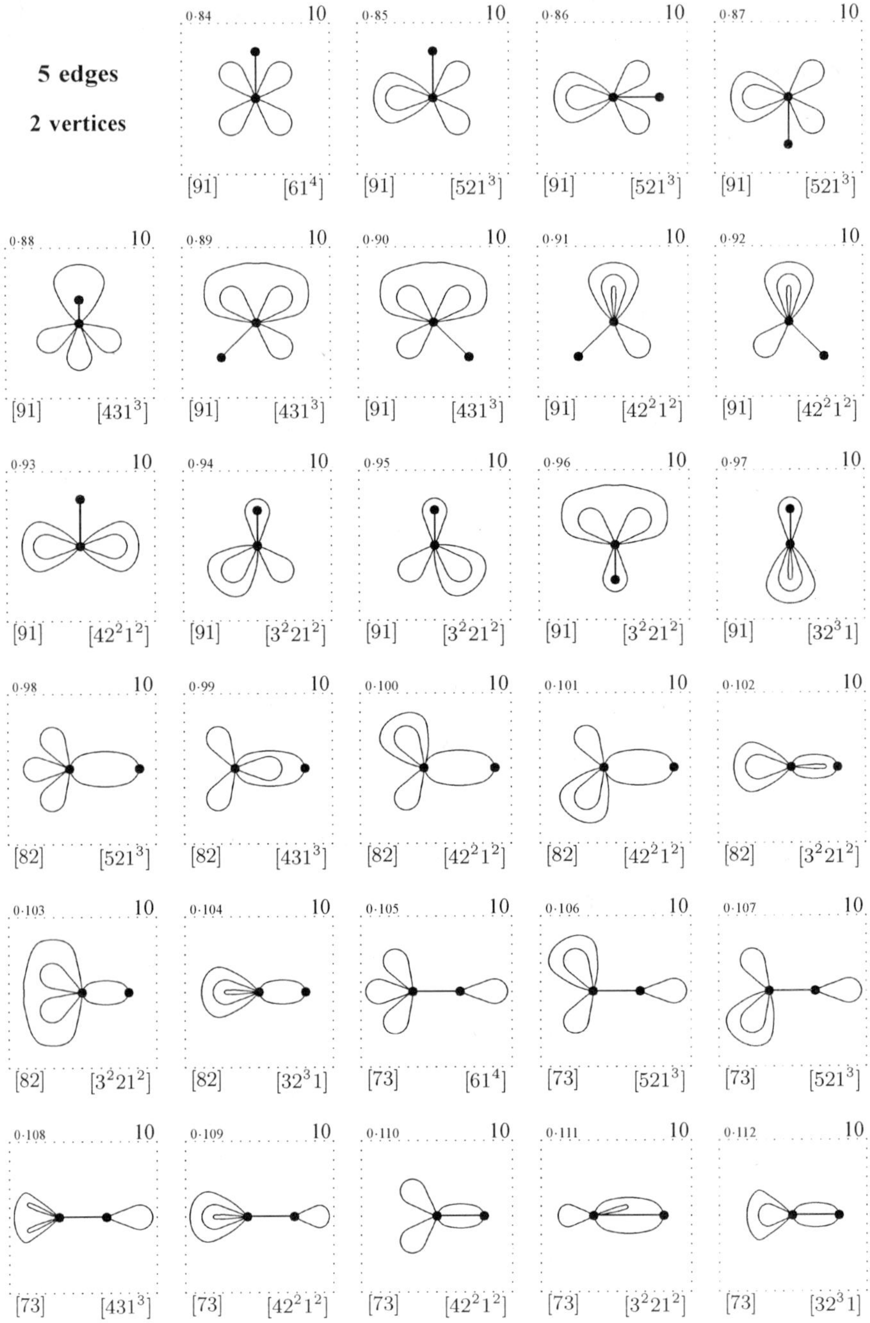

Genus 0: 5e 2v

0·113 10
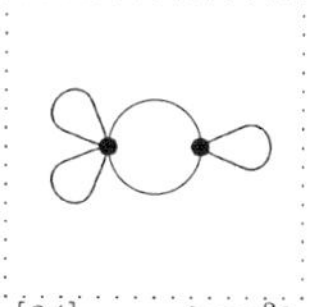
$[64]$ $[521^3]$

0·114 10
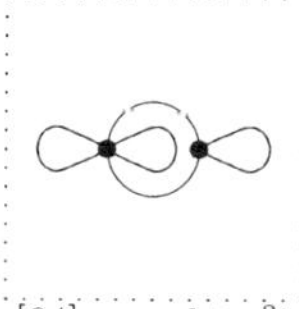
$[64]$ $[431^3]$

0·115 10
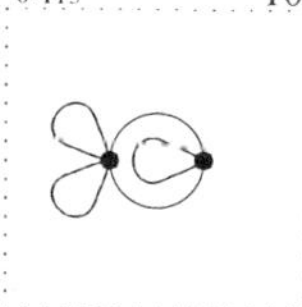
$[64]$ $[431^3]$

0·116 10
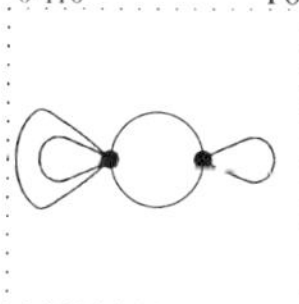
$[64]$ $[42^21^2]$

0·117 10
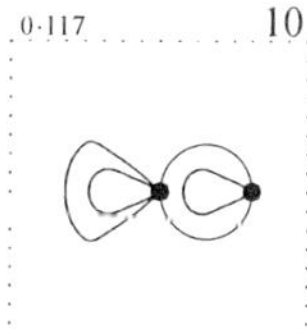
$[64]$ $[3^221^2]$

0·118 10
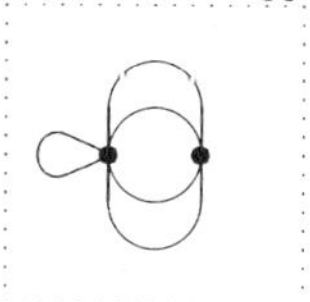
$[64]$ $[32^31]$

0·119 5
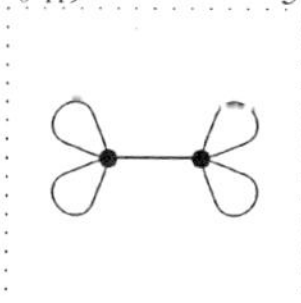
$[5^2]$ $[61^4]$

0·120 10
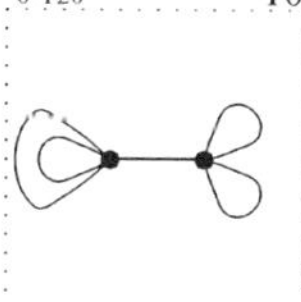
$[5^2]$ $[521^3]$

0·121 5
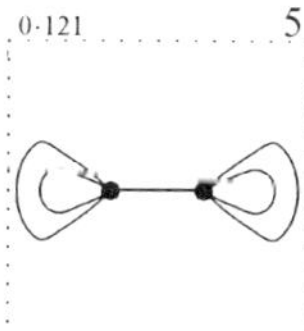
$[5^2]$ $[42^21^2]$

0·122 5
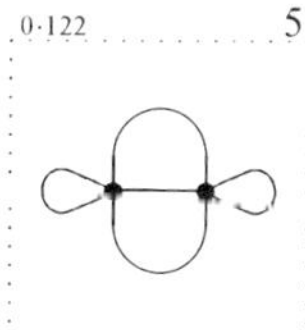
$[5^2]$ $[42^21^2]$

0·123 5
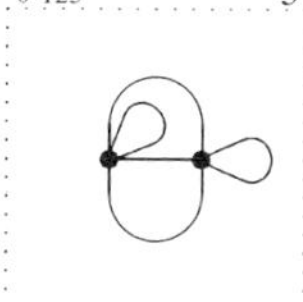
$[5^2]$ $[3^221^2]$

0·124 5
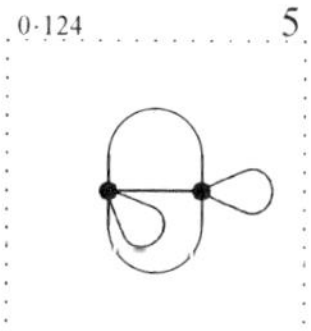
$[5^2]$ $[3^221^2]$

0·125 1
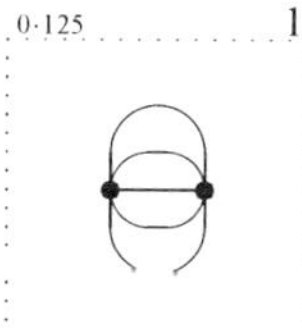
$[5^2]$ $[2^5]$

5 edges

3 vertices

0·126 10
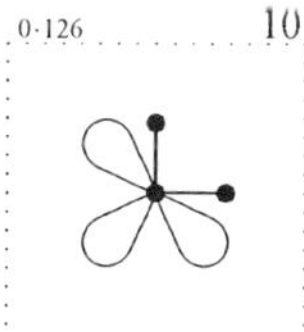
$[81^2]$ $[71^3]$

0·127 10
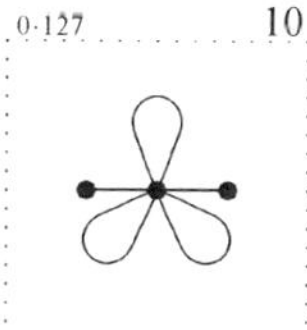
$[81^2]$ $[71^3]$

0·128 10
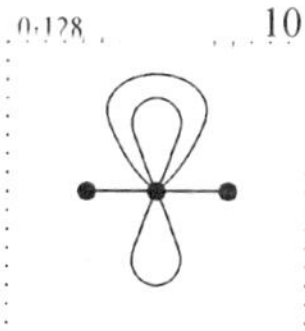
$[81^2]$ $[621^2]$

0·129 10
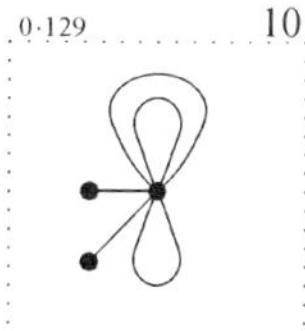
$[81^2]$ $[621^2]$

0·130 10
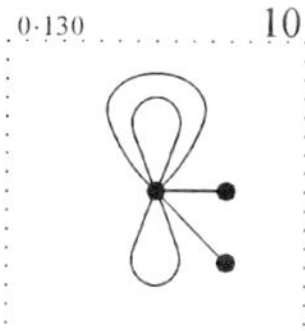
$[81^2]$ $[621^2]$

0·131 10
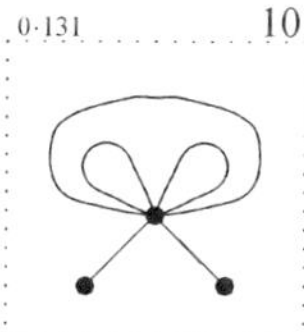
$[81^2]$ $[531^2]$

0·132 10
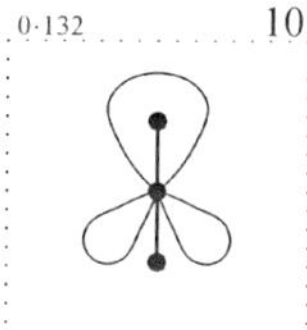
$[81^2]$ $[531^2]$

0·133 10
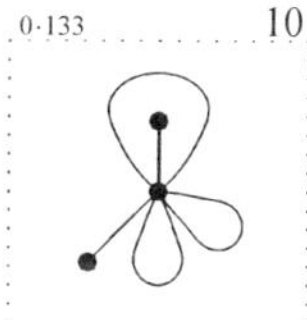
$[81^2]$ $[531^2]$

0·134 10
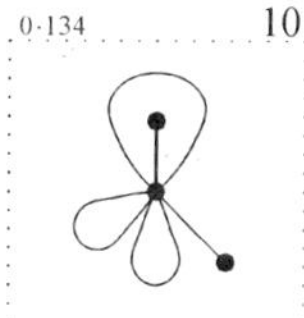
$[81^2]$ $[531^2]$

0·135 10
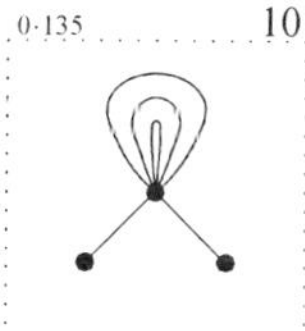
$[81^2]$ $[52^21]$

0·136 5
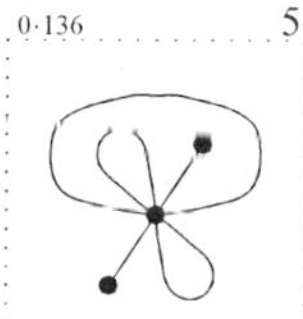
$[81^2]$ $[4^21^2]$

0·137 5
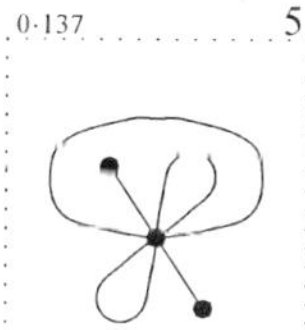
$[81^2]$ $[4^21^2]$

0·138 10
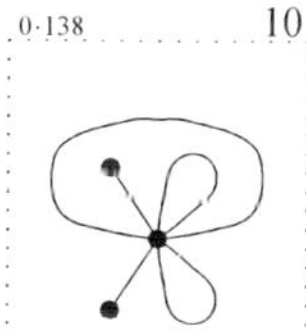
$[81^2]$ $[4^21^2]$

0·139 10
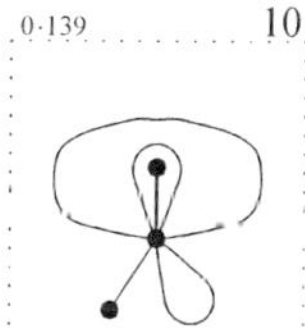
$[81^2]$ $[4321]$

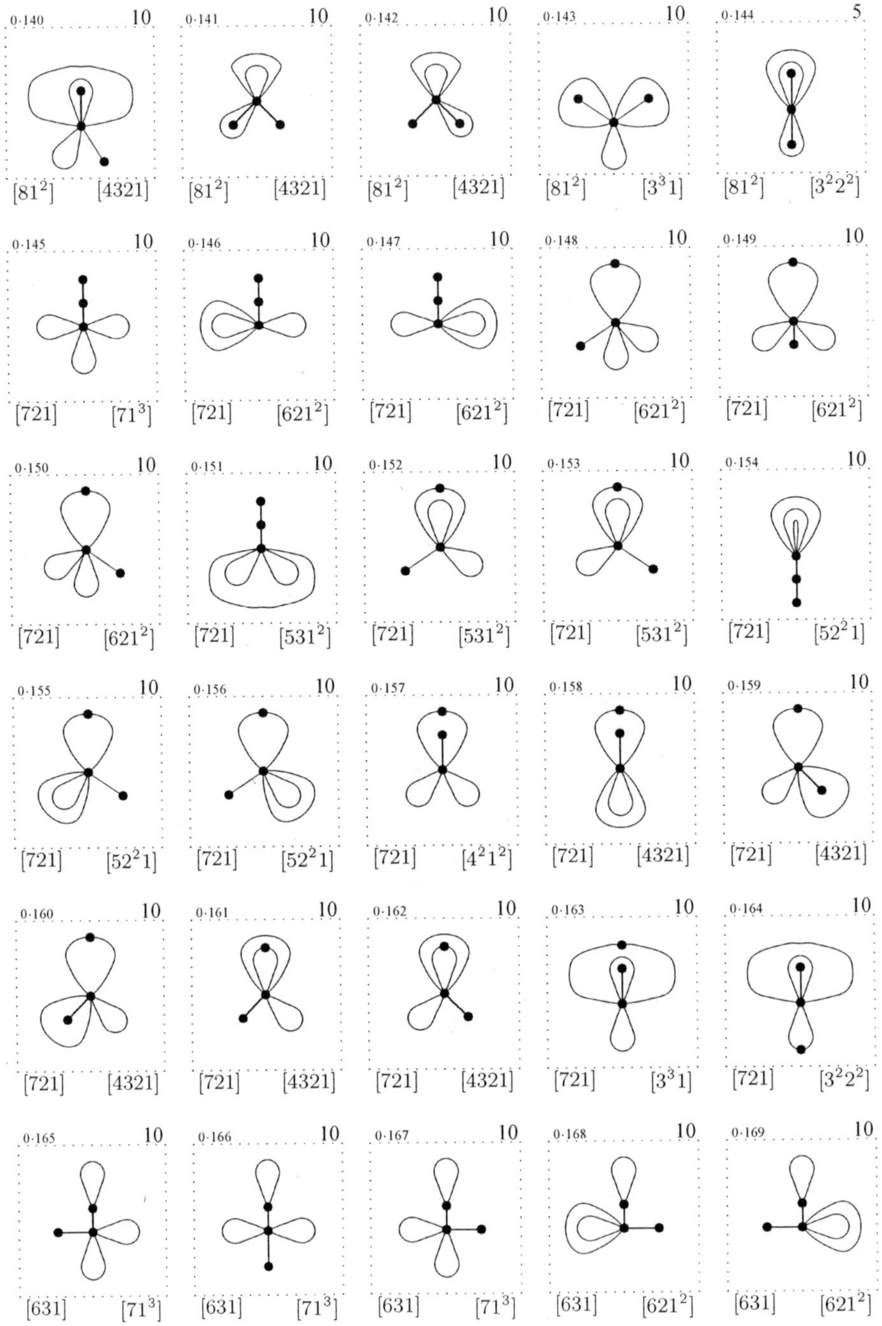
0·140 10 $[81^2]$ $[4321]$
0·141 10 $[81^2]$ $[4321]$
0·142 10 $[81^2]$ $[4321]$
0·143 10 $[81^2]$ $[3^31]$
0·144 5 $[81^2]$ $[3^22^2]$
0·145 10 $[721]$ $[71^3]$
0·146 10 $[721]$ $[621^2]$
0·147 10 $[721]$ $[621^2]$
0·148 10 $[721]$ $[621^2]$
0·149 10 $[721]$ $[621^2]$
0·150 10 $[721]$ $[621^2]$
0·151 10 $[721]$ $[531^2]$
0·152 10 $[721]$ $[531^2]$
0·153 10 $[721]$ $[531^2]$
0·154 10 $[721]$ $[52^21]$
0·155 10 $[721]$ $[52^21]$
0·156 10 $[721]$ $[52^21]$
0·157 10 $[721]$ $[4^21^2]$
0·158 10 $[721]$ $[4321]$
0·159 10 $[721]$ $[4321]$
0·160 10 $[721]$ $[4321]$
0·161 10 $[721]$ $[4321]$
0·162 10 $[721]$ $[4321]$
0·163 10 $[721]$ $[3^31]$
0·164 10 $[721]$ $[3^22^2]$
0·165 10 $[631]$ $[71^3]$
0·166 10 $[631]$ $[71^3]$
0·167 10 $[631]$ $[71^3]$
0·168 10 $[631]$ $[621^2]$
0·169 10 $[631]$ $[621^2]$

Genus 0: 5e 3v

0·170 10
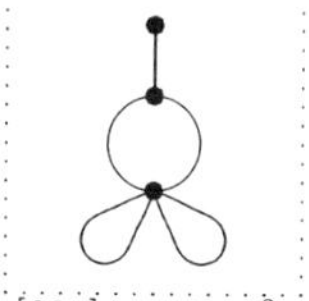
$[631]$ $[621^2]$

0·171 10
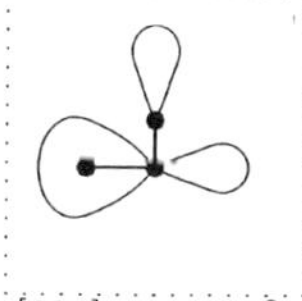
$[631]$ $[531^2]$

0·172 10
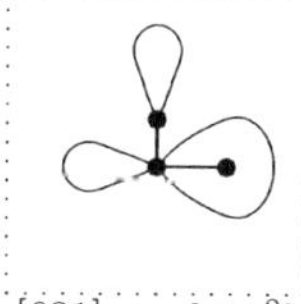
$[631]$ $[531^2]$

0·173 10
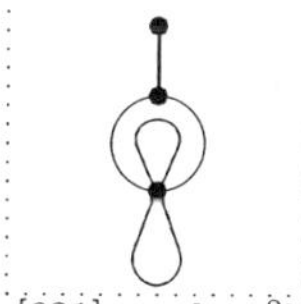
$[631]$ $[531^2]$

0·174 10
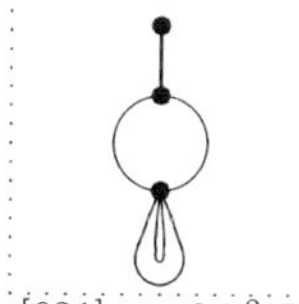
$[631]$ $[52^2 1]$

0·175 10
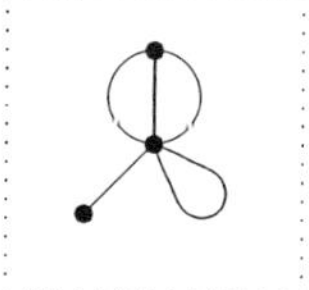
$[631]$ $[52^2 1]$

0·176 10
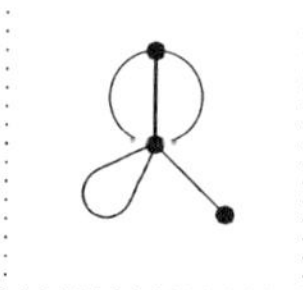
$[631]$ $[52^2 1]$

0·177 10
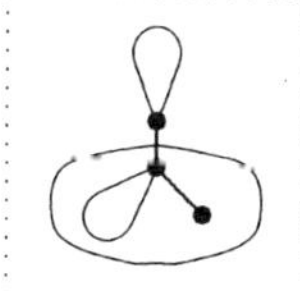
$[631]$ $[4^2 1^2]$

0·178 10
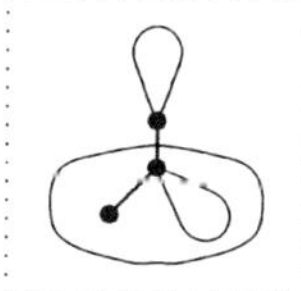
$[631]$ $[4^2 1^2]$

0·179 10
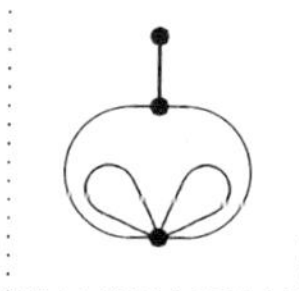
$[631]$ $[4^2 1^2]$

0·180 10
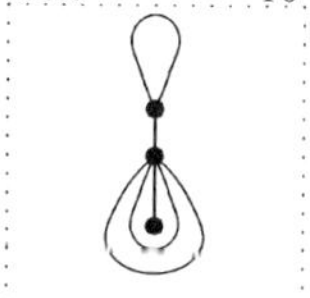
$[631]$ $[4321]$

0·181 10
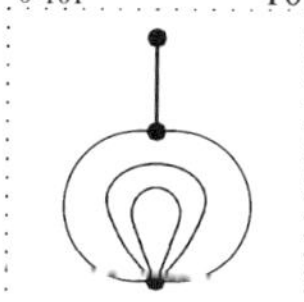
$[631]$ $[4321]$

0·182 10
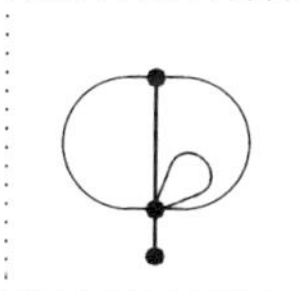
$[631]$ $[4321]$

0·183 10
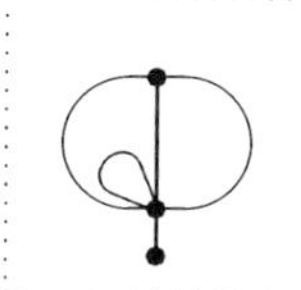
$[631]$ $[4321]$

0·184 10
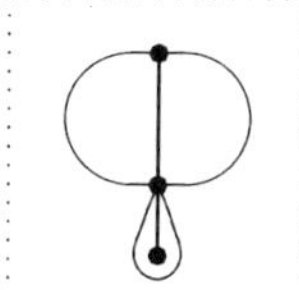
$[631]$ $[3^2 2^2]$

0·185 10
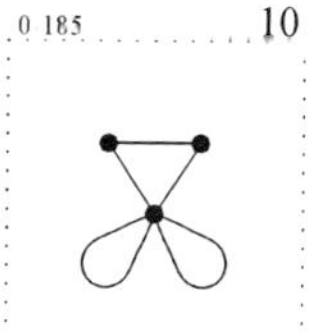
$[62^2]$ $[531^2]$

0·186 10
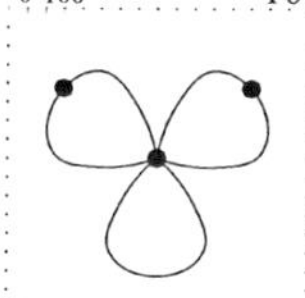
$[62^2]$ $[52^2 1]$

0·187 5
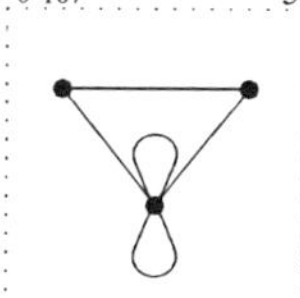
$[62^2]$ $[4^2 1^2]$

0·188 10
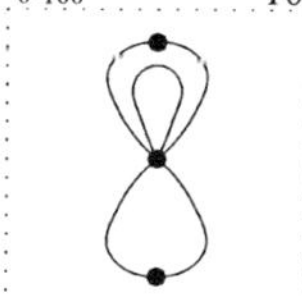
$[62^2]$ $[4321]$

0·189 10
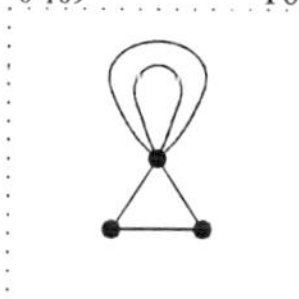
$[62^2]$ $[4321]$

0·190 5
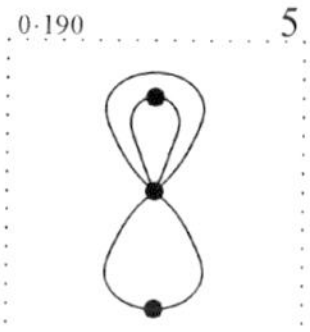
$[62^2]$ $[3^2 2^2]$

0·191 10
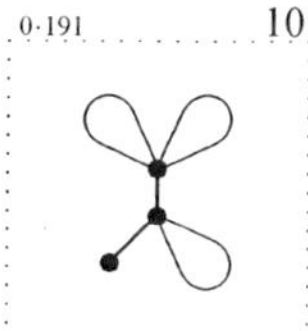
$[541]$ $[71^3]$

0·192 10
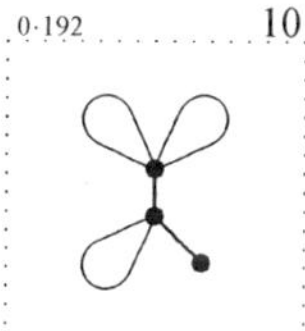
$[541]$ $[71^3]$

0·193 10
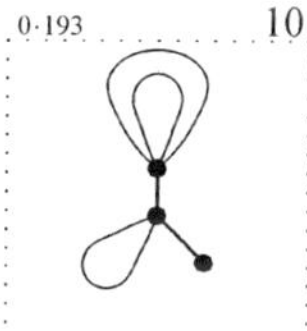
$[541]$ $[621^2]$

0·194 10
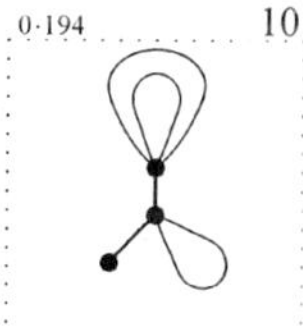
$[541]$ $[621^2]$

0·195 10
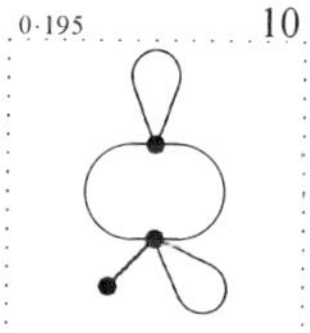
$[541]$ $[621^2]$

0·196 10
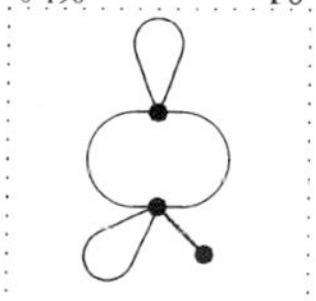
$[541]$ $[621^2]$

0·197 10
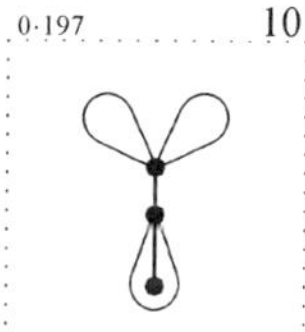
$[541]$ $[531^2]$

0·198 10
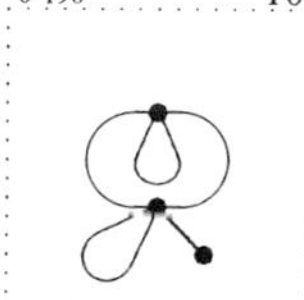
$[541]$ $[531^2]$

0·199 10
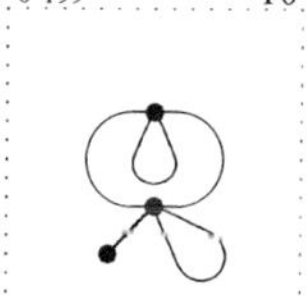
$[541]$ $[531^2]$

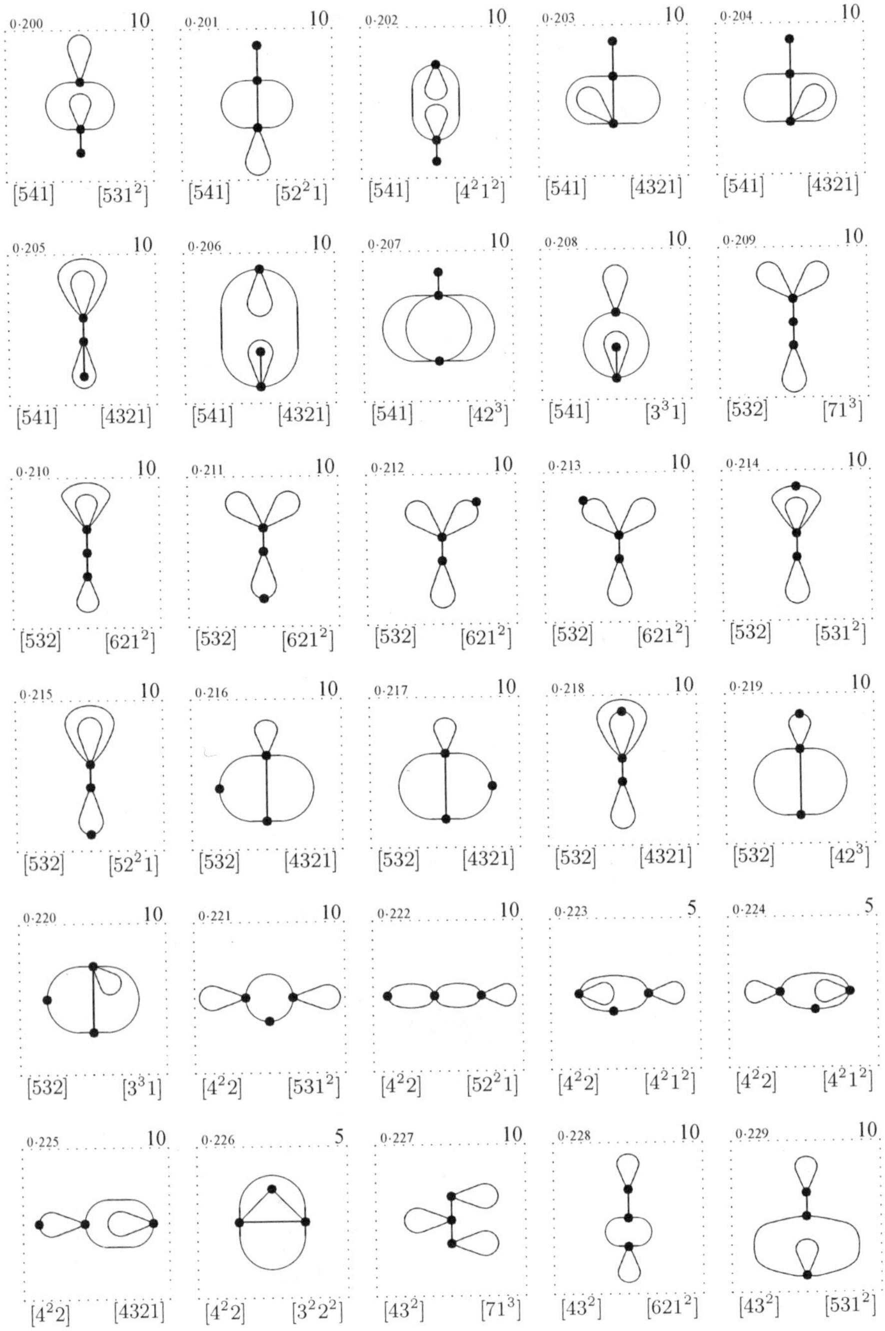
0·200 10 [541] [531^2]
0·201 10 [541] [52^{2}1]
0·202 10 [541] [4^{2}1^2]
0·203 10 [541] [4321]
0·204 10 [541] [4321]
0·205 10 [541] [4321]
0·206 10 [541] [4321]
0·207 10 [541] [42^3]
0·208 10 [541] [3^{3}1]
0·209 10 [532] [71^3]
0·210 10 [532] [621^2]
0·211 10 [532] [621^2]
0·212 10 [532] [621^2]
0·213 10 [532] [621^2]
0·214 10 [532] [531^2]
0·215 10 [532] [52^{2}1]
0·216 10 [532] [4321]
0·217 10 [532] [4321]
0·218 10 [532] [4321]
0·219 10 [532] [42^3]
0·220 10 [532] [3^{3}1]
0·221 10 [4^{2}2] [531^2]
0·222 10 [4^{2}2] [52^{2}1]
0·223 5 [4^{2}2] [4^{2}1^2]
0·224 5 [4^{2}2] [4^{2}1^2]
0·225 10 [4^{2}2] [4321]
0·226 5 [4^{2}2] [3^{2}2^2]
0·227 10 [43^2] [71^3]
0·228 10 [43^2] [621^2]
0·229 10 [43^2] [531^2]

Genus 0: 5e 4v

0·230 10
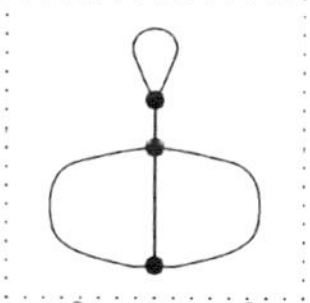
$[43^2]$ $[52^21]$

0·231 5
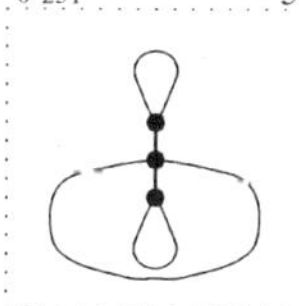
$[43^2]$ $[4^21^2]$

0·232 10
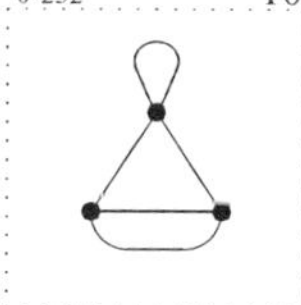
$[43^2]$ $[4321]$

0·233 5
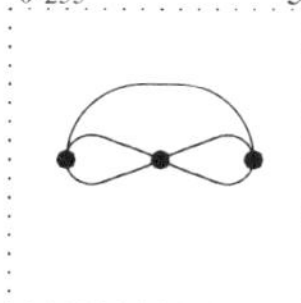
$[43^2]$ $[3^22^2]$

5 edges

4 vertices

0·234 10
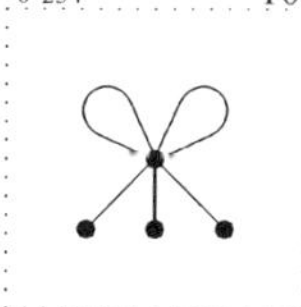
$[71^3]$ $[81^2]$

0·235 10
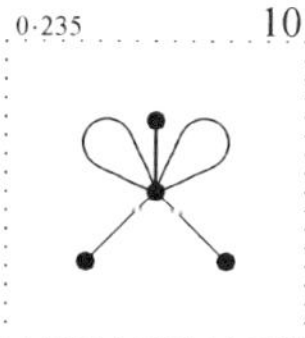
$[71^3]$ $[81^2]$

0·236 10
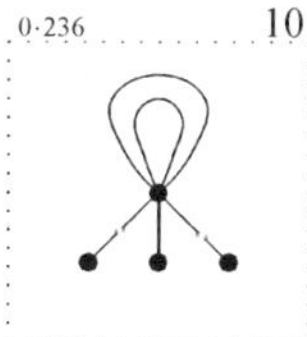
$[71^3]$ $[721]$

0·237 10
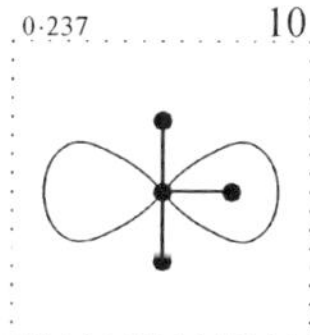
$[71^3]$ $[631]$

0·238 10
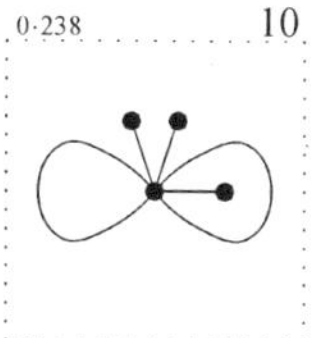
$[71^3]$ $[631]$

0·239 10
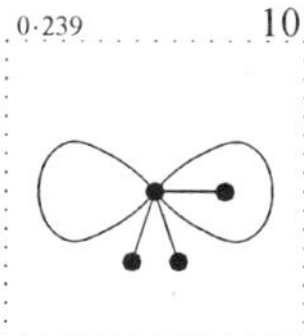
$[71^3]$ $[631]$

0·240 10
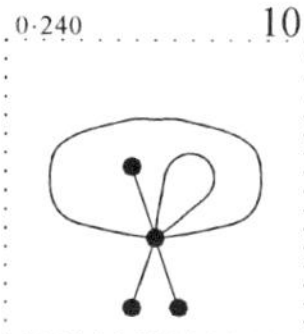
$[71^3]$ $[541]$

0·241 10
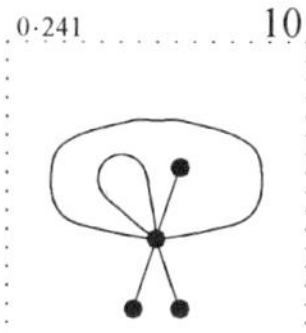
$[71^3]$ $[541]$

0·242 10
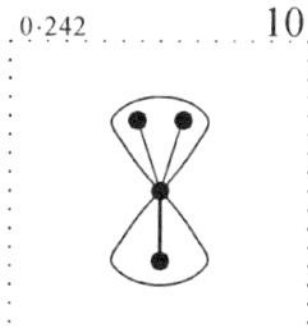
$[71^3]$ $[532]$

0·243 10
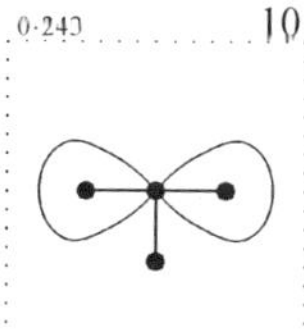
$[71^3]$ $[43^2]$

0·244 10
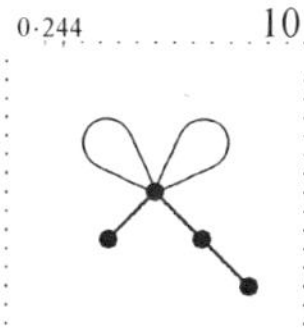
$[621^2]$ $[81^2]$

0·245 10
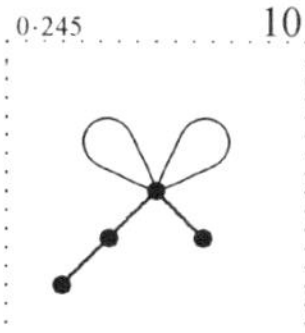
$[621^2]$ $[81^2]$

0·246 10
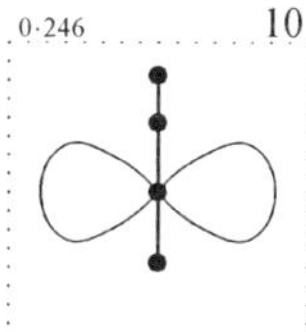
$[621^2]$ $[81^2]$

0·247 10
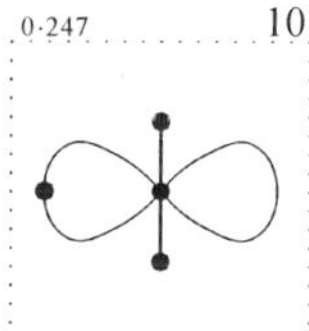
$[621^2]$ $[721]$

0·248 10
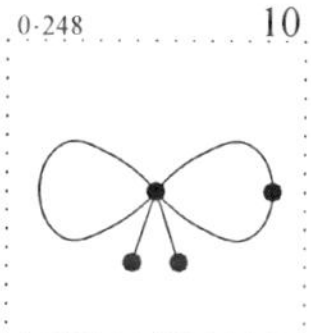
$[621^2]$ $[721]$

0·249 10
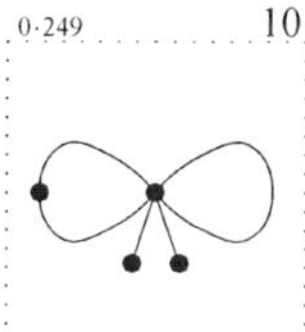
$[621^2]$ $[721]$

0·250 10
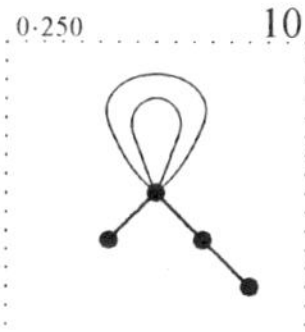
$[621^2]$ $[721]$

0·251 10
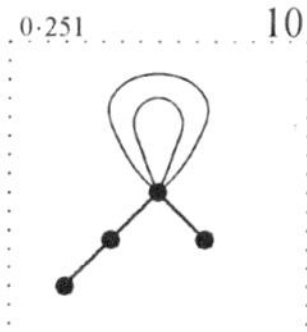
$[621^2]$ $[721]$

0·252 10
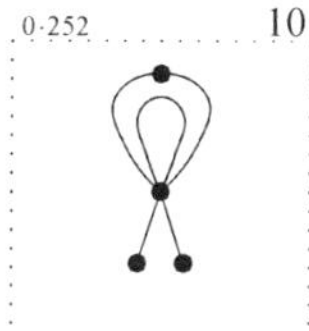
$[621^2]$ $[631]$

0·253 10
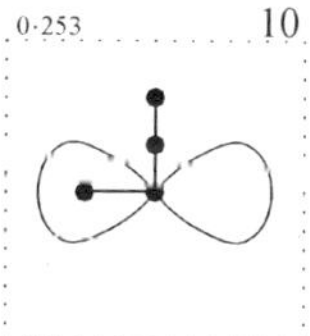
$[621^2]$ $[631]$

0·254 10
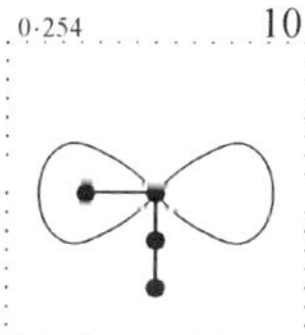
$[621^2]$ $[631]$

0·255 10
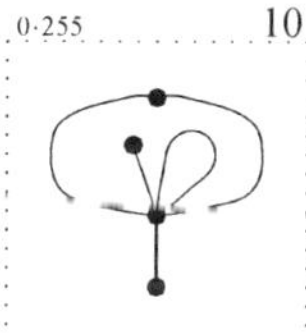
$[621^2]$ $[541]$

0·256 10
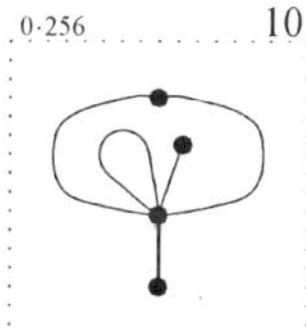
$[621^2]$ $[541]$

0·257 10
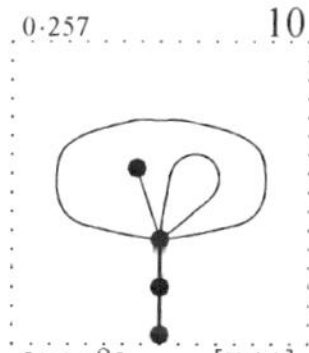
$[621^2]$ $[541]$

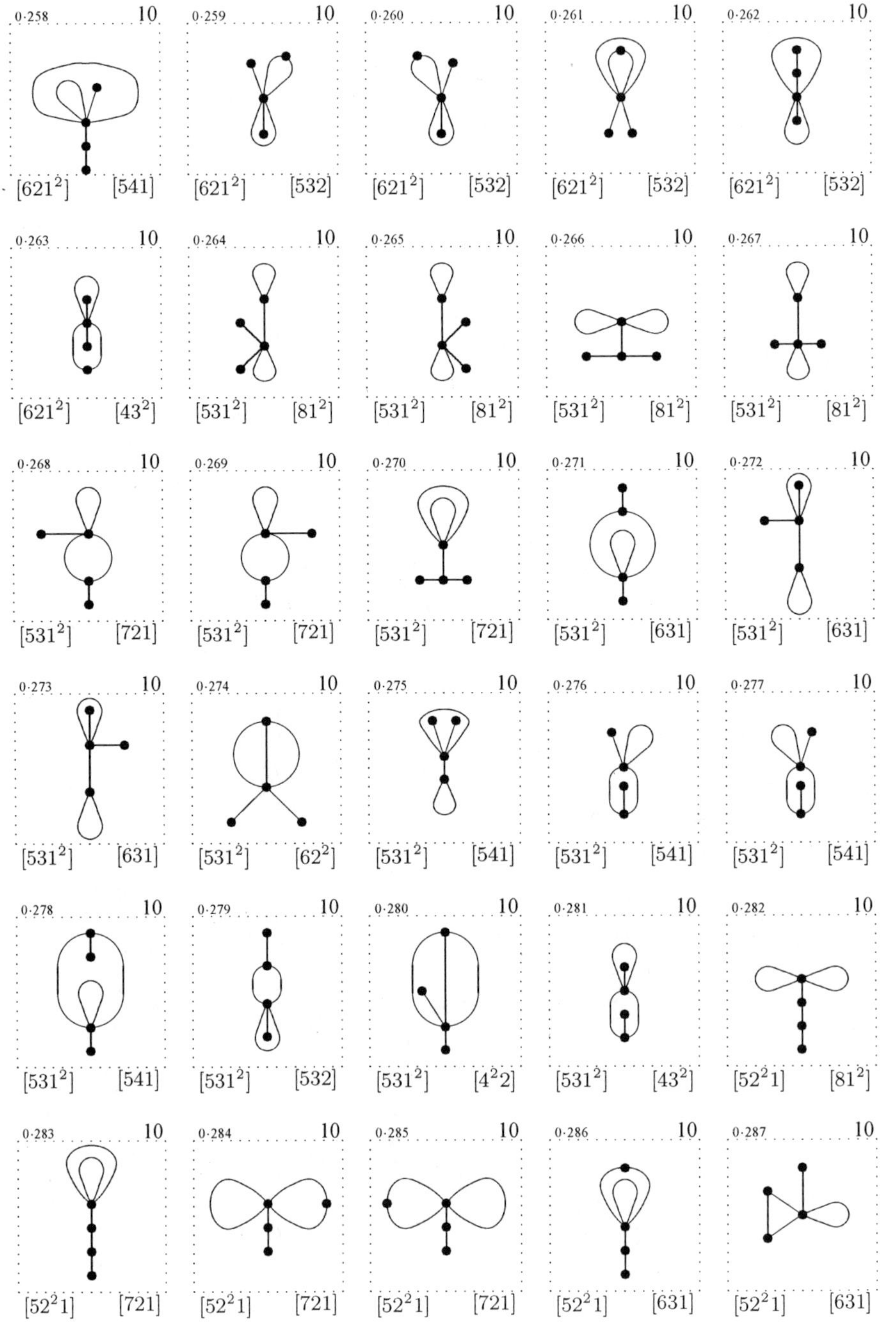
0·258 10 $[621^2]$ $[541]$
0·259 10 $[621^2]$ $[532]$
0·260 10 $[621^2]$ $[532]$
0·261 10 $[621^2]$ $[532]$
0·262 10 $[621^2]$ $[532]$
0·263 10 $[621^2]$ $[43^2]$
0·264 10 $[531^2]$ $[81^2]$
0·265 10 $[531^2]$ $[81^2]$
0·266 10 $[531^2]$ $[81^2]$
0·267 10 $[531^2]$ $[81^2]$
0·268 10 $[531^2]$ $[721]$
0·269 10 $[531^2]$ $[721]$
0·270 10 $[531^2]$ $[721]$
0·271 10 $[531^2]$ $[631]$
0·272 10 $[531^2]$ $[631]$
0·273 10 $[531^2]$ $[631]$
0·274 10 $[531^2]$ $[62^2]$
0·275 10 $[531^2]$ $[541]$
0·276 10 $[531^2]$ $[541]$
0·277 10 $[531^2]$ $[541]$
0·278 10 $[531^2]$ $[541]$
0·279 10 $[531^2]$ $[532]$
0·280 10 $[531^2]$ $[4^22]$
0·281 10 $[531^2]$ $[43^2]$
0·282 10 $[52^21]$ $[81^2]$
0·283 10 $[52^21]$ $[721]$
0·284 10 $[52^21]$ $[721]$
0·285 10 $[52^21]$ $[721]$
0·286 10 $[52^21]$ $[631]$
0·287 10 $[52^21]$ $[631]$

Genus 0: 5e 4v

0·288 10
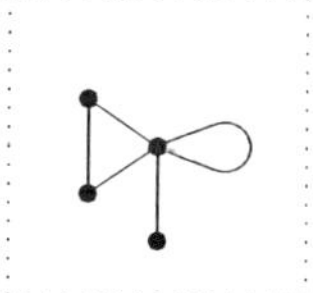
$[52^21]$ $[631]$

0·289 10
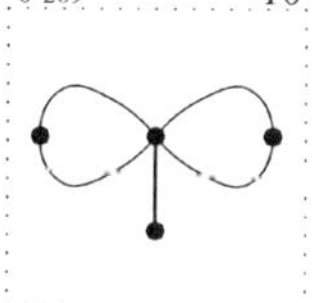
$[52^21]$ $[62^2]$

0·290 10
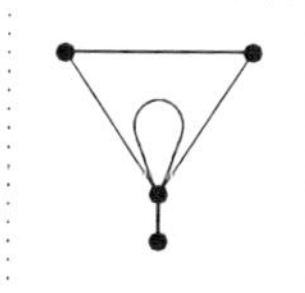
$[52^21]$ $[541]$

0·291 10
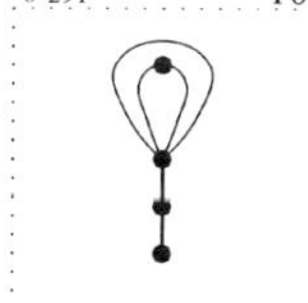
$[52^21]$ $[532]$

0·292 10
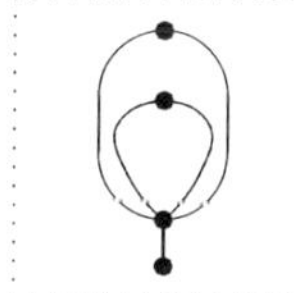
$[52^21]$ $[4^22]$

0·293 10
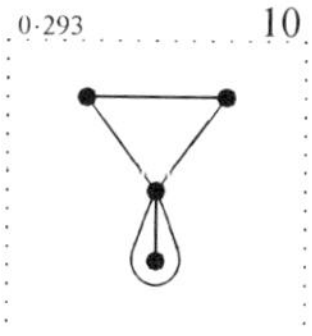
$[52^21]$ $[43^2]$

0·294 5
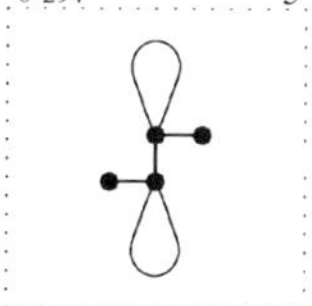
$[4^21^2]$ $[81^2]$

0·295 5
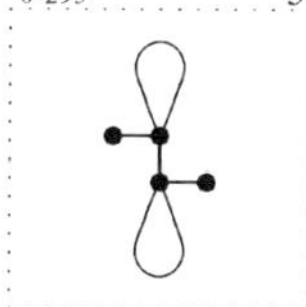
$[4^21^2]$ $[81^2]$

0·296 10
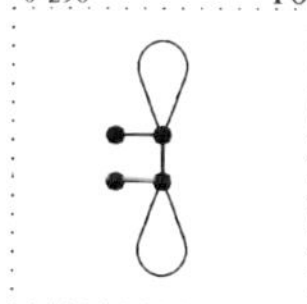
$[4^21^2]$ $[81^2]$

0·297 10
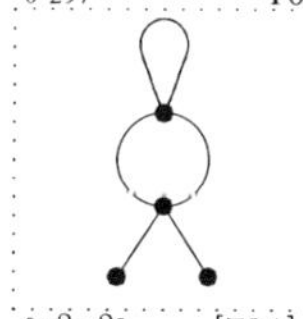
$[4^21^2]$ $[721]$

0·298 10
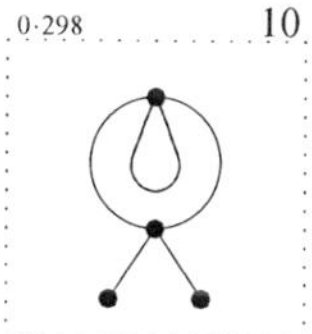
$[4^21^2]$ $[631]$

0·299 10
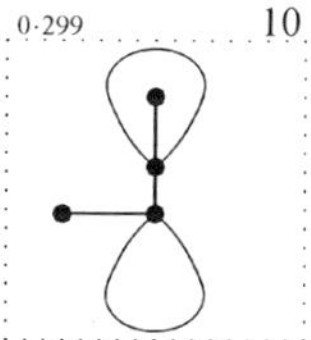
$[4^21^2]$ $[631]$

0·300 10
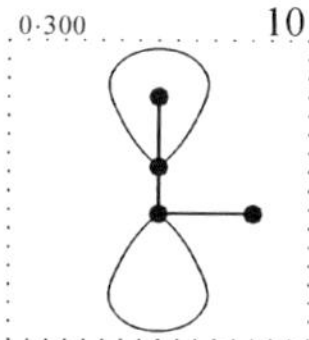
$[4^21^2]$ $[631]$

0·301 5
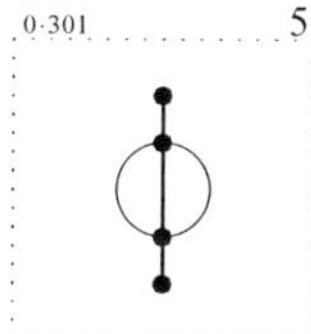
$[4^21^2]$ $[62^2]$

0·302 10
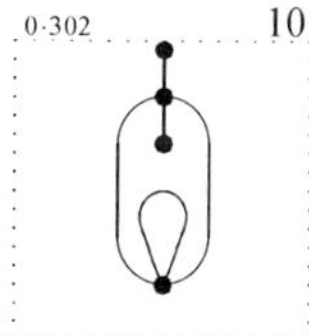
$[4^21^2]$ $[541]$

0·303 5
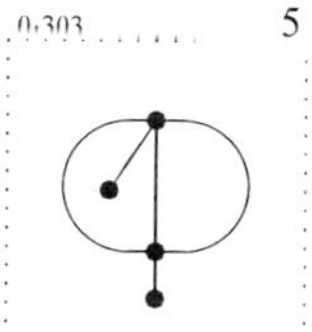
$[4^21^2]$ $[4^22]$

0·304 5
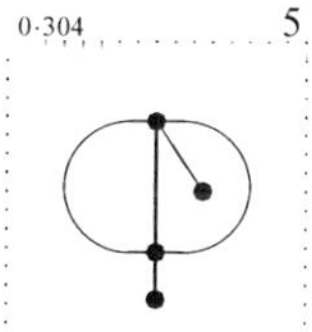
$[4^21^2]$ $[4^22]$

0·305 5
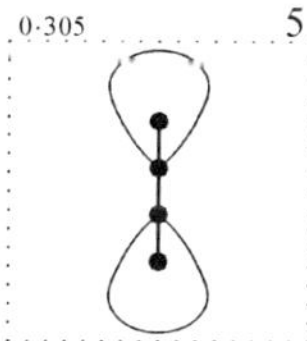
$[4^21^2]$ $[43^2]$

0·306 10
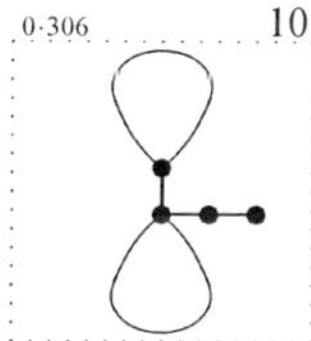
$[4321]$ $[81^2]$

0·307 10
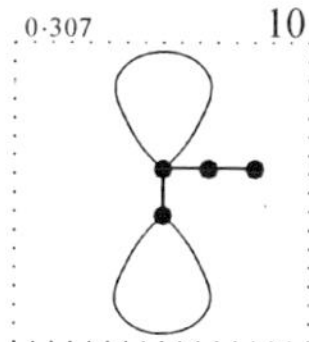
$[4321]$ $[81^2]$

0·308 10
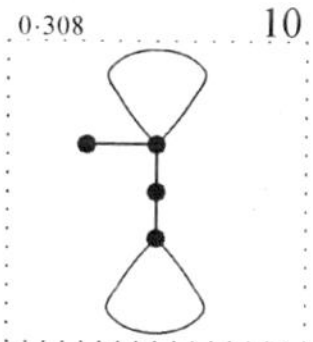
$[4321]$ $[81^2]$

0·309 10
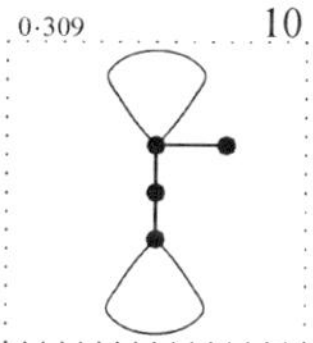
$[4321]$ $[81^2]$

0·310 10
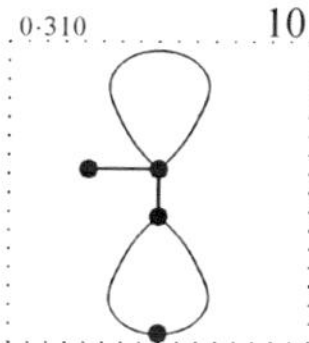
$[4321]$ $[721]$

0·311 10
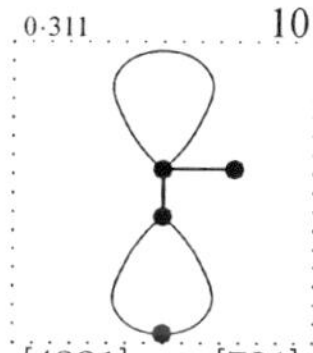
$[4321]$ $[721]$

0·312 10
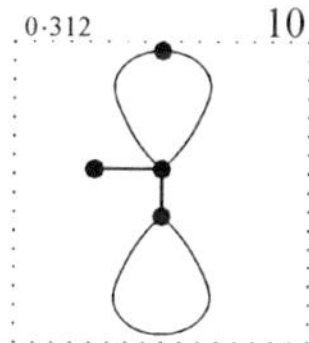
$[4321]$ $[721]$

0·313 10
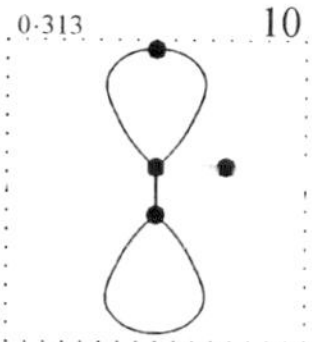
$[4321]$ $[721]$

0·314 10
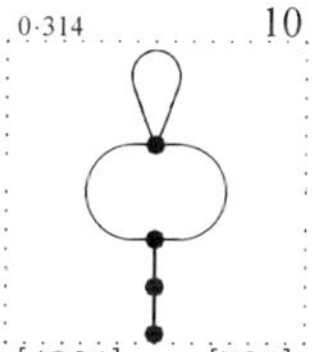
$[4321]$ $[721]$

0·315 10
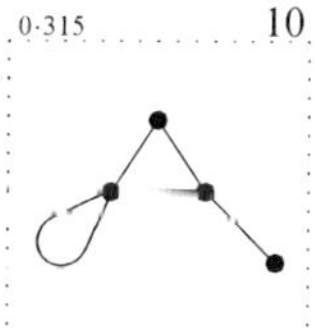
$[4321]$ $[631]$

0·316 10
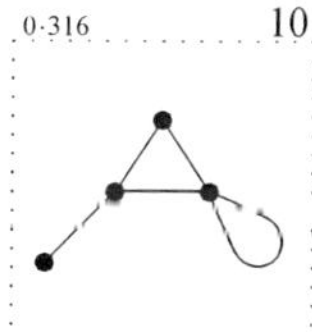
$[4321]$ $[631]$

0·317 10
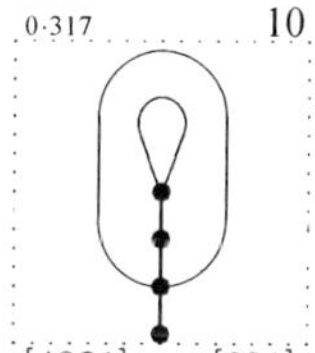
$[4321]$ $[631]$

0·318 10

[4321] [631]

0·319 10

$[4321]$ $[62^2]$

0·320 10

$[4321]$ $[62^2]$

0·321 10

[4321] [541]

0·322 10
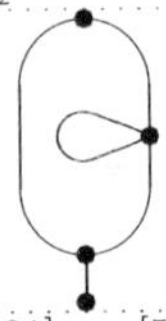
[4321] [541]

0·323 10

[4321] [541]

0·324 10

[4321] [541]

0·325 10

[4321] [532]

0·326 10

[4321] [532]

0·327 10

[4321] [532]

0·328 10

$[4321]$ $[4^2 2]$

0·329 10

$[4321]$ $[43^2]$

0·330 10

$[42^3]$ $[541]$

0·331 10

$[42^3]$ $[532]$

0·332 10

$[3^3 1]$ $[81^2]$

0·333 10

$[3^3 1]$ $[721]$

0·334 10

$[3^3 1]$ $[541]$

0·335 10
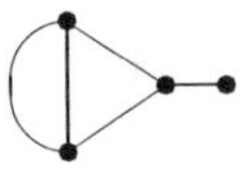
$[3^3 1]$ $[532]$

0·336 5

$[3^2 2^2]$ $[81^2]$

0·337 10

$[3^2 2^2]$ $[721]$

0·338 10

$[3^2 2^2]$ $[631]$

0·339 5

$[3^2 2^2]$ $[62^2]$

0·340 5

$[3^2 2^2]$ $[4^2 2]$

0·341 5

$[3^2 2^2]$ $[43^2]$

5 edges

5 vertices

0·342 10

$[61^4]$ $[91]$

0·343 10
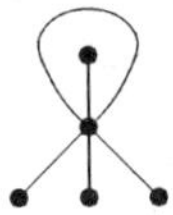
$[61^4]$ $[73]$

0·344 5
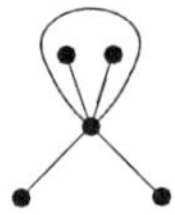
$[61^4]$ $[5^2]$

0·345 10

$[521^3]$ $[91]$

Genus 0: 5e 5v

0·346 10

$[521^3]$ $[91]$

0·347 10

$[521^3]$ $[91]$

0·348 10

$[521^3]$ $[82]$

0·349 10
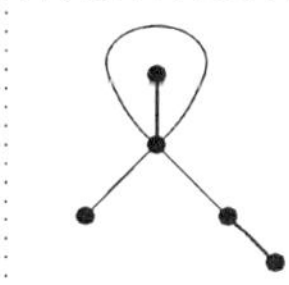
$[521^3]$ $[73]$

0·350 10
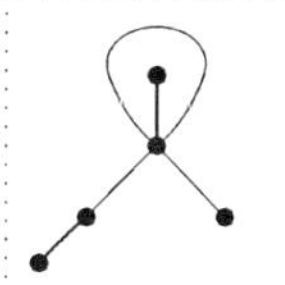
$[521^3]$ $[73]$

0·351 10
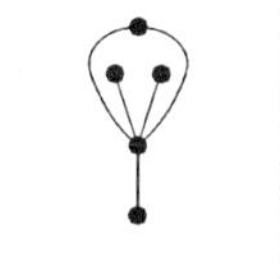
$[521^3]$ $[64]$

0·352 10

$[521^3]$ $[5^2]$

0·353 10

$[431^3]$ $[91]$

0·354 10
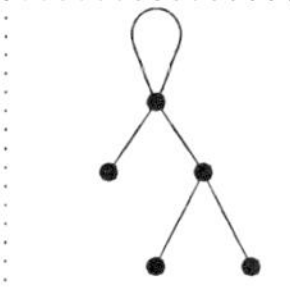
$[431^3]$ $[91]$

0·355 10
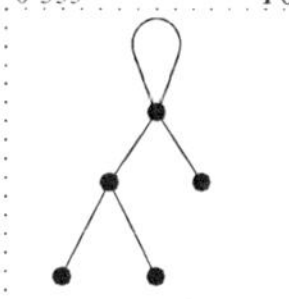
$[431^3]$ $[91]$

0·356 10
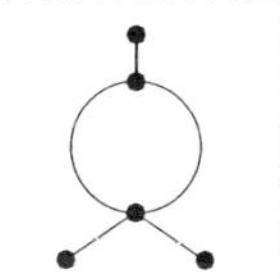
$[431^3]$ $[82]$

0·357 10

$[431^3]$ $[73]$

0·358 10
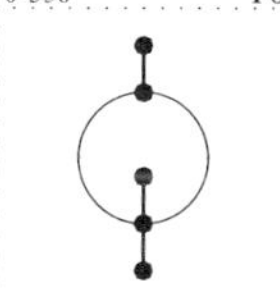
$[431^3]$ $[64]$

0·359 10
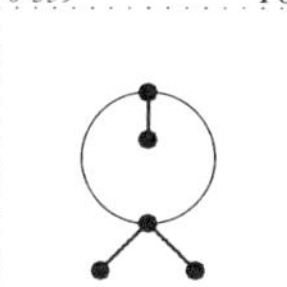
$[431^3]$ $[64]$

0·360 10
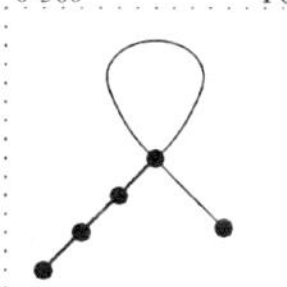
$[42^2 1^2]$ $[91]$

0·361 10
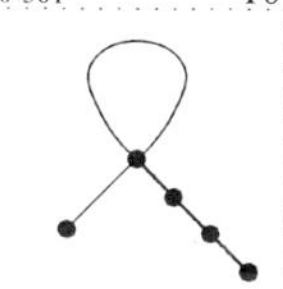
$[42^2 1^2]$ $[91]$

0·362 10
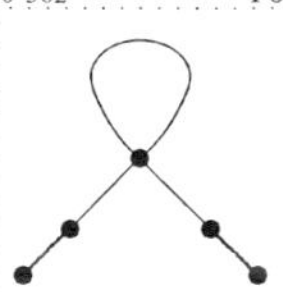
$[42^2 1^2]$ $[91]$

0·363 10
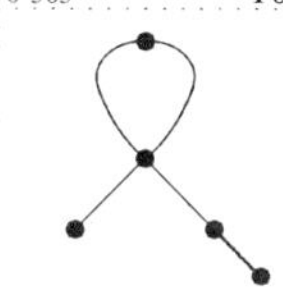
$[42^2 1^2]$ $[82]$

0·364 10
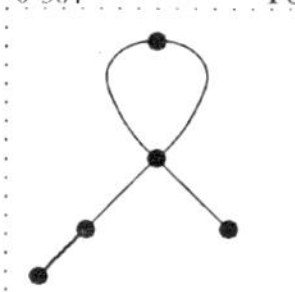
$[42^2 1^2]$ $[82]$

0·365 10
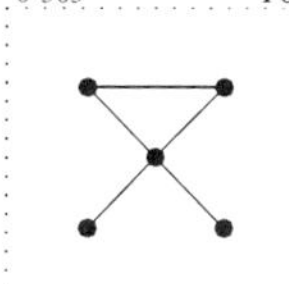
$[42^2 1^2]$ $[73]$

0·366 10

$[42^2 1^2]$ $[73]$

0·367 10
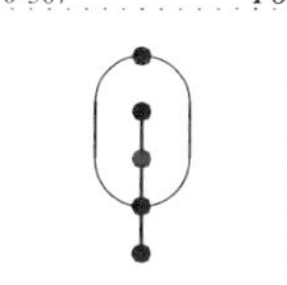
$[42^2 1^2]$ $[64]$

0·368 5
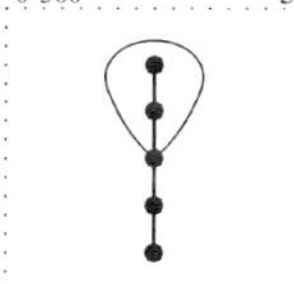
$[42^2 1^2]$ $[5^2]$

0·369 5
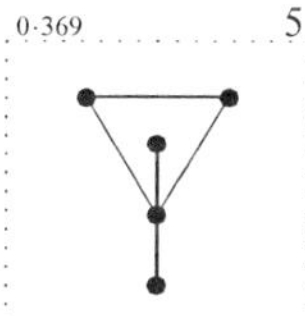
$[42^2 1^2]$ $[5^2]$

0·370 10
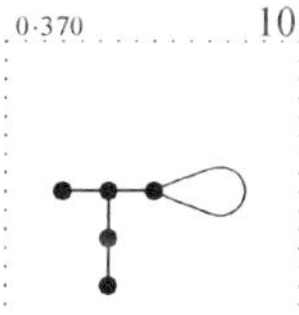
$[3^2 21^2]$ $[91]$

0·371 10
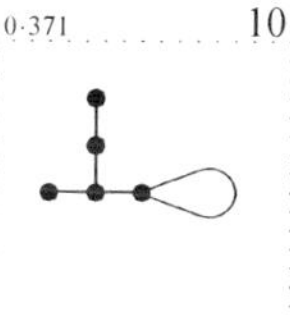
$[3^2 21^2]$ $[91]$

0·372 10
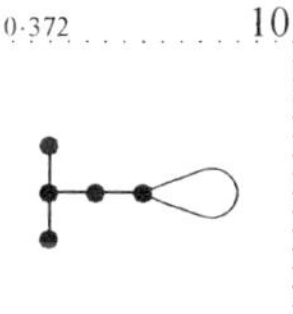
$[3^2 21^2]$ $[91]$

0·373 10
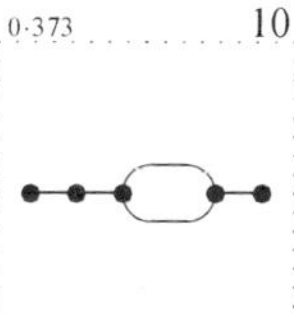
$[3^2 21^2]$ $[82]$

0·374 10
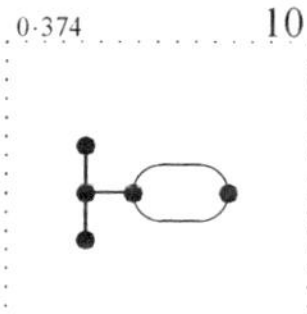
$[3^2 21^2]$ $[82]$

0·375 10
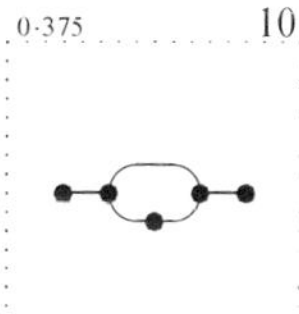
$[3^2 21^2]$ $[73]$

0·376 10
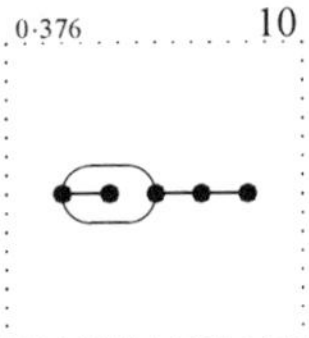
$[3^2 21^2]$ [64]

0·377 5
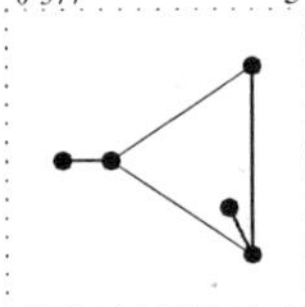
$[3^2 21^2]$ $[5^2]$

0·378 5
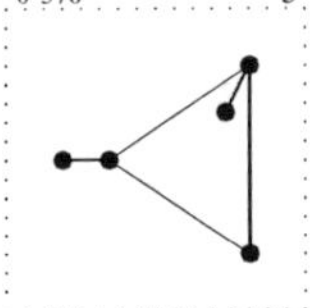
$[3^2 21^2]$ $[5^2]$

0·379 10

$[32^3 1]$ [91]

0·380 10
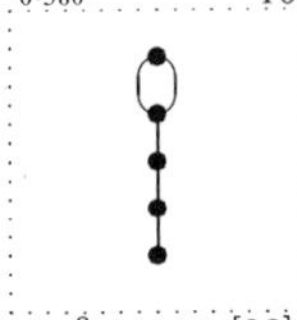
$[32^3 1]$ [82]

0·381 10
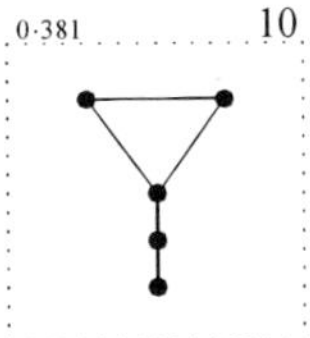
$[32^3 1]$ [73]

0·382 10
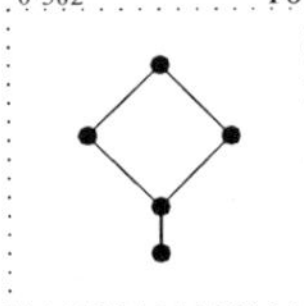
$[32^3 1]$ [64]

0·383 1
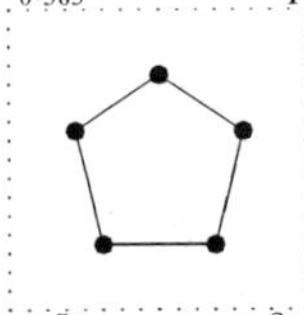
$[2^5]$ $[5^2]$

5 edges

6 vertices

0·384 2
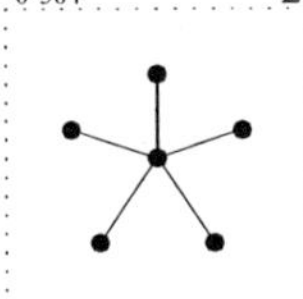
$[51^5]$ [10]

0·385 10
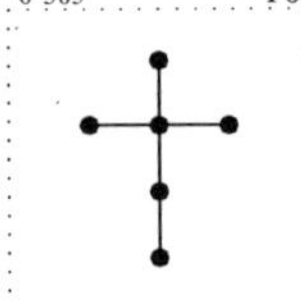
$[421^4]$ [10]

0·386 5
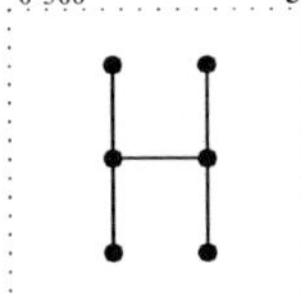
$[3^2 1^4]$ [10]

0·387 10
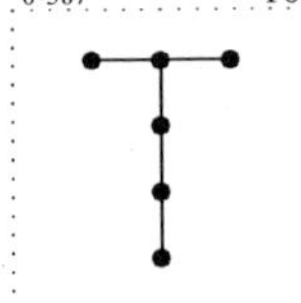
$[32^2 1^3]$ [10]

0·388 10
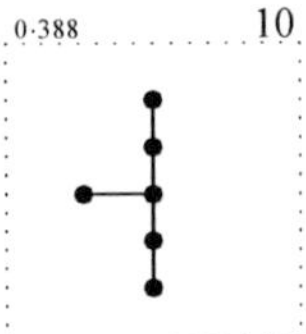
$[32^2 1^3]$ [10]

0·389 5
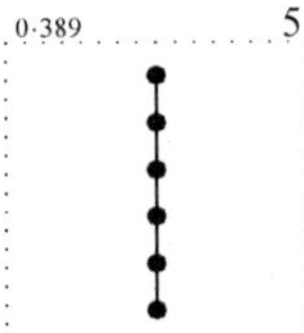
$[2^4 1^2]$ [10]

5.2 Genus 1 – the torus

2 edges

1 vertex

1·1 1
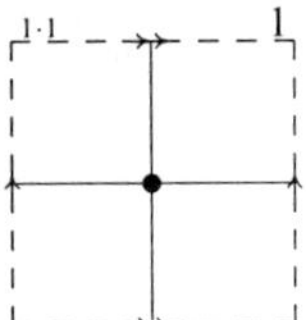
[4] [4]

Genus 1: 3e 2v

3 edges

1 vertex

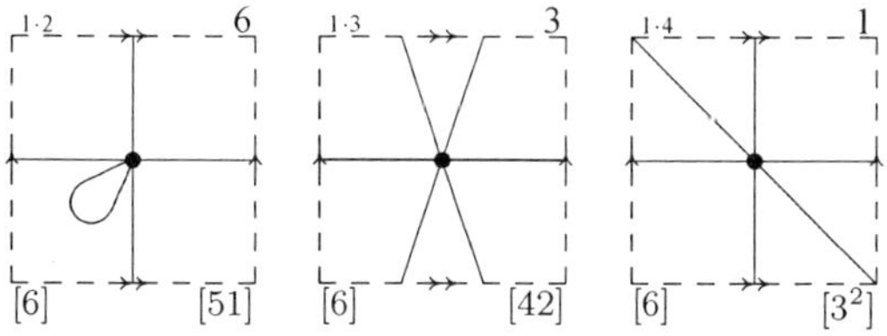

3 edges

2 vertices

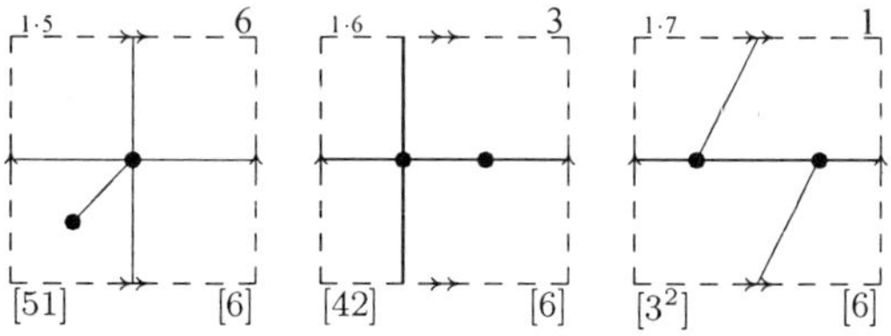

4 edges

1 vertex

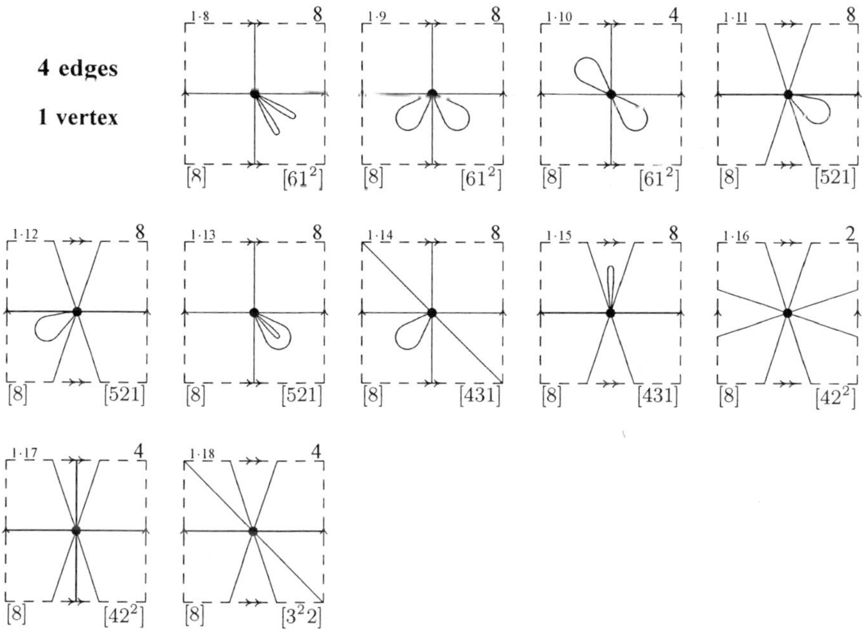

4 edges

2 vertices

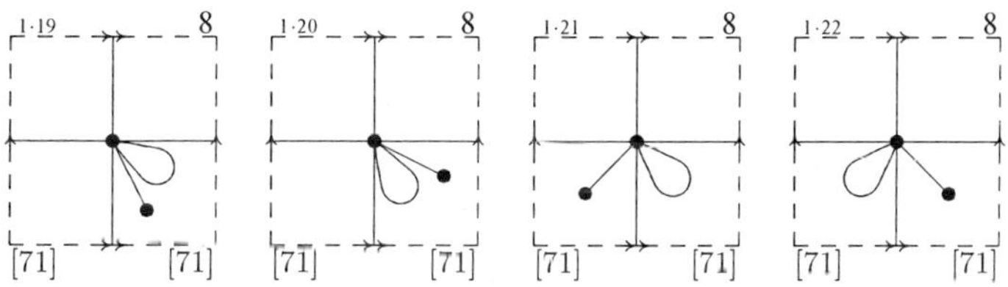

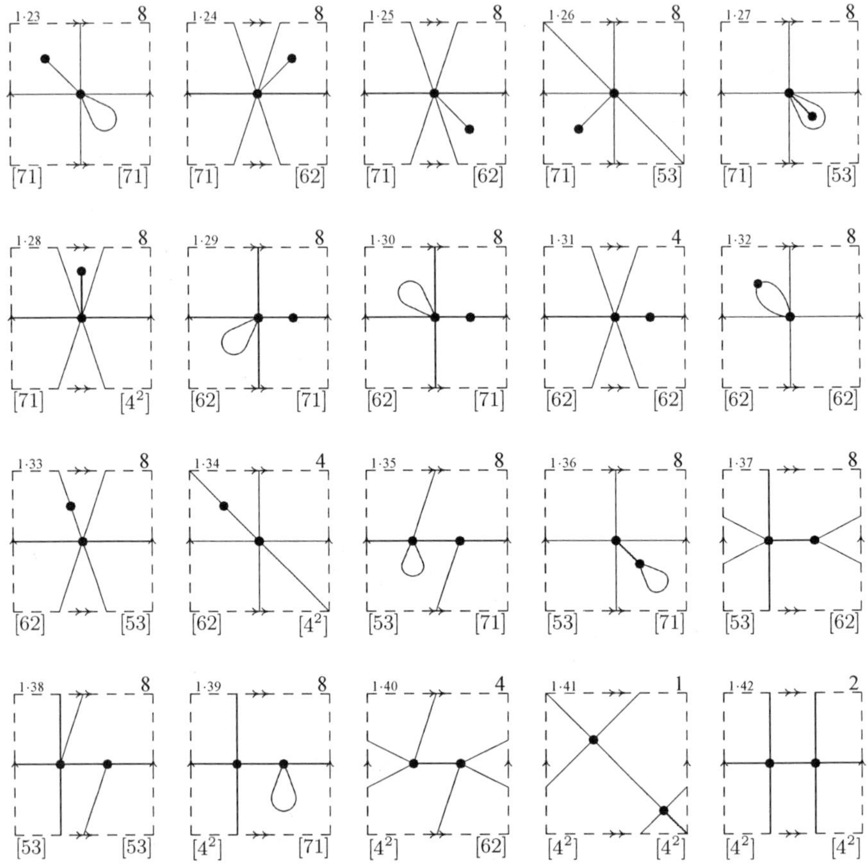

4 edges

3 vertices

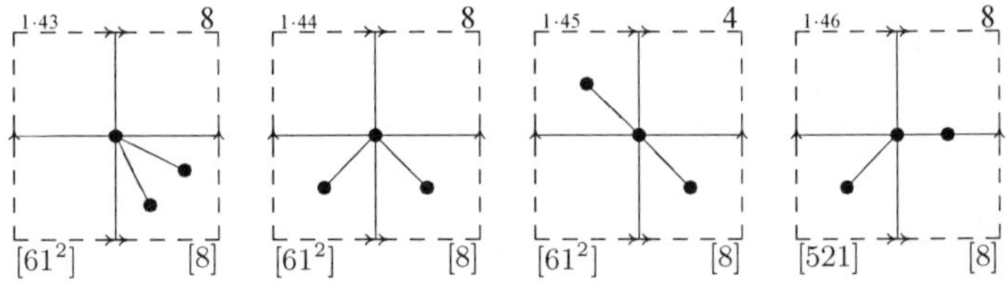

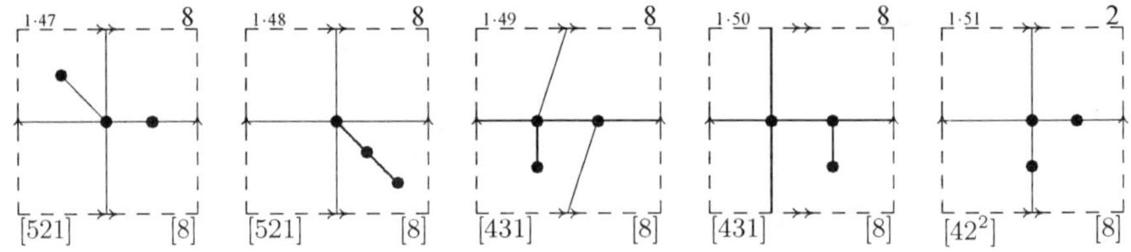

Genus 1: 5e 0v

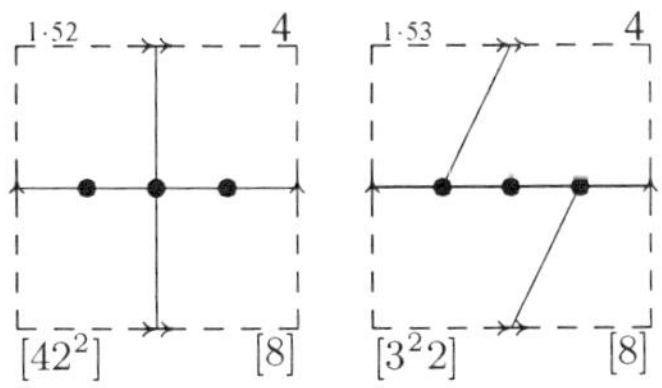

5 edges

1 vertex

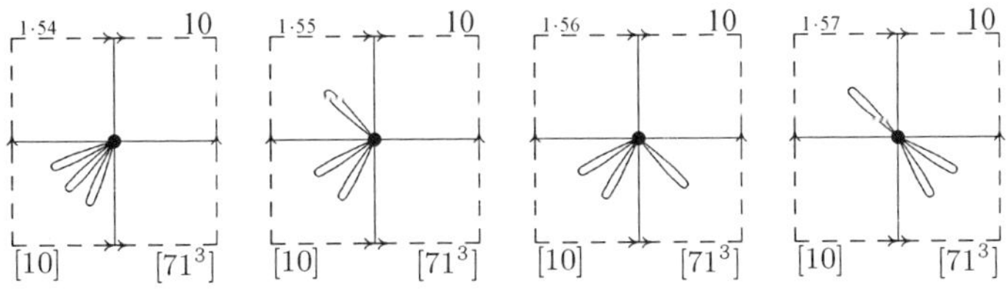

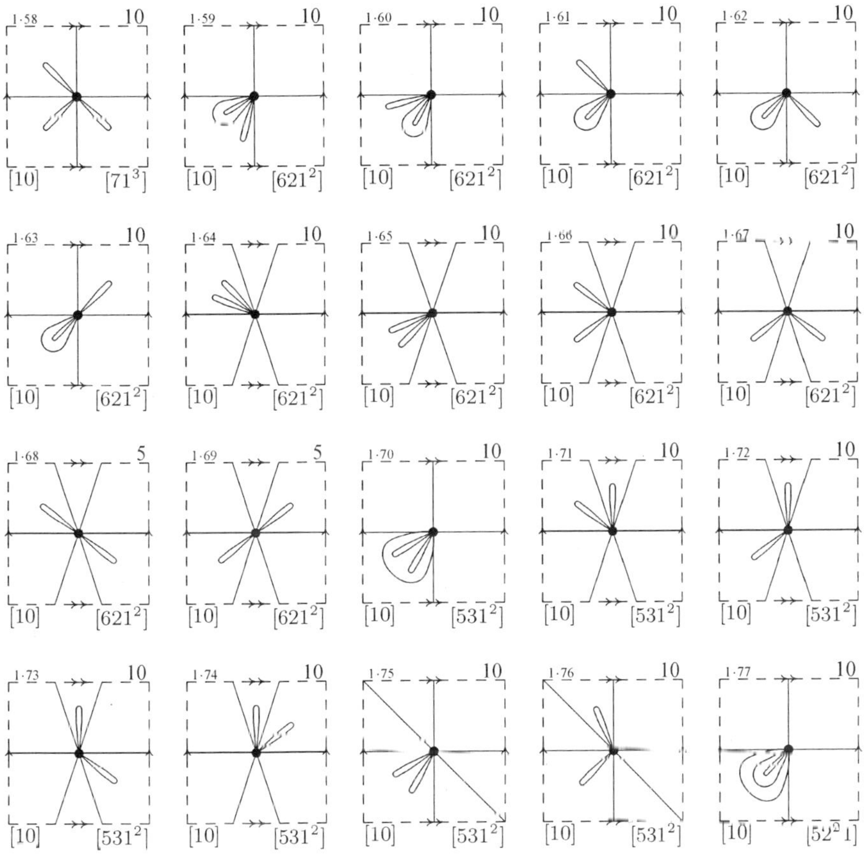

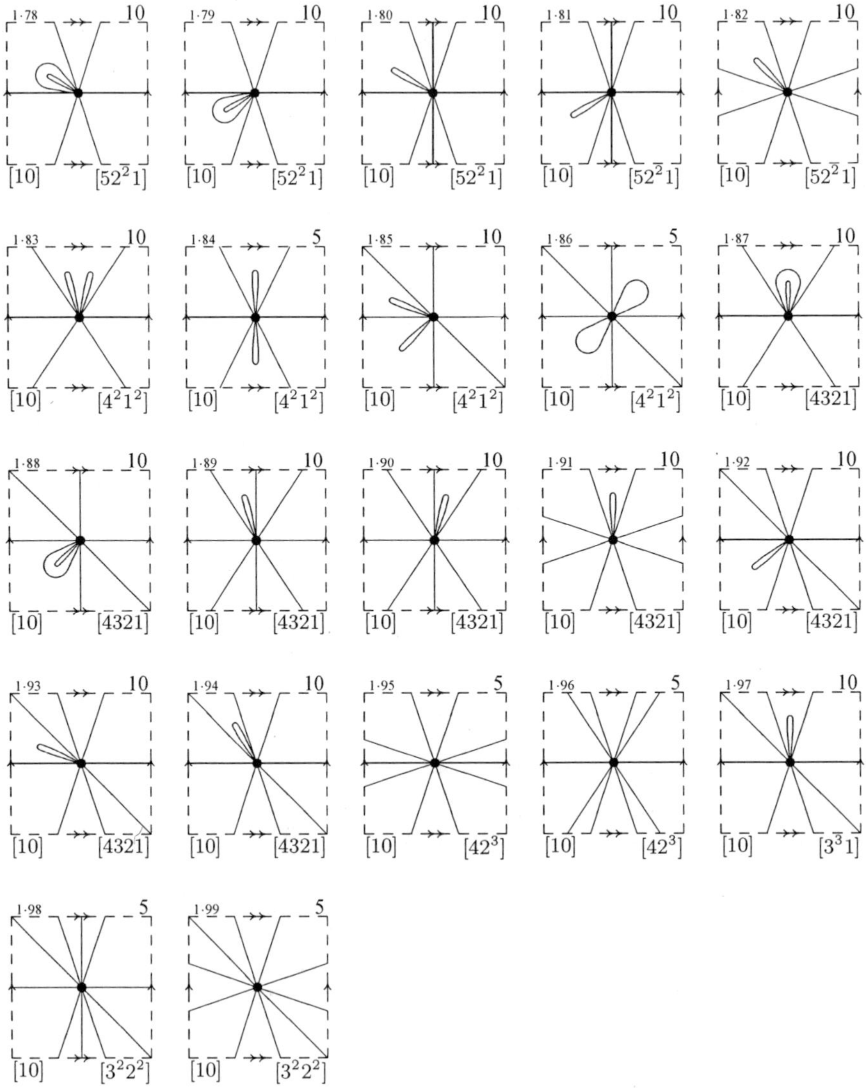

5 edges

2 vertices

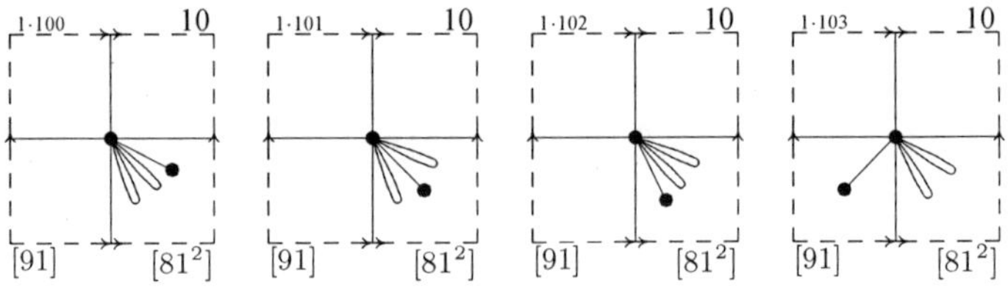

Genus 1: 5e 2v

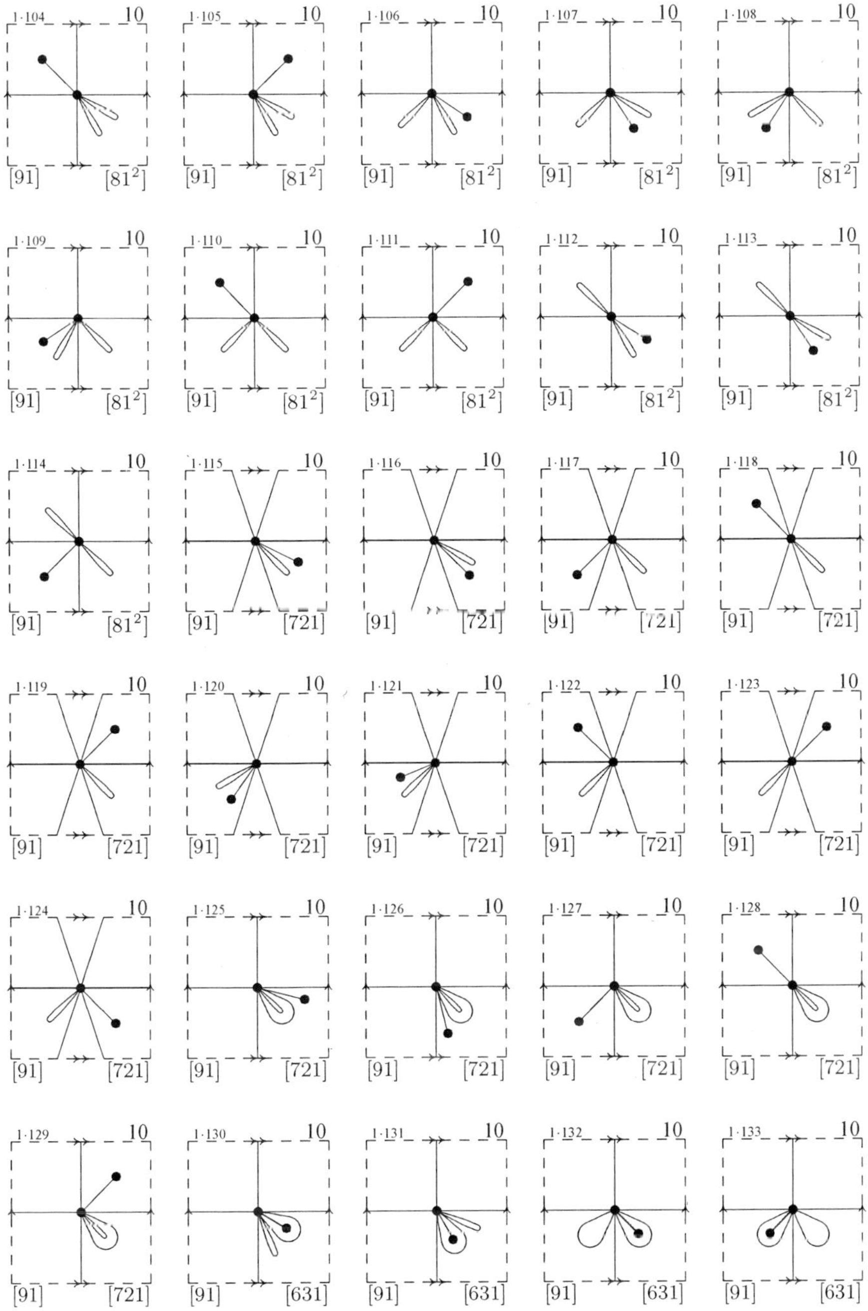
1·104 10 [91] [81²]
1·105 10 [91] [81²]
1·106 10 [91] [81²]
1·107 10 [91] [81²]
1·108 10 [91] [81²]
1·109 10 [91] [81²]
1·110 10 [91] [81²]
1·111 10 [91] [81²]
1·112 10 [91] [81²]
1·113 10 [91] [81²]
1·114 10 [91] [81²]
1·115 10 [91] [721]
1·116 10 [91] [721]
1·117 10 [91] [721]
1·118 10 [91] [721]
1·119 10 [91] [721]
1·120 10 [91] [721]
1·121 10 [91] [721]
1·122 10 [91] [721]
1·123 10 [91] [721]
1·124 10 [91] [721]
1·125 10 [91] [721]
1·126 10 [91] [721]
1·127 10 [91] [721]
1·128 10 [91] [721]
1·129 10 [91] [721]
1·130 10 [91] [631]
1·131 10 [91] [631]
1·132 10 [91] [631]
1·133 10 [91] [631]

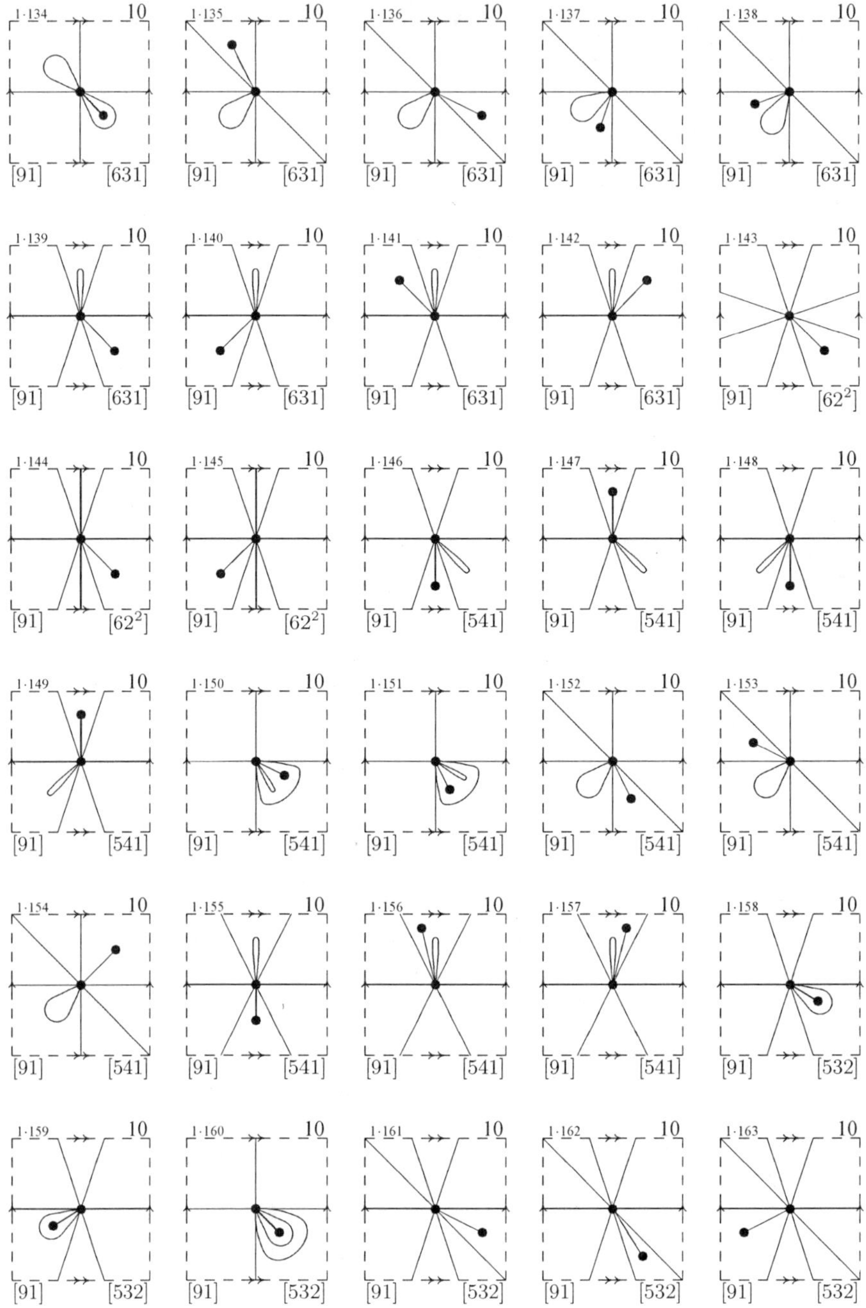
1·134 10 $[9\bar{1}]$ $[63\bar{1}]$
1·135 10 $[9\bar{1}]$ $[63\bar{1}]$
1·136 10 $[9\bar{1}]$ $[63\bar{1}]$
1·137 10 $[9\bar{1}]$ $[63\bar{1}]$
1·138 10 $[9\bar{1}]$ $[63\bar{1}]$
1·139 10 $[9\bar{1}]$ $[63\bar{1}]$
1·140 10 $[9\bar{1}]$ $[63\bar{1}]$
1·141 10 $[9\bar{1}]$ $[63\bar{1}]$
1·142 10 $[9\bar{1}]$ $[63\bar{1}]$
1·143 10 $[9\bar{1}]$ $[62^2]$
1·144 10 $[9\bar{1}]$ $[62^2]$
1·145 10 $[9\bar{1}]$ $[62^2]$
1·146 10 $[9\bar{1}]$ $[54\bar{1}]$
1·147 10 $[9\bar{1}]$ $[54\bar{1}]$
1·148 10 $[9\bar{1}]$ $[54\bar{1}]$
1·149 10 $[9\bar{1}]$ $[54\bar{1}]$
1·150 10 $[9\bar{1}]$ $[54\bar{1}]$
1·151 10 $[9\bar{1}]$ $[54\bar{1}]$
1·152 10 $[9\bar{1}]$ $[54\bar{1}]$
1·153 10 $[9\bar{1}]$ $[54\bar{1}]$
1·154 10 $[9\bar{1}]$ $[54\bar{1}]$
1·155 10 $[9\bar{1}]$ $[54\bar{1}]$
1·156 10 $[9\bar{1}]$ $[54\bar{1}]$
1·157 10 $[9\bar{1}]$ $[54\bar{1}]$
1·158 10 $[9\bar{1}]$ $[53\bar{2}]$
1·159 10 $[9\bar{1}]$ $[53\bar{2}]$
1·160 10 $[9\bar{1}]$ $[53\bar{2}]$
1·161 10 $[9\bar{1}]$ $[53\bar{2}]$
1·162 10 $[9\bar{1}]$ $[53\bar{2}]$
1·163 10 $[9\bar{1}]$ $[53\bar{2}]$

Genus 1: 5e 2v

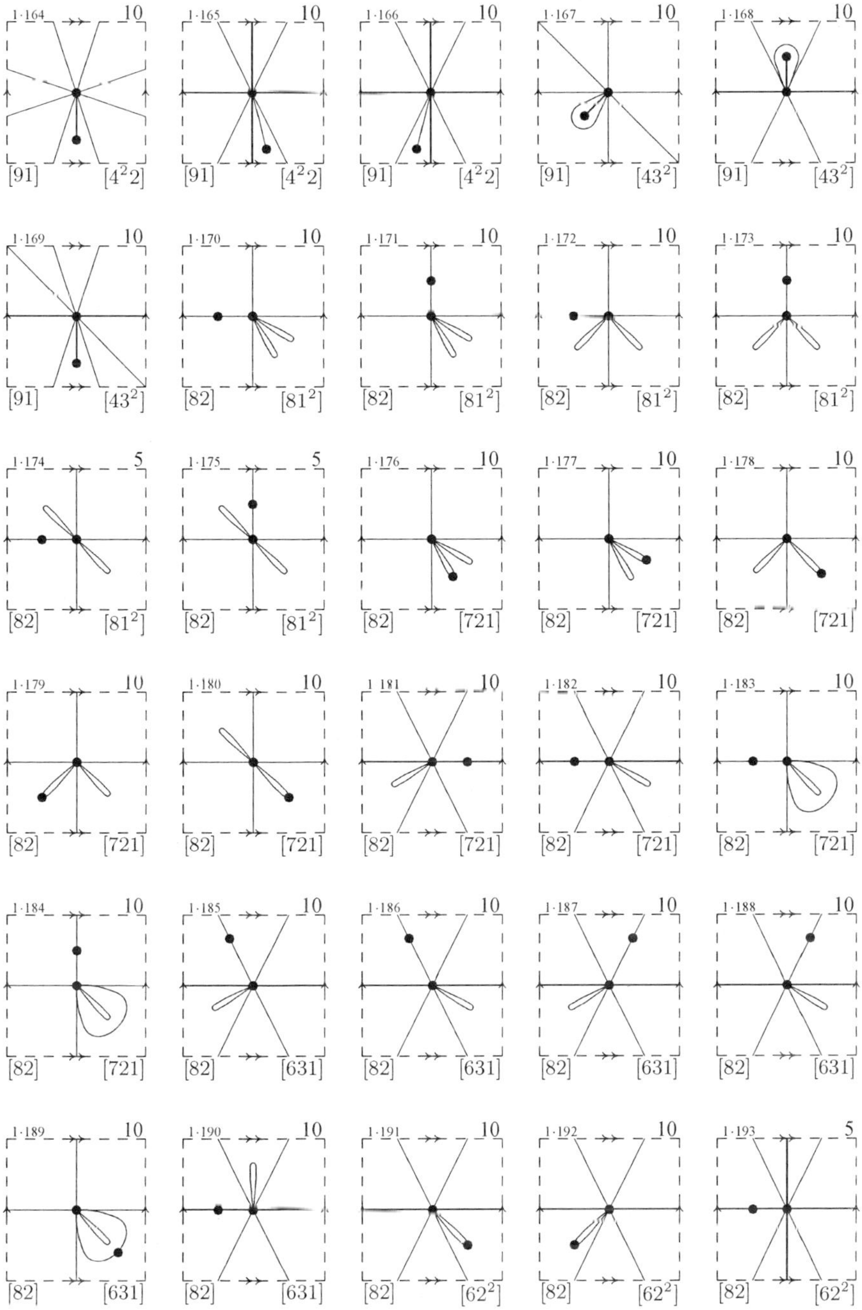
1·164 10 [91] $[4^22]$
1·165 10 [91] $[4^22]$
1·166 10 [91] $[4^22]$
1·167 10 [91] $[43^2]$
1·168 10 [91] $[43^2]$
1·169 10 [91] $[43^2]$
1·170 10 [82] $[81^2]$
1·171 10 [82] $[81^2]$
1·172 10 [82] $[81^2]$
1·173 10 [82] $[81^2]$
1·174 5 [82] $[81^2]$
1·175 5 [82] $[81^2]$
1·176 10 [82] [721]
1·177 10 [82] [721]
1·178 10 [82] [721]
1·179 10 [82] [721]
1·180 10 [82] [721]
1·181 10 [82] [721]
1·182 10 [82] [721]
1·183 10 [82] [721]
1·184 10 [82] [721]
1·185 10 [82] [631]
1·186 10 [82] [631]
1·187 10 [82] [631]
1·188 10 [82] [631]
1·189 10 [82] [631]
1·190 10 [82] [631]
1·191 10 [82] $[62^2]$
1·192 10 [82] $[62^2]$
1·193 5 [82] $[62^2]$

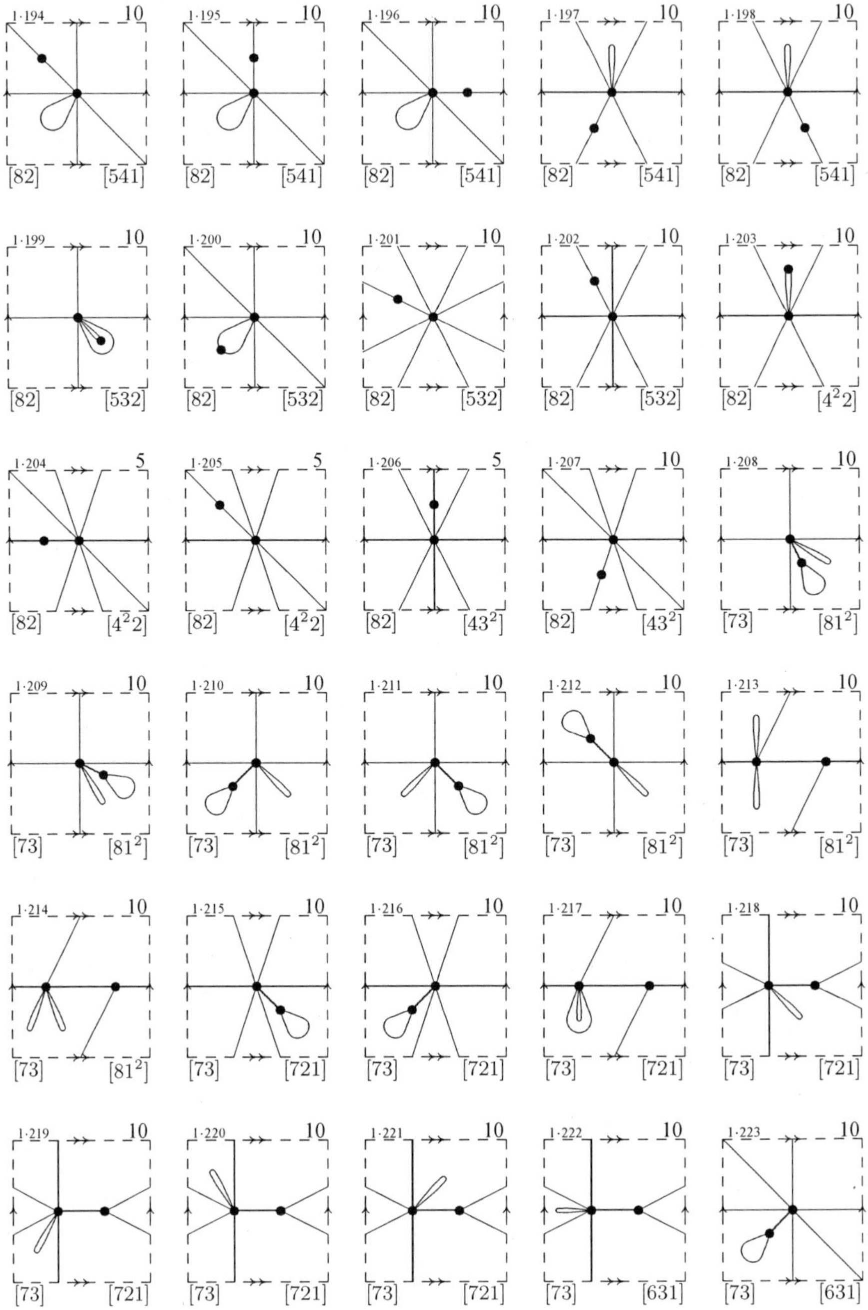
1·194 10 [82] [541]
1·195 10 [82] [541]
1·196 10 [82] [541]
1·197 10 [82] [541]
1·198 10 [82] [541]
1·199 10 [82] [532]
1·200 10 [82] [532]
1·201 10 [82] [532]
1·202 10 [82] [532]
1·203 10 [82] [4²2]
1·204 5 [82] [4²2]
1·205 5 [82] [4²2]
1·206 5 [82] [43²]
1·207 10 [82] [43²]
1·208 10 [73] [81²]
1·209 10 [73] [81²]
1·210 10 [73] [81²]
1·211 10 [73] [81²]
1·212 10 [73] [81²]
1·213 10 [73] [81²]
1·214 10 [73] [81²]
1·215 10 [73] [721]
1·216 10 [73] [721]
1·217 10 [73] [721]
1·218 10 [73] [721]
1·219 10 [73] [721]
1·220 10 [73] [721]
1·221 10 [73] [721]
1·222 10 [73] [631]
1·223 10 [73] [631]

Genus 1: 5e 2v

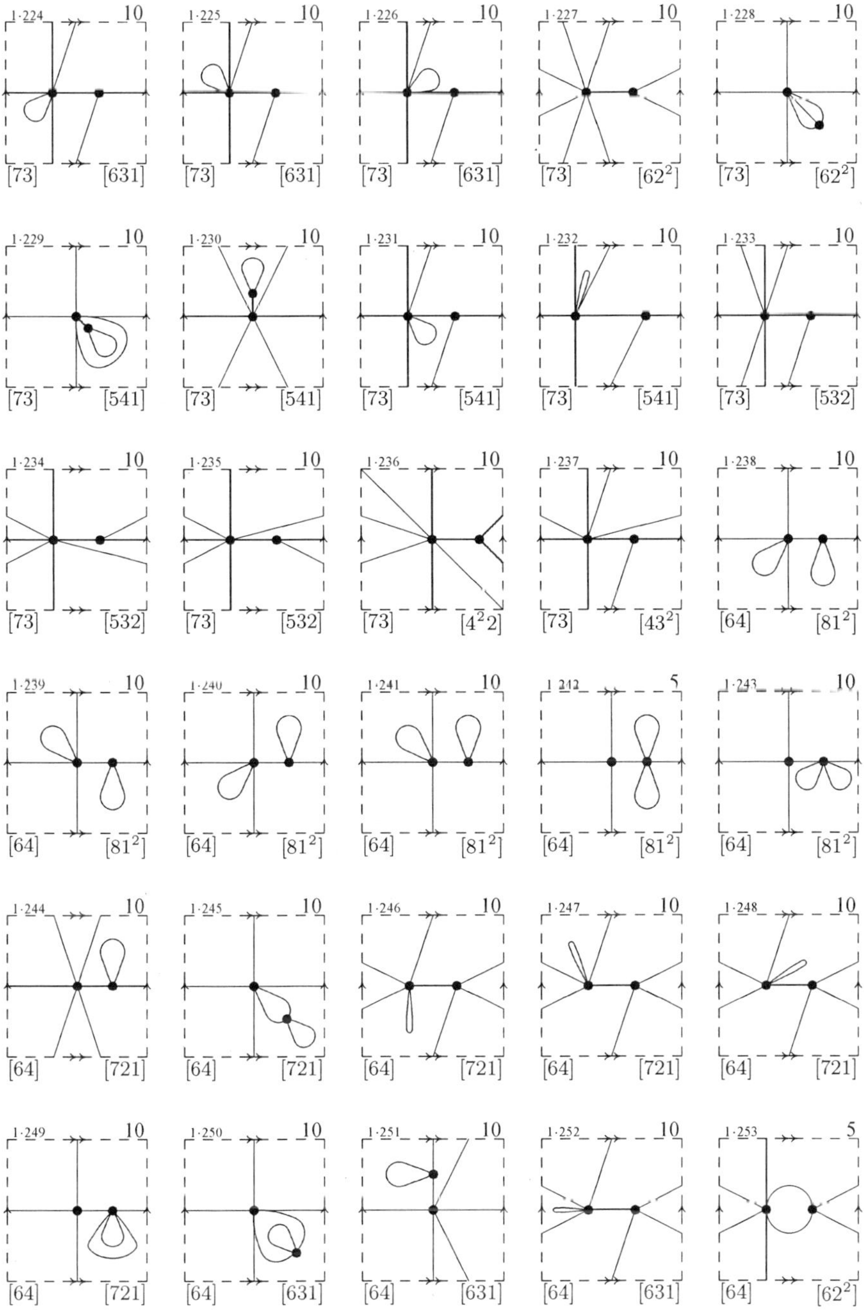
1·224 10 [73] [631]
1·225 10 [73] [631]
1·226 10 [73] [631]
1·227 10 [73] [62^2]
1·228 10 [73] [62^2]
1·229 10 [73] [541]
1·230 10 [73] [541]
1·231 10 [73] [541]
1·232 10 [73] [541]
1·233 10 [73] [532]
1·234 10 [73] [532]
1·235 10 [73] [532]
1·236 10 [73] [4^2 2]
1·237 10 [73] [43^2]
1·238 10 [64] [81^2]
1·239 10 [64] [81^2]
1·240 10 [64] [81^2]
1·241 10 [64] [81^2]
1·242 5 [64] [81^2]
1·243 10 [64] [81^2]
1·244 10 [64] [721]
1·245 10 [64] [721]
1·246 10 [64] [721]
1·247 10 [64] [721]
1·248 10 [64] [721]
1·249 10 [64] [721]
1·250 10 [64] [631]
1·251 10 [64] [631]
1·252 10 [64] [631]
1·253 5 [64] [62^2]

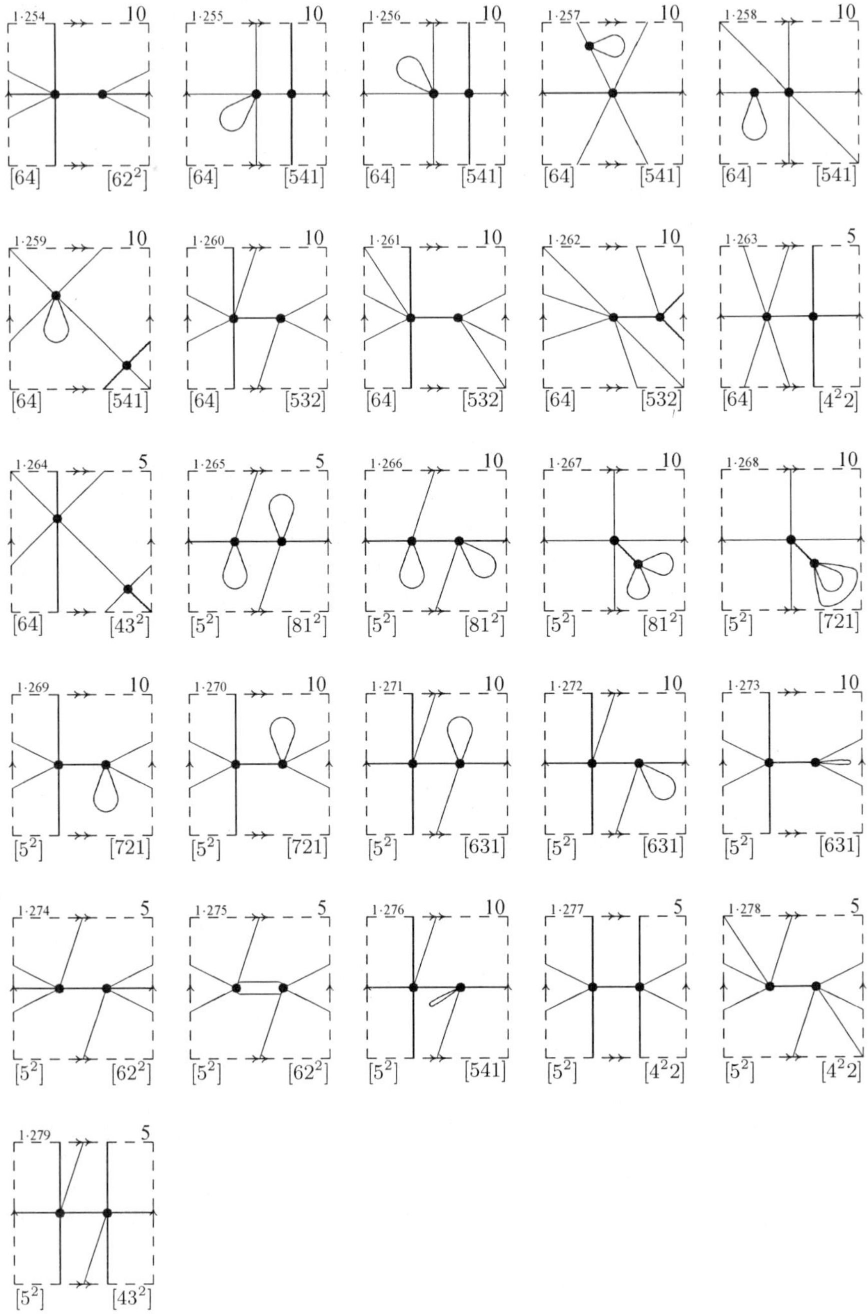
1·254 10 [64] [62²]
1·255 10 [64] [541]
1·256 10 [64] [541]
1·257 10 [64] [541]
1·258 10 [64] [541]
1·259 10 [64] [541]
1·260 10 [64] [532]
1·261 10 [64] [532]
1·262 10 [64] [532]
1·263 5 [64] [4²2]
1·264 5 [64] [43²]
1·265 5 [5²] [81²]
1·266 10 [5²] [81²]
1·267 10 [5²] [81²]
1·268 10 [5²] [721]
1·269 10 [5²] [721]
1·270 10 [5²] [721]
1·271 10 [5²] [631]
1·272 10 [5²] [631]
1·273 10 [5²] [631]
1·274 5 [5²] [62²]
1·275 5 [5²] [62²]
1·276 10 [5²] [541]
1·277 5 [5²] [4²2]
1·278 5 [5²] [4²2]
1·279 5 [5²] [43²]

Genus 1: 5e 3v

5 edges

3 vertices

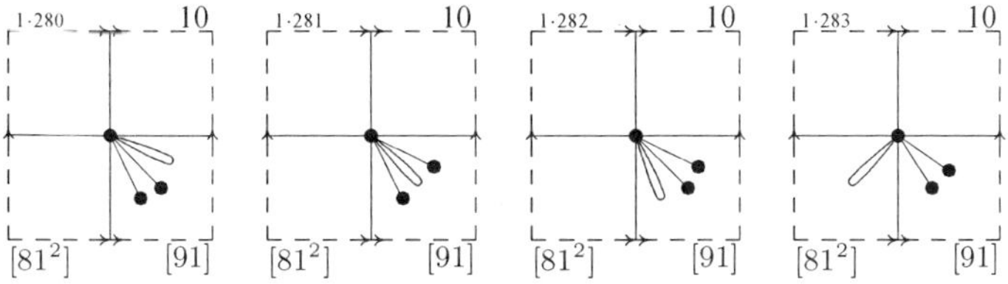

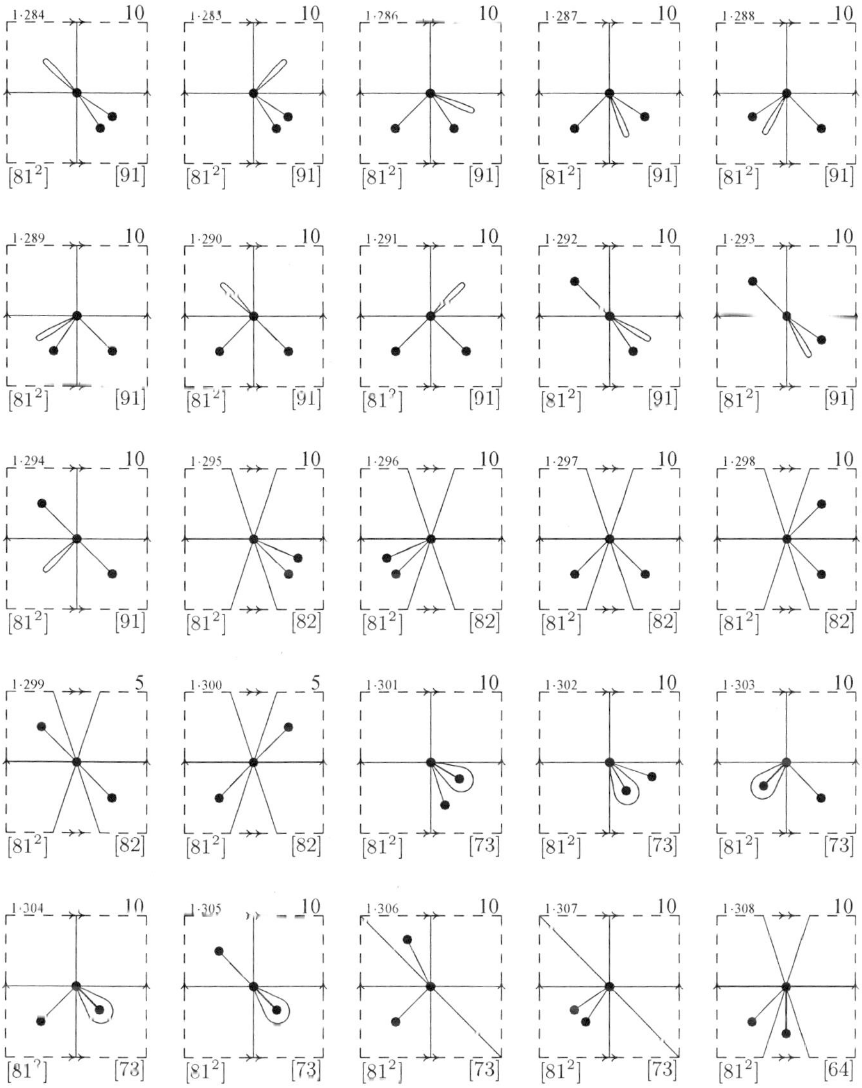

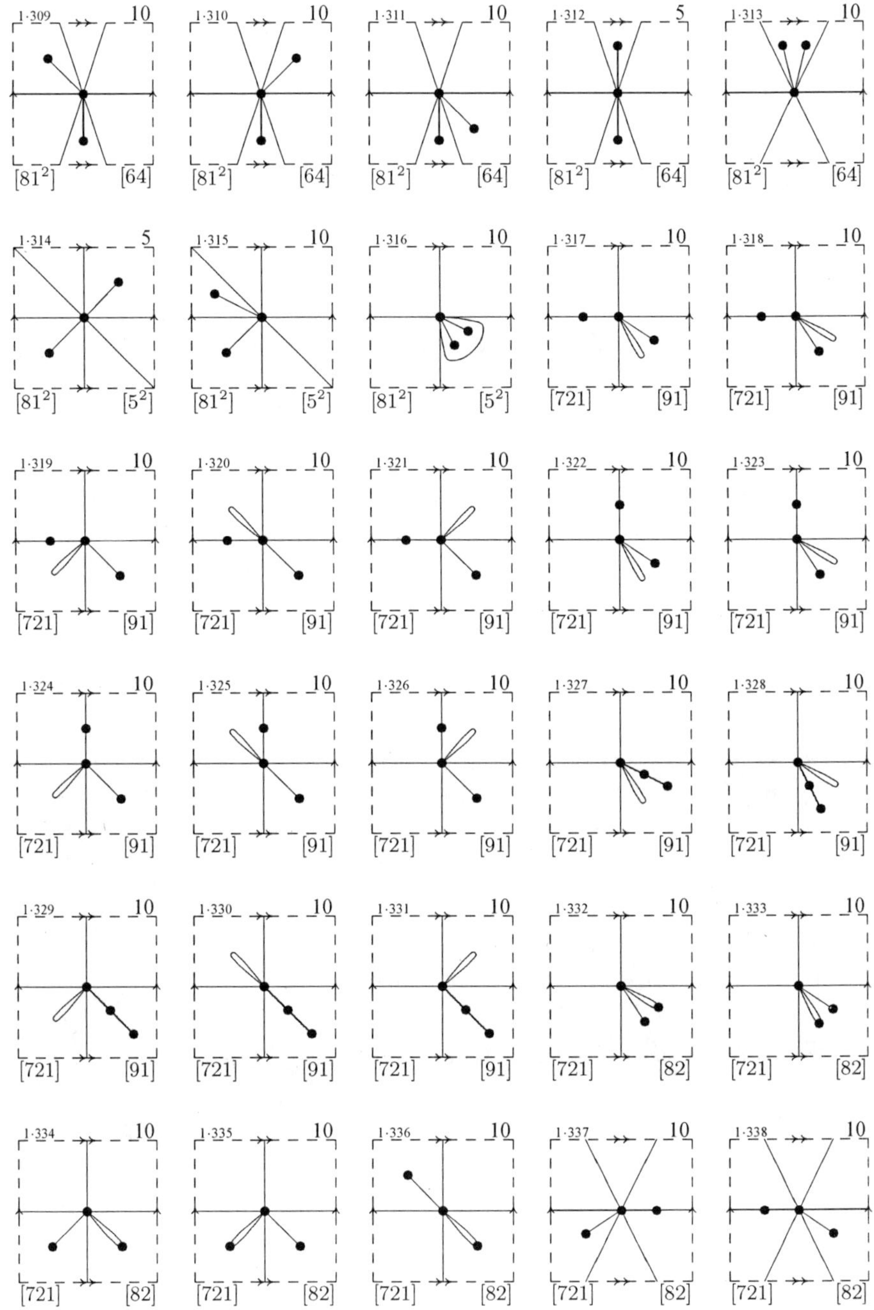
1·309 10 $[81^2]$ [64]
1·310 10 $[81^2]$ [64]
1·311 10 $[81^2]$ [64]
1·312 5 $[81^2]$ [64]
1·313 10 $[81^2]$ [64]
1·314 5 $[81^2]$ $[5^2]$
1·315 10 $[81^2]$ $[5^2]$
1·316 10 $[81^2]$ $[5^2]$
1·317 10 [721] [91]
1·318 10 [721] [91]
1·319 10 [721] [91]
1·320 10 [721] [91]
1·321 10 [721] [91]
1·322 10 [721] [91]
1·323 10 [721] [91]
1·324 10 [721] [91]
1·325 10 [721] [91]
1·326 10 [721] [91]
1·327 10 [721] [91]
1·328 10 [721] [91]
1·329 10 [721] [91]
1·330 10 [721] [91]
1·331 10 [721] [91]
1·332 10 [721] [82]
1·333 10 [721] [82]
1·334 10 [721] [82]
1·335 10 [721] [82]
1·336 10 [721] [82]
1·337 10 [721] [82]
1·338 10 [721] [82]

Genus 1: 5e 3v

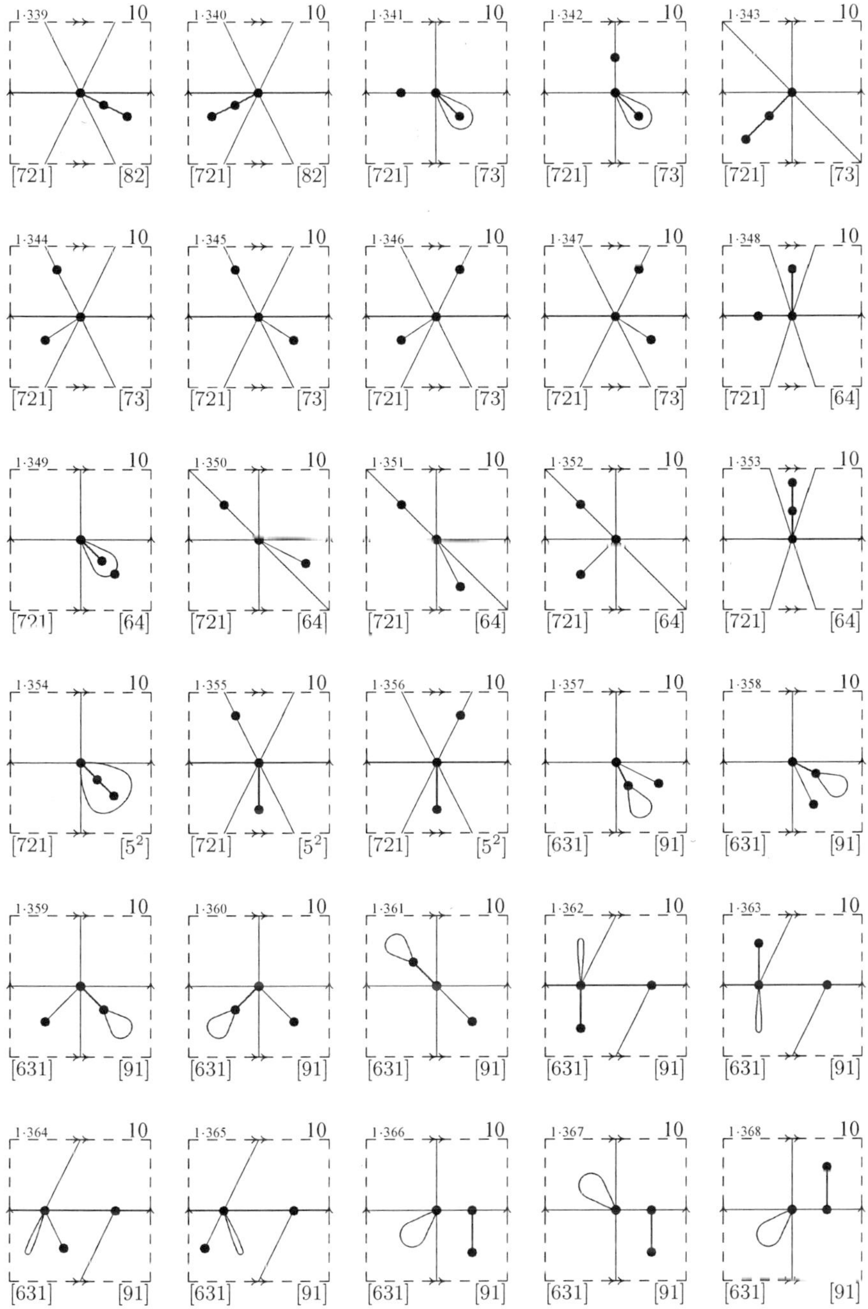
1·339 10 [721] [82]
1·340 10 [721] [82]
1·341 10 [721] [73]
1·342 10 [721] [73]
1·343 10 [721] [73]
1·344 10 [721] [73]
1·345 10 [721] [73]
1·346 10 [721] [73]
1·347 10 [721] [73]
1·348 10 [721] [64]
1·349 10 [721] [64]
1·350 10 [721] [64]
1·351 10 [721] [64]
1·352 10 [721] [64]
1·353 10 [721] [64]
1·354 10 [721] [5^2]
1·355 10 [721] [5^2]
1·356 10 [721] [5^2]
1·357 10 [631] [91]
1·358 10 [631] [91]
1·359 10 [631] [91]
1·360 10 [631] [91]
1·361 10 [631] [91]
1·362 10 [631] [91]
1·363 10 [631] [91]
1·364 10 [631] [91]
1·365 10 [631] [91]
1·366 10 [631] [91]
1·367 10 [631] [91]
1·368 10 [631] [91]

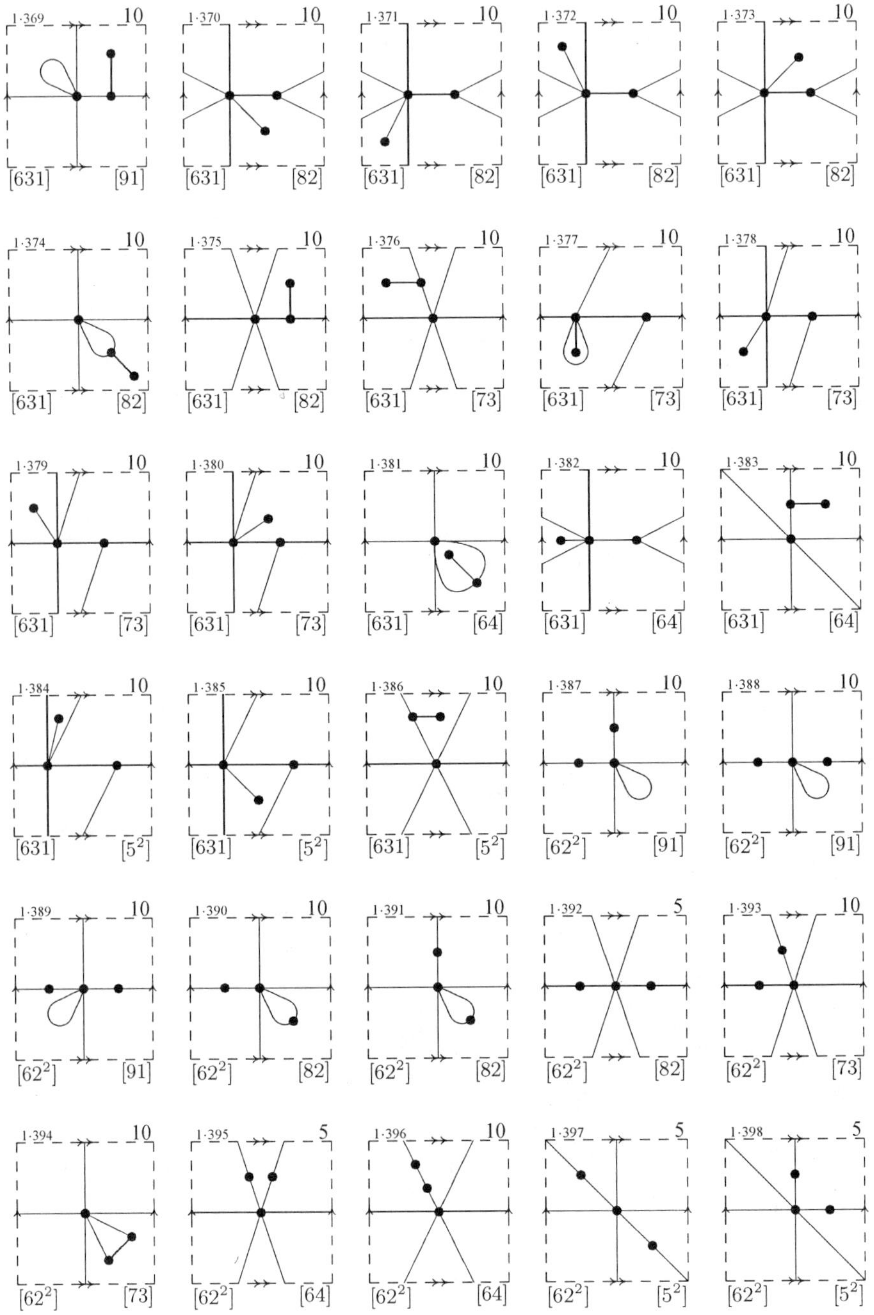
1·369 10 [631] [91]
1·370 10 [631] [82]
1·371 10 [631] [82]
1·372 10 [631] [82]
1·373 10 [631] [82]
1·374 10 [631] [82]
1·375 10 [631] [82]
1·376 10 [631] [73]
1·377 10 [631] [73]
1·378 10 [631] [73]
1·379 10 [631] [73]
1·380 10 [631] [73]
1·381 10 [631] [64]
1·382 10 [631] [64]
1·383 10 [631] [64]
1·384 10 [631] [5^2]
1·385 10 [631] [5^2]
1·386 10 [631] [5^2]
1·387 10 [62^2] [91]
1·388 10 [62^2] [91]
1·389 10 [62^2] [91]
1·390 10 [62^2] [82]
1·391 10 [62^2] [82]
1·392 5 [62^2] [82]
1·393 10 [62^2] [73]
1·394 10 [62^2] [73]
1·395 5 [62^2] [64]
1·396 10 [62^2] [64]
1·397 5 [62^2] [5^2]
1·398 5 [62^2] [5^2]

Genus 1: 5e 3v

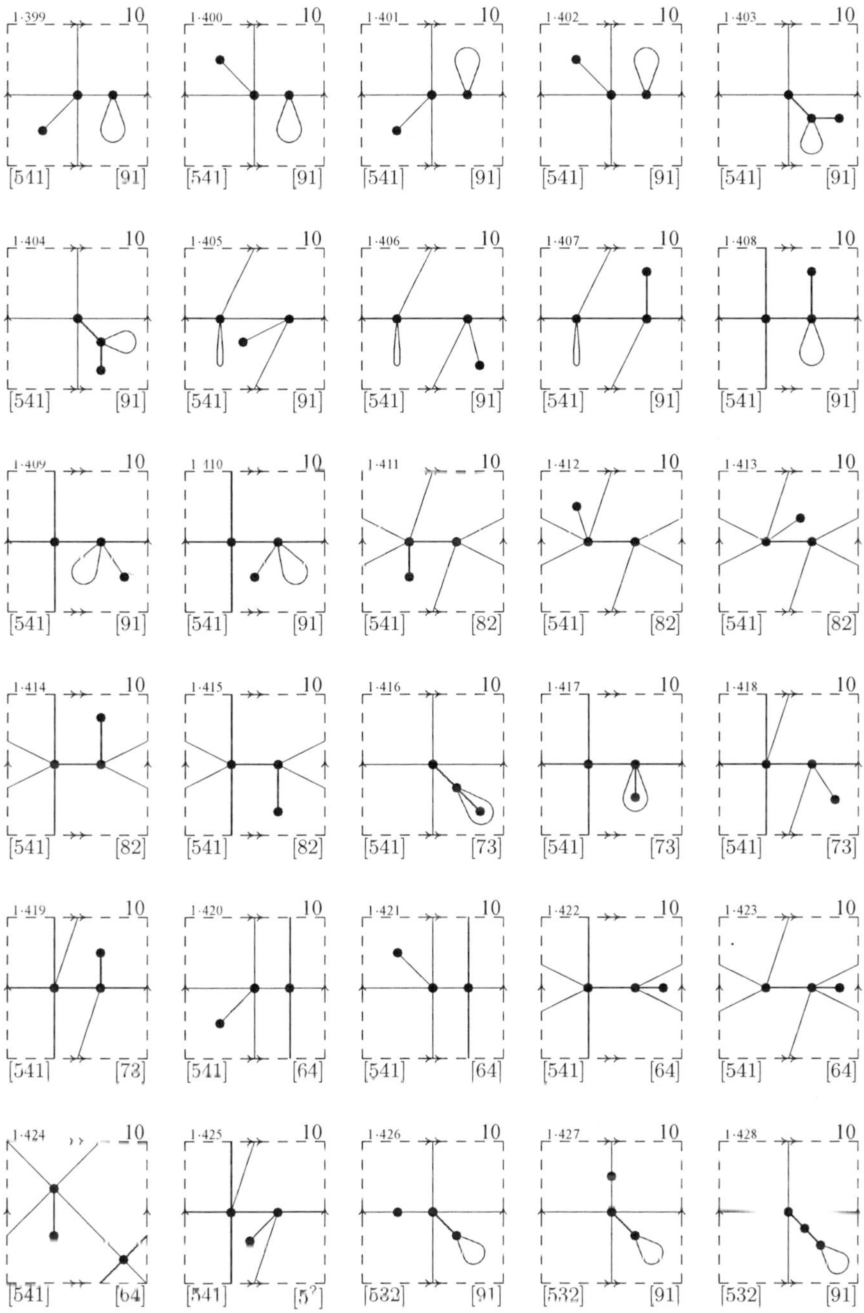
1·399 10 [541] [91]
1·400 10 [541] [91]
1·401 10 [541] [91]
1·402 10 [541] [91]
1·403 10 [541] [91]
1·404 10 [541] [91]
1·405 10 [541] [91]
1·406 10 [541] [91]
1·407 10 [541] [91]
1·408 10 [541] [91]
1·409 10 [541] [91]
1·410 10 [541] [91]
1·411 10 [541] [82]
1·412 10 [541] [82]
1·413 10 [541] [82]
1·414 10 [541] [82]
1·415 10 [541] [82]
1·416 10 [541] [73]
1·417 10 [541] [73]
1·418 10 [541] [73]
1·419 10 [541] [73]
1·420 10 [541] [64]
1·421 10 [541] [64]
1·422 10 [541] [64]
1·423 10 [541] [64]
1·424 10 [541] [64]
1·425 10 [541]
1·426 10 [532] [91]
1·427 10 [532] [91]
1·428 10 [532] [91]

1·429 10
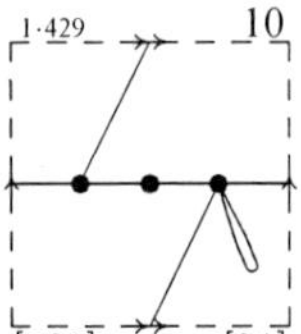
[532] [91]

1·430 10
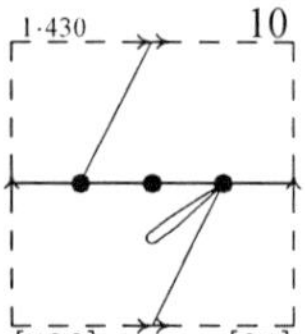
[532] [91]

1·431 10
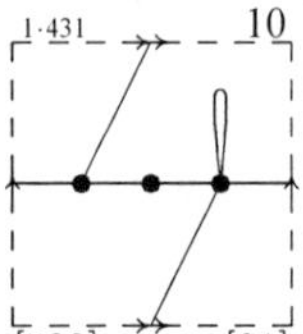
[532] [91]

1·432 10
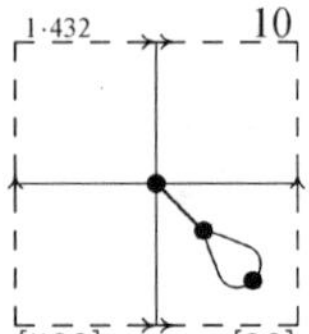
[532] [82]

1·433 10
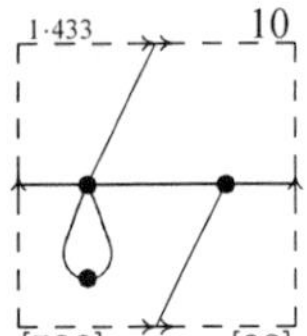
[532] [82]

1·434 10
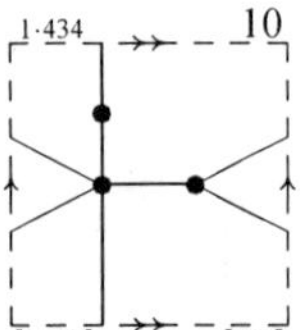
[532] [82]

1·435 10
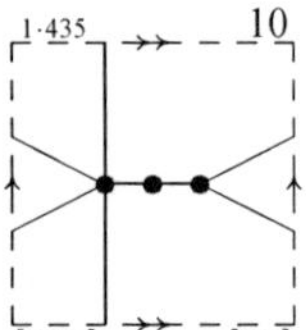
[532] [82]

1·436 10
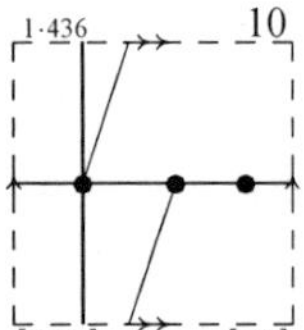
[532] [73]

1·437 10
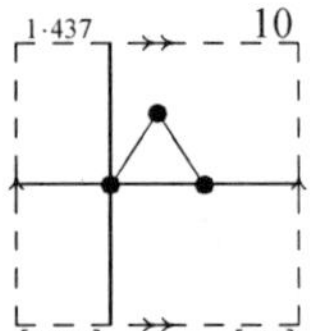
[532] [73]

1·438 10
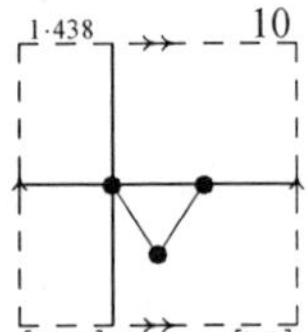
[532] [73]

1·439 10
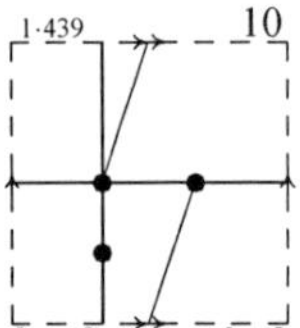
[532] [64]

1·440 10
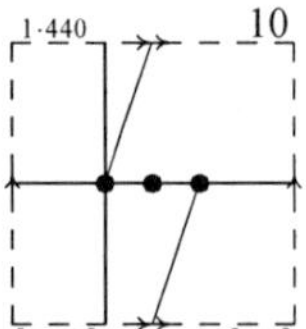
[532] [64]

1·441 10
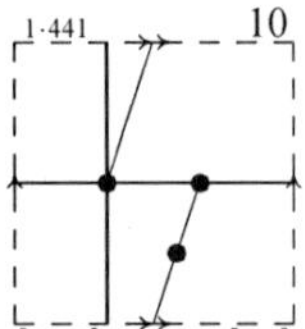
[532] [64]

1·442 10
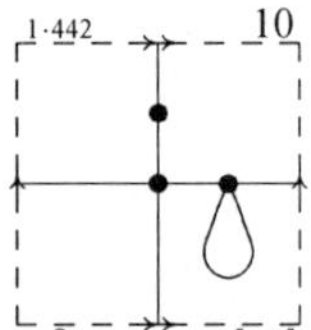
$[4^2 2]$ [91]

1·443 10
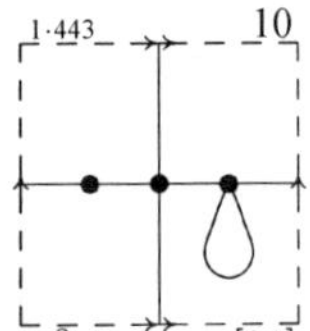
$[4^2 2]$ [91]

1·444 10
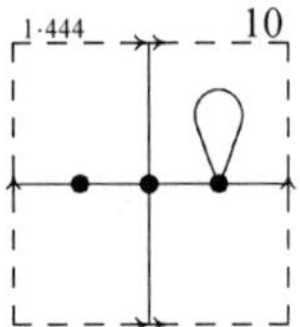
$[4^2 2]$ [91]

1·445 10
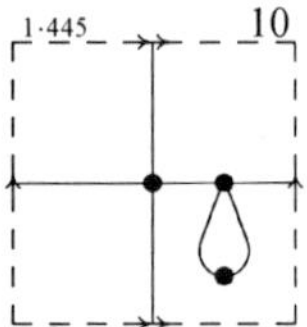
$[4^2 2]$ [82]

1·446 5
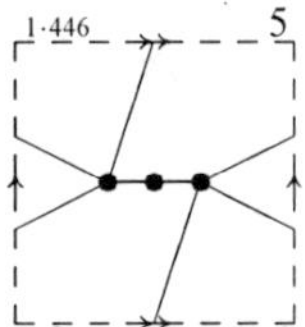
$[4^2 2]$ [82]

1·447 5
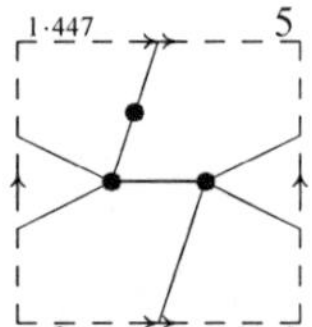
$[4^2 2]$ [82]

1·448 10
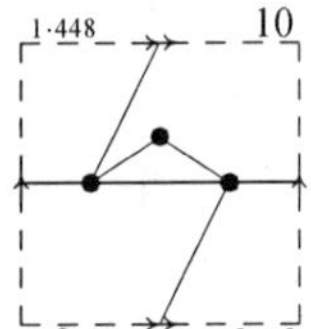
$[4^2 2]$ [73]

1·449 5
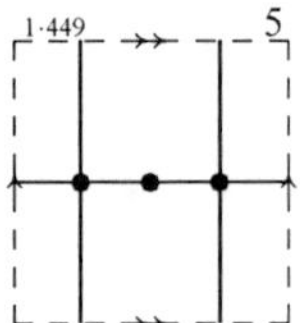
$[4^2 2]$ [64]

1·450 5
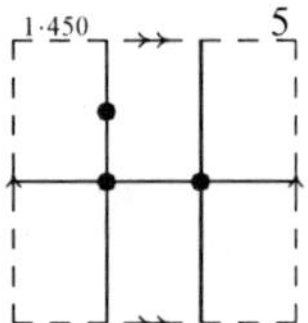
$[4^2 2]$ $[5^2]$

1·451 5
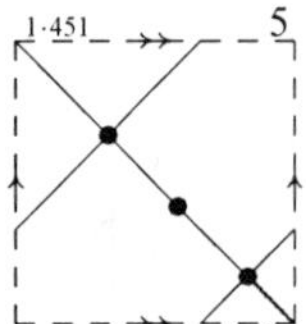
$[4^2 2]$ $[5^2]$

1·452 10
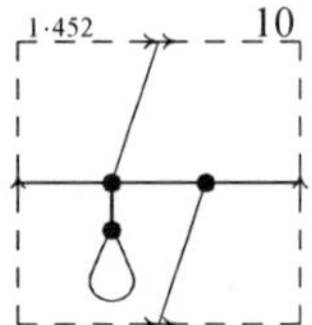
$[43^2]$ [91]

1·453 10
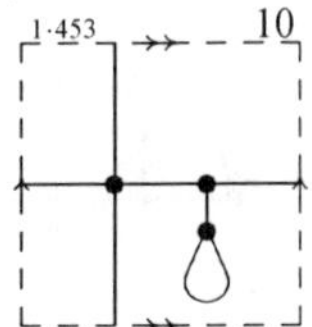
$[43^2]$ [91]

1·454 10
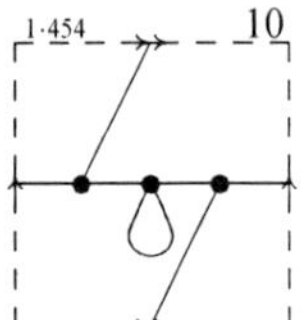
$[43^2]$ [91]

1·455 5
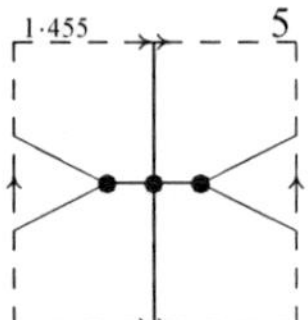
$[43^2]$ [82]

1·456 10
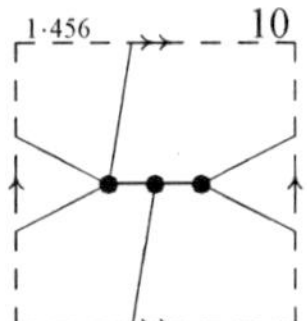
$[43^2]$ [82]

1·457 10
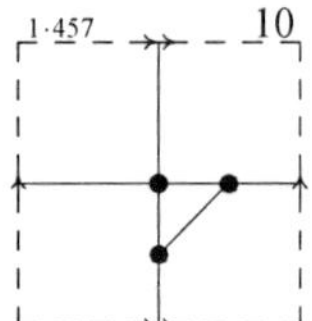
$[43^2]$ [73]

1·458 5
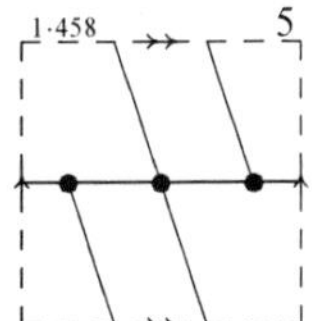
$[43^2]$ [64]

Genus 1: 5e 4v

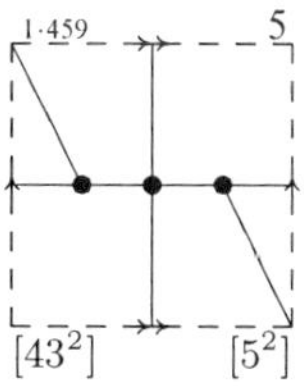

5 edges

4 vertices

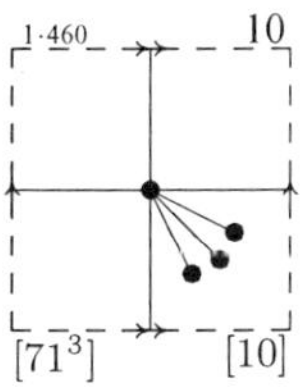

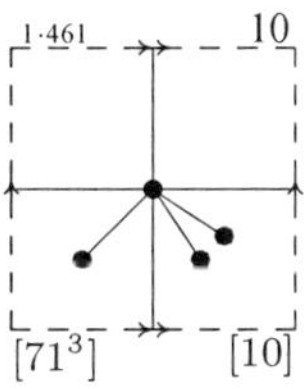

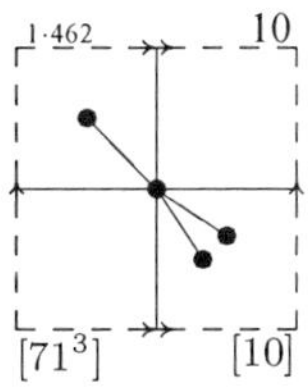

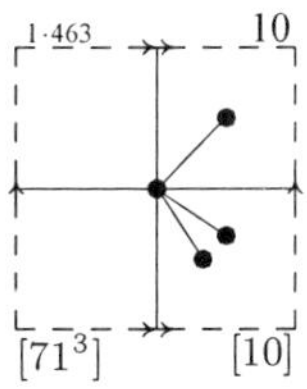

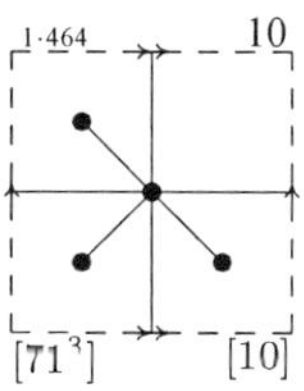

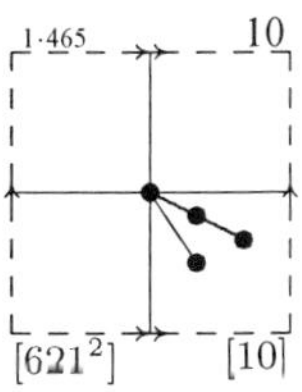

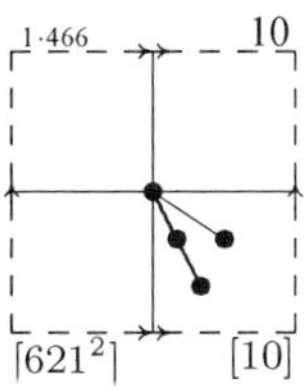

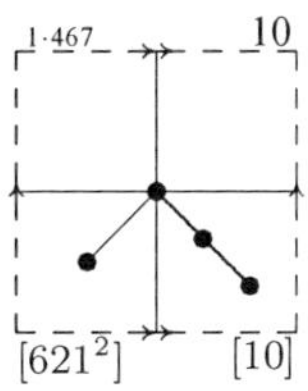

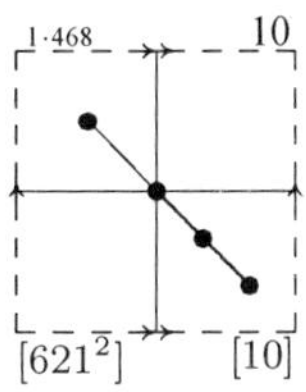

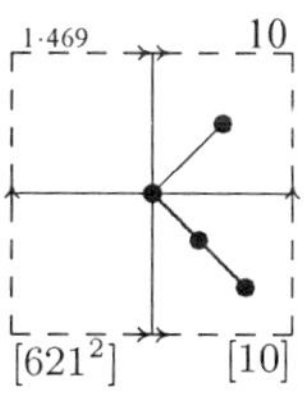

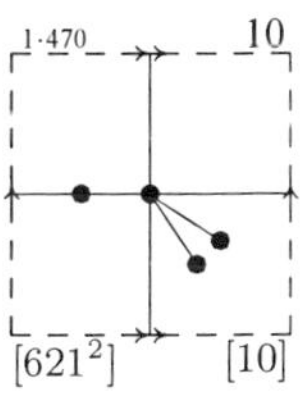

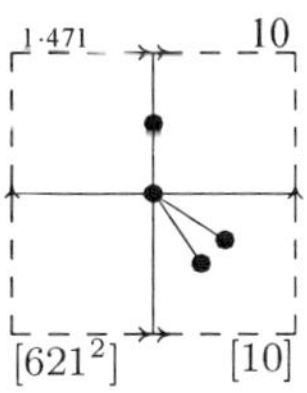

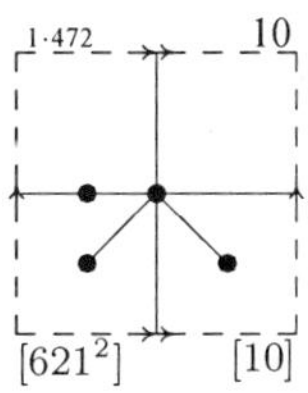

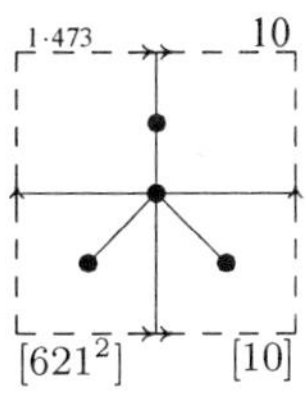

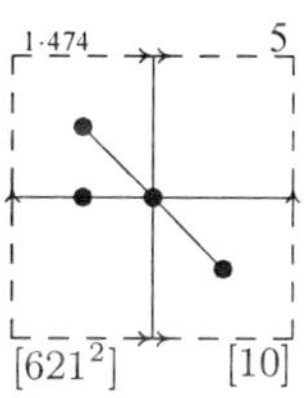

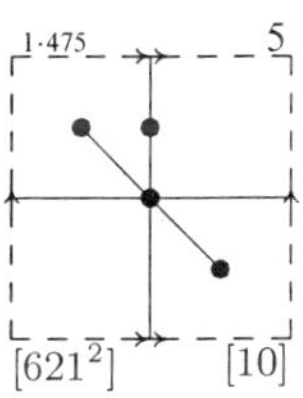

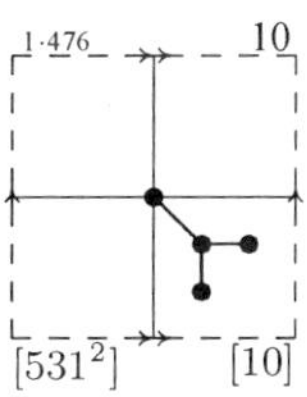

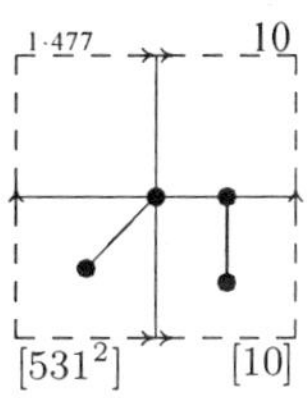

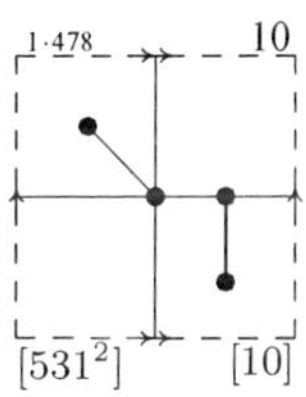

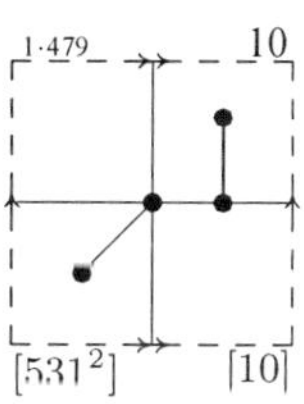

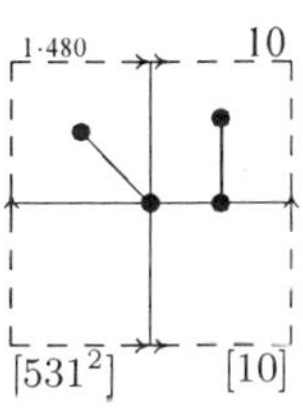

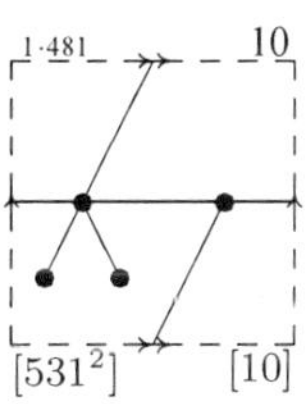

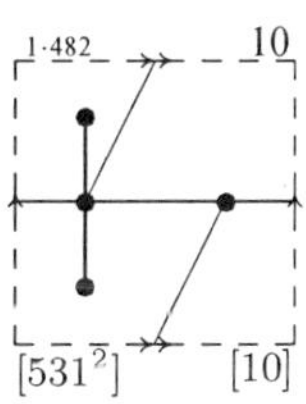

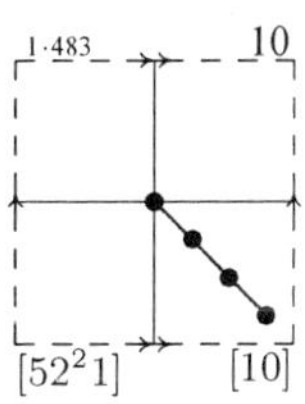

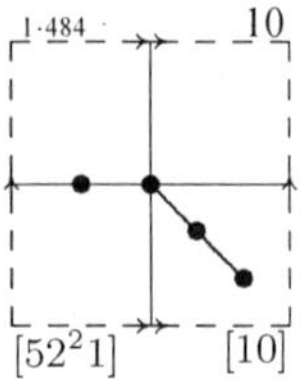
1·484
10
[52²1]
[1̄0]

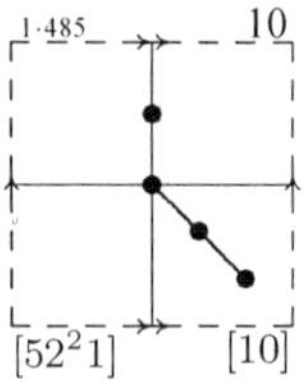
1·485
10
[52²1]
[1̄0]

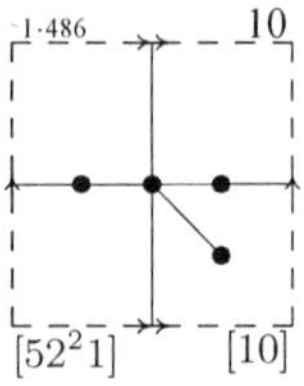
1·486
10
[52²1]
[1̄0]

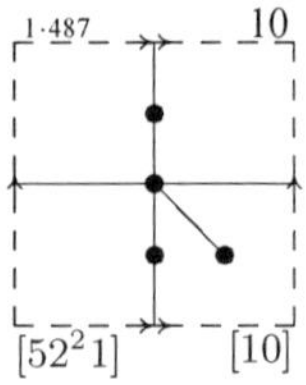
1·487
10
[52²1]
[1̄0]

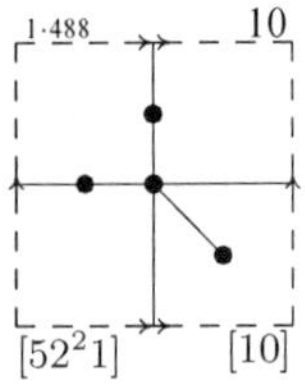
1·488
10
[52²1]
[1̄0]

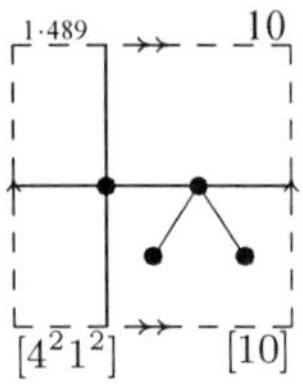
1·489
10
[4²1²]
[1̄0]

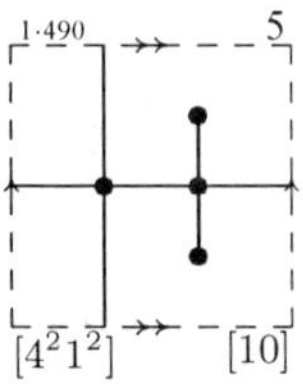
1·490
5
[4²1²]
[1̄0]

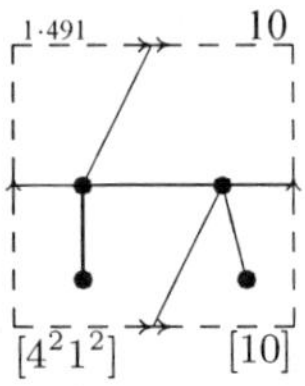
1·491
10
[4²1²]
[1̄0]

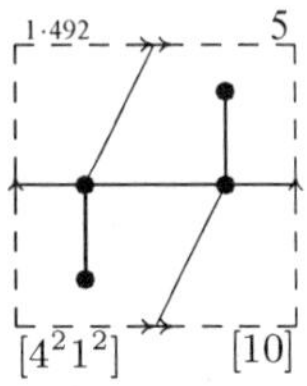
1·492
5
[4²1²]
[1̄0]

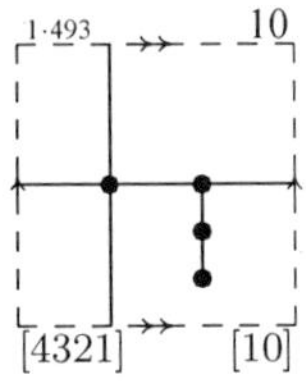
1·493
10
[43̄21̄]
[1̄0]

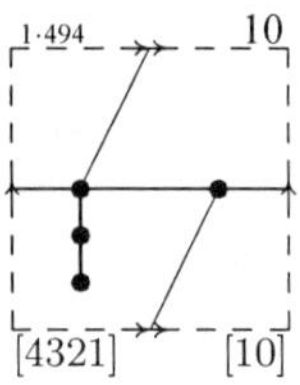
1·494
10
[43̄21̄]
[1̄0]

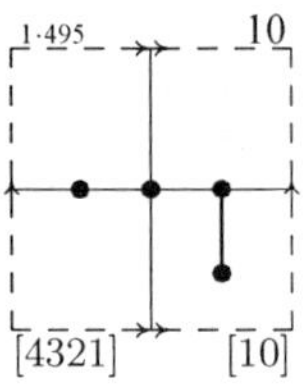
1·495
10
[43̄21̄]
[1̄0]

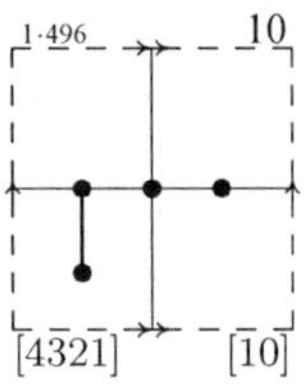
1·496
10
[43̄21̄]
[1̄0]

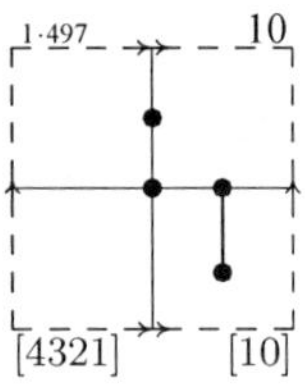
1·497
10
[43̄21̄]
[1̄0]

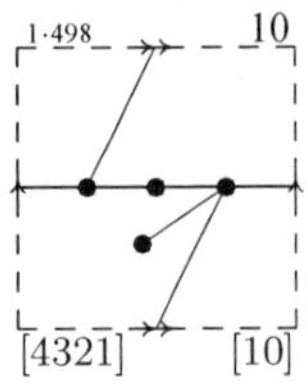
1·498
10
[43̄21̄]
[1̄0]

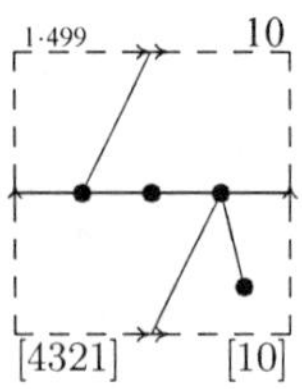
1·499
10
[43̄21̄]
[1̄0]

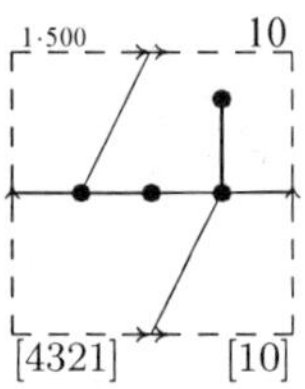
1·500
10
[43̄21̄]
[1̄0]

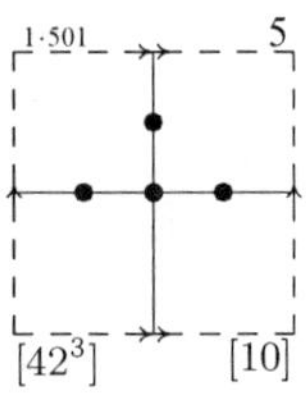
1·501
5
[42³]
[1̄0]

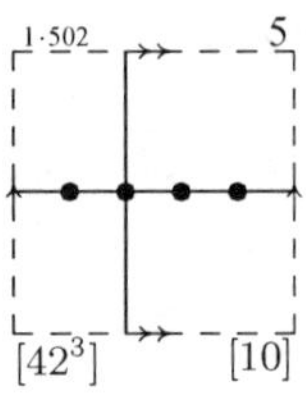
1·502
5
[42³]
[1̄0]

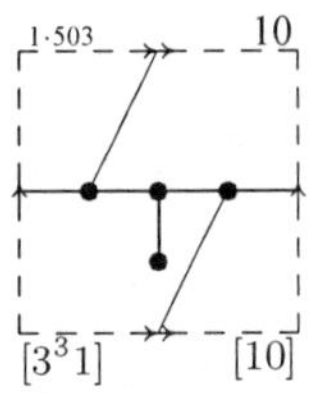
1·503
10
[3³1]
[1̄0]

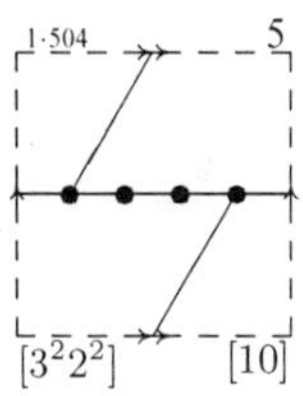
1·504
5
[3²2²]
[1̄0]

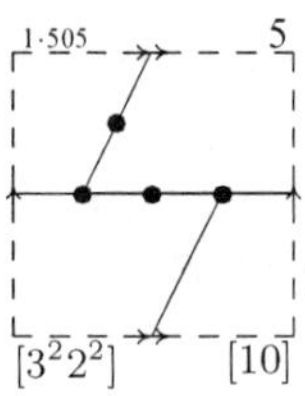
1·505
5
[3²2²]
[1̄0]

Genus 2: 4e 1v

5.3 Genus 2 – the double torus

4 edges

1 vertex

2·1 1

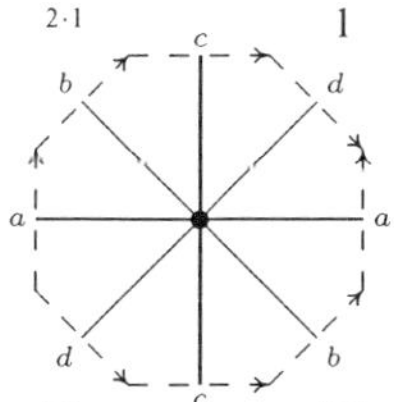

[8] [8]

2·2 4

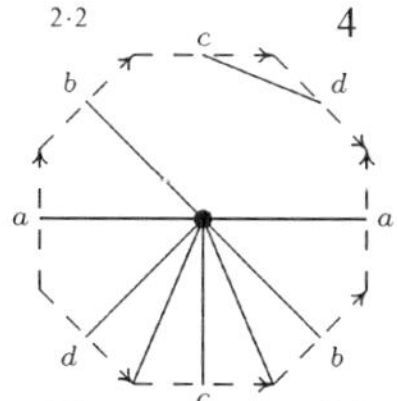

[8] [8]

2·3 8

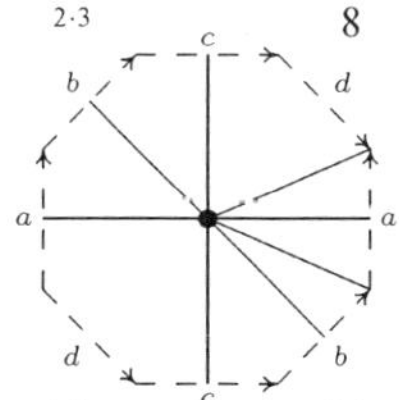

[8] [8]

2·4 8

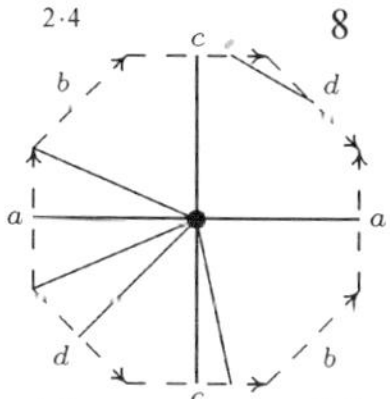

[8] [8]

5 edges

1 vertex

2·5 10

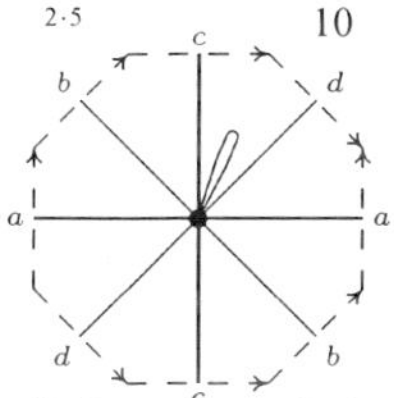

[10] [91]

2·6 10

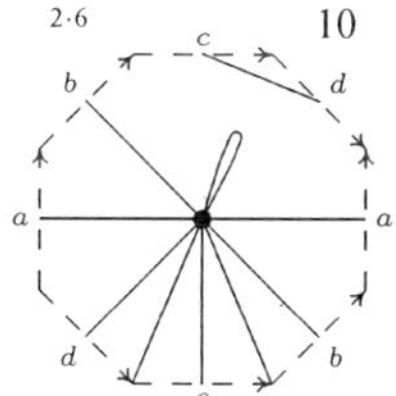

[10] [91]

2·7 10

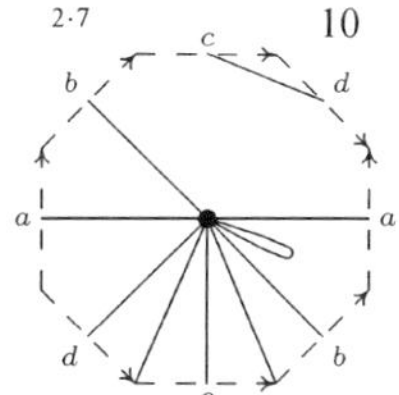

[10] [91]

2·8 10

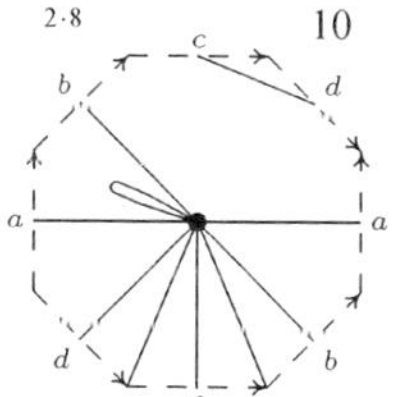

[10] [01]

2·9 10

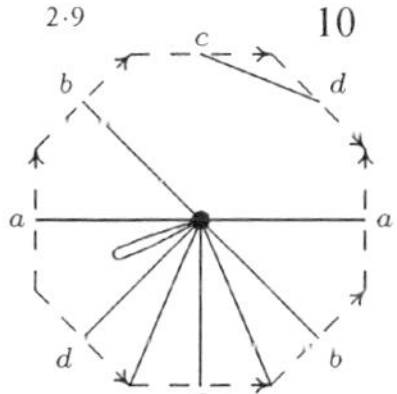

[10] [91]

2·10 10

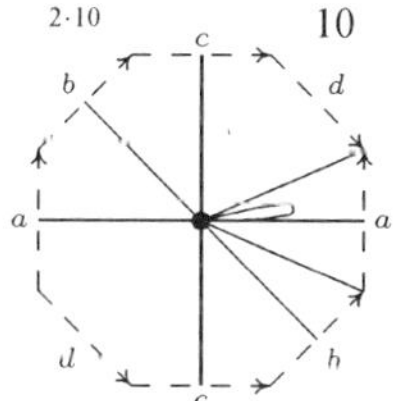

[10] [91]

2·11 10

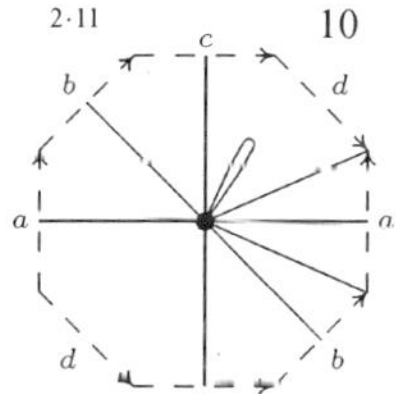

[10] [91]

2·12 10

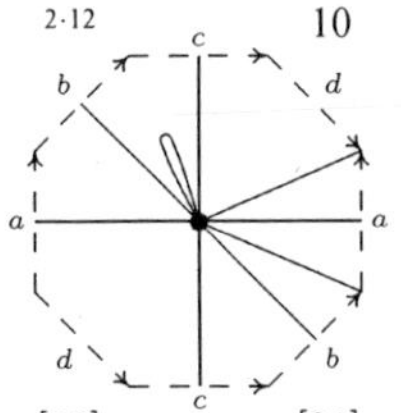

[10] [91]

2·13 10

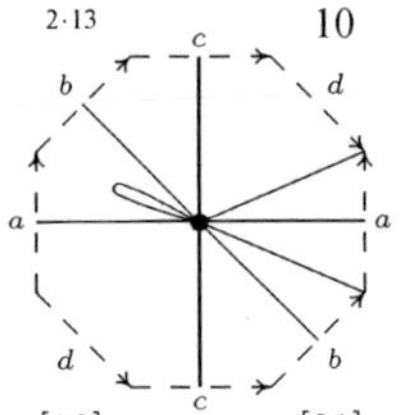

[10] [91]

2·14 10

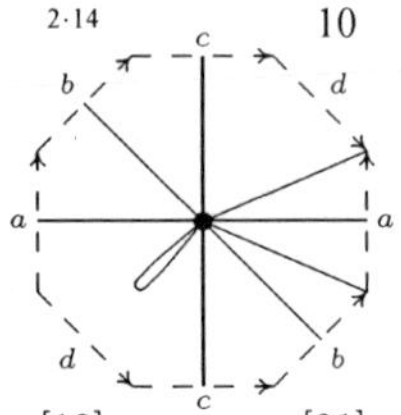

[10] [91]

2·15 10

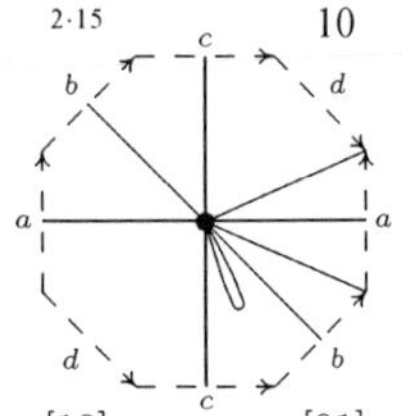

[10] [91]

2·16 10

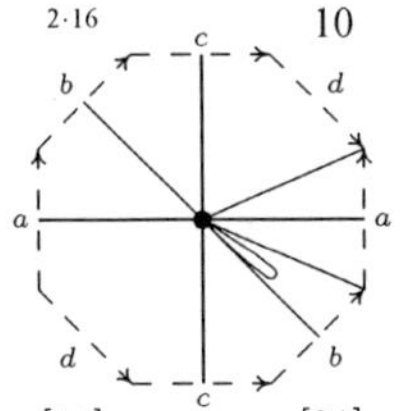

[10] [91]

2·17 10

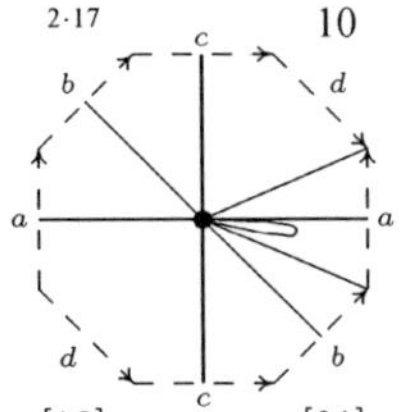

[10] [91]

2·18 10

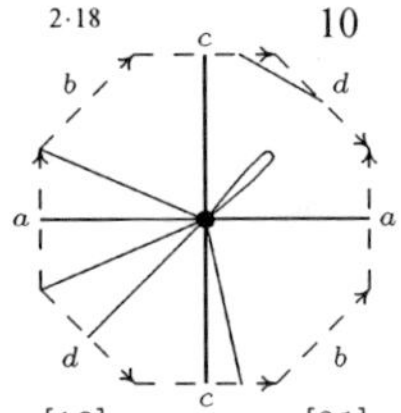

[10] [91]

2·19 10

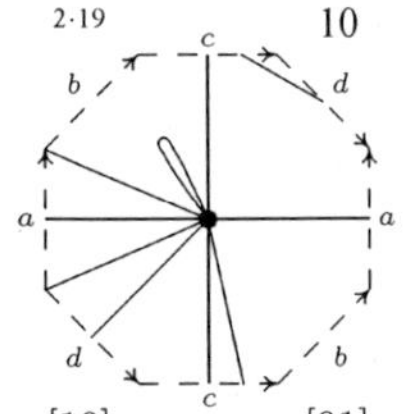

[10] [91]

2·20 10

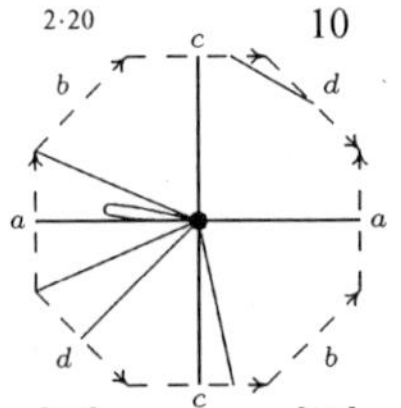

[10] [91]

2·21 10

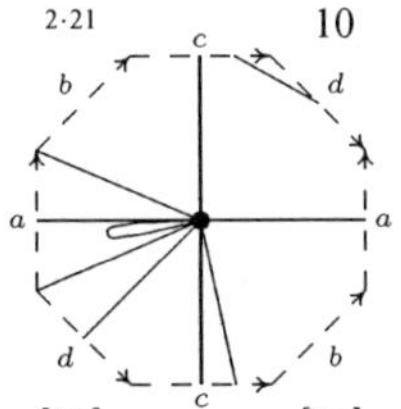

[10] [91]

2·22 10

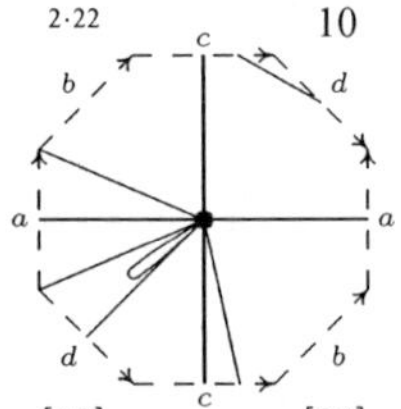

[10] [91]

2·23 10

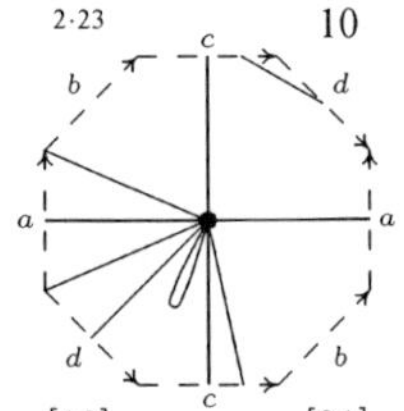

[10] [91]

2·24 10

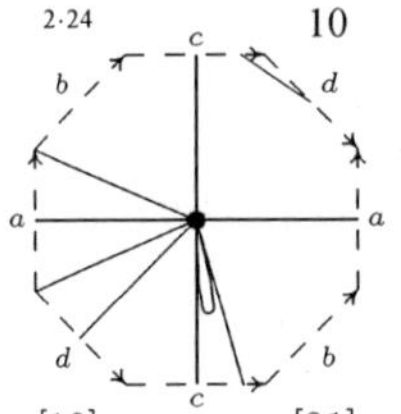

[10] [91]

2·25 10

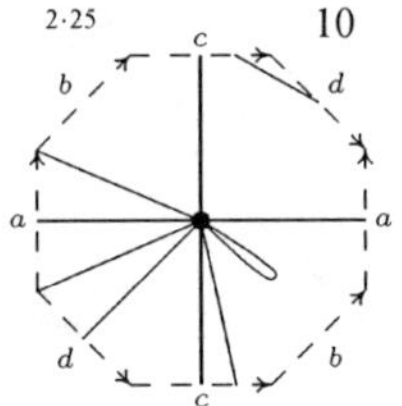

[10] [91]

2·26 5

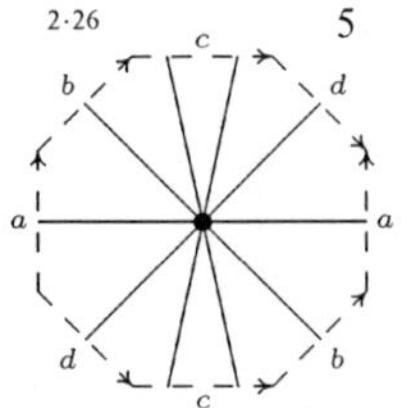

[10] [82]

2·27 10

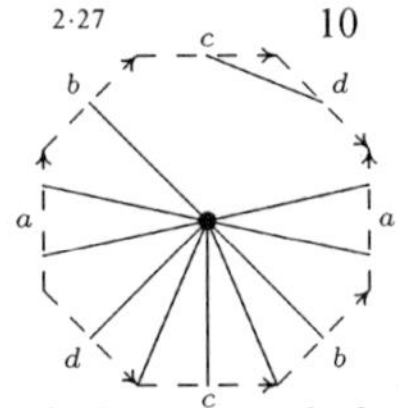

[10] [82]

2·28 10

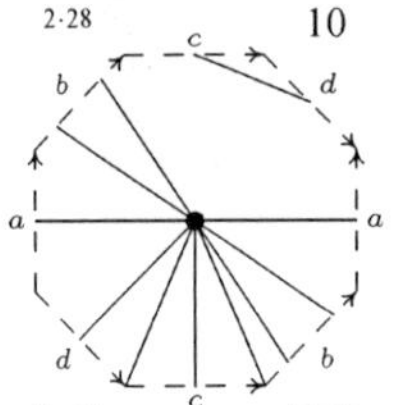

[10] [82]

2·29 10

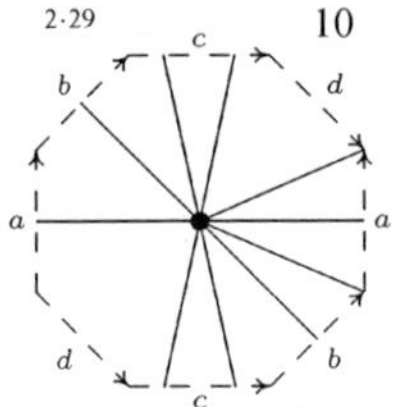

[10] [82]

2·30 10

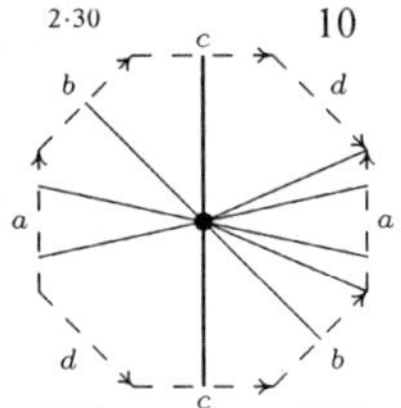

[10] [82]

2·31 10

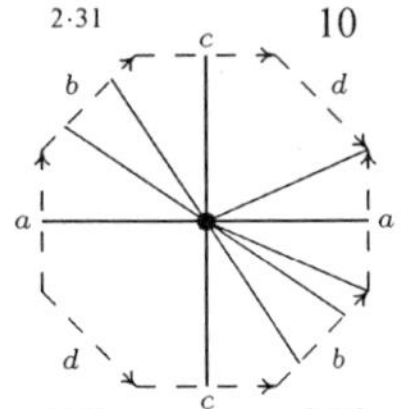

[10] [82]

Genus 2: 5e 1v

2·32 10

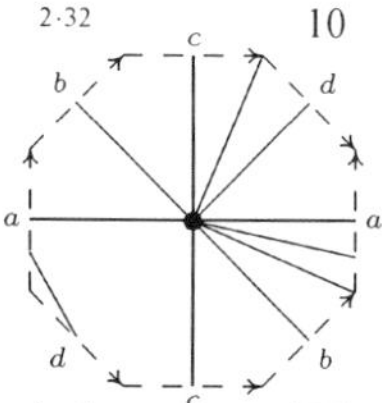

[10] [82]

2·33 10

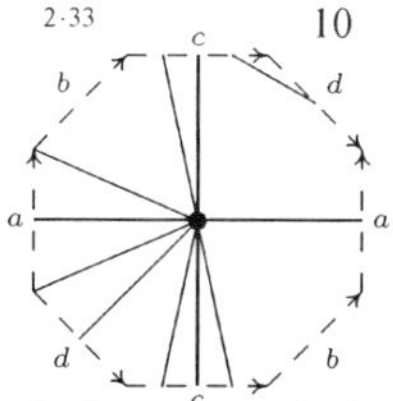

[10] [82]

2·34 10

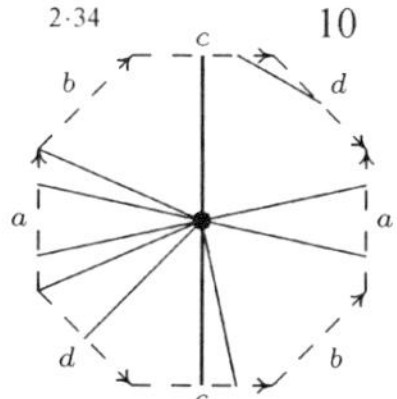

[10] [82]

2·35 10

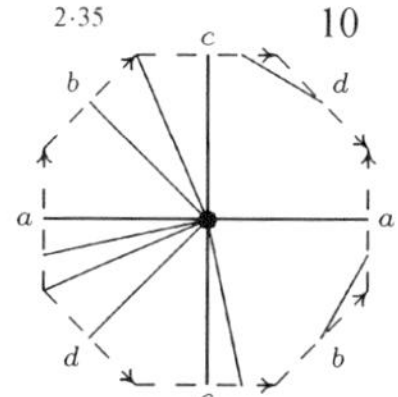

[10] [82]

2·36 10

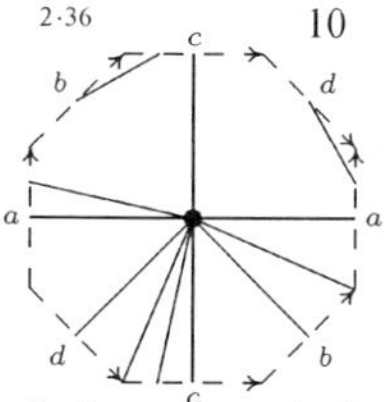

[10] [82]

2·37 10

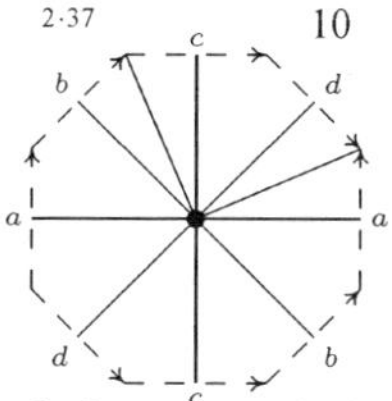

[10] [73]

2·38 10

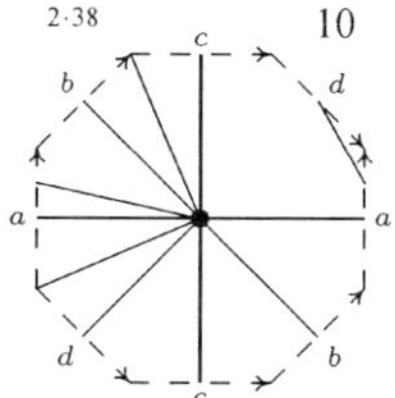

[10] [73]

2·39 10

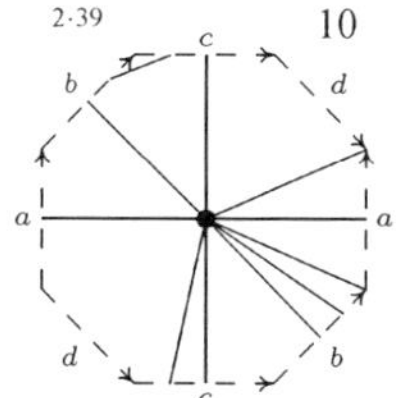

[10] [73]

2·40 10

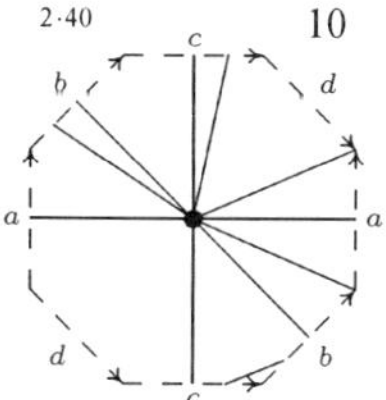

[10] [73]

2·41 10

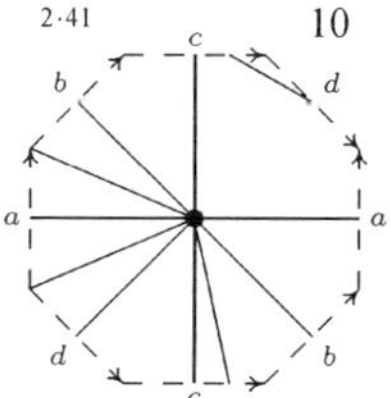

[10] [73]

2·42 10

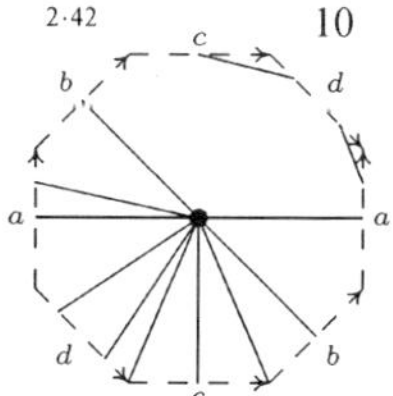

[10] [73]

2·43 10

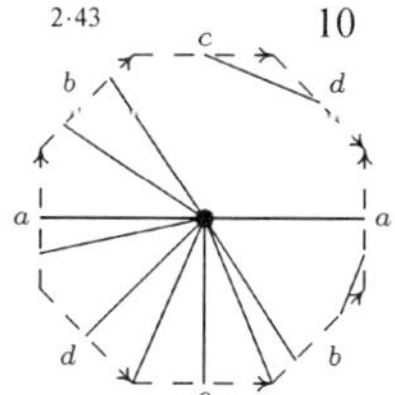

[10] [73]

2·44 5

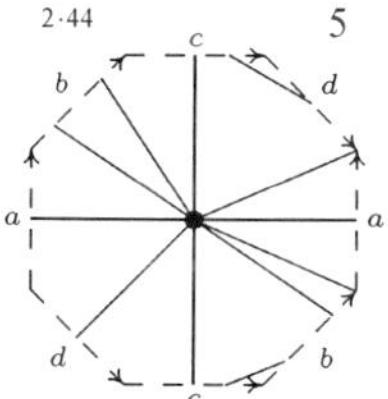

[10] [64]

2·45 10

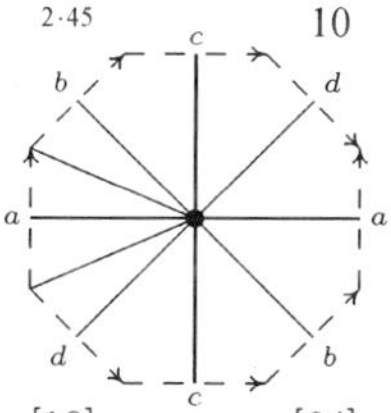

[10] [64]

2·46 10

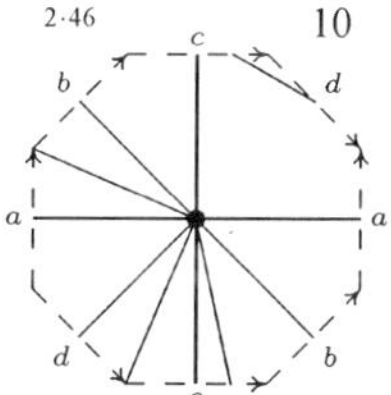

[10] [64]

2·47 10

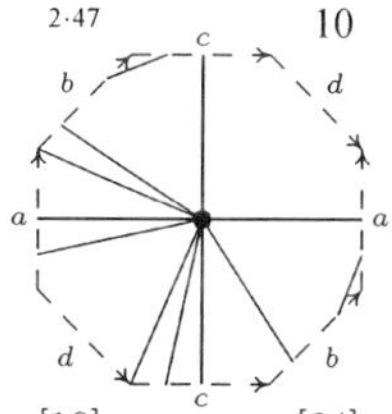

[10] [64]

2·48 10

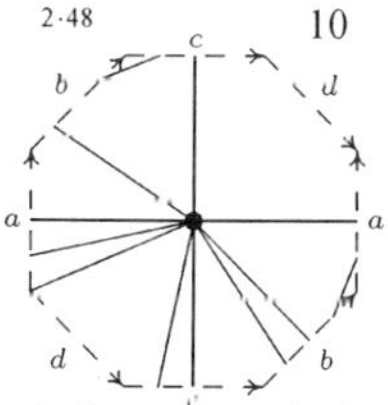

[10] [64]

2·49 10

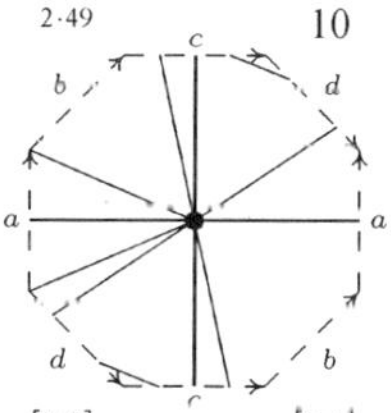

[10] [64]

2·50 10

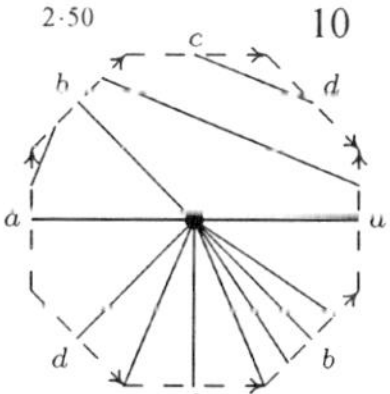

[10] [64]

2·51 1

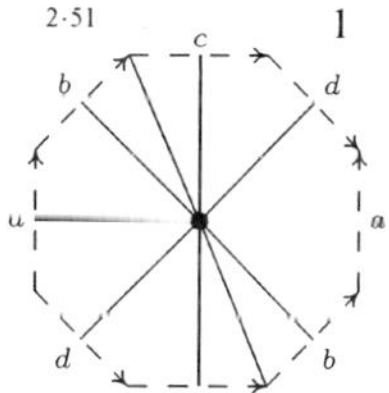

[10] [5^2]

2·52 2

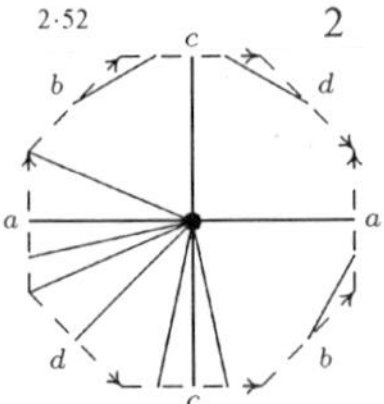

[10] $[5^2]$

2·53 10

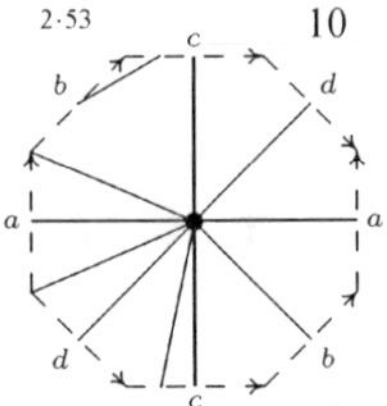

[10] $[5^2]$

2·54 5

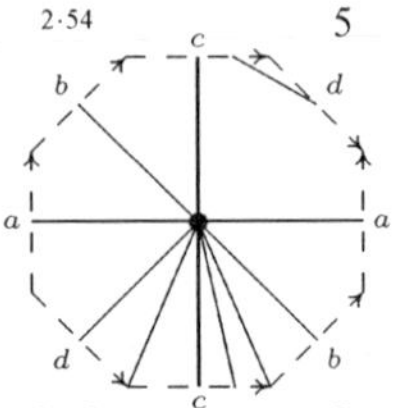

[10] $[5^2]$

2·55 5

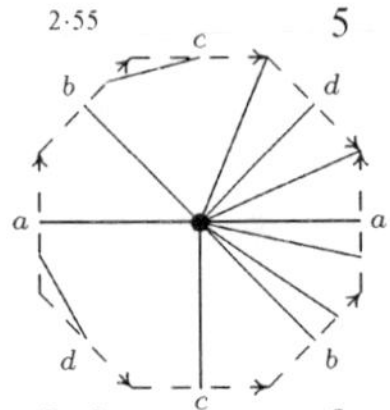

[10] $[5^2]$

2·56 5

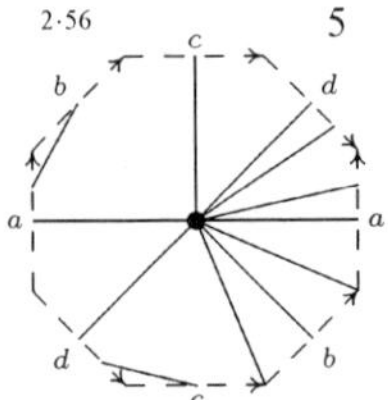

[10] $[5^2]$

2·57 5

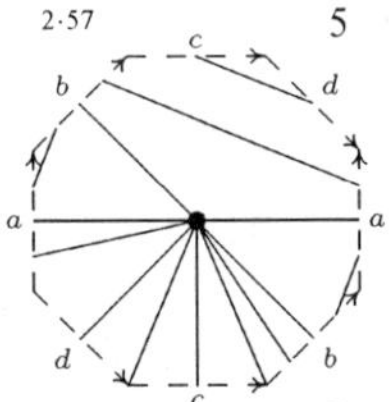

[10] $[5^2]$

5 edges

2 vertices

2·58 10

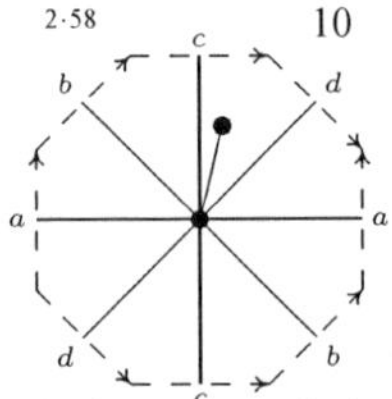

[91] [10]

2·59 10

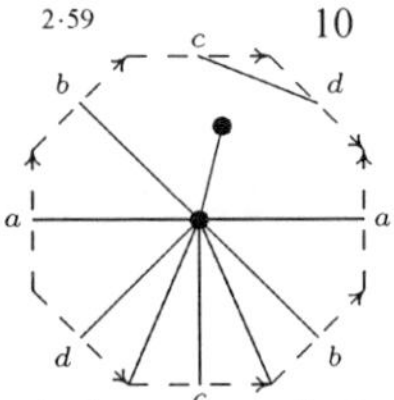

[91] [10]

2·60 10

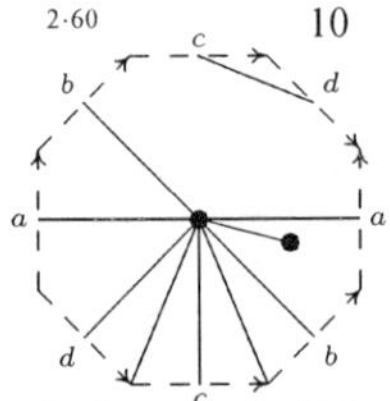

[91] [10]

2·61 10

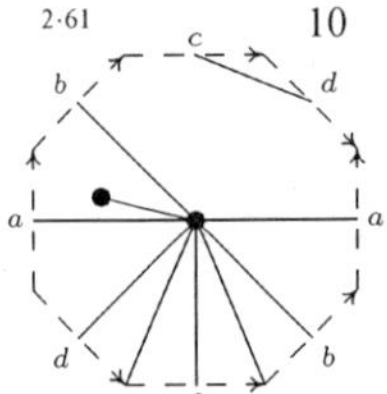

[91] [10]

2·62 10

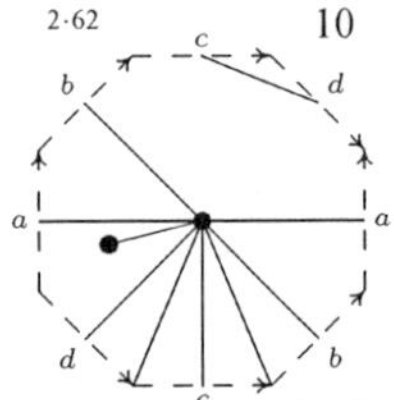

[91] [10]

2·63 10

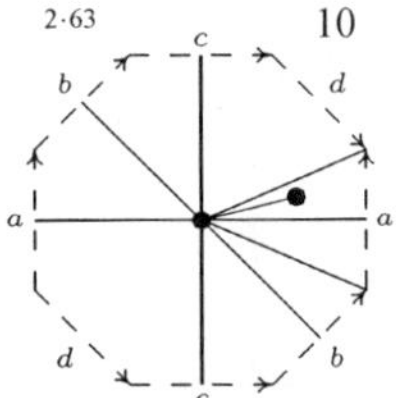

[91] [10]

2·64 10

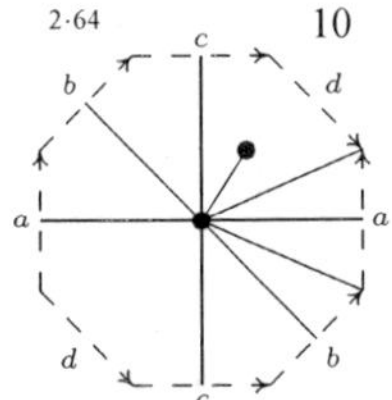

[91] [10]

2·65 10

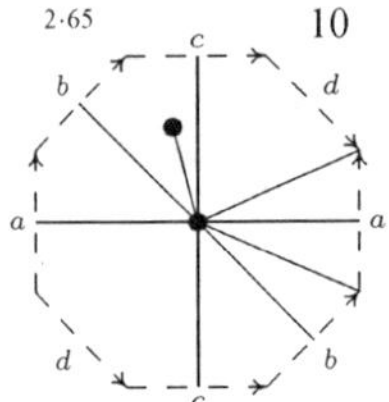

[91] [10]

2·66 10

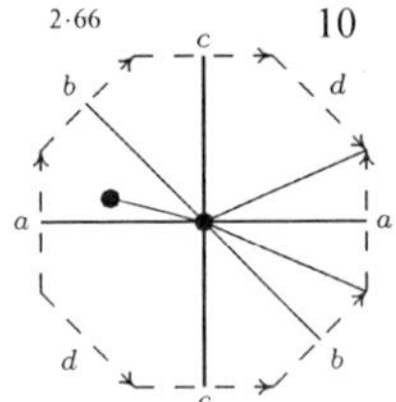

[91] [10]

2·67 10

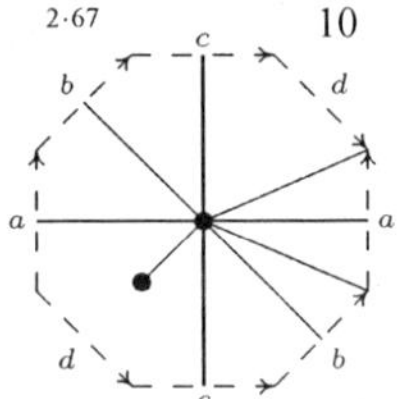

[91] [10]

2·68 10

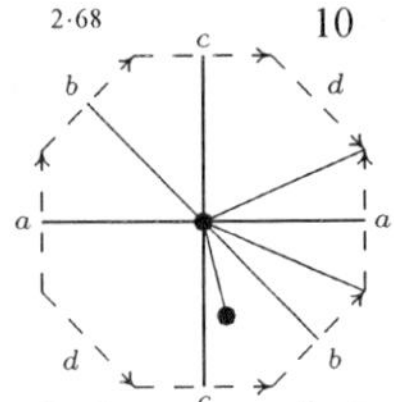

[91] [10]

Genus 2: 5e 2v

2·69 10
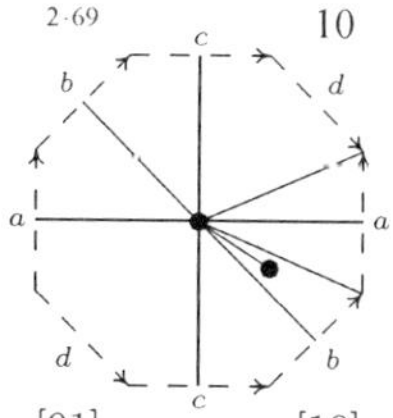
[91] [10]

2·70 10
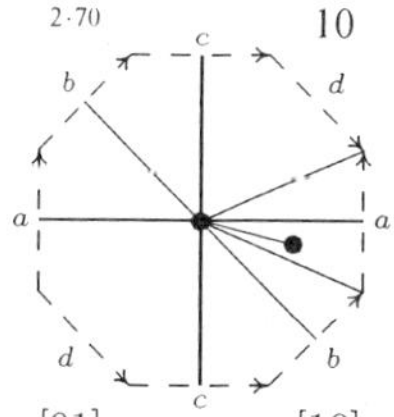
[91] [10]

2·71 10
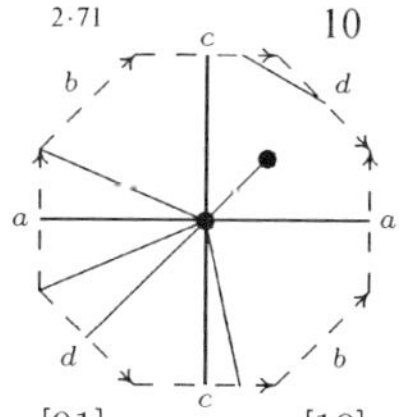
[91] [10]

2·72 10
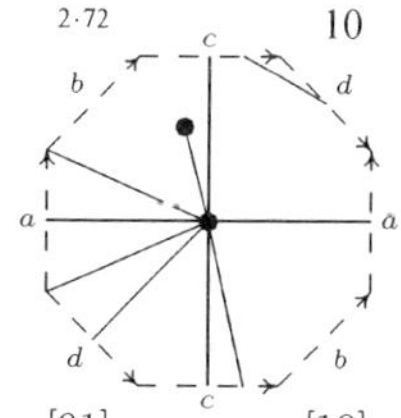
[91] [10]

2·73 10
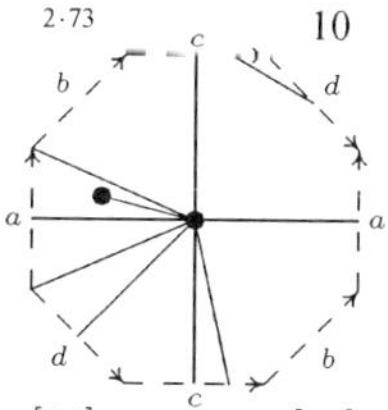
[91] [10]

2·74 10
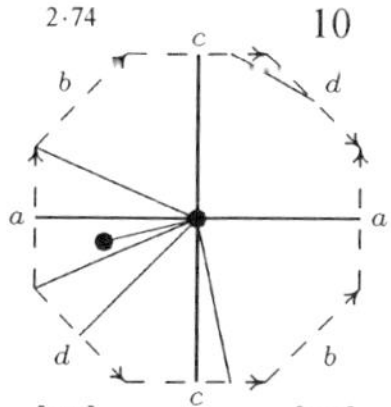
[91] [10]

2·75 10
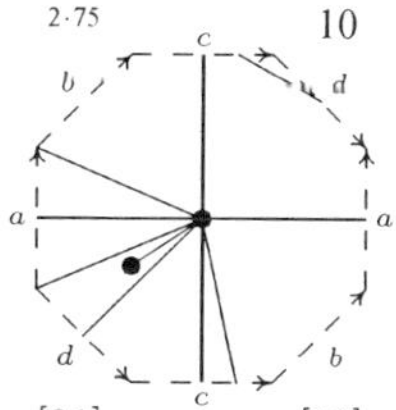
[91] [10]

2·76 10
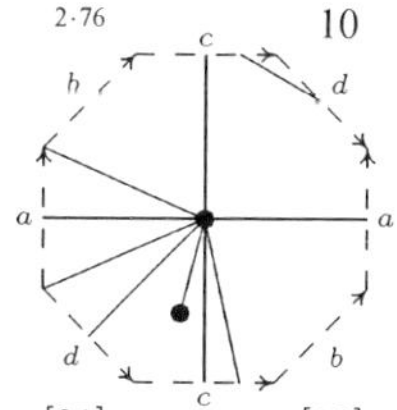
[91] [10]

2·77 10
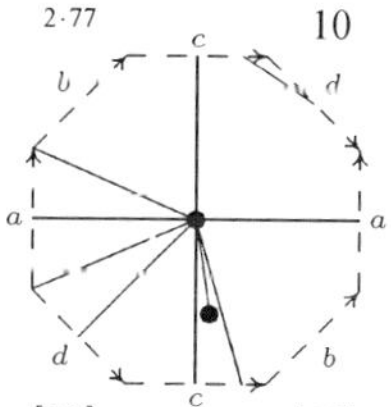
[91] [10]

2·78 10
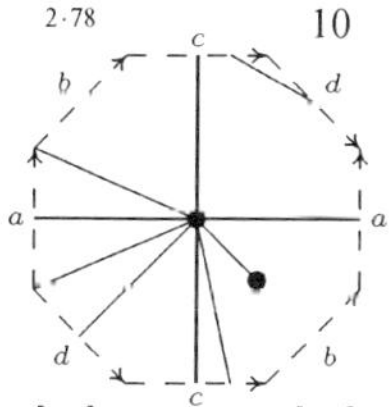
[91] [10]

2·79 5
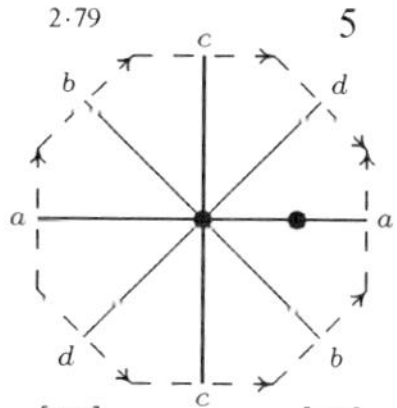
[82] [10]

2·80 10
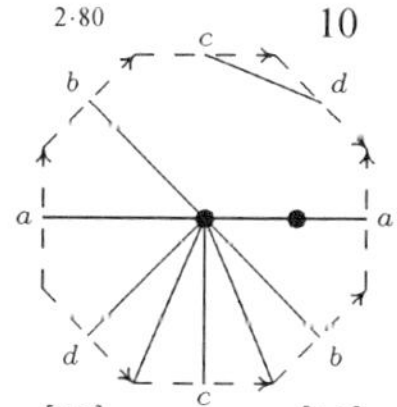
[82] [10]

2·81 10
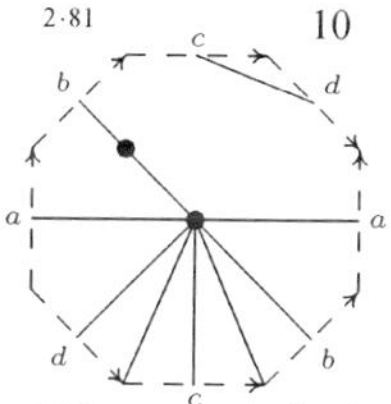
[82] [10]

2·82 10
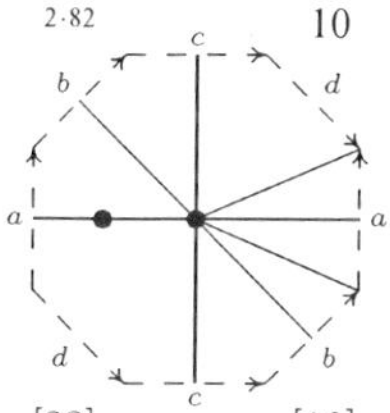
[82] [10]

2·83 10
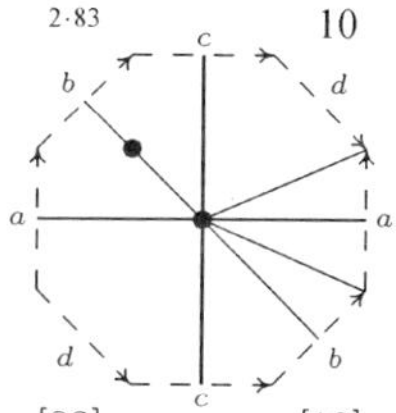
[82] [10]

2·84 10
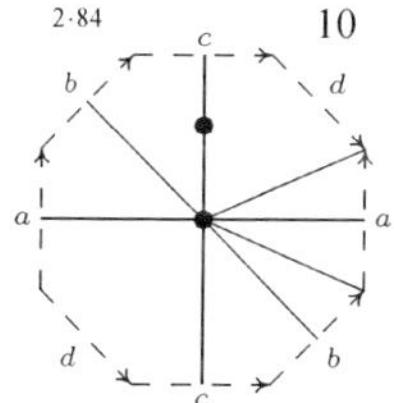
[82] [10]

2·85 10
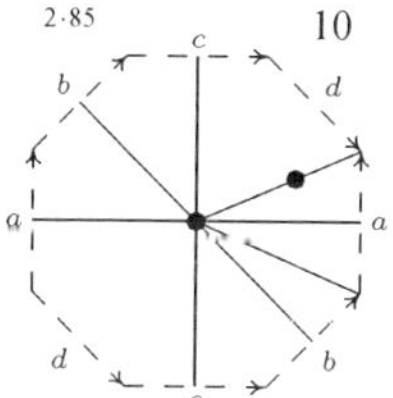
[82] [10]

2·86 10
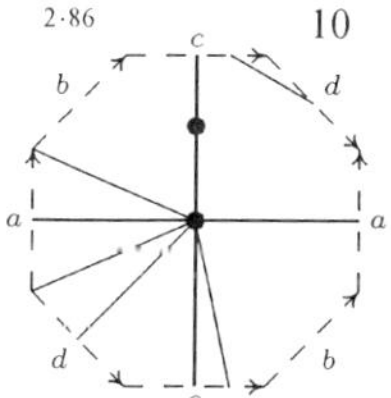
[82] [10]

2·87 10
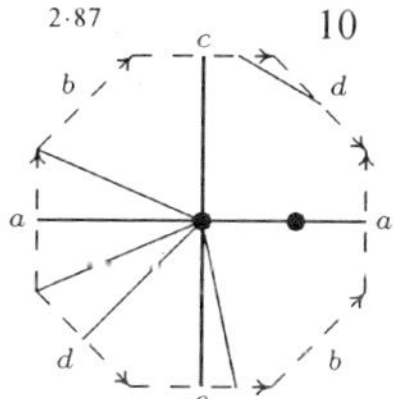
[82] [10]

2·88 10
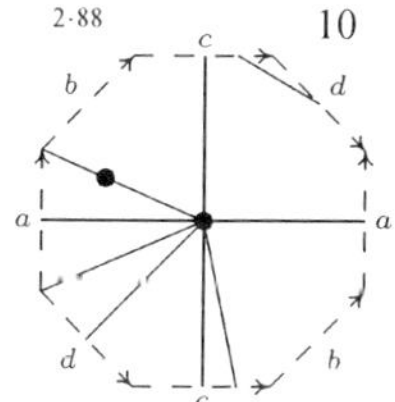
[82] [10]

2·89 10
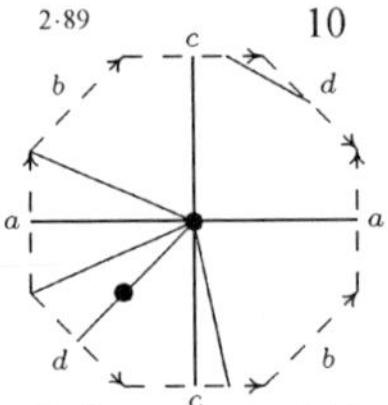
[82] [10]

2·90 10
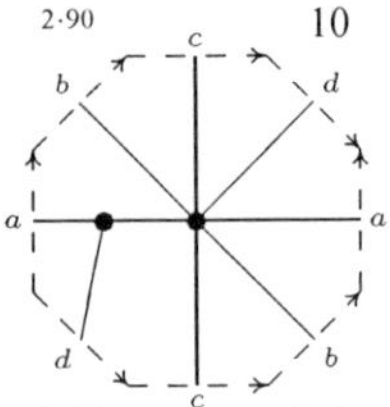
[73] [10]

2·91 10
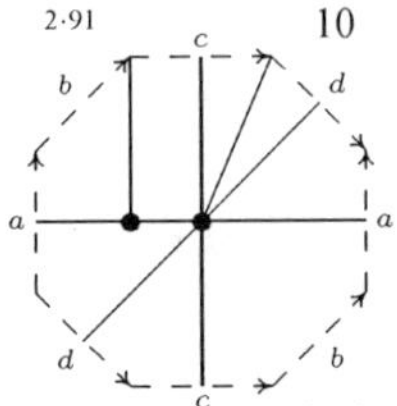
[73] [10]

2·92 10
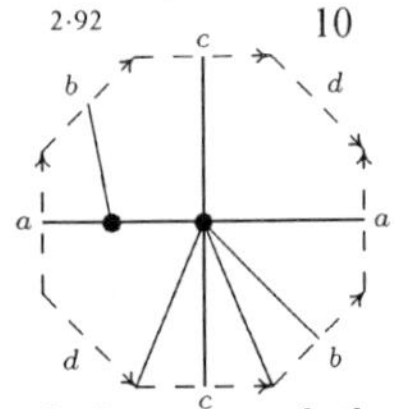
[73] [10]

2·93 10
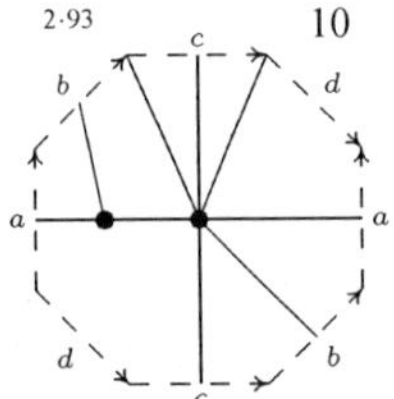
[73] [10]

2·94 10
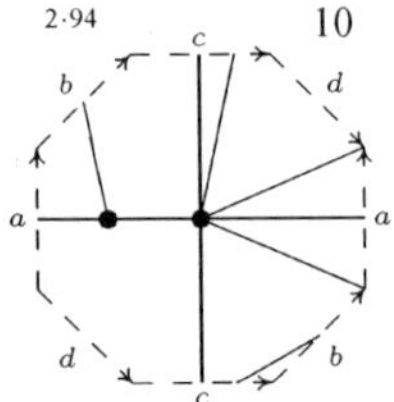
[73] [10]

2·95 10
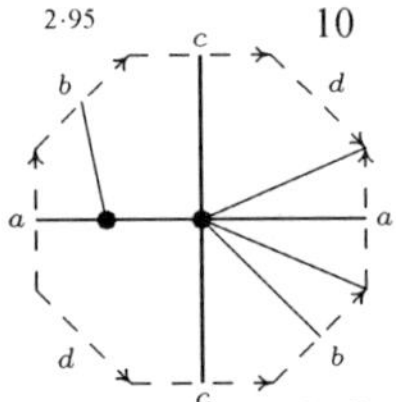
[73] [10]

2·96 10
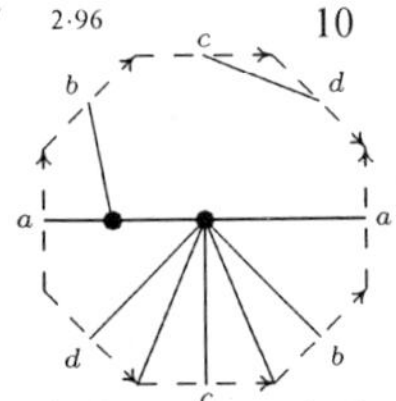
[73] [10]

2·97 5
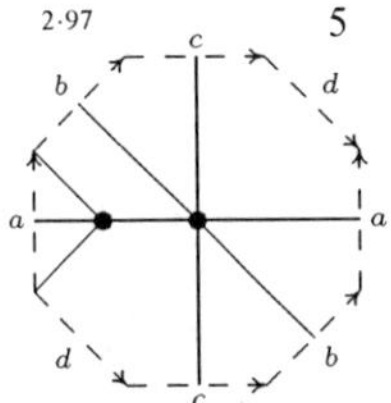
[64] [10]

2·98 10
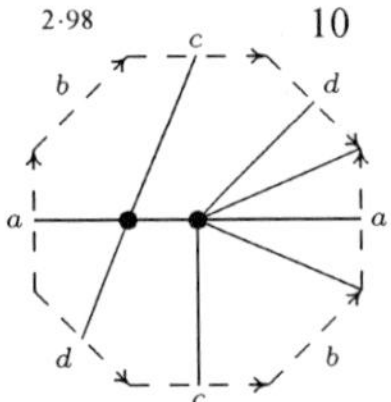
[64] [10]

2·99 10
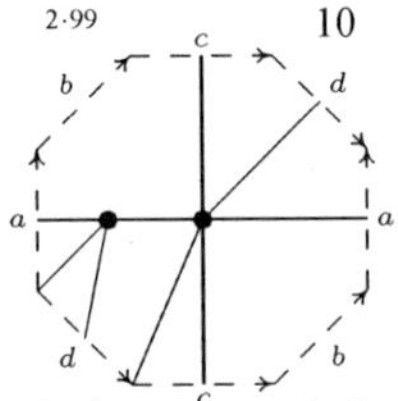
[64] [10]

2·100 10
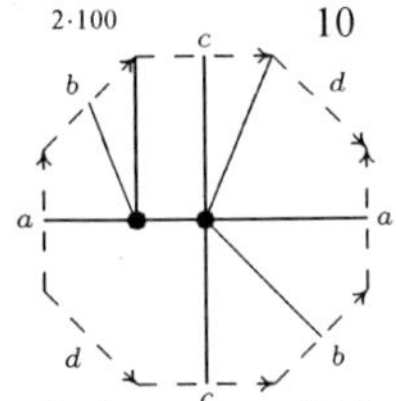
[64] [10]

2·101 10
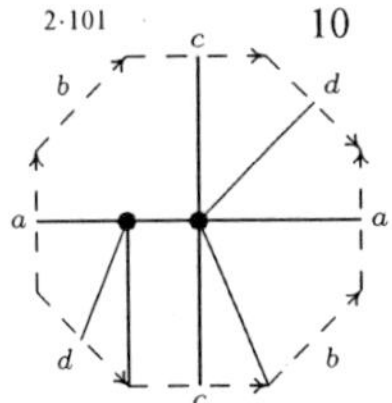
[64] [10]

2·102 10
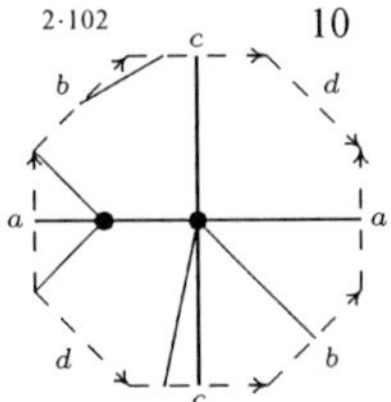
[64] [10]

2·103 10
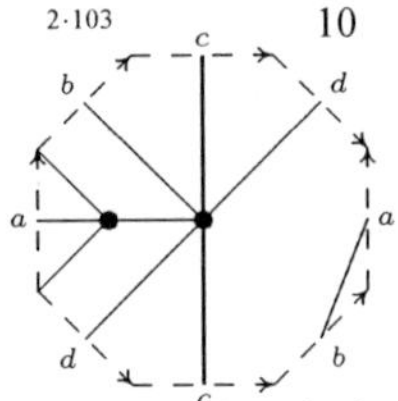
[64] [10]

2·104 1
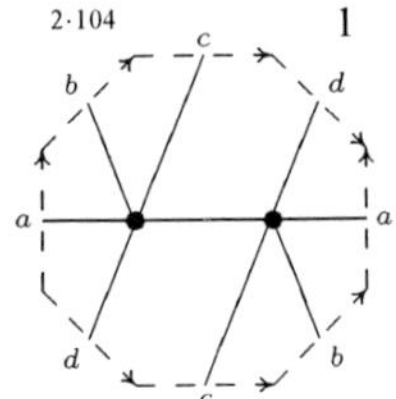
$[5^2]$ [10]

2·105 2
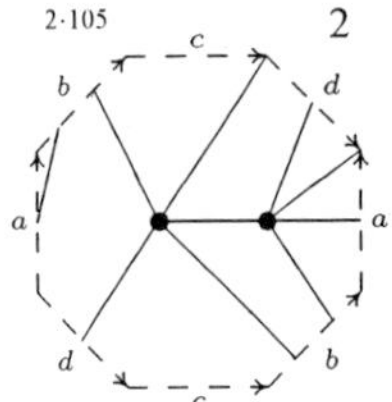
$[5^2]$ [10]

2·106 10
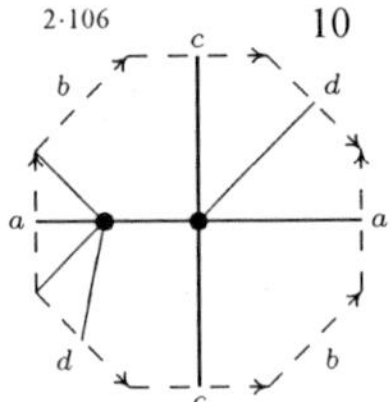
$[5^2]$ [10]

2·107 5
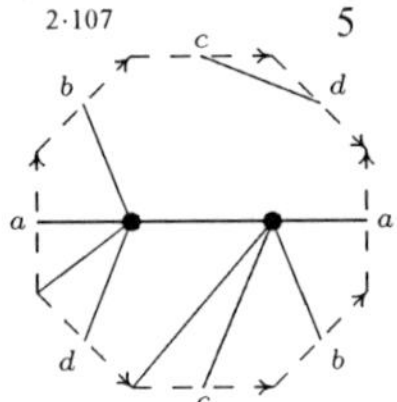
$[5^2]$ [10]

2·108 5
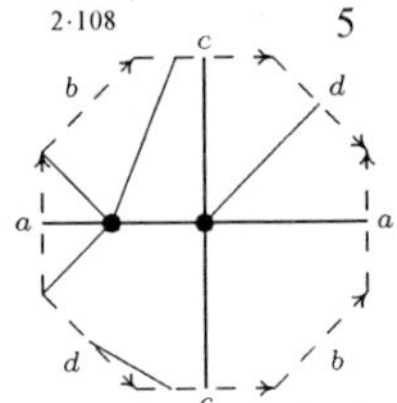
$[5^2]$ [10]

Genus 2: 5e 2v

2·109 5

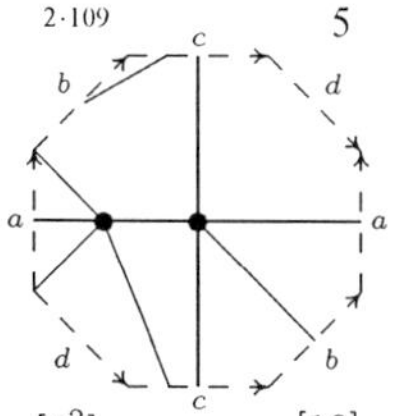

$[5^2]$ [10]

2·110 5

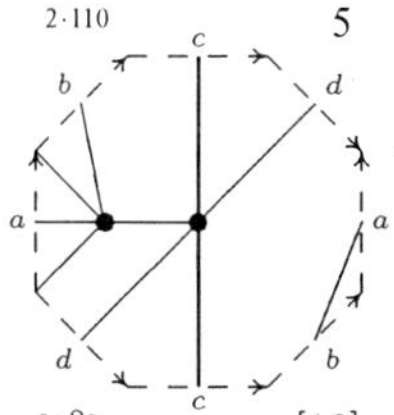

$[5^2]$ [10]

Chapter 6

Maps in nonorientable surfaces

In this chapter we give the polygonal representation of all maps on up to four edges in nonorientable surfaces. The maps are listed by *genus, number of edges* and *number of vertices* where these are written in decreasing order of subordination, with the maps arranged in decreasing (lexicographic) order of *vertex partition* and then *edge partition*. When the number of edges or vertices changes, a title indicating this appears at the beginning of a line.

At the foot of each map, the left-hand partition is the vertex partition and the right-hand partition is the face partition. The integer in the top right-hand corner of each map is the number of distinct rootings of the map and the serial number is given in the top left-hand corner.

An annotation is given at the foot of odd pages indicating the genus, number of edges and number of vertices of the first map on the page. The annotation at the foot of the next even page gives the corresponding information for the last map on the page.

6.1 Genus 1 – the projective plane

1 edge

1 vertex

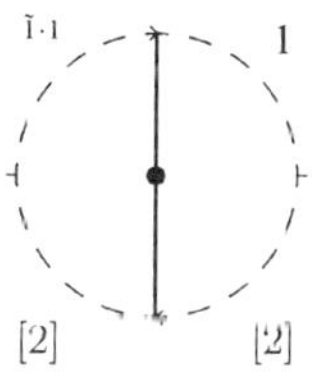

2 edges
1 vertex

Ĩ·2 4
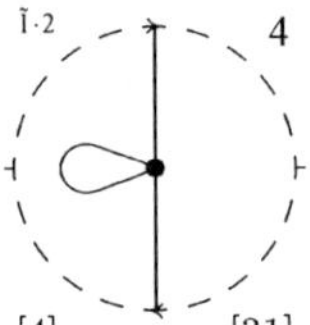
[4] [31]

Ĩ·3 1
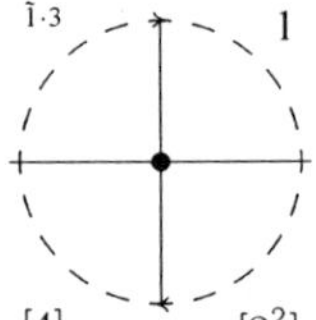
[4] [2²]

2 edges
2 vertices

Ĩ·4 4
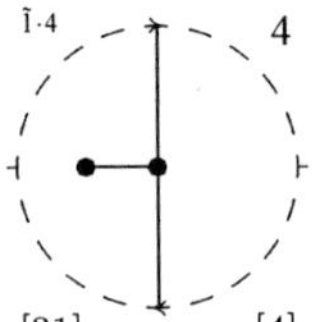
[31] [4]

Ĩ·5 1
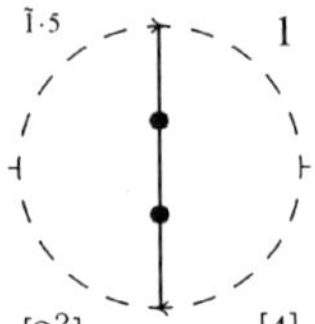
[2²] [4]

3 edges
1 vertex

Ĩ·6 6
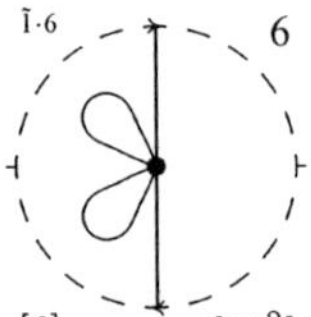
[6] [41²]

Ĩ·7 3
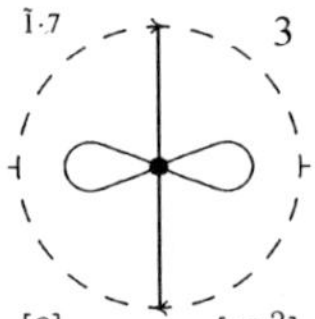
[6] [41²]

Ĩ·8 6
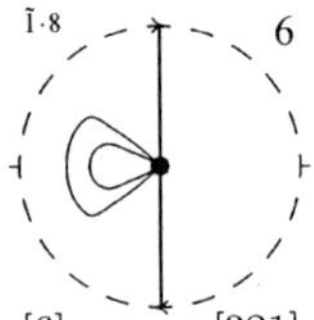
[6] [321]

Ĩ·9 6
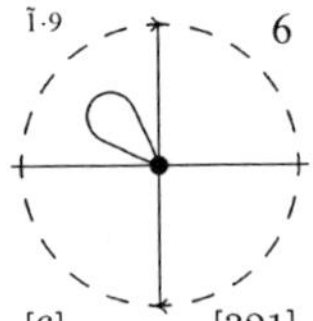
[6] [321]

Ĩ·10 1
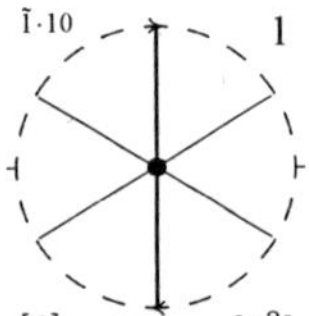
[6] [2³]

3 edges
2 vertices

Ĩ·11 12
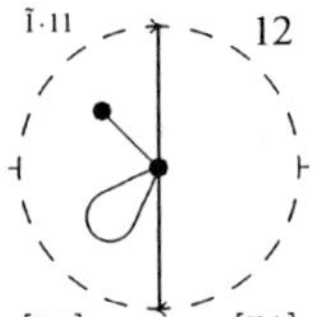
[51] [51]

Ĩ·12 6
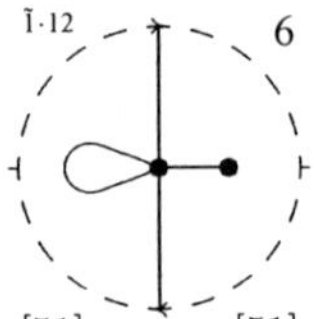
[51] [51]

Ĩ·13 6
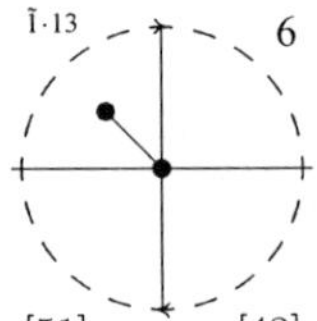
[51] [42]

Ĩ·14 6
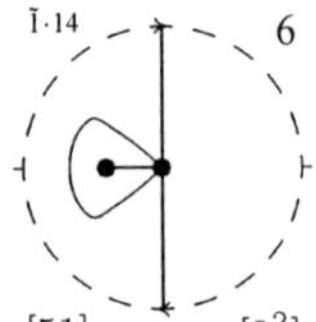
[51] [3²]

Ĩ·15 6
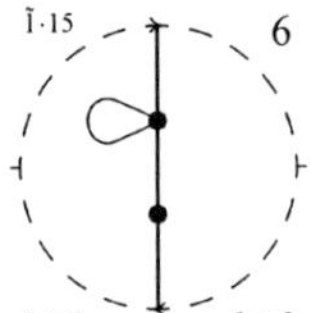
[42] [51]

Ĩ·16 6
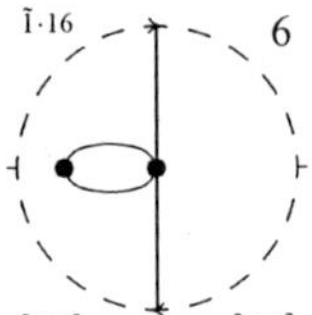
[42] [42]

Ĩ·17 3
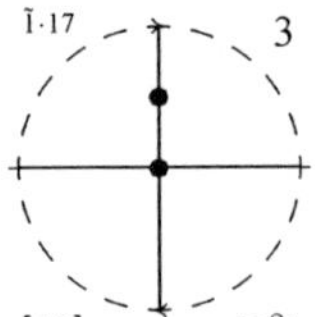
[42] [3²]

Ĩ·18 6
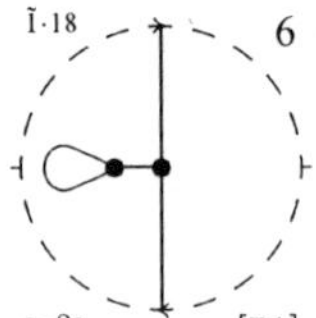
[3²] [51]

Ĩ·19 3
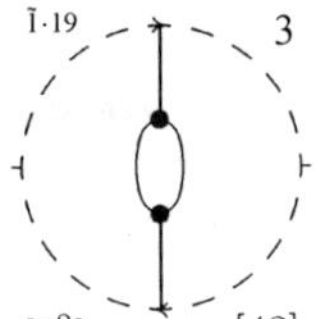
[3²] [42]

Genus Ĩ: 3e 3v

3 edges

3 vertices

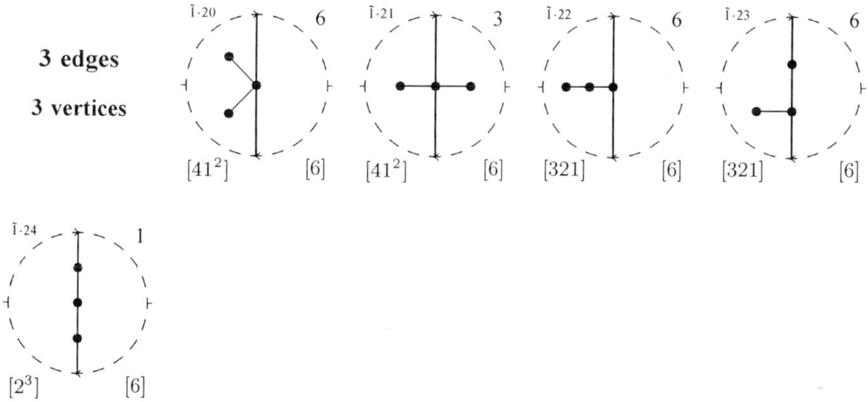

4 edges

1 vertex

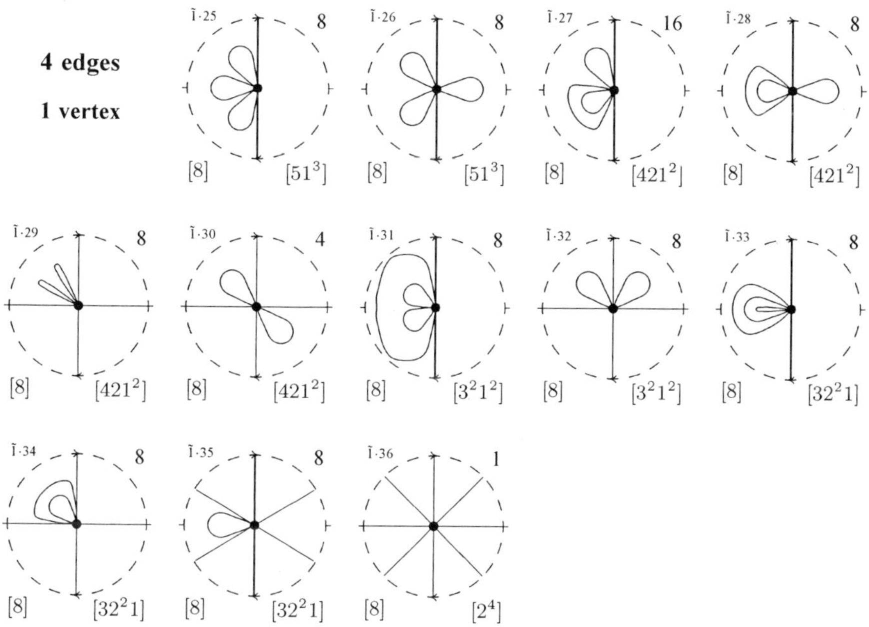

4 edges

2 vertices

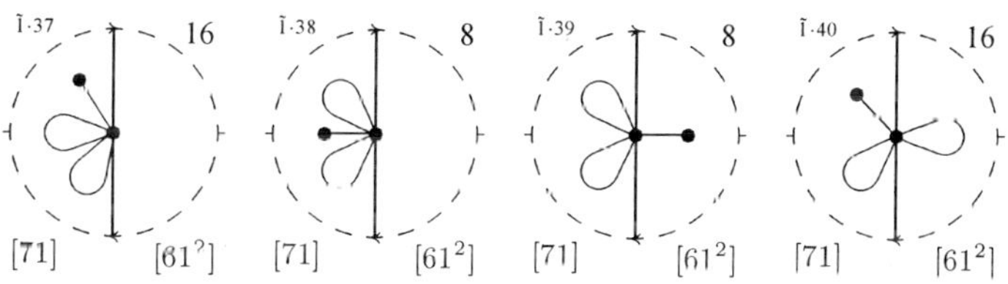

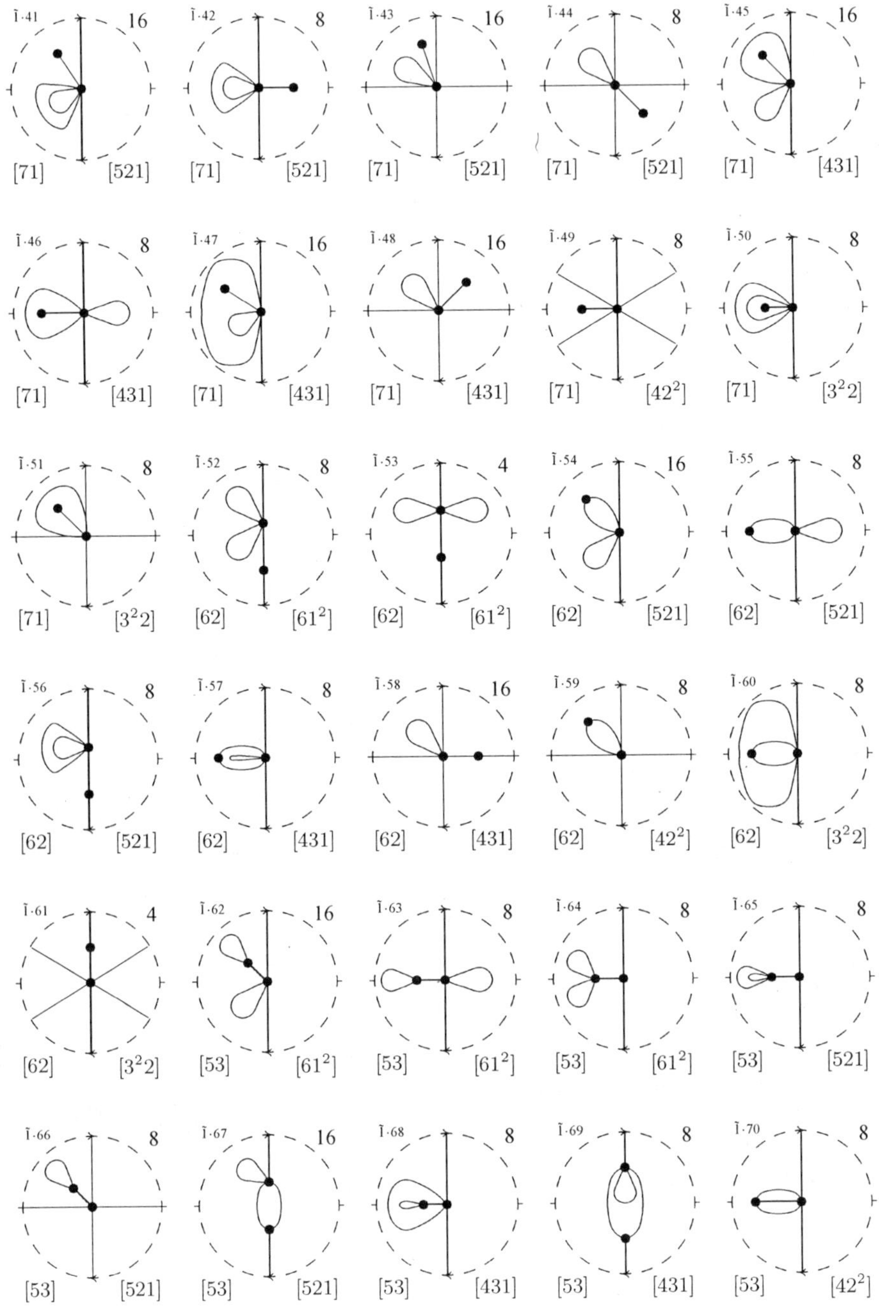
Ĩ·41 16 [71] [521]
Ĩ·42 8 [71] [521]
Ĩ·43 16 [71] [521]
Ĩ·44 8 [71] [521]
Ĩ·45 16 [71] [431]
Ĩ·46 8 [71] [431]
Ĩ·47 16 [71] [431]
Ĩ·48 16 [71] [431]
Ĩ·49 8 [71] [42^2]
Ĩ·50 8 [71] [$3^2 2$]
Ĩ·51 8 [71] [$3^2 2$]
Ĩ·52 8 [62] [61^2]
Ĩ·53 4 [62] [61^2]
Ĩ·54 16 [62] [521]
Ĩ·55 8 [62] [521]
Ĩ·56 8 [62] [521]
Ĩ·57 8 [62] [431]
Ĩ·58 16 [62] [431]
Ĩ·59 8 [62] [42^2]
Ĩ·60 8 [62] [$3^2 2$]
Ĩ·61 4 [62] [$3^2 2$]
Ĩ·62 16 [53] [61^2]
Ĩ·63 8 [53] [61^2]
Ĩ·64 8 [53] [61^2]
Ĩ·65 8 [53] [521]
Ĩ·66 8 [53] [521]
Ĩ·67 16 [53] [521]
Ĩ·68 8 [53] [431]
Ĩ·69 8 [53] [431]
Ĩ·70 8 [53] [42^2]

Genus Ĩ: 4e 2v

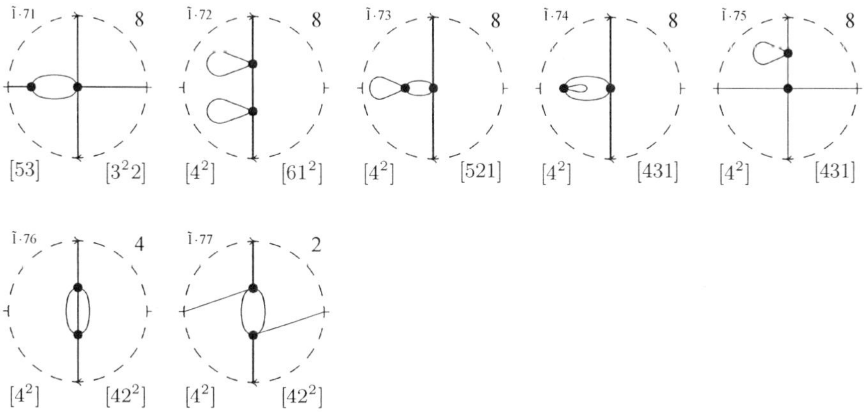

4 edges

3 vertices

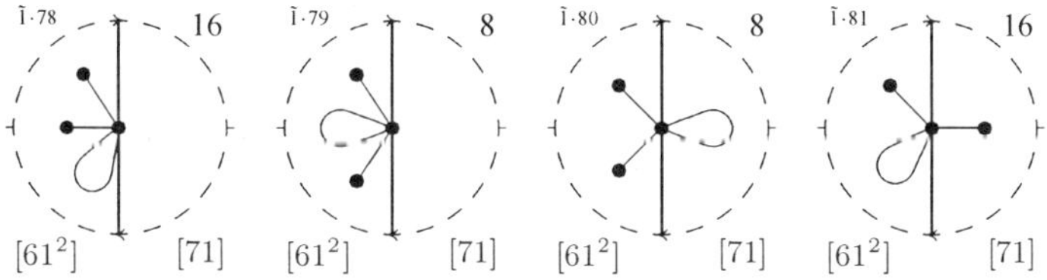

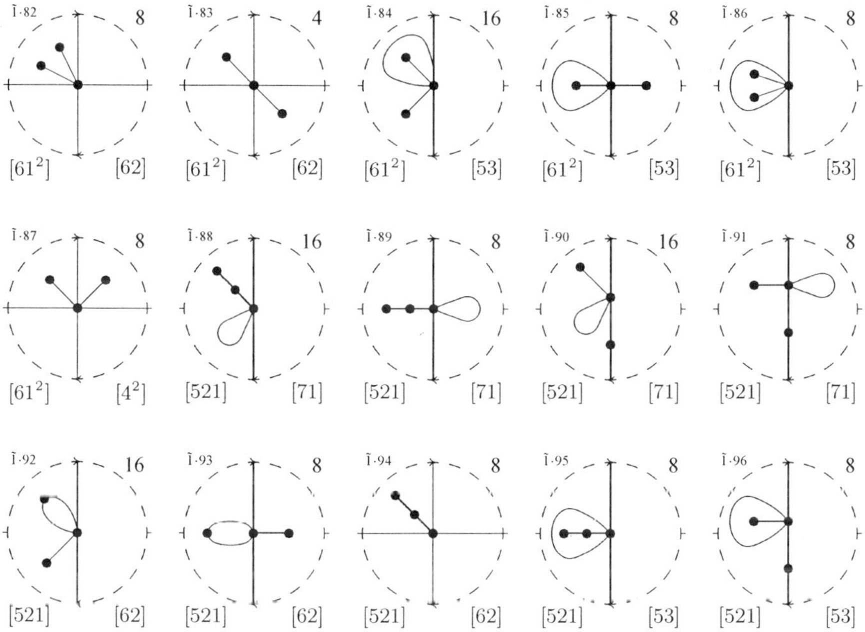

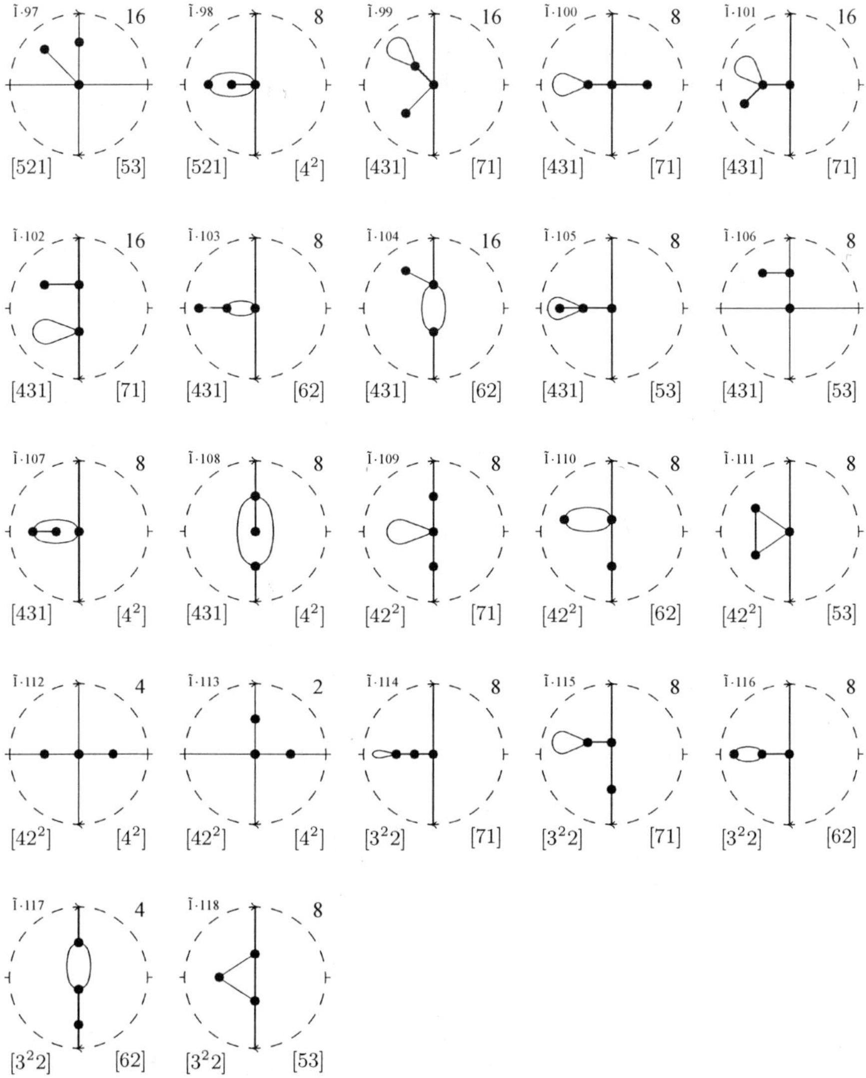

4 edges

4 vertices

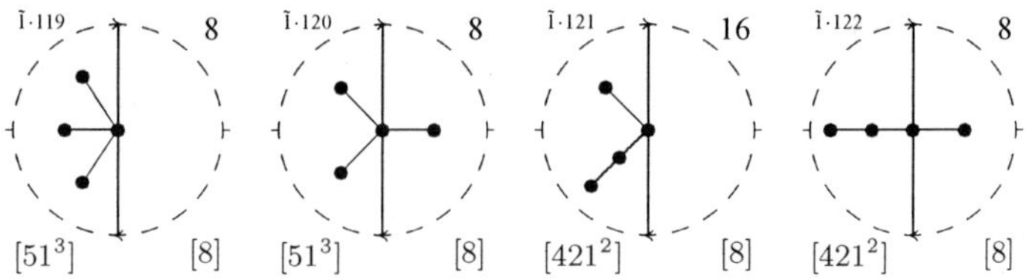

Genus Ĩ: 4e 4v

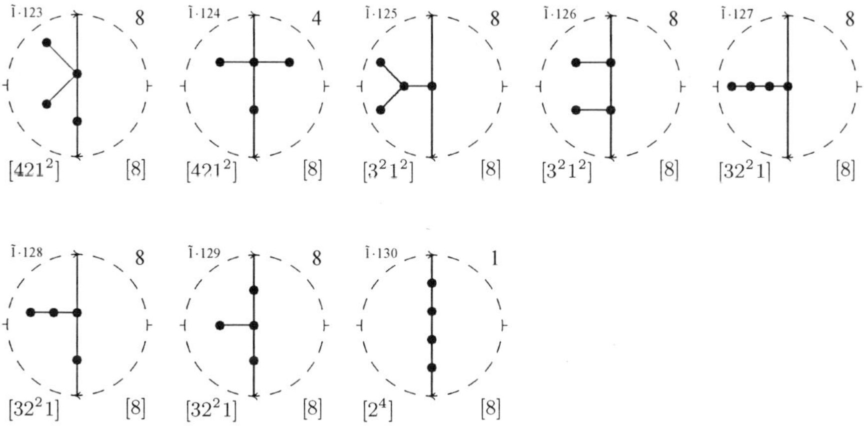

6.2 Genus 2 – the Klein bottle

2 edges

1 vertex

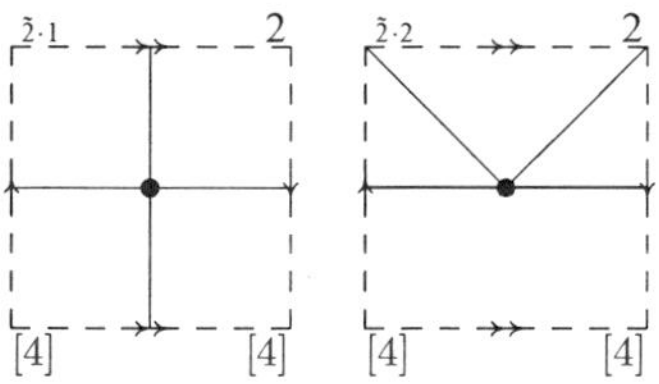

3 edges

1 vertex

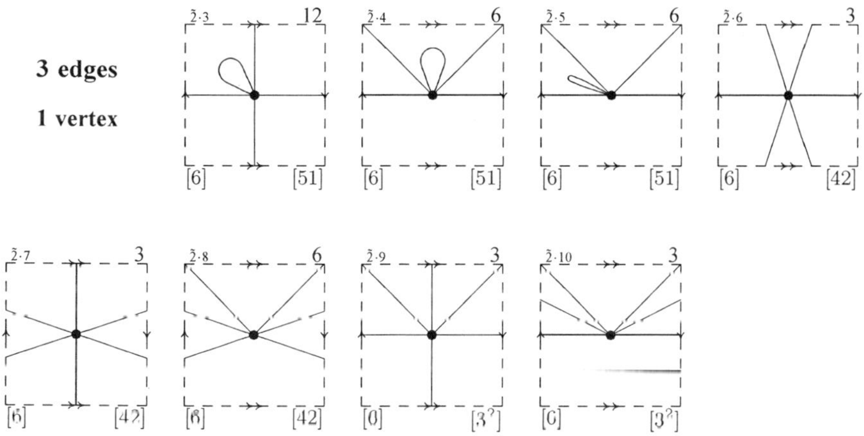

3 edges

2 vertices

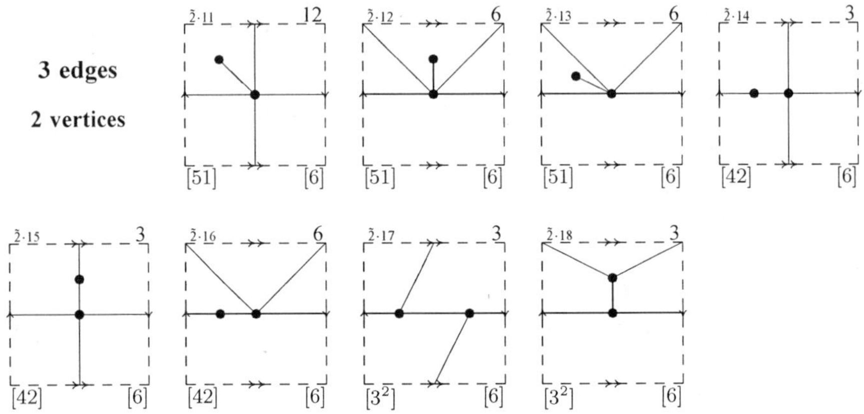

4 edges

1 vertex

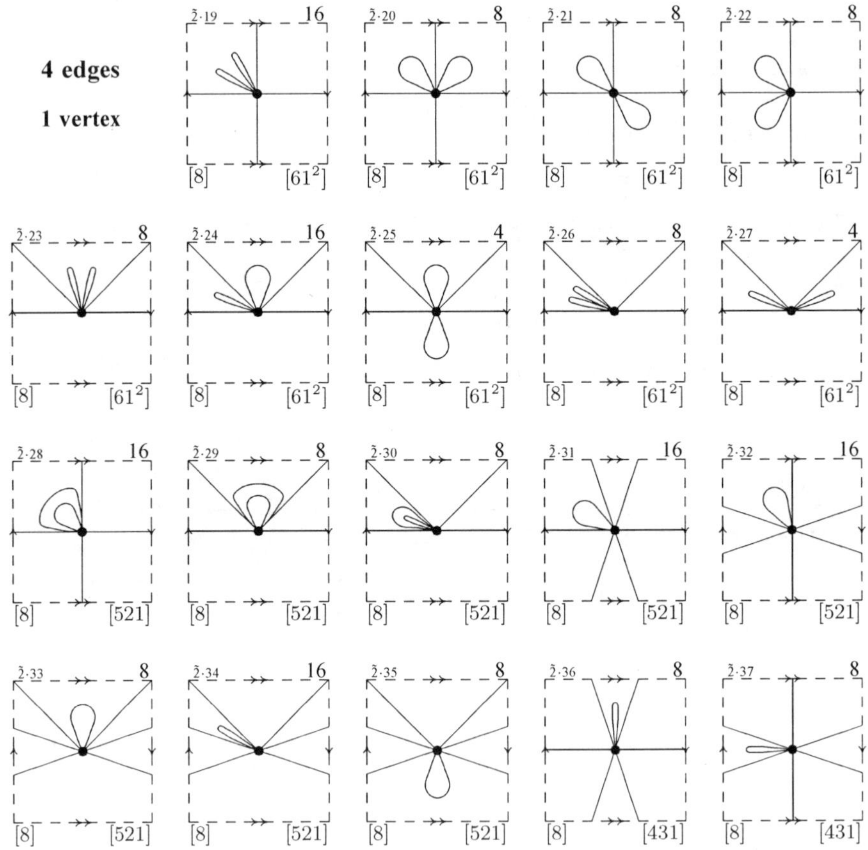

Genus $\tilde{2}$: 4e 1v

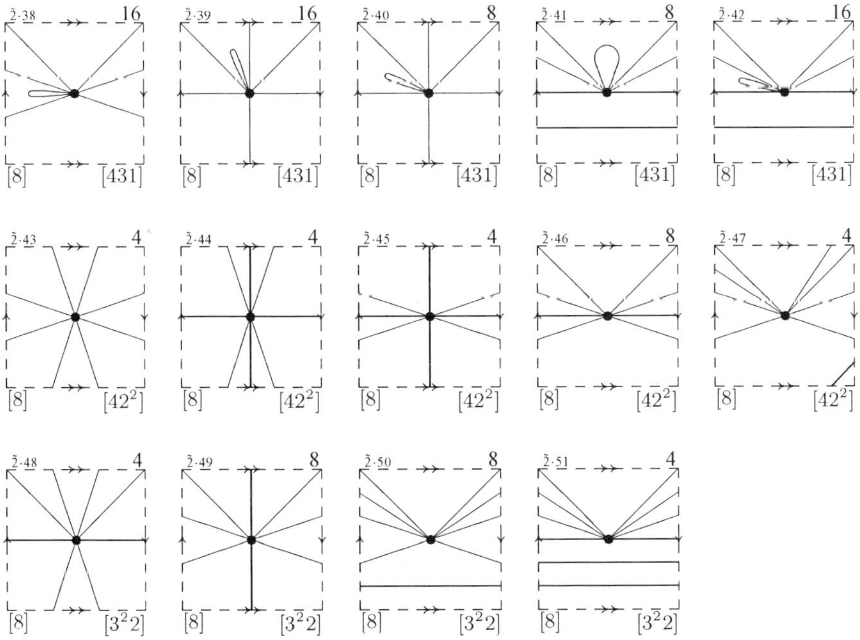

4 edges

2 vertices

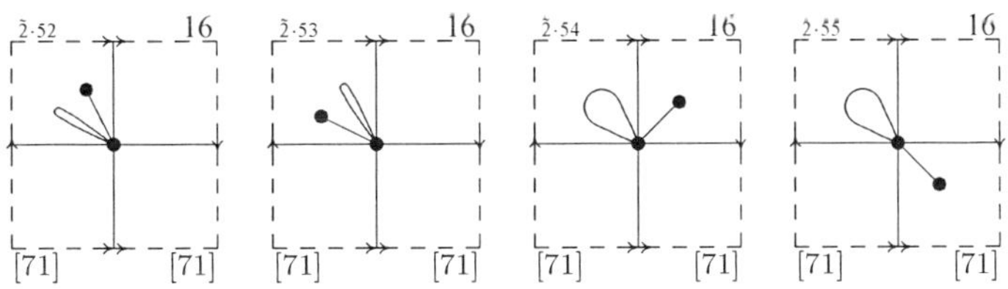

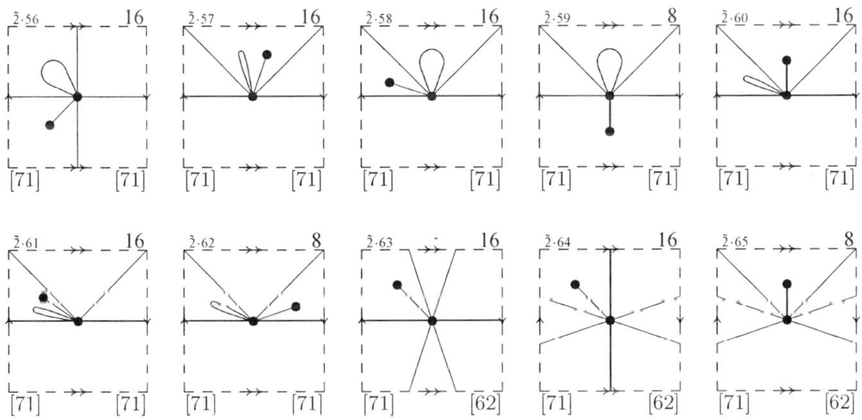

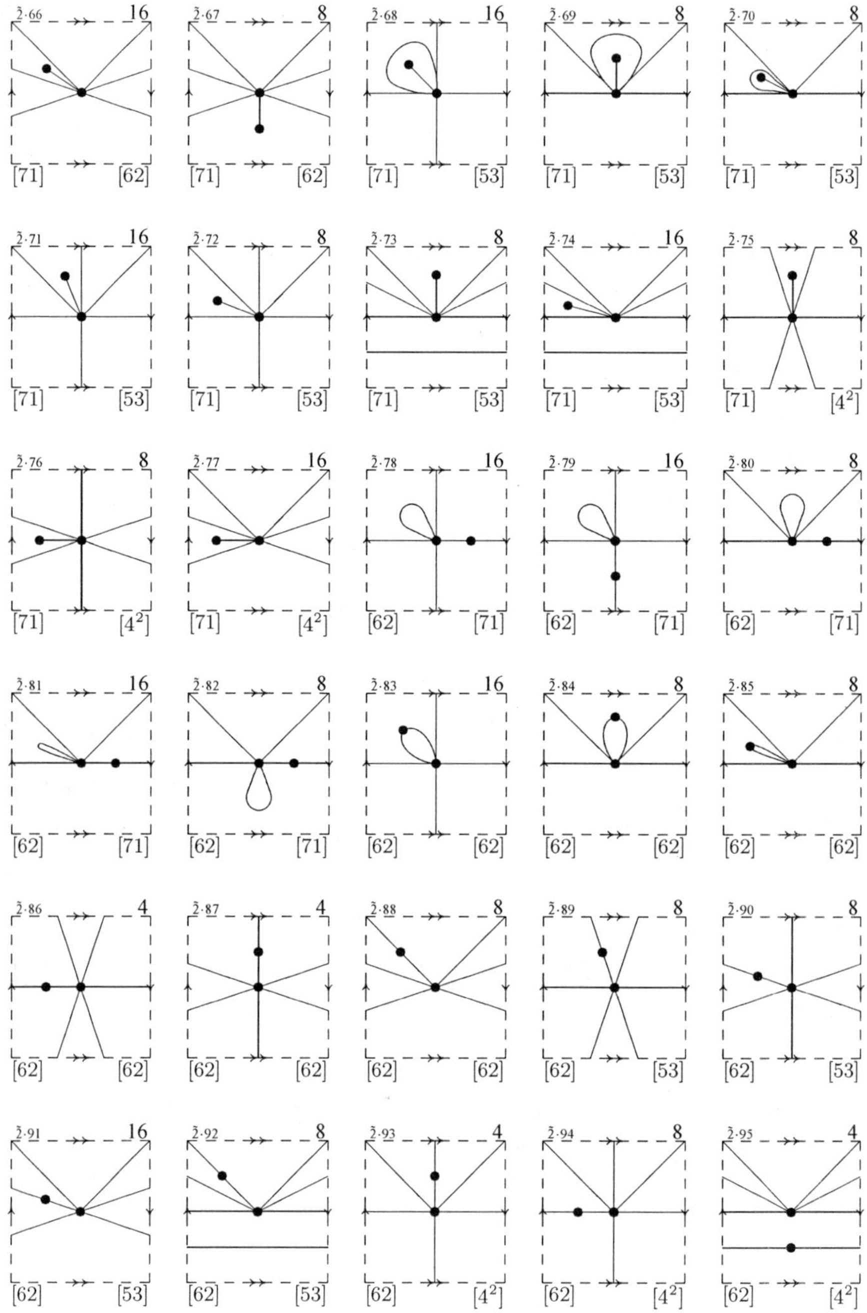

Genus $\tilde{2}$: 4e 2v

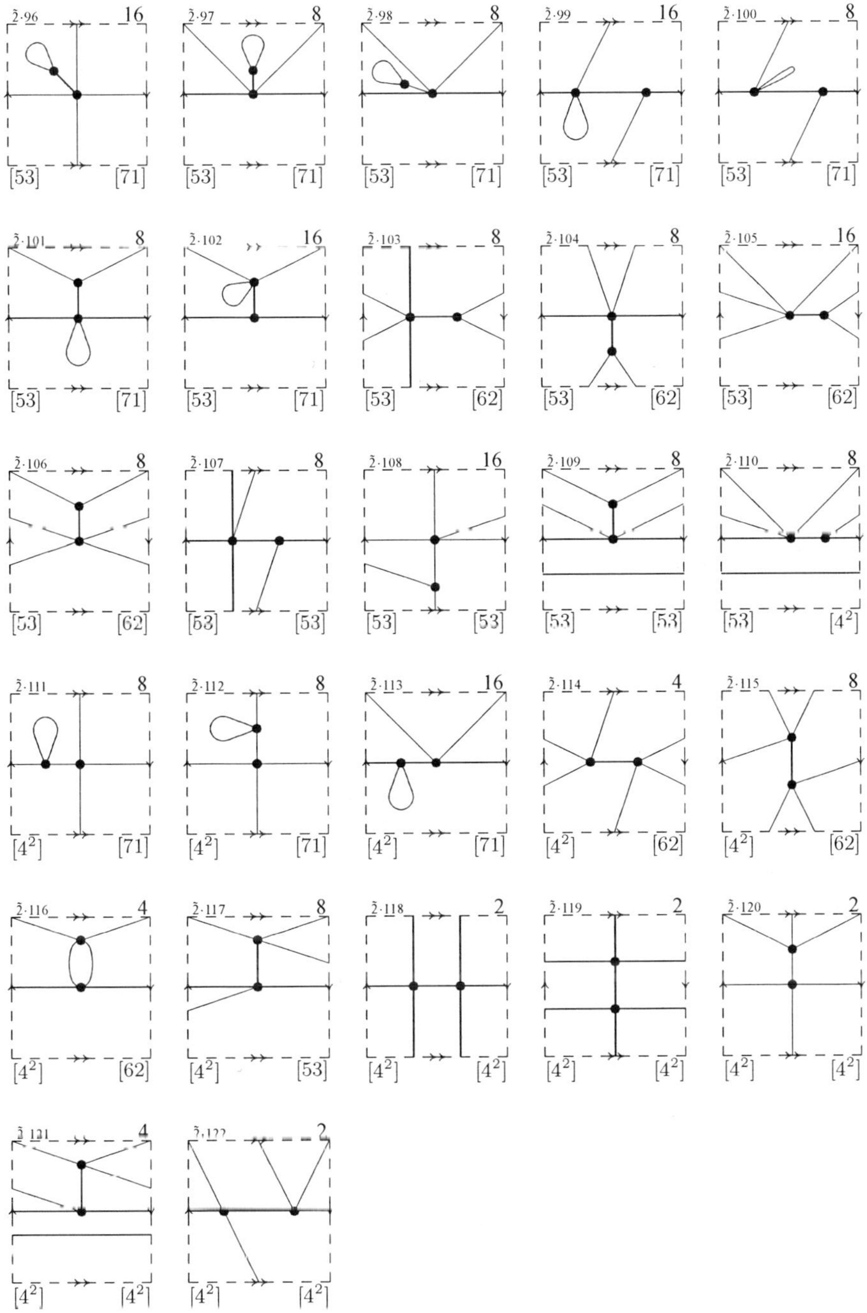

2̃·96 16 [53] [71]
2̃·97 8 [53] [71]
2̃·98 8 [53] [71]
2̃·99 16 [53] [71]
2̃·100 8 [53] [71]
2̃·101 8 [53] [71]
2̃·102 16 [53] [71]
2̃·103 8 [53] [62]
2̃·104 8 [53] [62]
2̃·105 16 [53] [62]
2̃·106 8 [53] [62]
2̃·107 8 [53] [53]
2̃·108 16 [53] [53]
2̃·109 8 [53] [53]
2̃·110 8 [53] [4²]
2̃·111 8 [4²] [71]
2̃·112 8 [4²] [71]
2̃·113 16 [4²] [71]
2̃·114 4 [4²] [62]
2̃·115 8 [4²] [62]
2̃·116 4 [4²] [62]
2̃·117 8 [4²] [53]
2̃·118 2 [4²] [4²]
2̃·119 2 [4²] [4²]
2̃·120 2 [4²] [4²]
2̃·121 4 [4²] [4²]
2̃·122 2 [4²] [4²]

4 edges

3 vertices

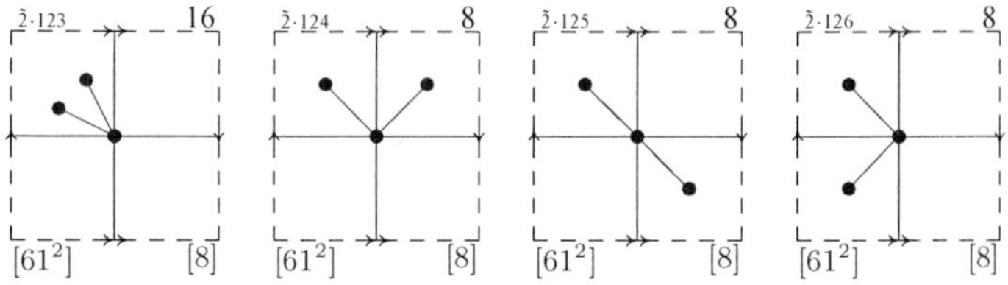

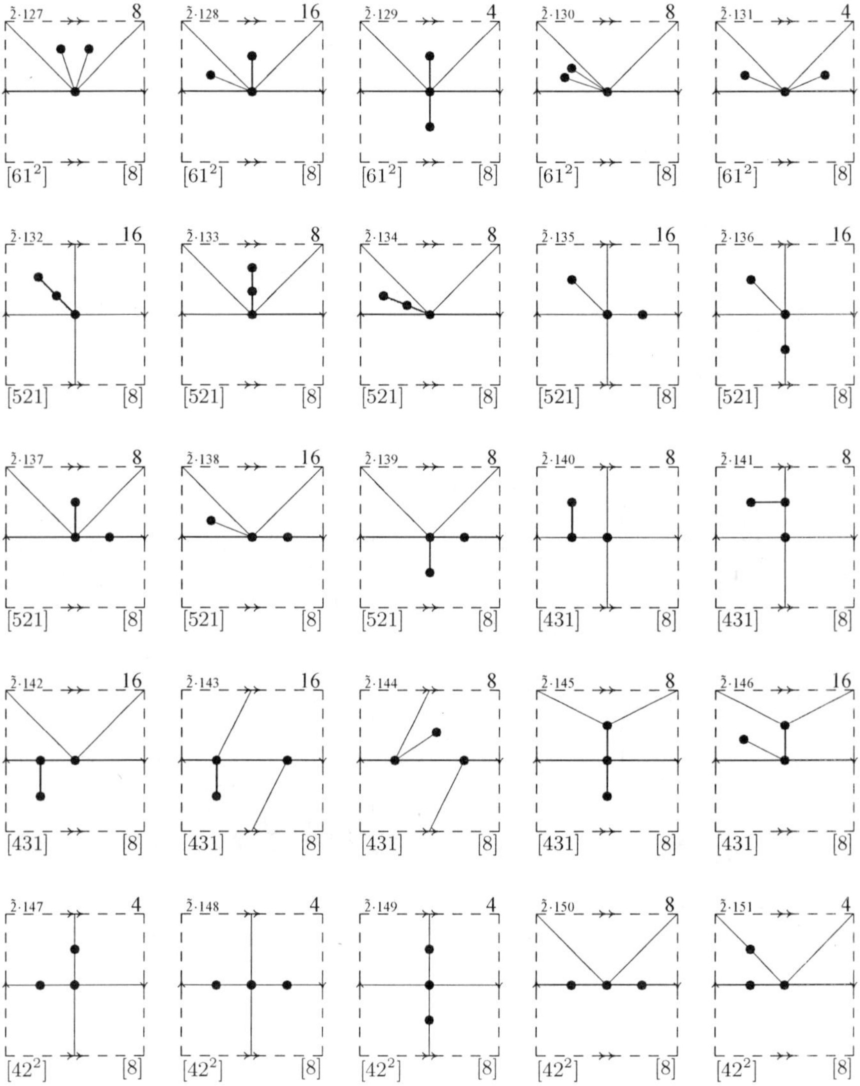

Genus $\tilde{2}$: 4e 3v

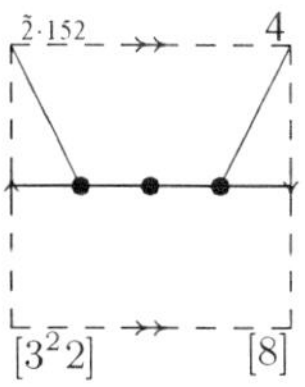

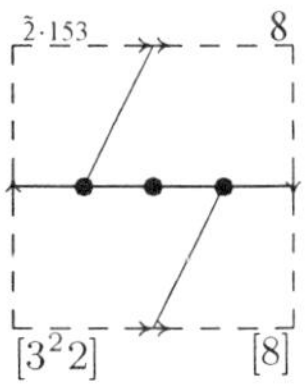

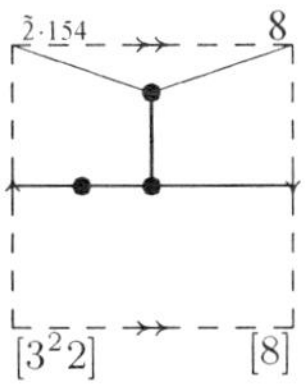

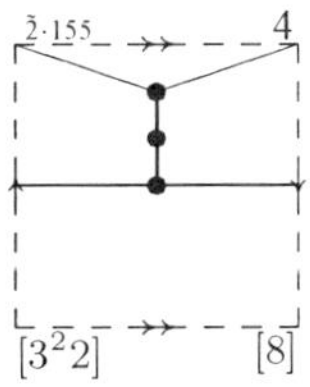

6.3 Genus 3 – the crosscapped torus

3 edges

1 vertex

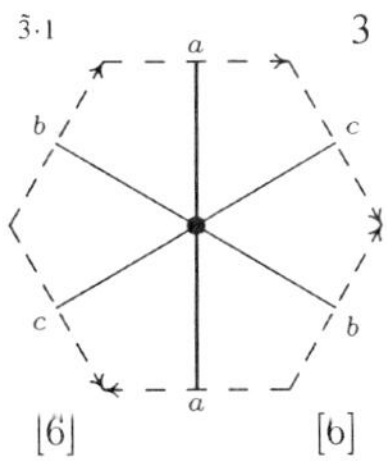

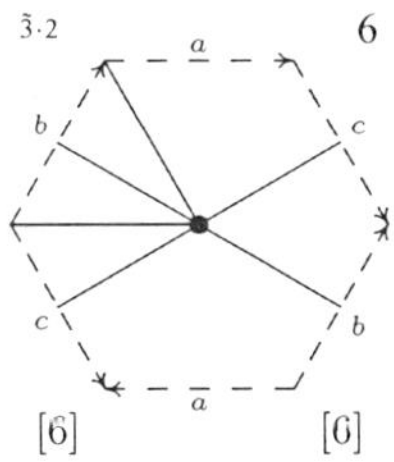

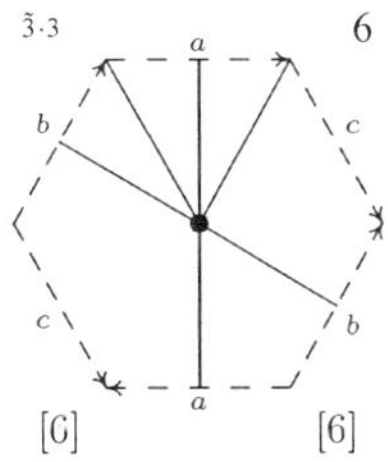

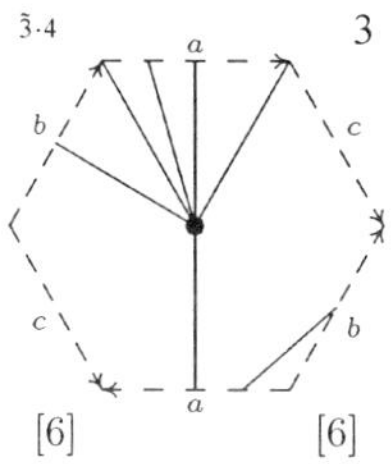

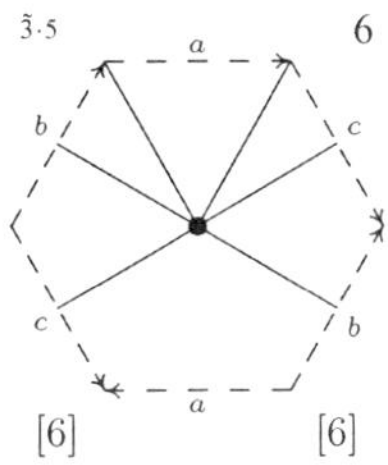

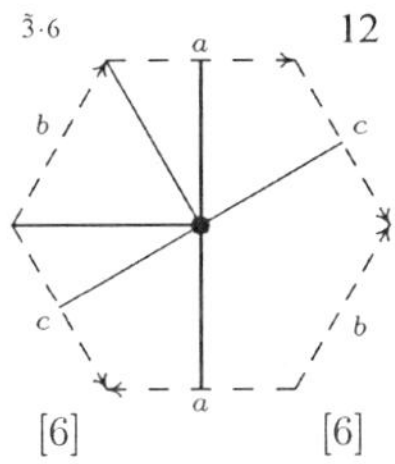

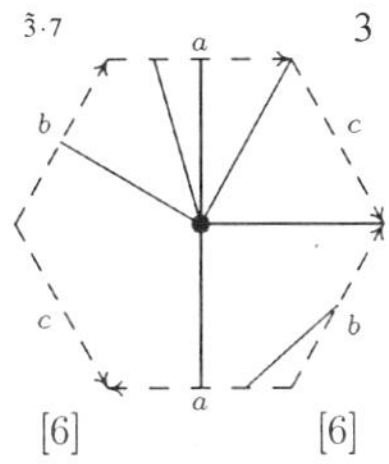

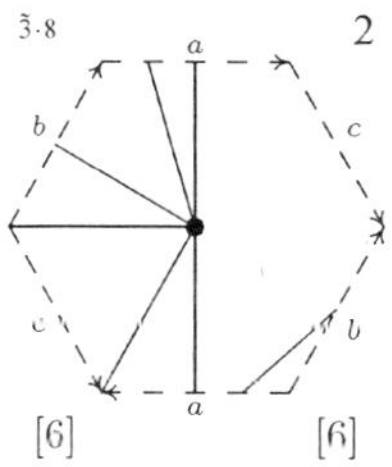

4 edges

1 vertex

$\tilde{3}\cdot 9$ 16

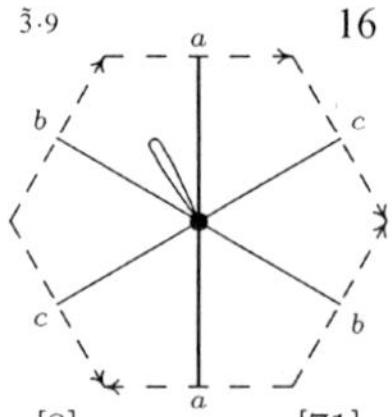

[8] [71]

$\tilde{3}\cdot 10$ 8

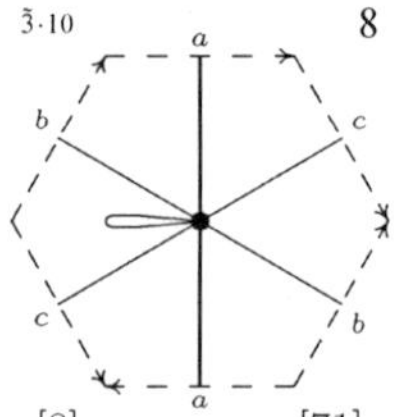

[8] [71]

$\tilde{3}\cdot 11$ 16

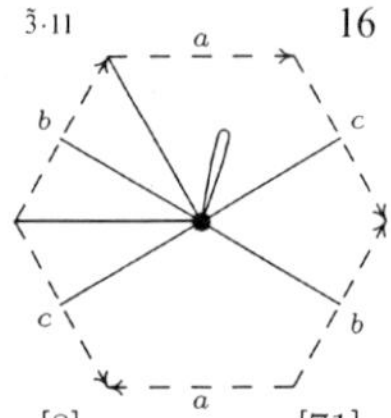

[8] [71]

$\tilde{3}\cdot 12$ 16

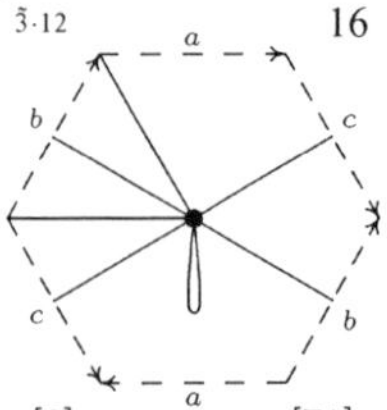

[8] [71]

$\tilde{3}\cdot 13$ 16

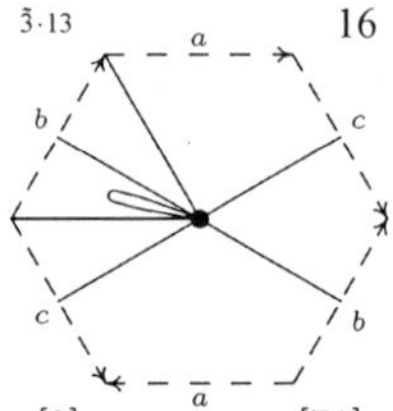

[8] [71]

$\tilde{3}\cdot 14$ 16

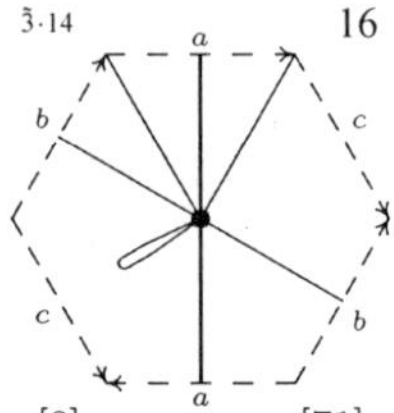

[8] [71]

$\tilde{3}\cdot 15$ 16

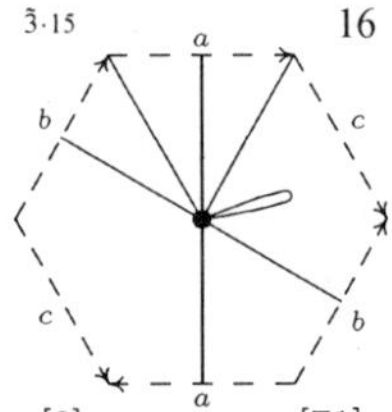

[8] [71]

$\tilde{3}\cdot 16$ 16

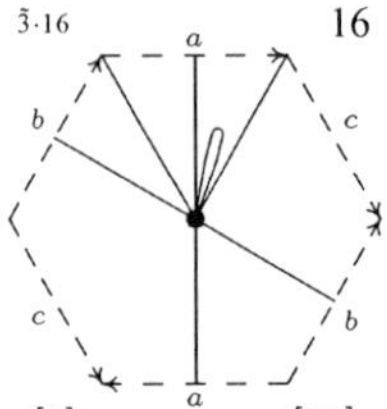

[8] [71]

$\tilde{3}\cdot 17$ 8

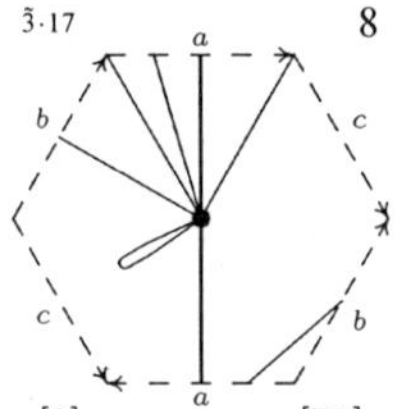

[8] [71]

$\tilde{3}\cdot 18$ 16

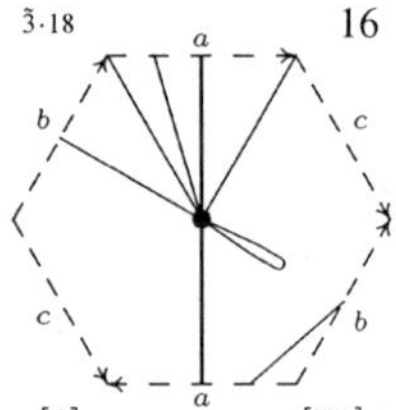

[8] [71]

$\tilde{3}\cdot 19$ 8

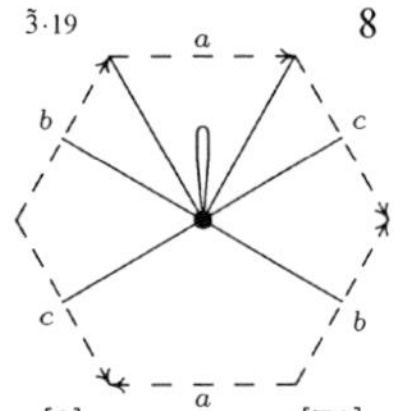

[8] [71]

$\tilde{3}\cdot 20$ 16

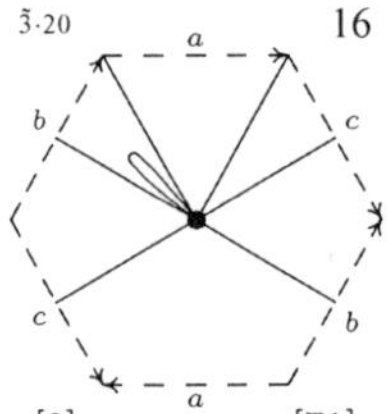

[8] [71]

$\tilde{3}\cdot 21$ 16

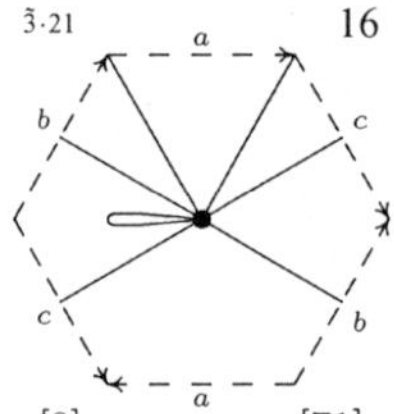

[8] [71]

$\tilde{3}\cdot 22$ 8

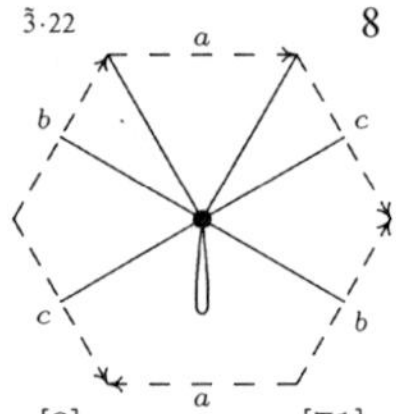

[8] [71]

$\tilde{3}\cdot 23$ 16

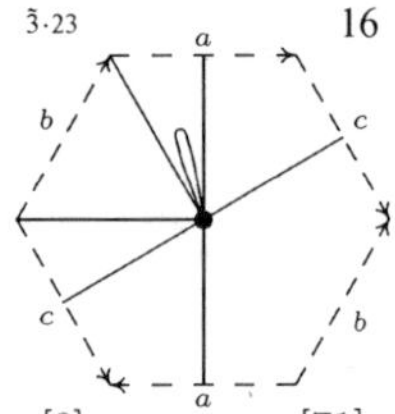

[8] [71]

$\tilde{3}\cdot 24$ 16

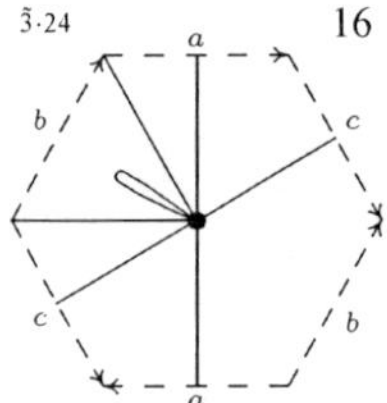

[8] [71]

$\tilde{3}\cdot 25$ 16

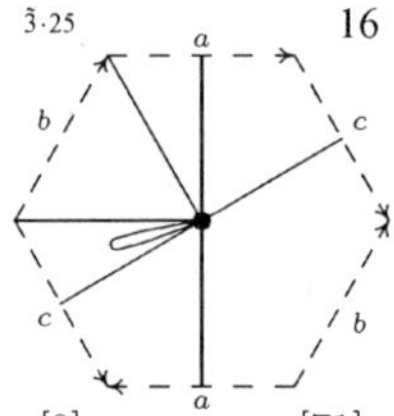

[8] [71]

$\tilde{3}\cdot 26$ 16

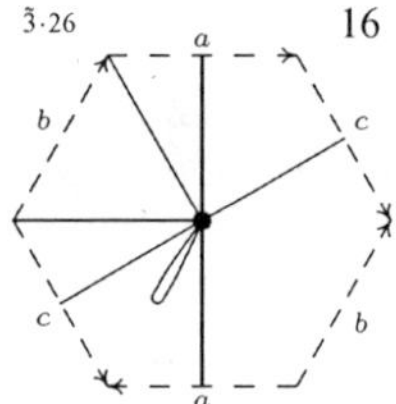

[8] [71]

$\tilde{3}\cdot 27$ 16

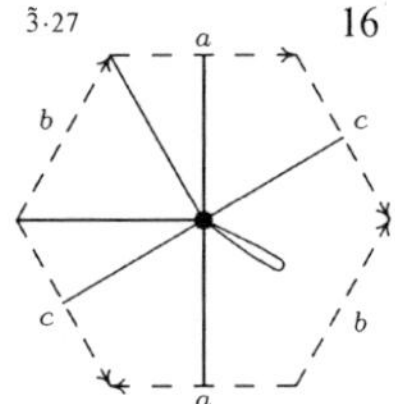

[8] [71]

Genus $\tilde{3}$: 4e 1v

$\tilde{3}$·28 16

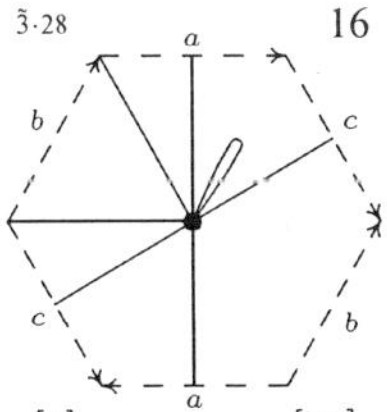

[8] [71]

$\tilde{3}$·29 16

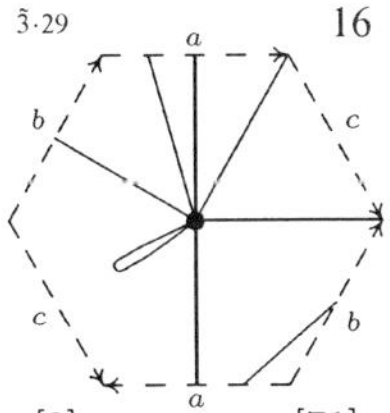

[8] [71]

$\tilde{3}$·30 8

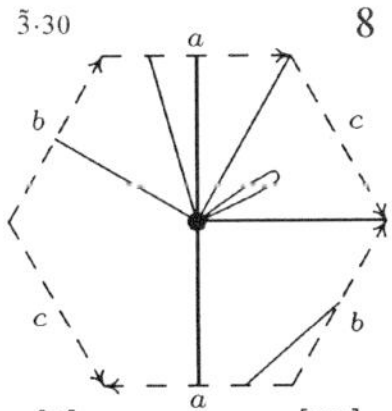

[8] [71]

$\tilde{3}$·31 8

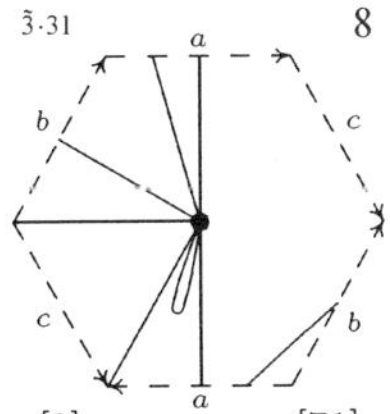

[8] [71]

$\tilde{3}$·32 8

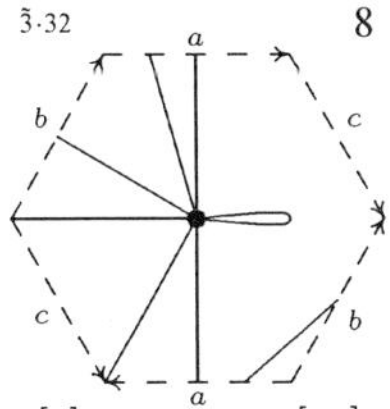

[8] [71]

$\tilde{3}$·33 4

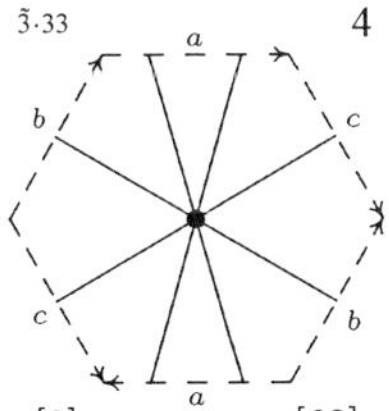

[8] [62]

$\tilde{3}$·34 8

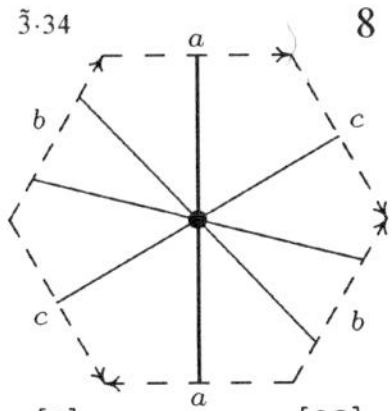

[8] [62]

$\tilde{3}$·35 8

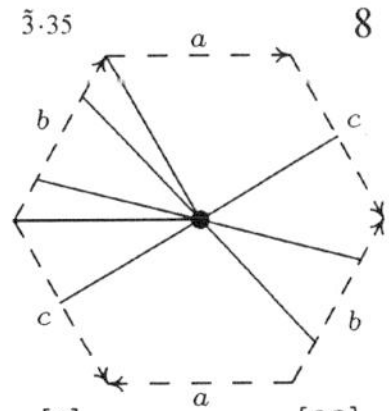

[8] [62]

$\tilde{3}$·36 8

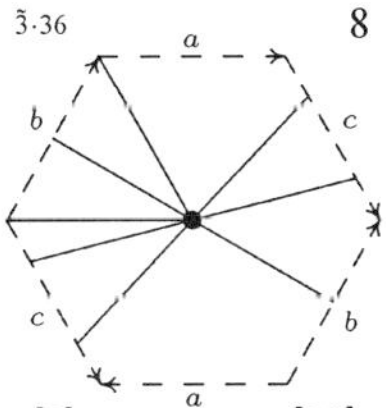

[8] [62]

$\tilde{3}$·37 8

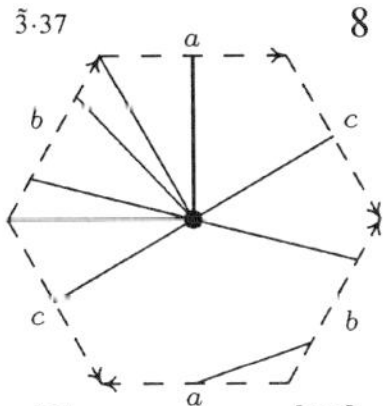

[8] [62]

$\tilde{3}$·38 8

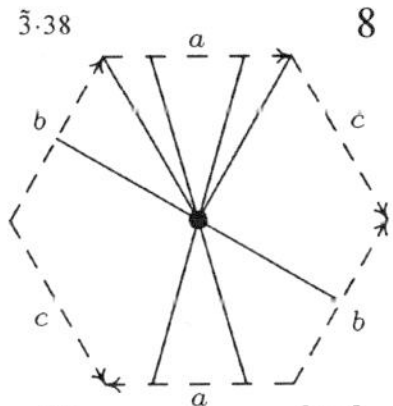

[8] [62]

$\tilde{3}$·39 8

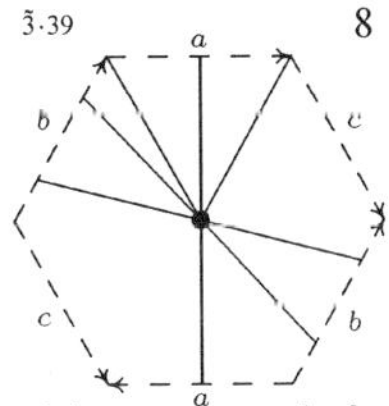

[8] [62]

$\tilde{3}$·40 8

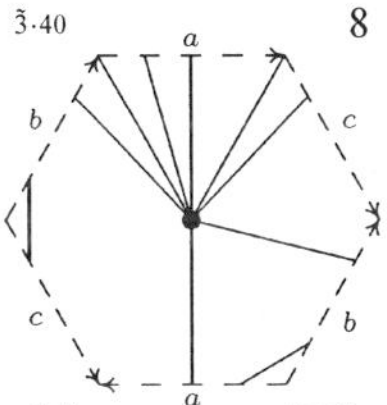

[8] [62]

$\tilde{3}$·41 4

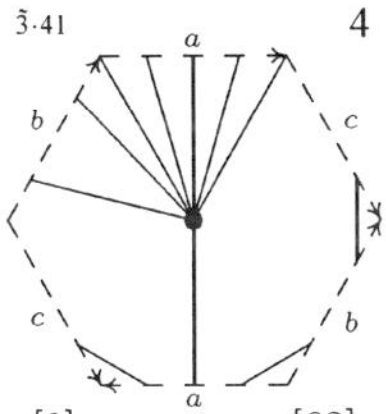

[8] [62]

$\tilde{3}$·42 8

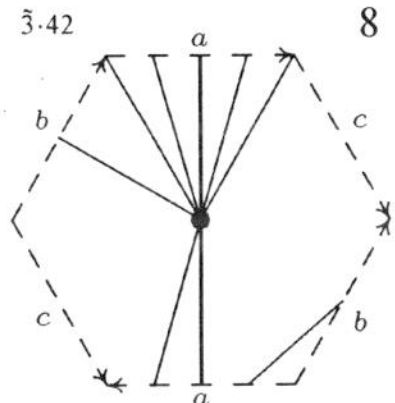

[8] [62]

$\tilde{3}$·43 8

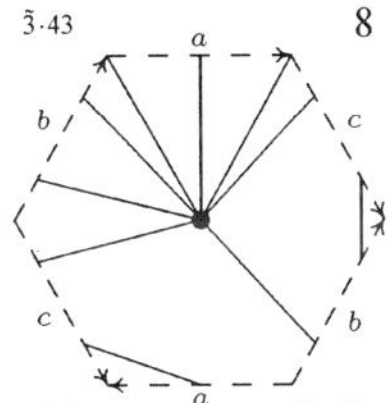

[8] [62]

$\tilde{3}$·44 16

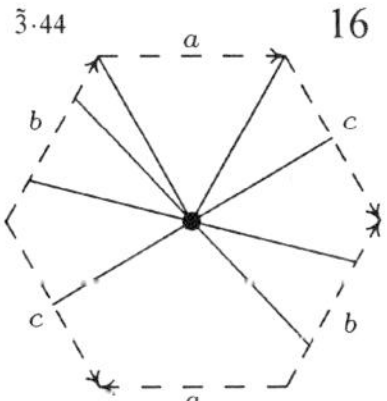

[8] [62]

$\tilde{3}$·45 16

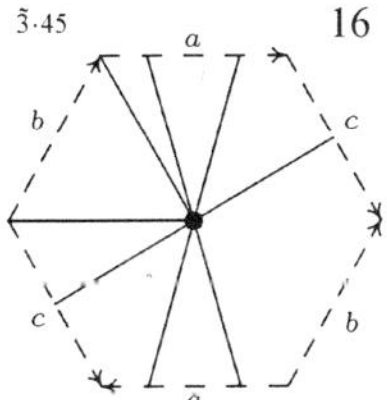

[8] [62]

$\tilde{3}$·46 16

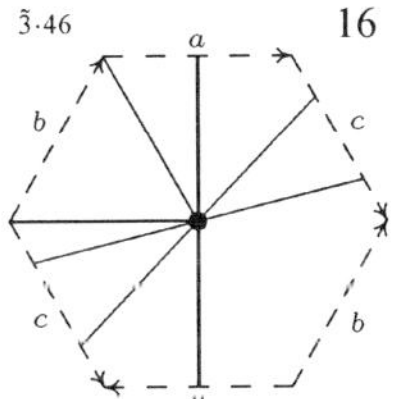

[8] [62]

$\tilde{3}$·47 16

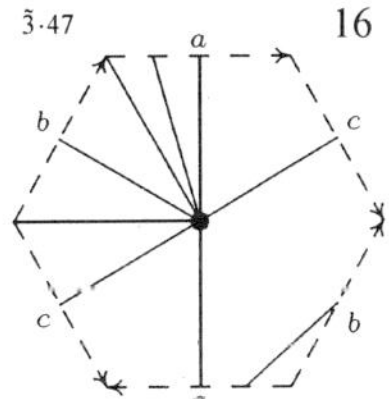

[8] [62]

$\tilde{3}$·48 8
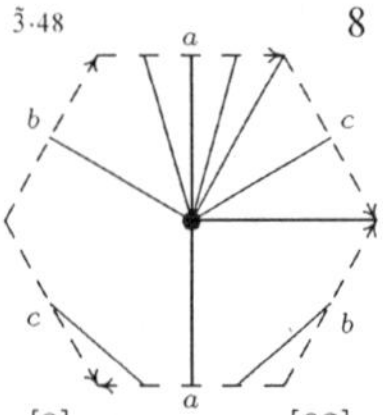
[8] [62]

$\tilde{3}$·49 4
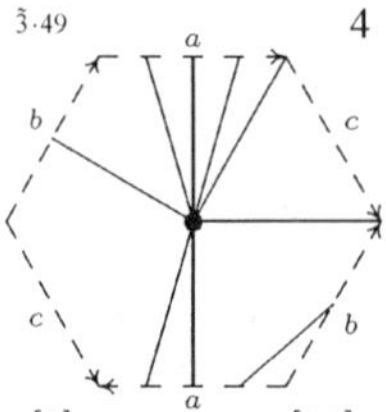
[8] [62]

$\tilde{3}$·50 8
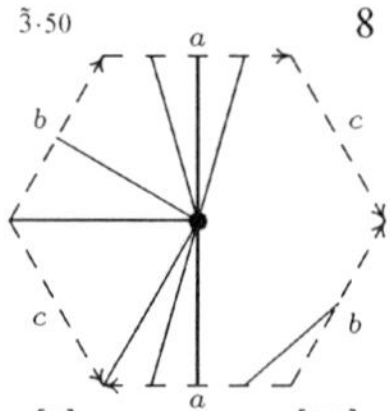
[8] [62]

$\tilde{3}$·51 8
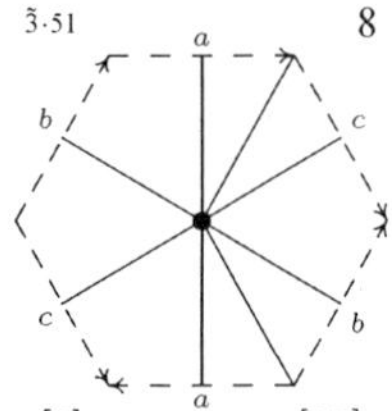
[8] [53]

$\tilde{3}$·52 8
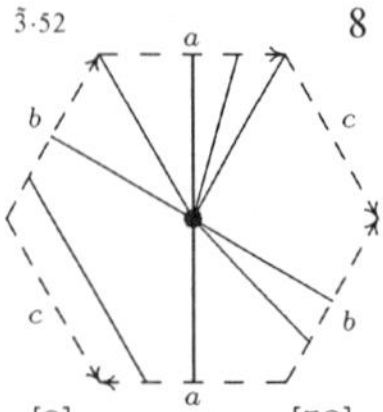
[8] [53]

$\tilde{3}$·53 8
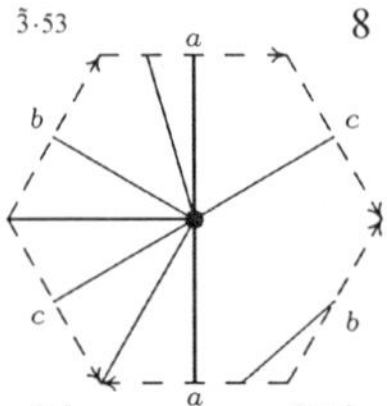
[8] [53]

$\tilde{3}$·54 8
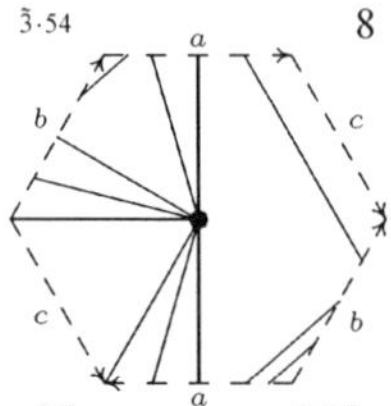
[8] [53]

$\tilde{3}$·55 16
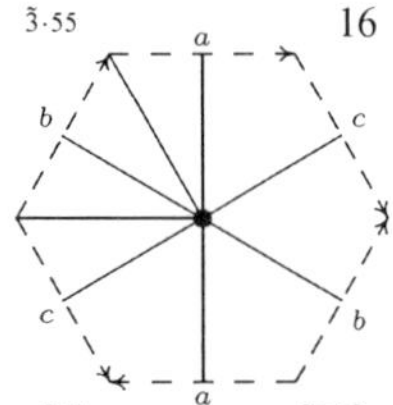
[8] [53]

$\tilde{3}$·56 16
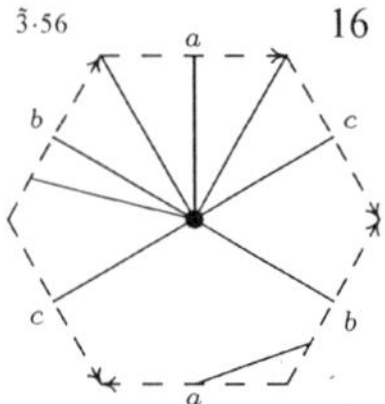
[8] [53]

$\tilde{3}$·57 16
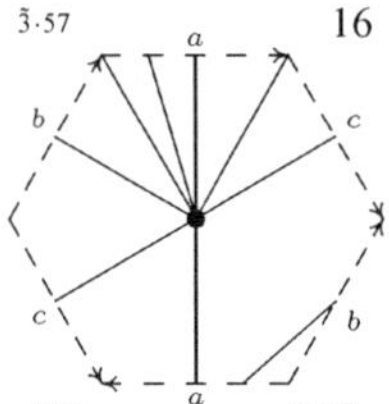
[8] [53]

$\tilde{3}$·58 8
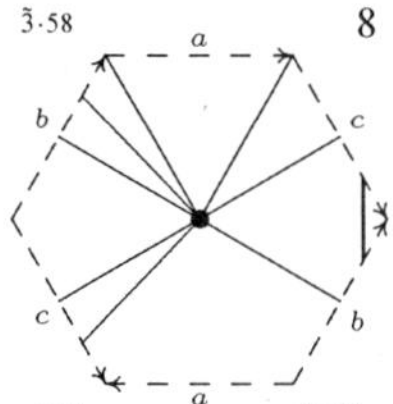
[8] [53]

$\tilde{3}$·59 16
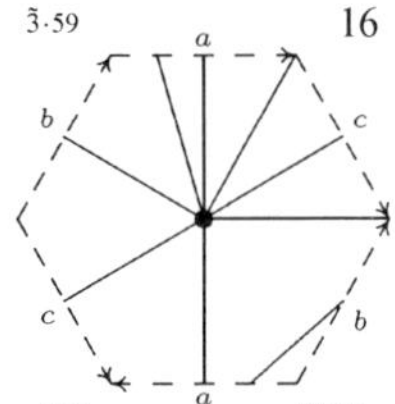
[8] [53]

$\tilde{3}$·60 8
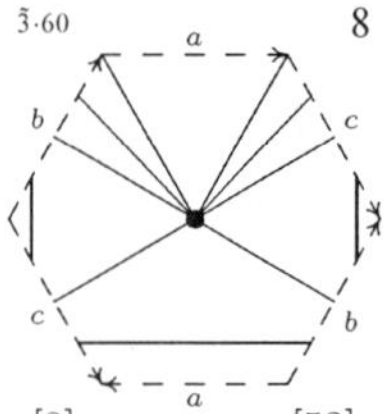
[8] [53]

$\tilde{3}$·61 16
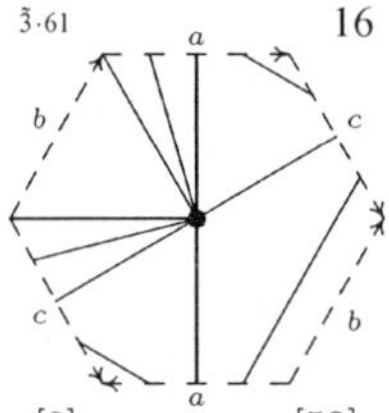
[8] [53]

$\tilde{3}$·62 8
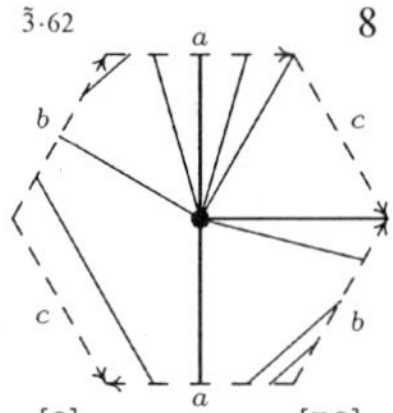
[8] [53]

$\tilde{3}$·63 2
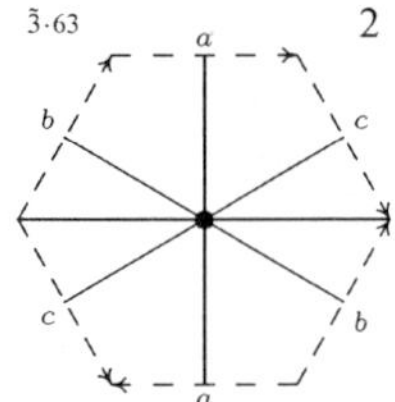
[8] [4^2]

$\tilde{3}$·64 8
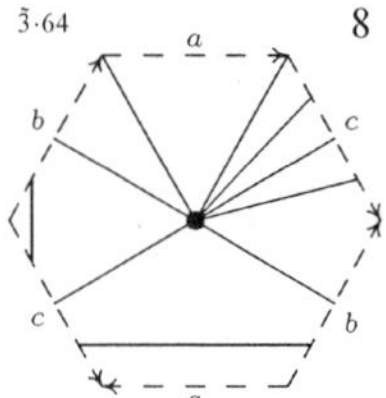
[8] [4^2]

$\tilde{3}$·65 8
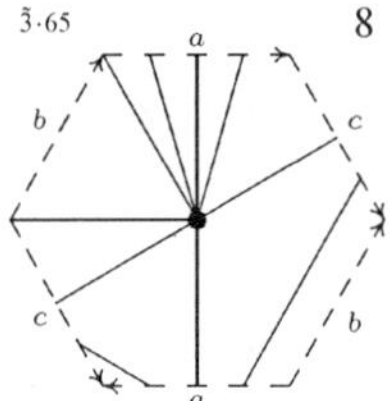
[8] [4^2]

$\tilde{3}$·66 8
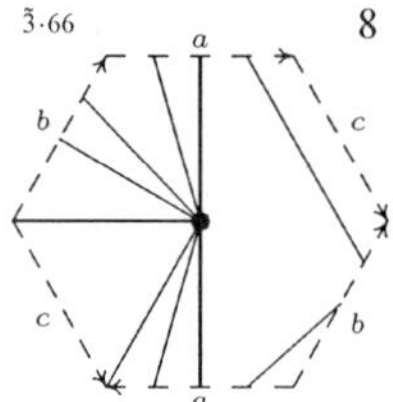
[8] [4^2]

$\tilde{3}$·67 8
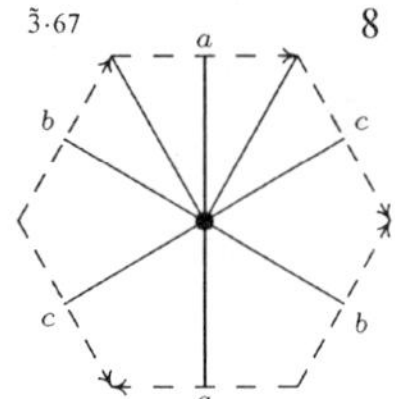
[8] [4^2]

Genus $\tilde{3}$: 4e 1v

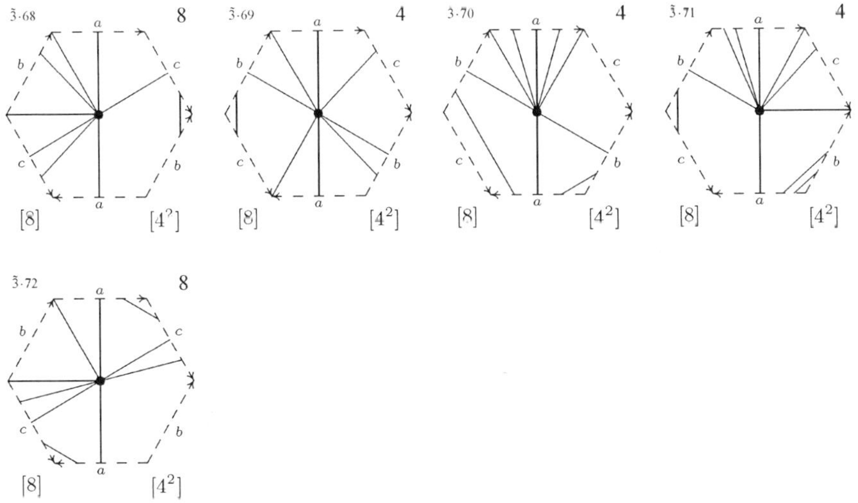

4 edges

2 vertices

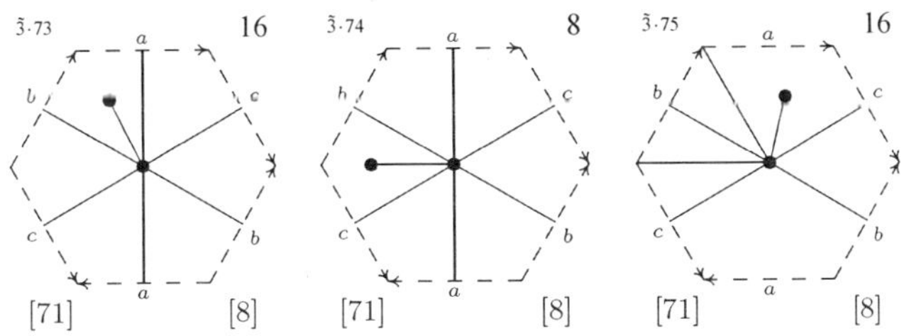

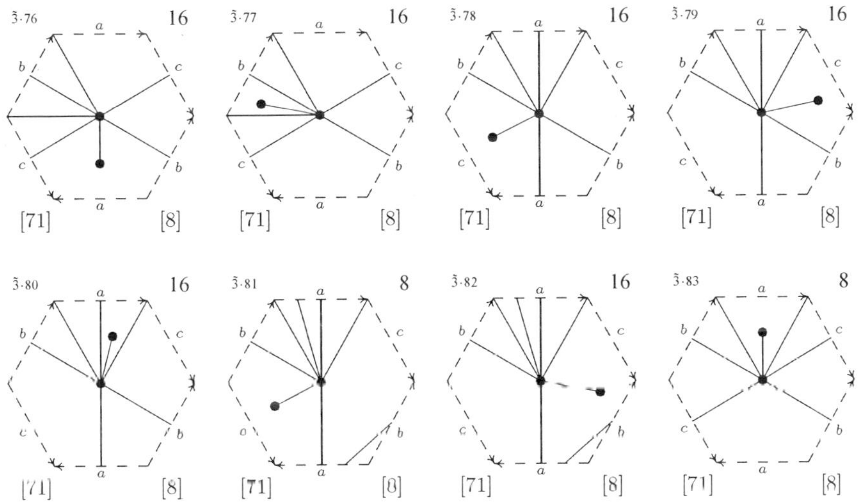

$\tilde{3}\cdot84$ 16
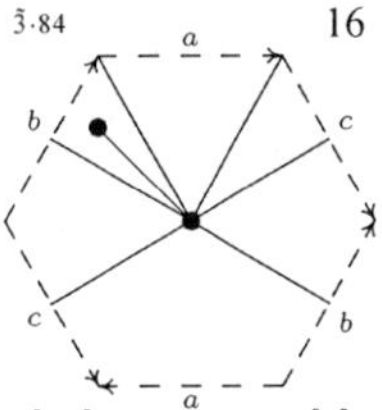

[71] [8]

$\tilde{3}\cdot85$ 16
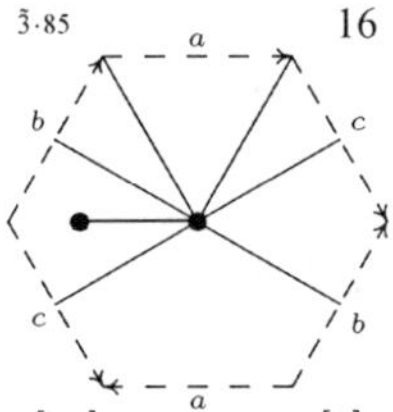
[71] [8]

$\tilde{3}\cdot86$ 8
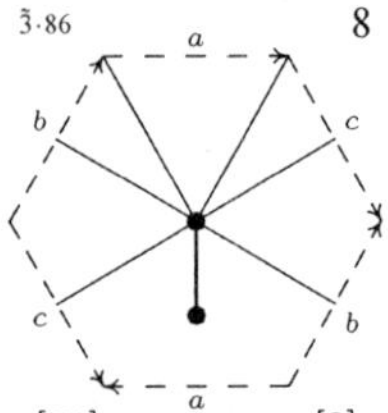
[71] [8]

$\tilde{3}\cdot87$ 16
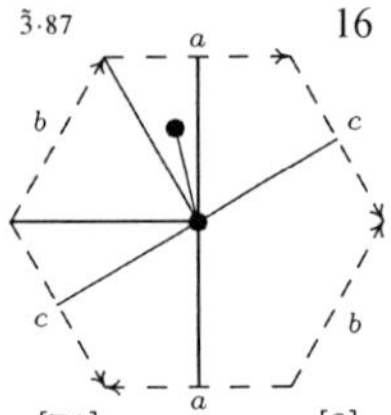
[71] [8]

$\tilde{3}\cdot88$ 16
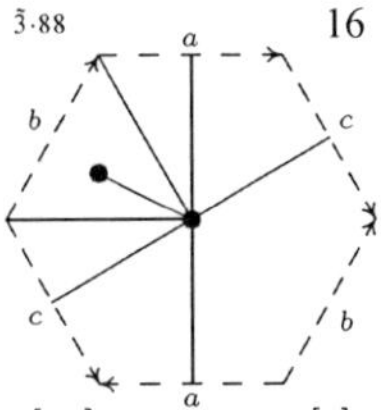
[71] [8]

$\tilde{3}\cdot89$ 16
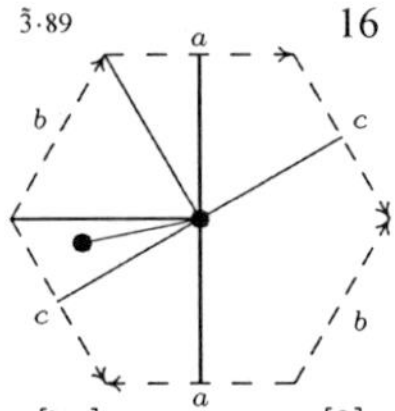
[71] [8]

$\tilde{3}\cdot90$ 16
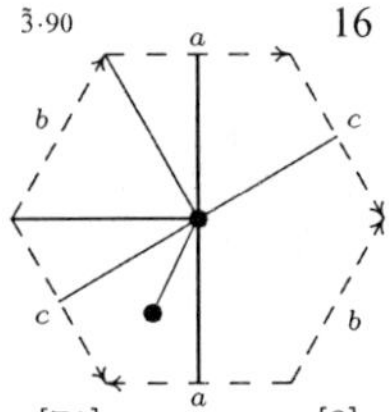
[71] [8]

$\tilde{3}\cdot91$ 16
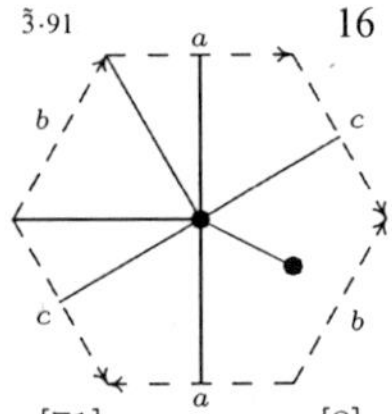
[71] [8]

$\tilde{3}\cdot92$ 16
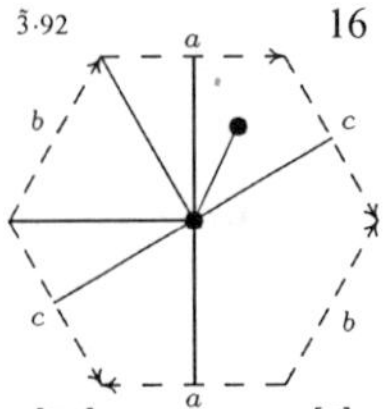
[71] [8]

$\tilde{3}\cdot93$ 16
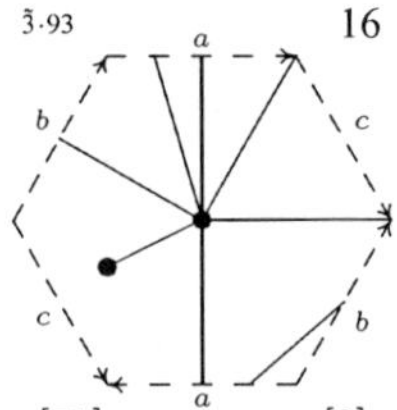
[71] [8]

$\tilde{3}\cdot94$ 8
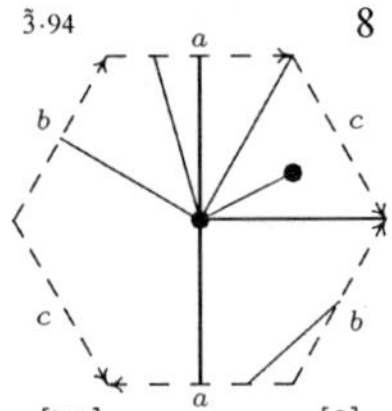
[71] [8]

$\tilde{3}\cdot95$ 8
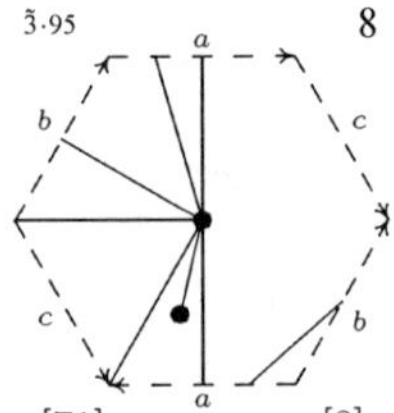
[71] [8]

$\tilde{3}\cdot96$ 8
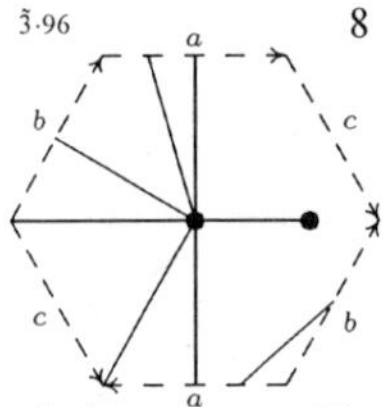
[71] [8]

$\tilde{3}\cdot97$ 4
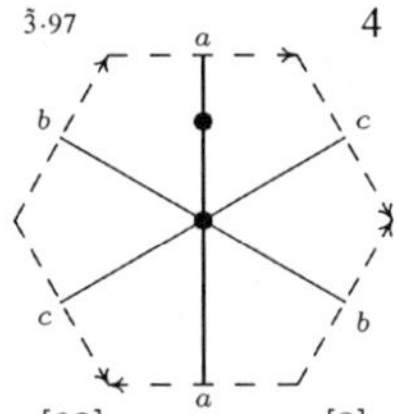
[62] [8]

$\tilde{3}\cdot98$ 8
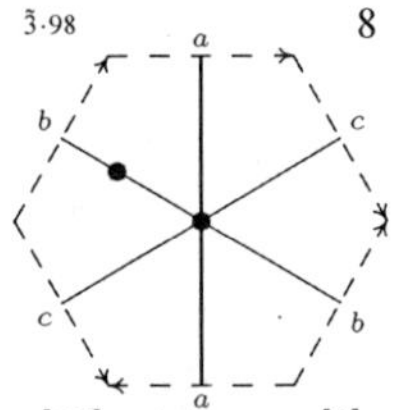
[62] [8]

$\tilde{3}\cdot99$ 8
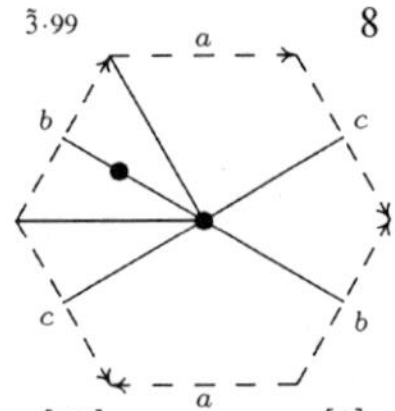
[62] [8]

$\tilde{3}\cdot100$ 8
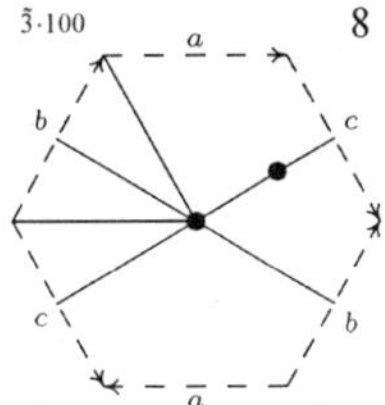
[62] [8]

$\tilde{3}\cdot101$ 8
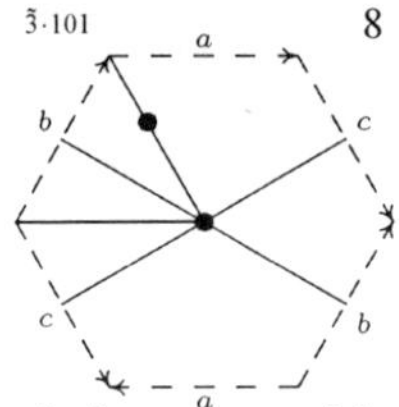
[62] [8]

$\tilde{3}\cdot102$ 8
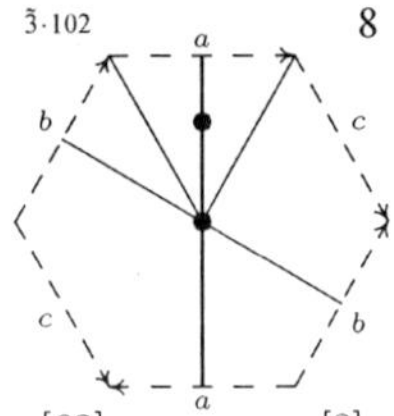
[62] [8]

$\tilde{3}\cdot103$ 8
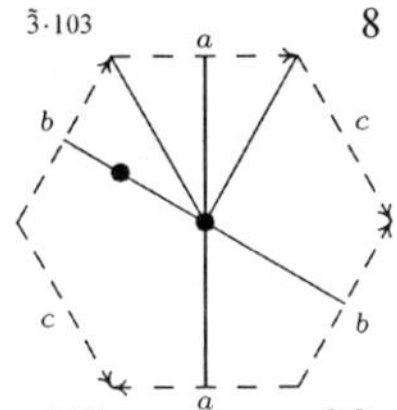
[62] [8]

Genus $\tilde{3}$: 4e 2v

$\tilde{3}$·104 8

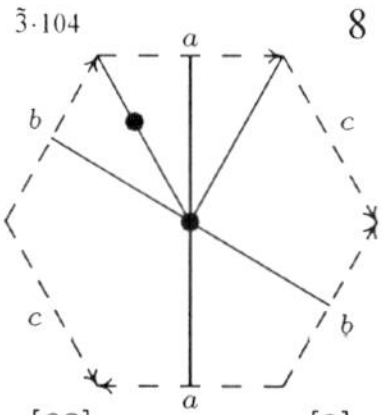

[62] [8]

$\tilde{3}$·105 4

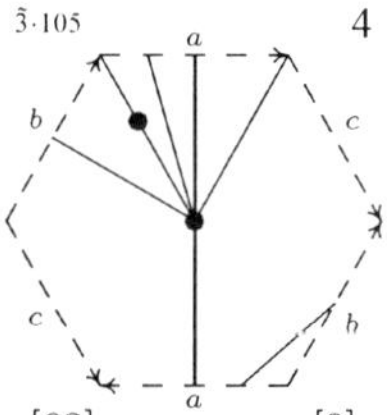

[62] [8]

$\tilde{3}$·106 8

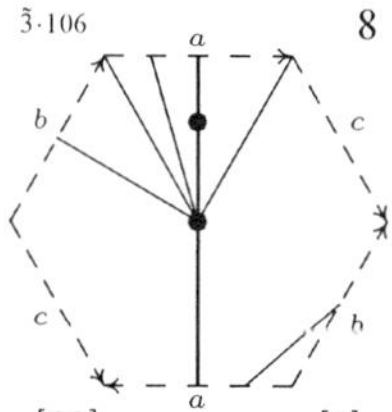

[62] [8]

$\tilde{3}$·107 8

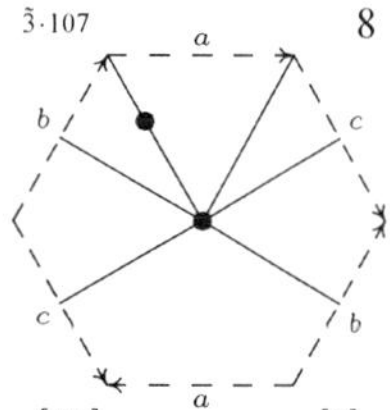

[62] [8]

$\tilde{3}$·108 16

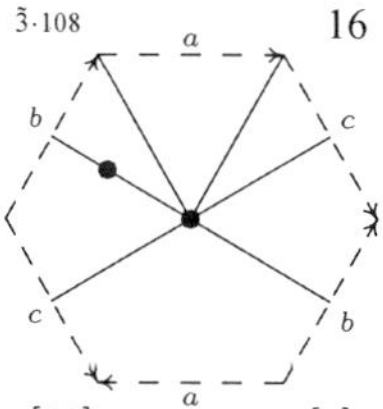

[62] [8]

$\tilde{3}$·109 16

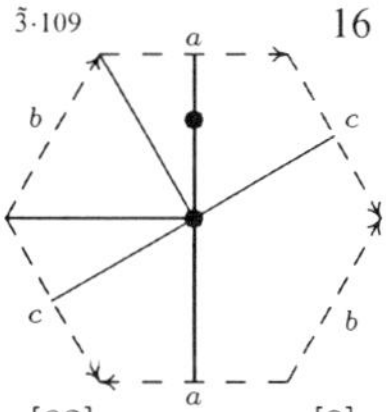

[62] [8]

$\tilde{3}$·110 16

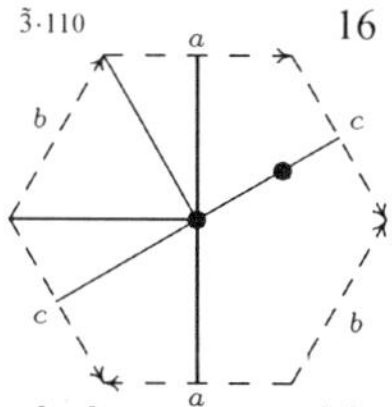

[62] [8]

$\tilde{3}$·111 16

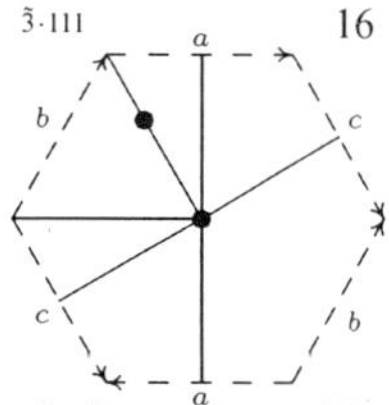

[62] [8]

$\tilde{3}$·112 8

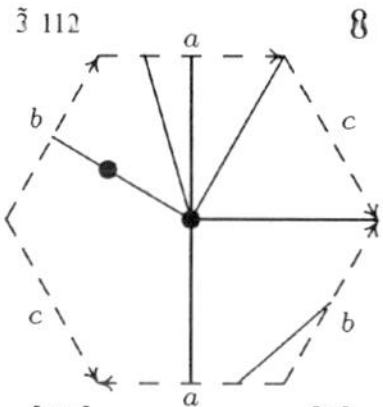

[62] [8]

$\tilde{3}$·113 4

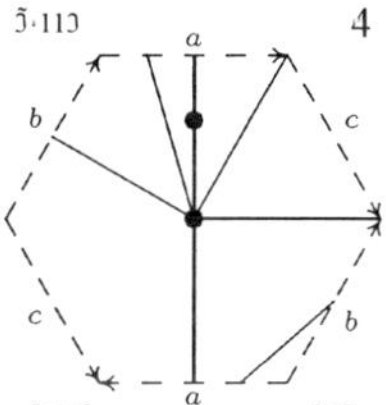

[62] [8]

$\tilde{3}$·114 8

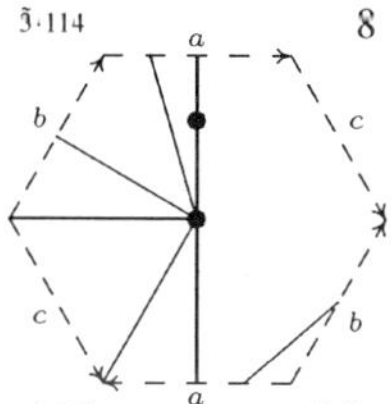

[62] [8]

$\tilde{3}$·115 8

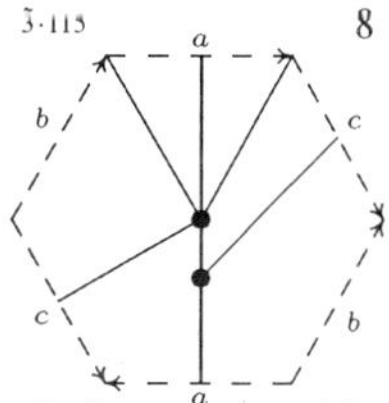

[53] [8]

$\tilde{3}$·116 8

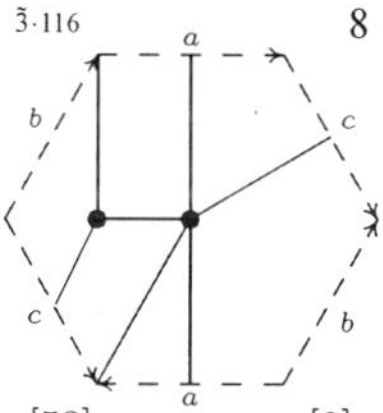

[53] [8]

$\tilde{3}$·117 8

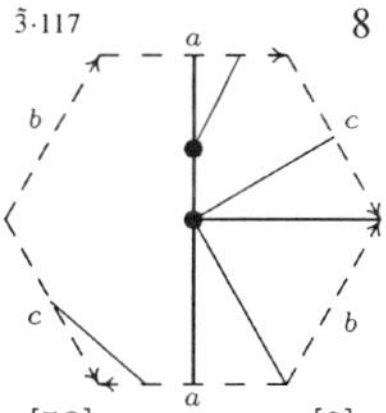

[53] [8]

$\tilde{3}$·118 8

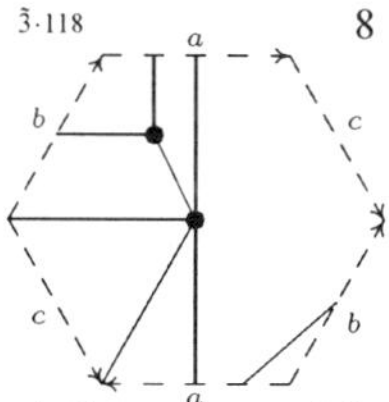

[53] [8]

$\tilde{3}$·119 16

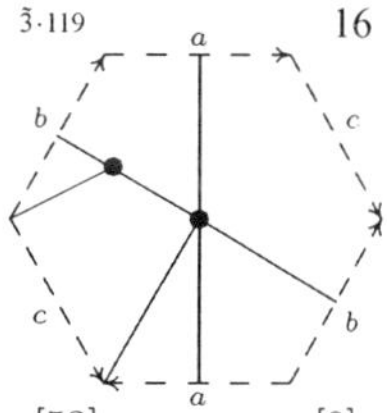

[53] [8]

$\tilde{3}$·120 16

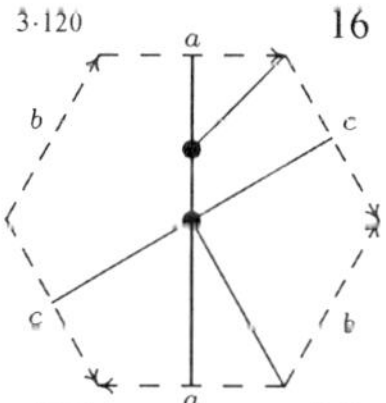

[53] [8]

$\tilde{3}$·121 16

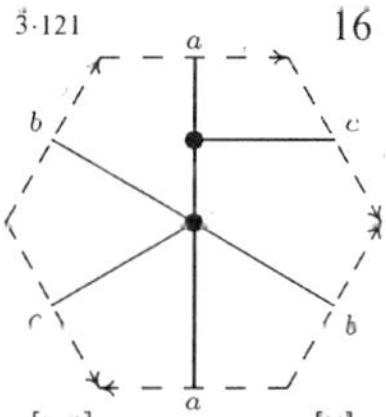

[53] [8]

$\tilde{3}$·122 8

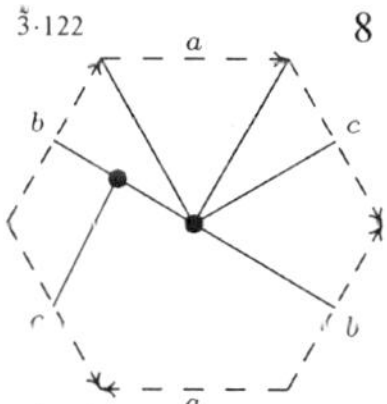

[53] [8]

$\tilde{3}$·123 16

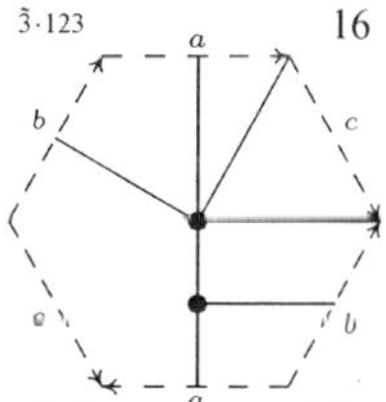

[53] [8]

Genus $\tilde{3}$: 4e 2v

$\tilde{3}\cdot124$ 8

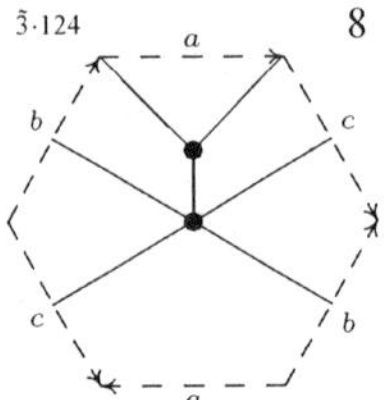

[53] [8]

$\tilde{3}\cdot125$ 16

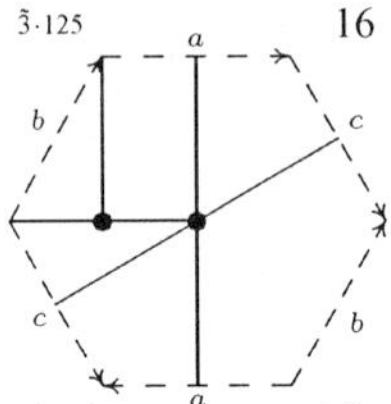

[53] [8]

$\tilde{3}\cdot126$ 8

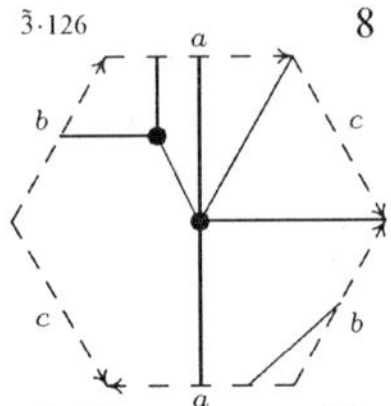

[53] [8]

$\tilde{3}\cdot127$ 2

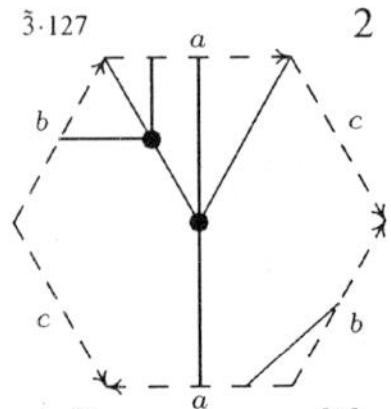

$[4^2]$ [8]

$\tilde{3}\cdot128$ 8

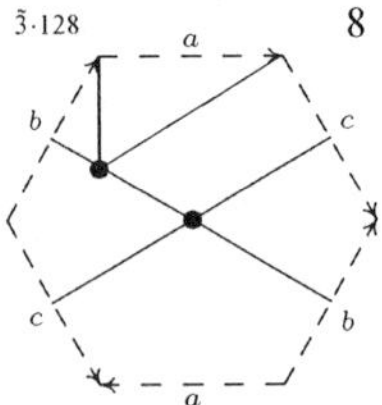

$[4^2]$ [8]

$\tilde{3}\cdot129$ 8

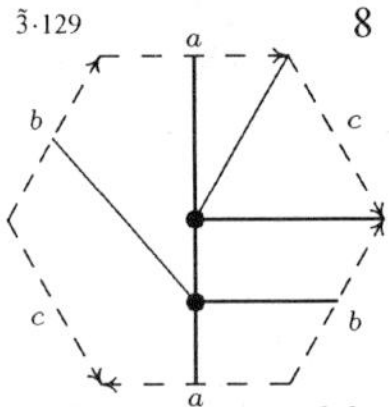

$[4^2]$ [8]

$\tilde{3}\cdot130$ 8

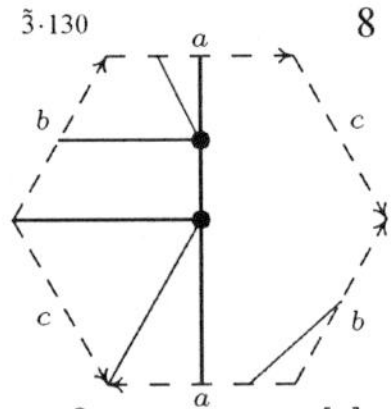

$[4^2]$ [8]

$\tilde{3}\cdot131$ 8

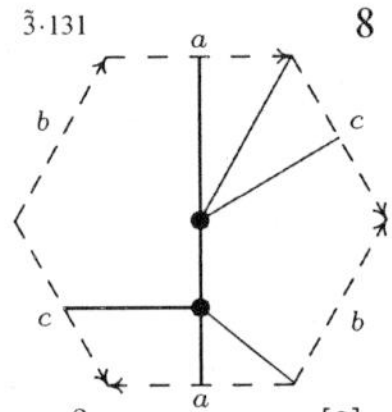

$[4^2]$ [8]

$\tilde{3}\cdot132$ 8

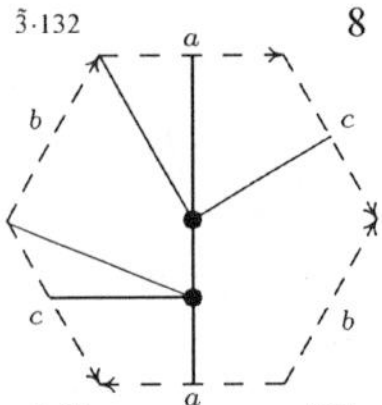

$[4^2]$ [8]

$\tilde{3}\cdot133$ 4

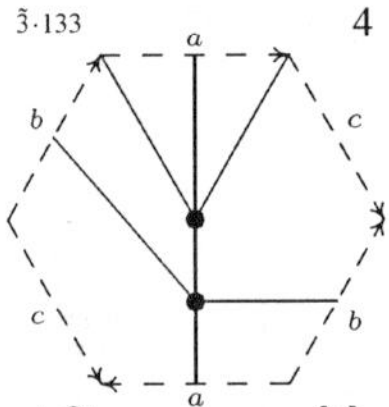

$[4^2]$ [8]

$\tilde{3}\cdot134$ 4

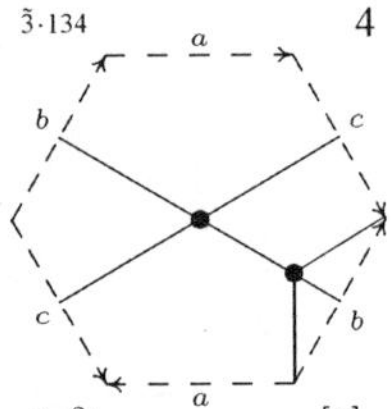

$[4^2]$ [8]

$\tilde{3}\cdot135$ 4

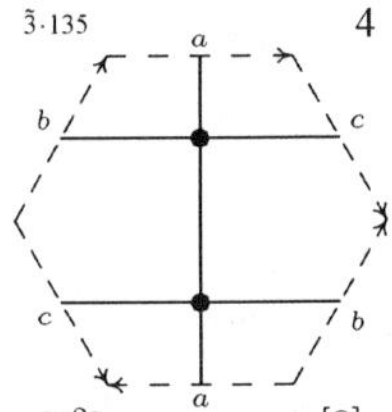

$[4^2]$ [8]

$\tilde{3}\cdot136$ 8

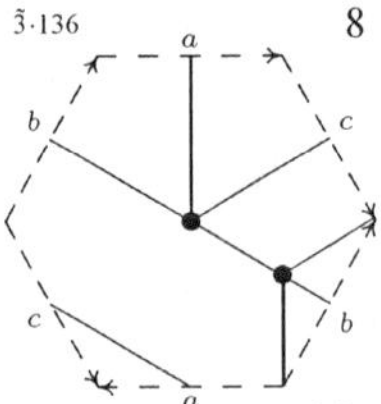

$[4^2]$ [8]

Genus $\tilde{4}$: 4e 1v

6.4 Genus 4 – the doubly crosscapped torus

4 edges

1 vertex

$\tilde{4}\cdot 1$ 4
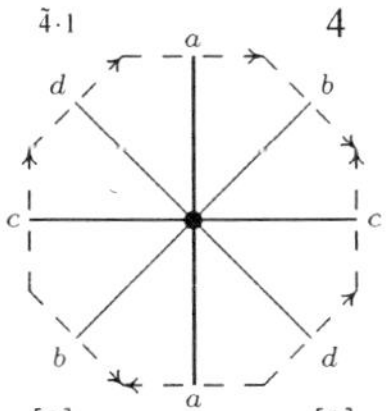
[8] [8]

$\tilde{4}\cdot 2$ 2
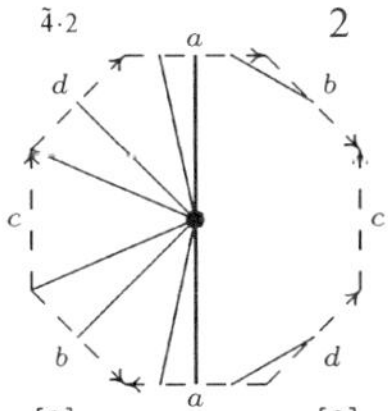
[8] [8]

$\tilde{4}\cdot 3$ 16
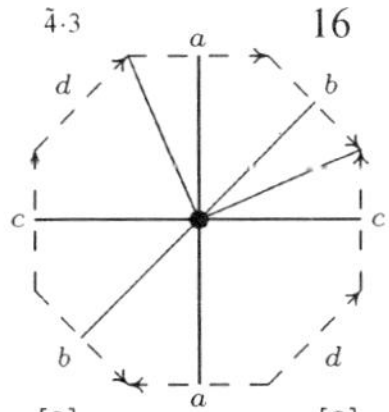
[8] [8]

$\tilde{4}\cdot 4$ 4
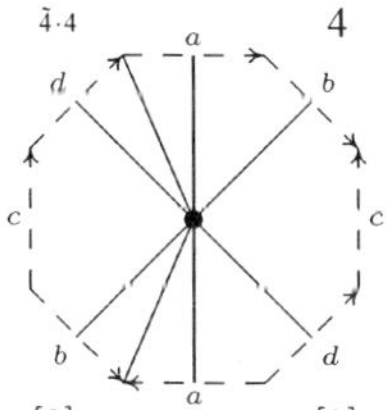
[8] [8]

$\tilde{4}\cdot 5$ 8
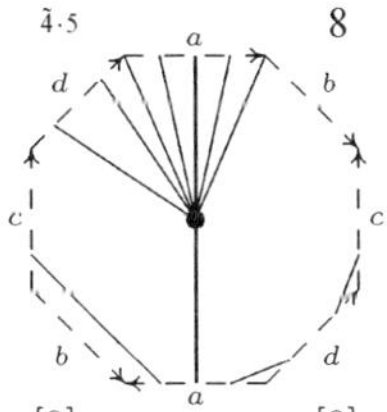
[8] [8]

$\tilde{4}\cdot 6$ 8
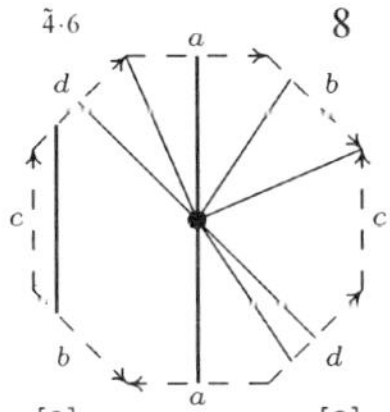
[8] [8]

$\tilde{4}\cdot 7$ 2
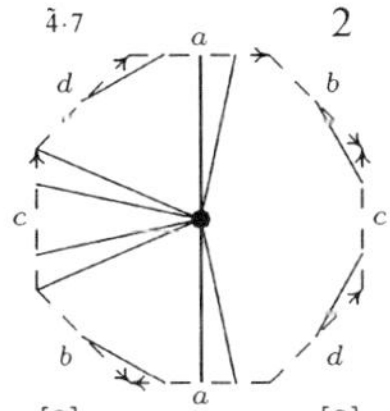
[8] [8]

$\tilde{4}\cdot 8$ 16
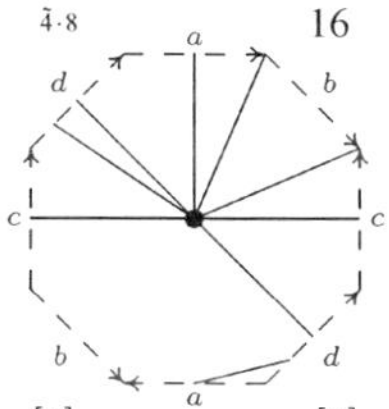
[8] [8]

$\tilde{4}\cdot 9$ 16
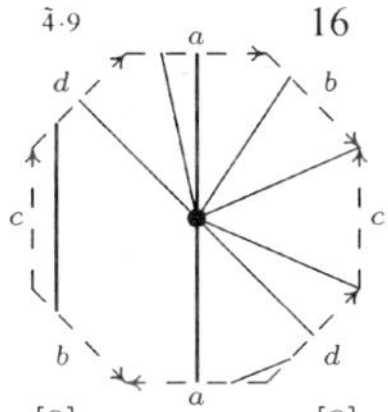
[8] [8]

$\tilde{4}\cdot 10$ 8
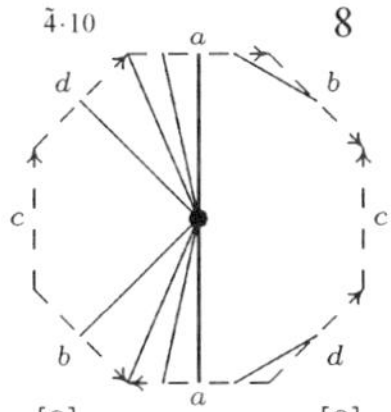
[8] [8]

$\tilde{4}\cdot 11$ 8
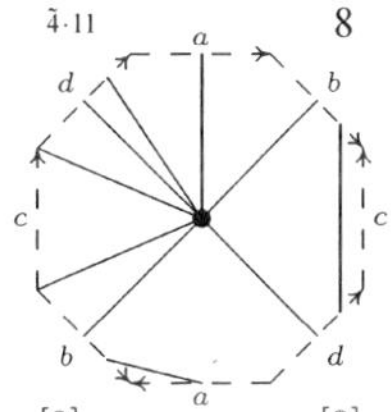
[8] [8]

$\tilde{4}\cdot 12$ 8
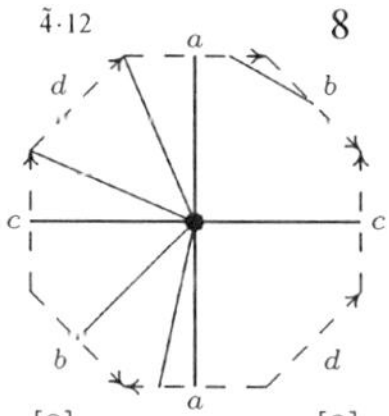
[8] [8]

$\tilde{4}\cdot 13$ 8
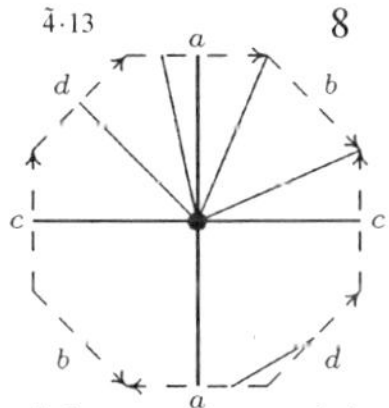
[8] [8]

$\tilde{4}\cdot 14$ 8
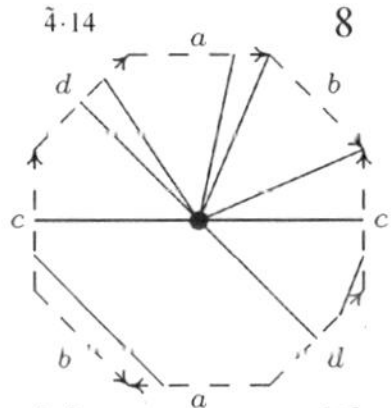
[8] [8]

$\tilde{4}\cdot 15$ 16
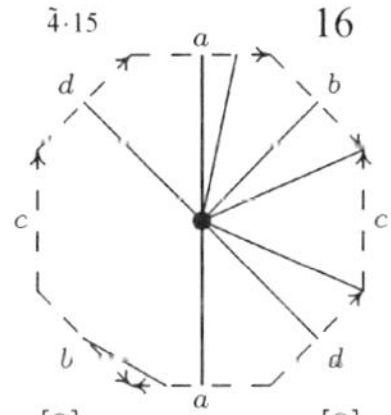
[8] [8]

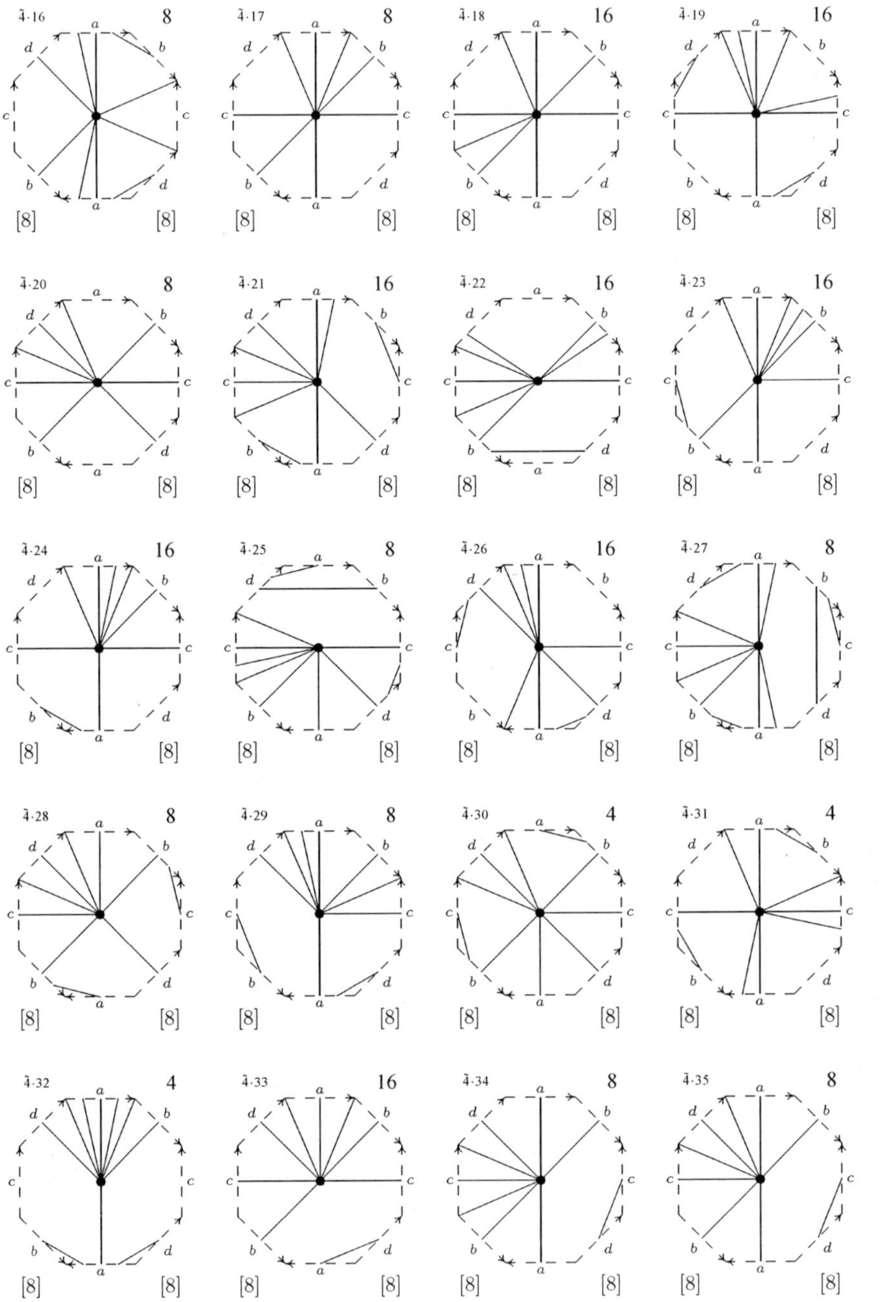

Genus $\tilde{4}$: 4e 1v

$\tilde{4}\cdot 36$ 16

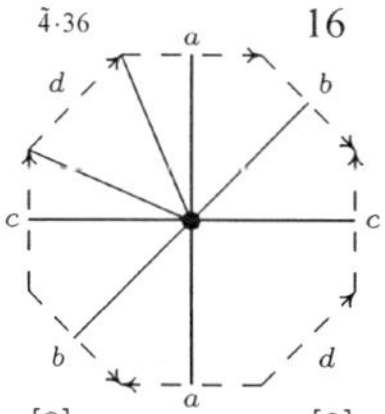

[8] [8]

$\tilde{4}\cdot 37$ 8

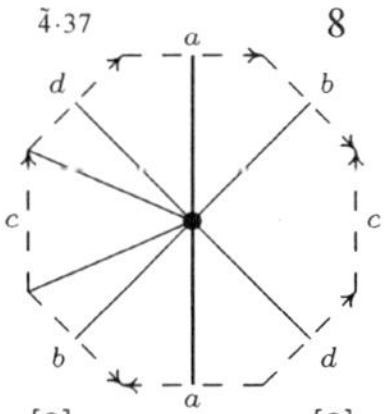

[8] [8]

$\tilde{4}\cdot 38$ 16

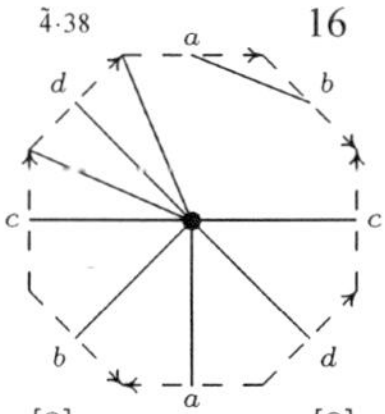

[8] [8]

$\tilde{4}\cdot 39$ 16

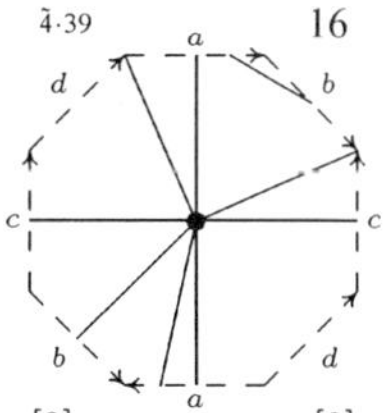

[8] [8]

$\tilde{4}\cdot 40$ 16

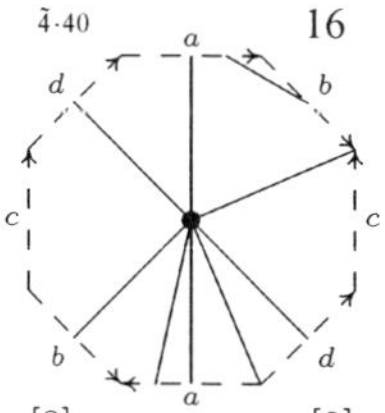

[8] [8]

$\tilde{4}\cdot 41$ 16

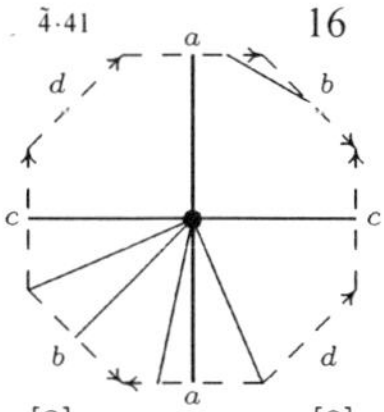

[8] [8]

$\tilde{4}\cdot 42$ 16

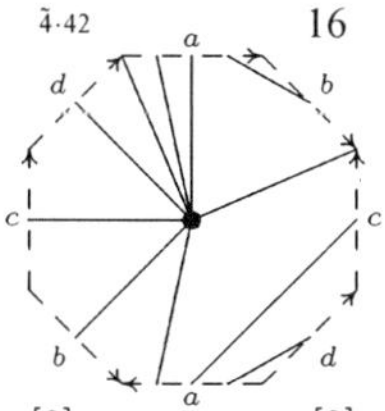

[8] [8]

$\tilde{4}\cdot 43$ 4

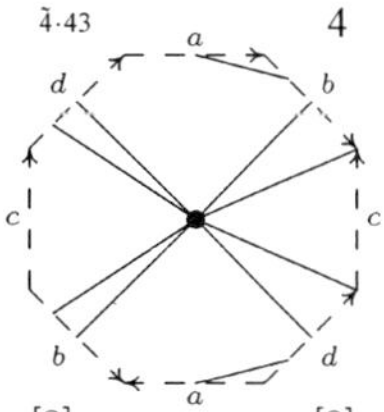

[8] [8]

$\tilde{4}\cdot 44$ 8

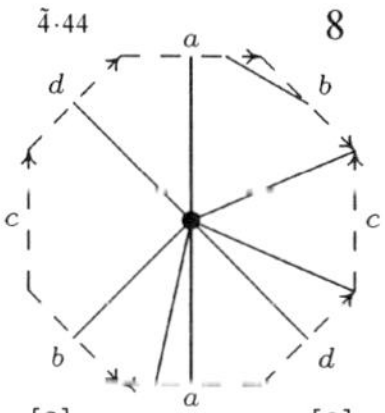

[8] [8]

$\tilde{4}\cdot 45$ 8

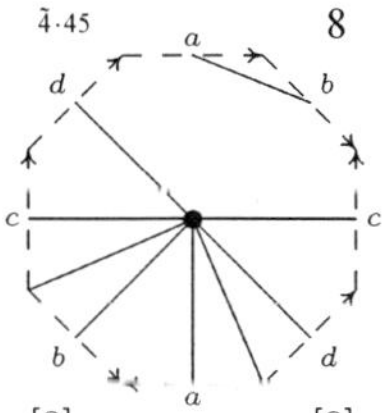

[8] [8]

$\tilde{4}\cdot 46$ 16

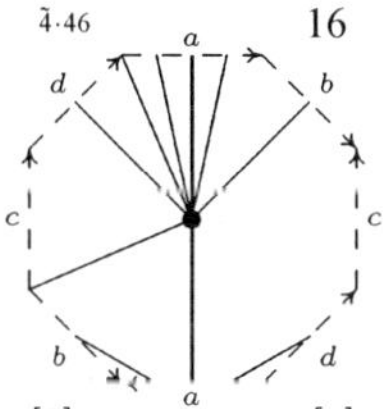

[8] [8]

$\tilde{4}\cdot 47$ 4

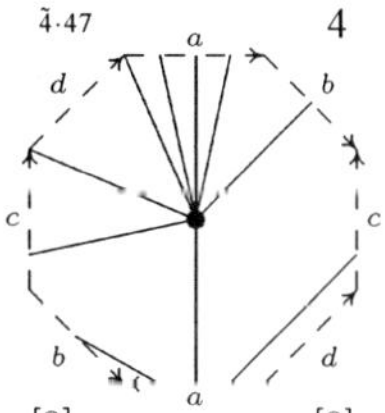

[8] [8]

Chapter 7

Face regular maps and hypermaps

Two classes of face regular maps are listed in this chapter. These are all the triangulations and quadrangulations on up to 6 edges in orientable surfaces and up to 4 edges in nonorientable surfaces. Triangulations are listed in Section 7.1 and quadrangulations are listed in Section 7.2.

7.1 Triangulations

7.1.1 Orientable of genus 0 and 1

Genus 0 – the sphere

3 edges

3 vertices

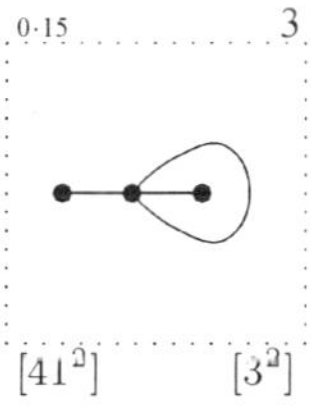

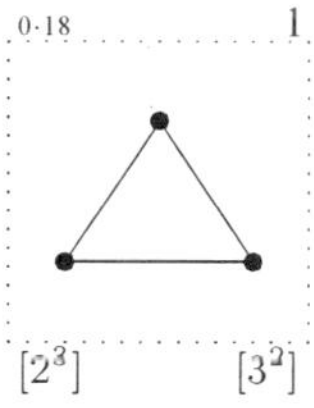

6 edges

4 vertices

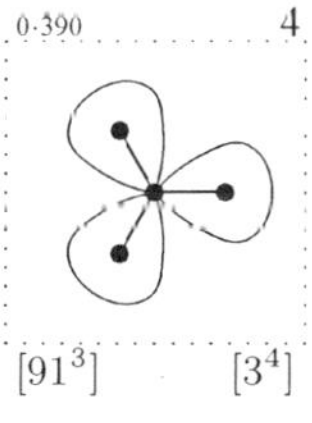

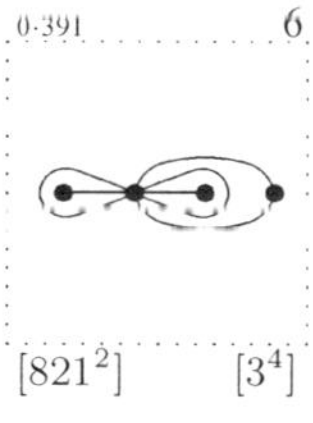

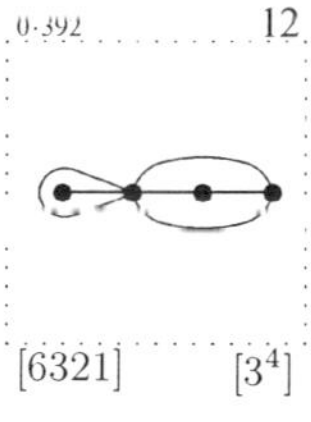

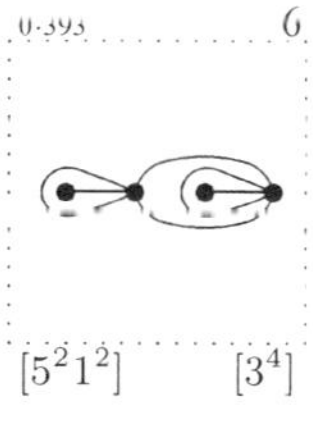

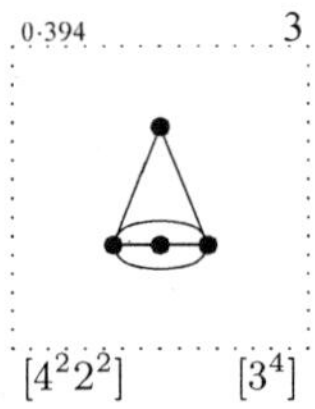

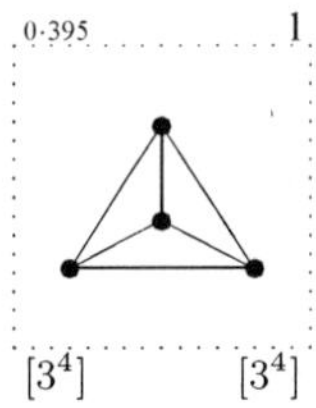

Genus 1 – the torus

3 edges

1 vertex

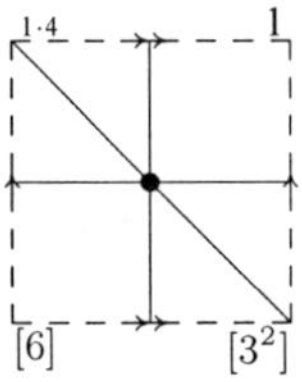

6 edges

2 vertices

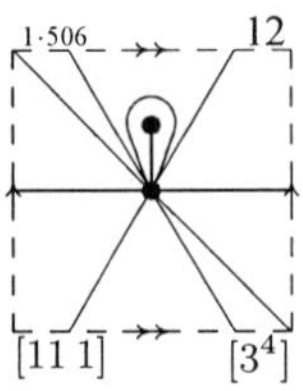

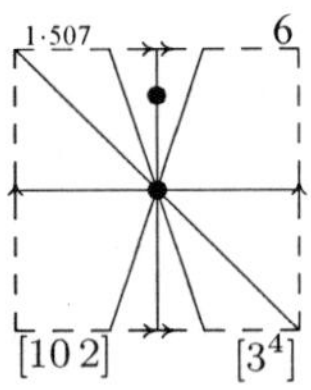

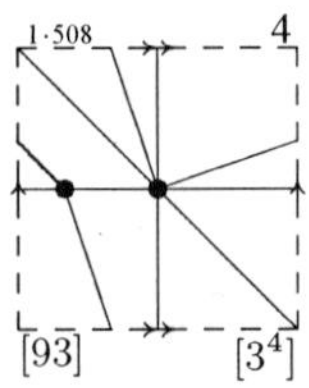

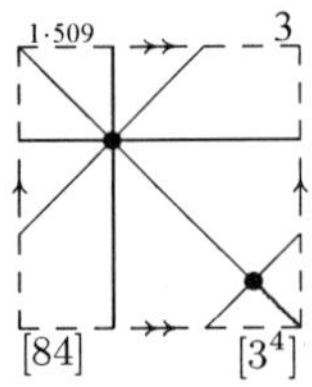

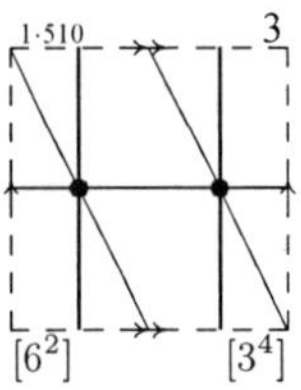

7.1.2 Nonorientable of genus 1 and 2

Genus 1 – the projective plane

3 edges

2 vertices

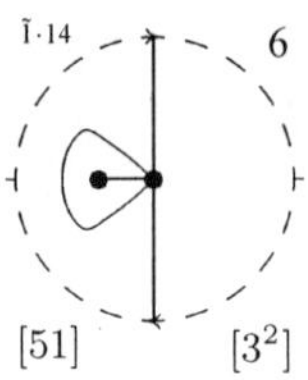

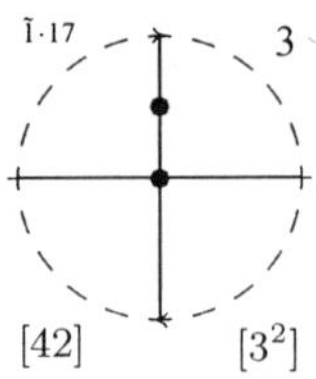

Genus 2 – the Klein bottle

3 edges

1 vertex

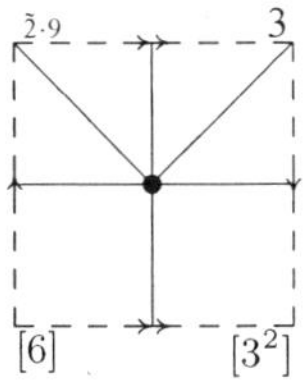

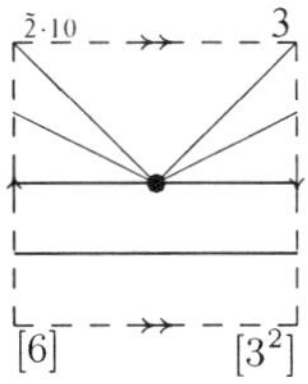

7.2 Quadrangulations

7.2.1 Orientable of genus 0, 1 and 2

Genus 0 – the sphere

2 edges

3 vertices

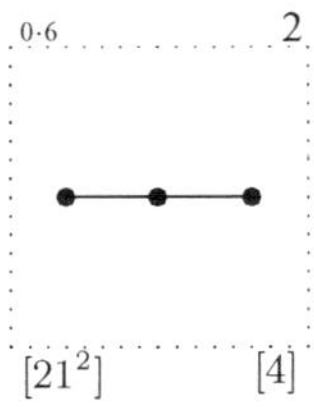

4 edges

4 vertices

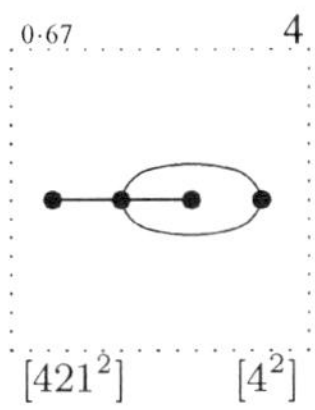

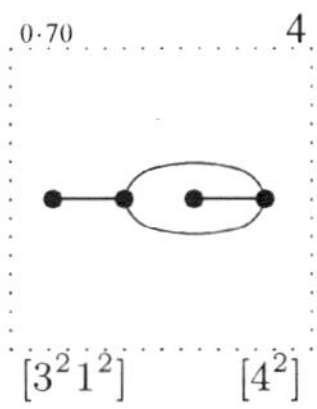

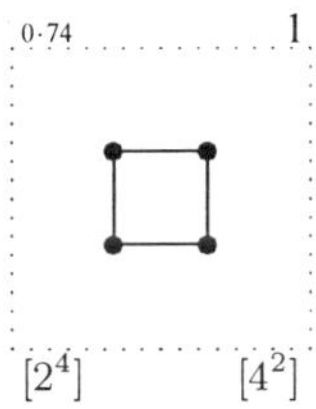

6 edges

5 vertices

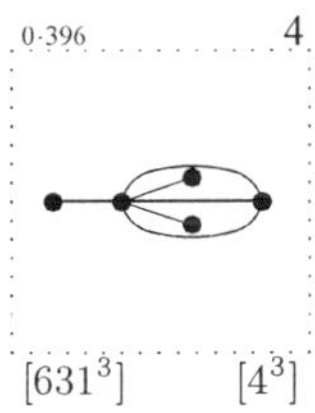

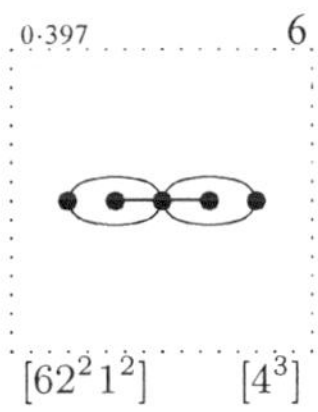

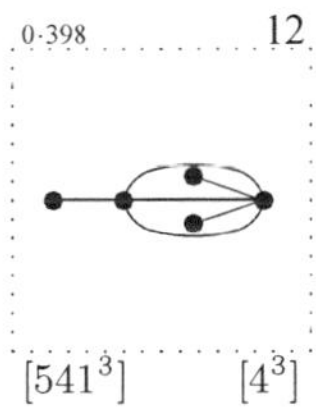

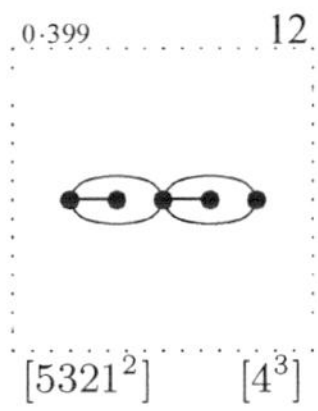

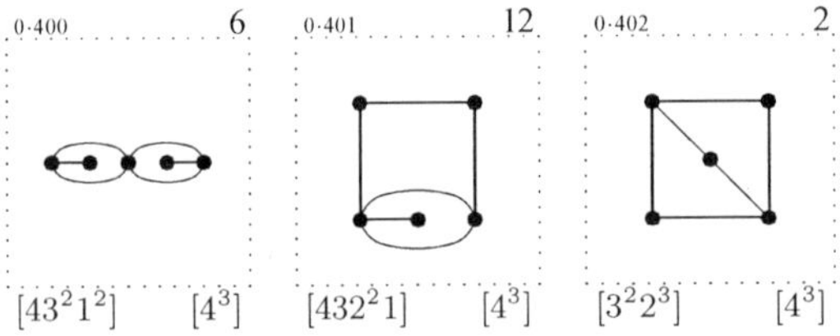

Genus 1 – the torus

2 edges

1 vertex

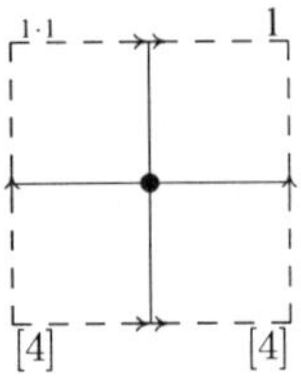

4 edges

2 vertices

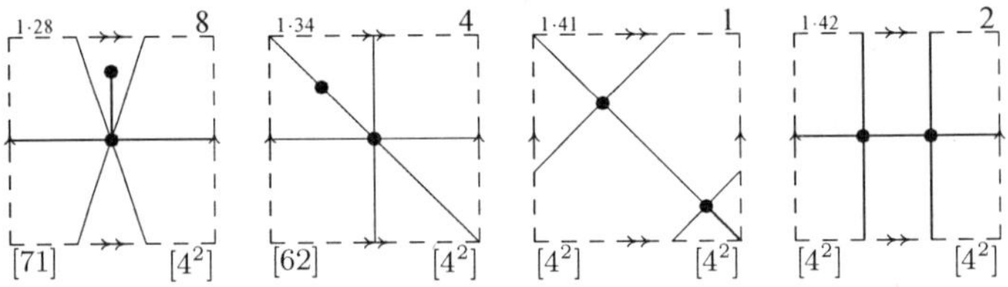

6 edges

3 vertices

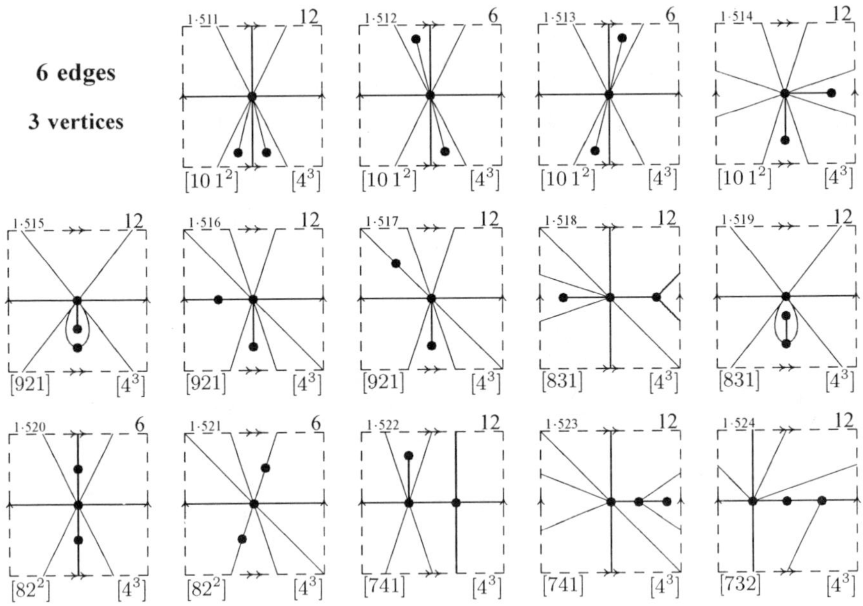

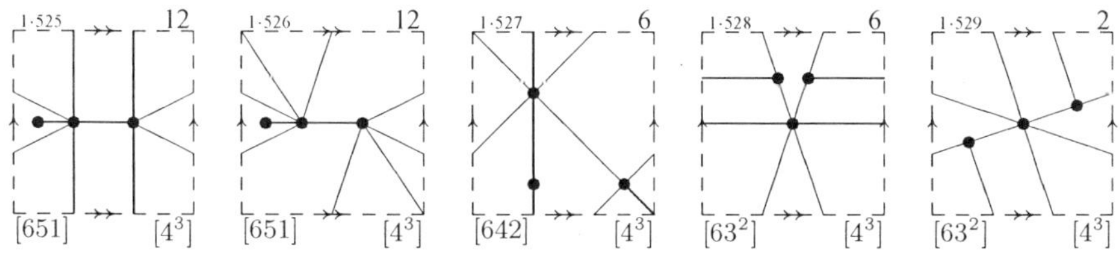

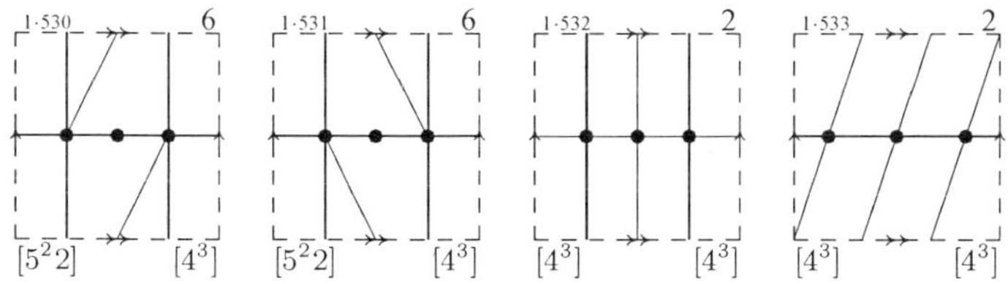

Genus 2 – the double torus

6 edges

1 vertex

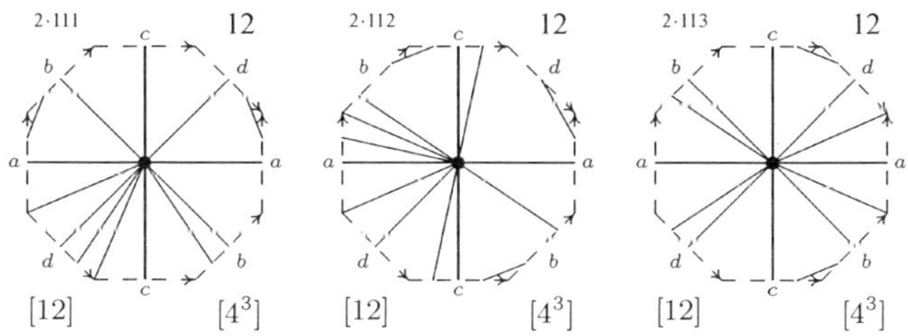

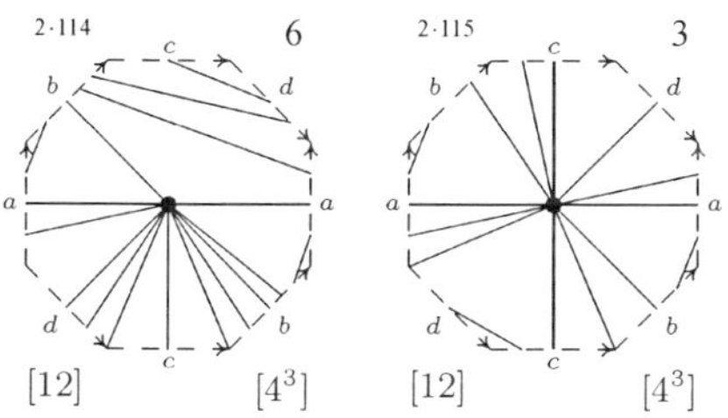

7.2.2 Nonorientable of genus 1, 2 and 3

Genus 1 – the projective plane

2 edges

2 vertices

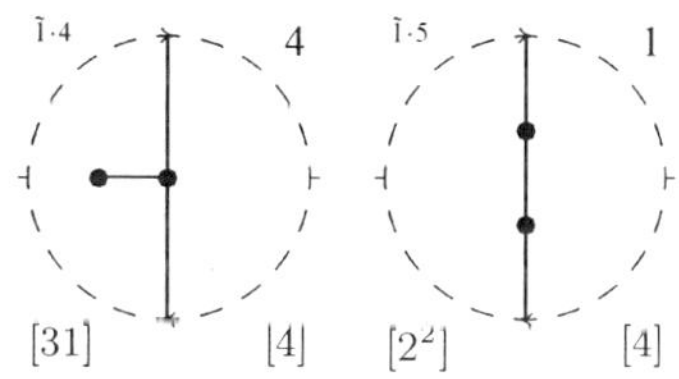

4 edges

3 vertices

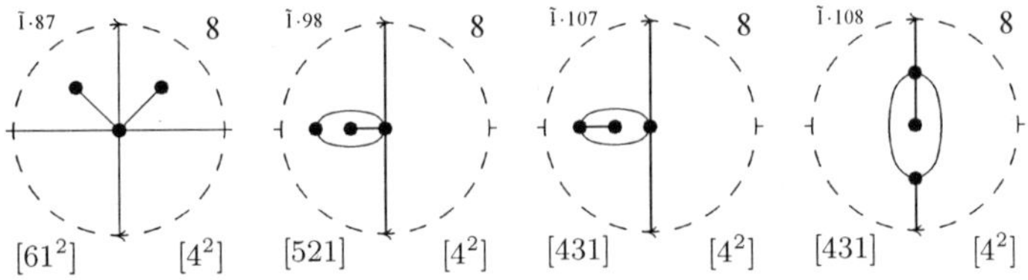

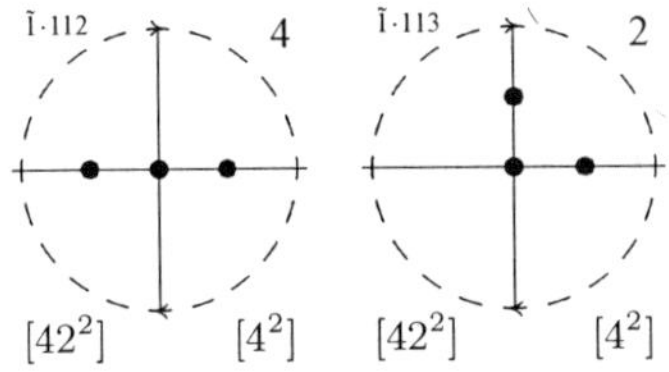

Genus 2 – the Klein bottle

2 edges

1 vertex

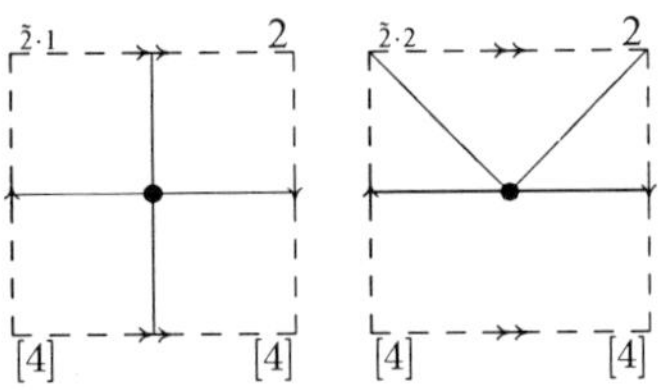

4 edges

2 vertices

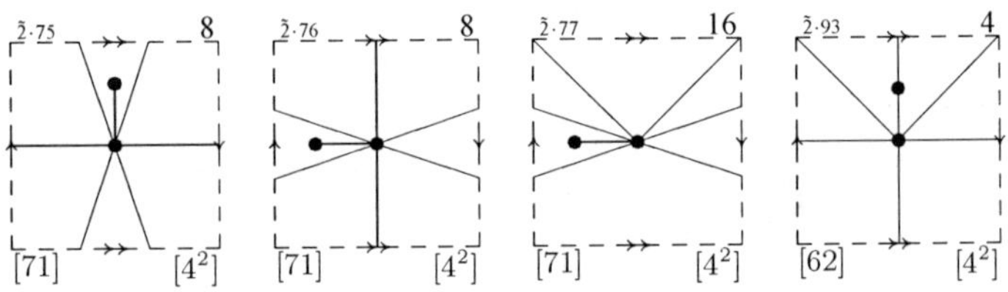

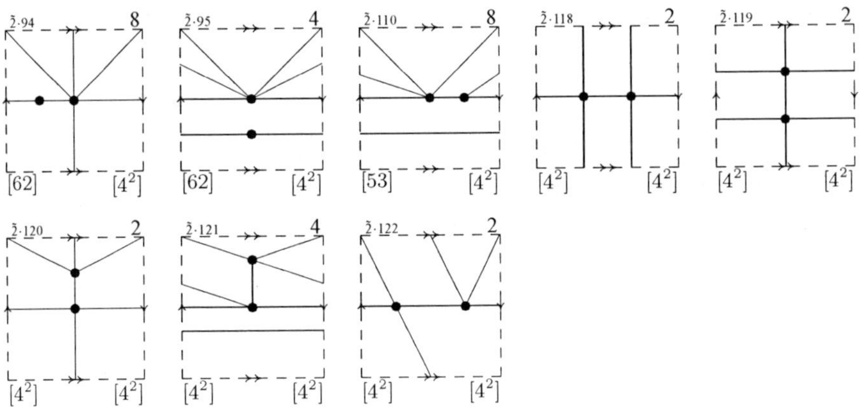

Genus 3 – the crosscapped torus

4 edges

1 vertex

$\tilde{3}\cdot 63$ 2

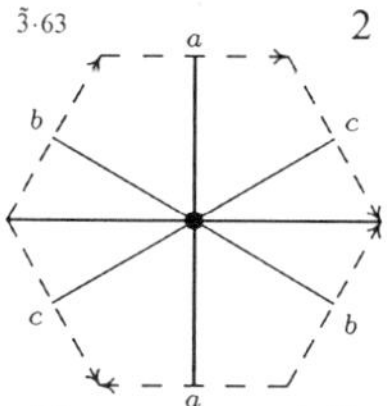

[8] $[4^2]$

$\tilde{3}\cdot 64$ 8

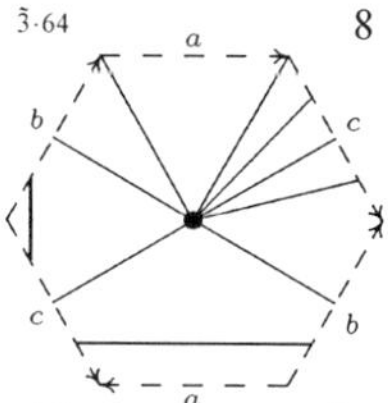

[8] $[4^2]$

$\tilde{3}\cdot 65$ 8

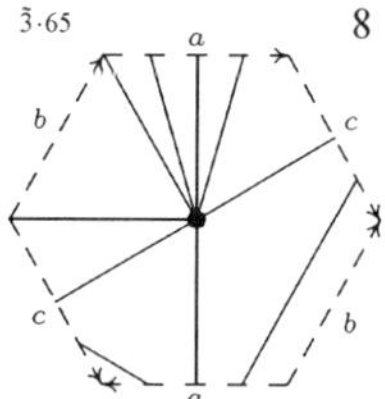

[8] $[4^2]$

$\tilde{3}\cdot 66$ 8

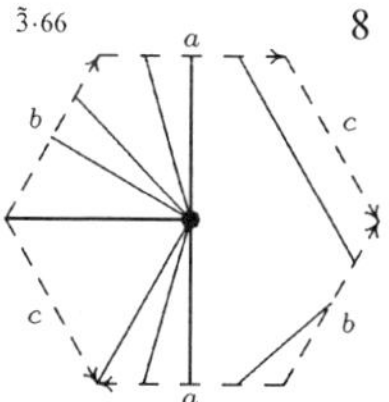

[8] $[4^2]$

$\tilde{3}\cdot 67$ 8

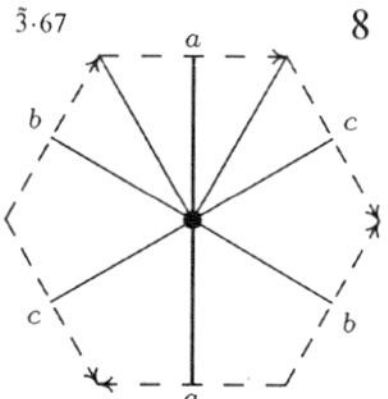

[8] $[4^2]$

$\tilde{3}\cdot 68$ 8

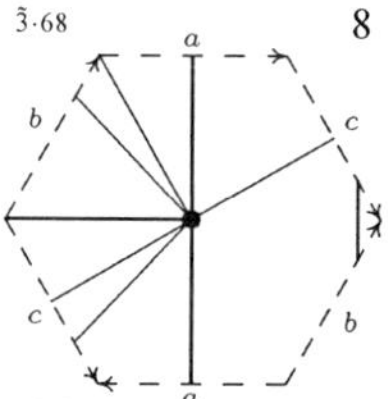

[8] $[4^2]$

$\tilde{3}\cdot 69$ 4

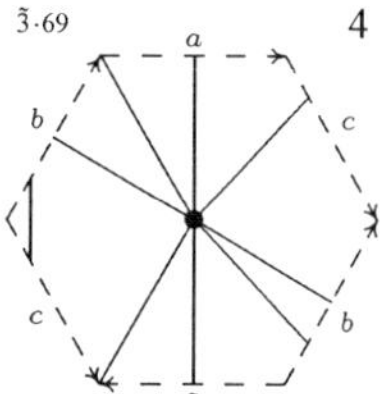

[8] $[4^2]$

$\tilde{3}\cdot 70$ 4

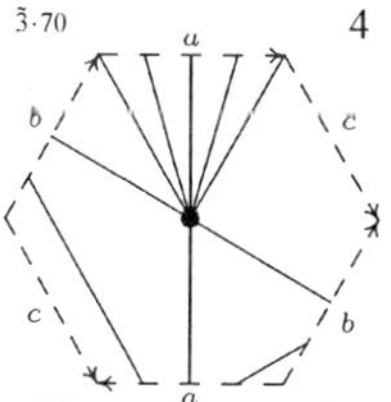

[8] $[4^2]$

$\tilde{3}\cdot 71$ 4

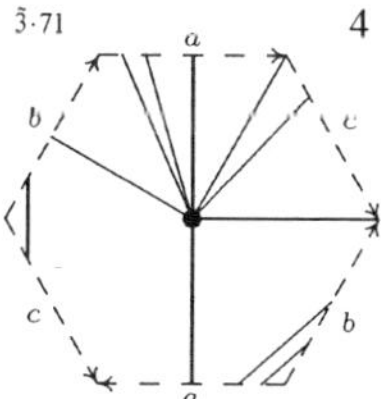

[8] $[4^2]$

$\tilde{3}\cdot 72$ 8

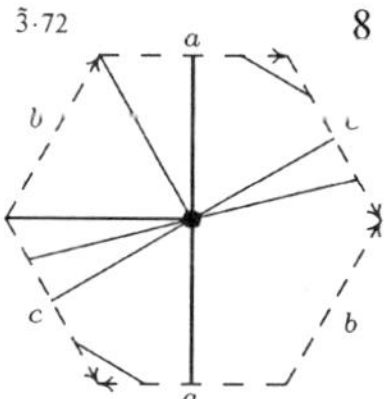

[8] $[4^2]$

7.3 Hypermaps

This collection includes all hypermaps on up to 5 edges in orientable surfaces and up to 4 edges in non-orientable surfaces. The pair of partitions in the bottom right-hand corner are the hyperedge- and hyperface-partitions. If a hypermap is symmetric, it has only one colouring, and the number of rootings is given in the top right-hand corner. If a hypermap is not symmetric, it has two distinct colourings, and the numbers of rootings for each are given in the top right-hand corner.

7.3.1 Orientable of genus 0, 1 and 2

Genus 0 – the sphere

1 edge

1 vertex

2 edges

1 vertex

2 edges

2 vertices

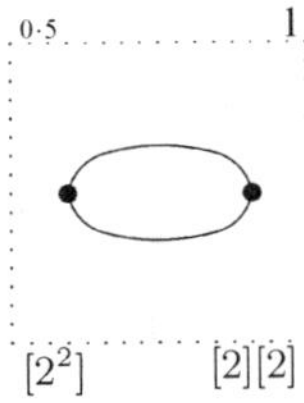

3 edges

1 vertex

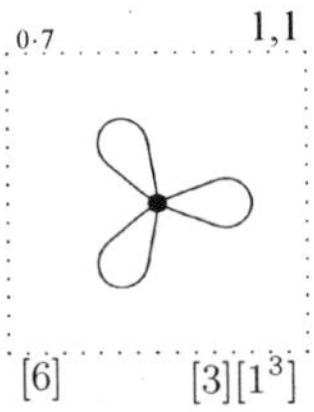

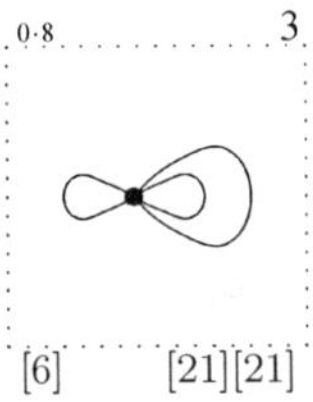

3 edges

2 vertices

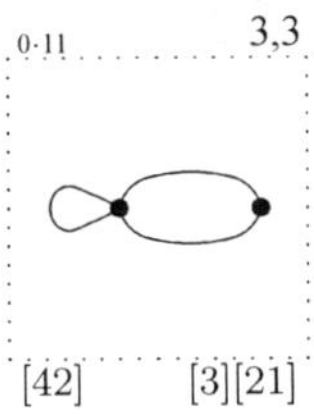

3 edges

3 vertices

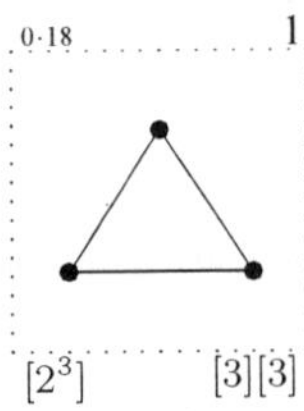

4 edges

1 vertex

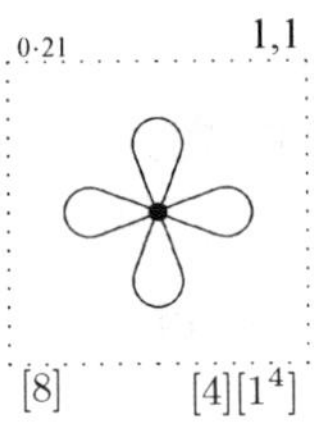

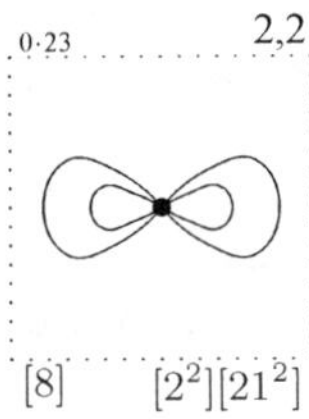

4 edges
2 vertices

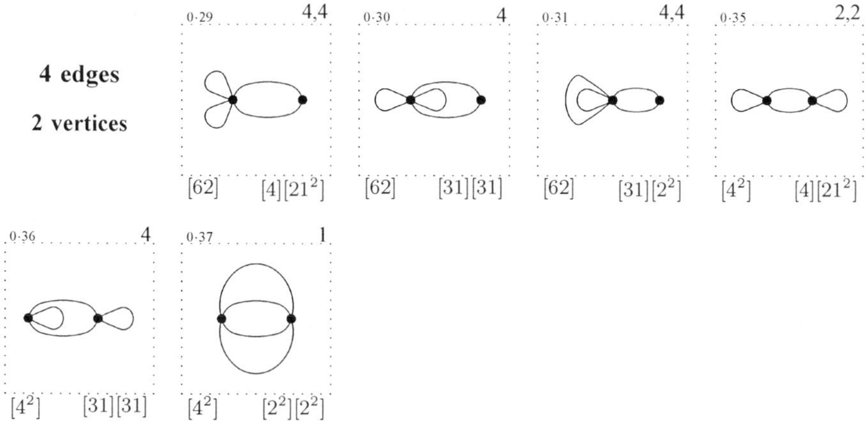

4 edges
3 vertices

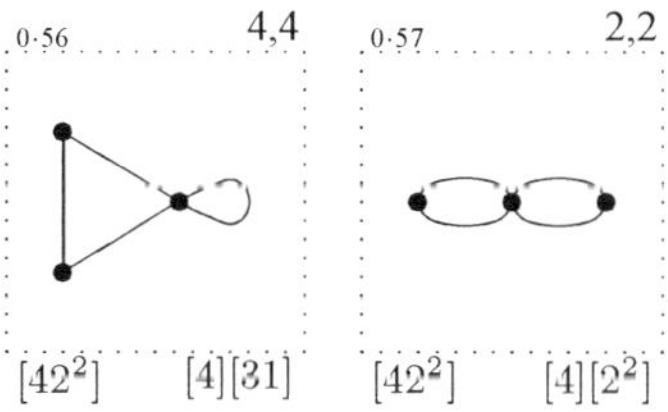

4 edges
4 vertices

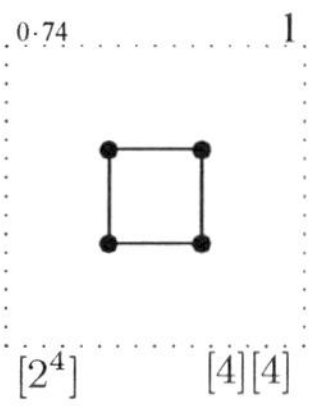

5 edges
1 vertex

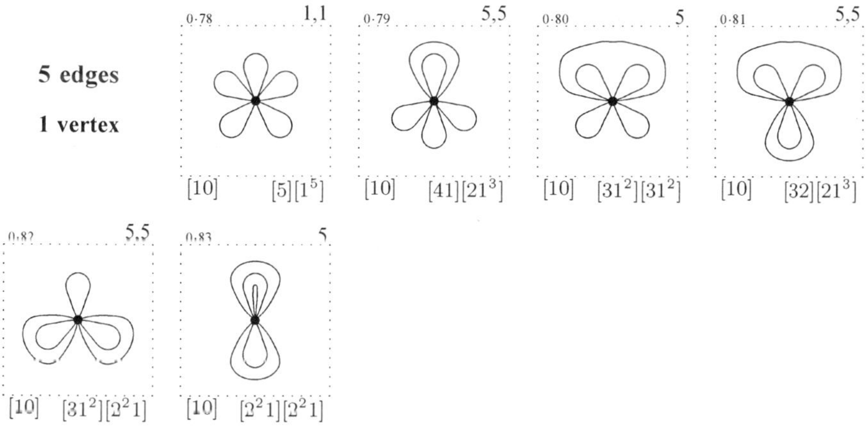

5 edges

2 vertices

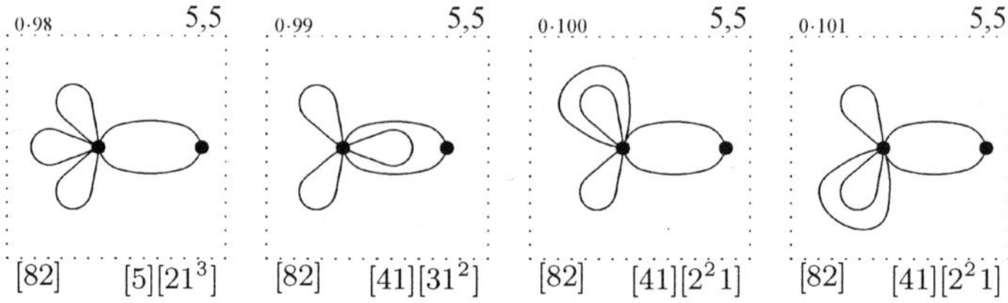

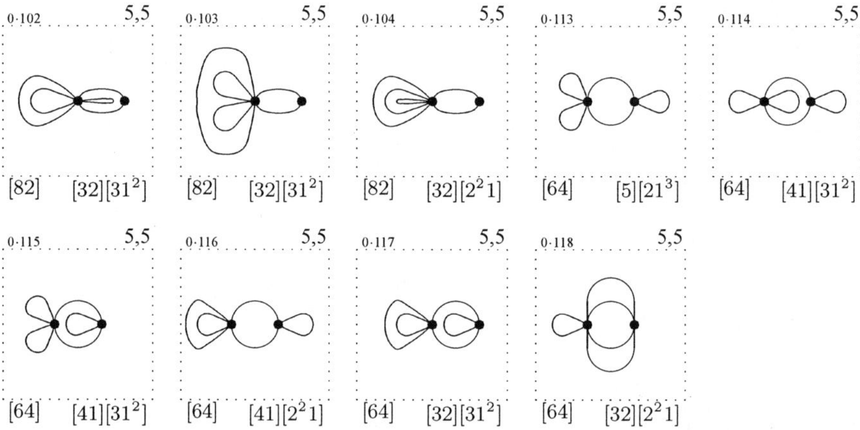

5 edges

3 vertices

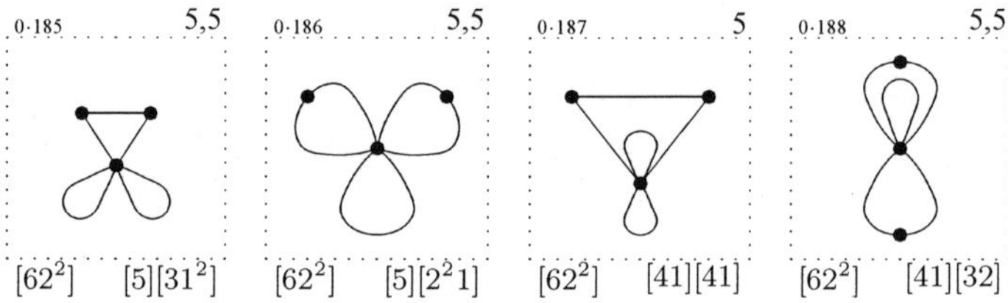

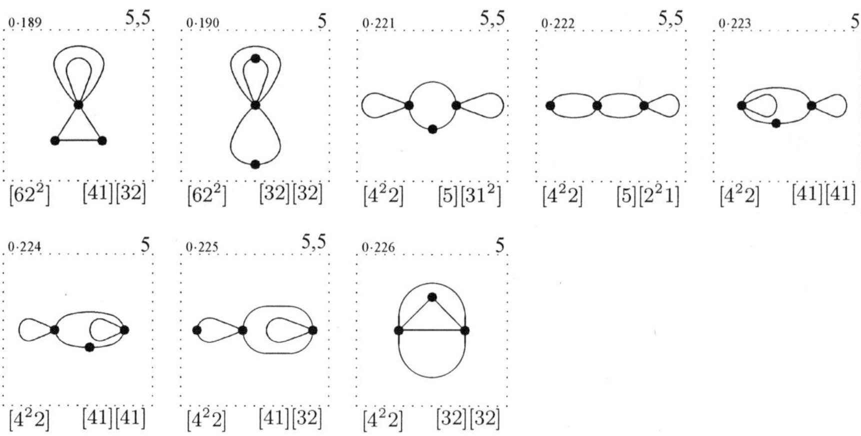

5 edges

4 vertices

0·330 5,5

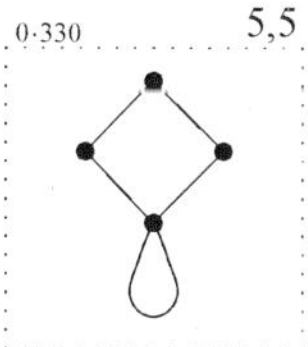

$[42^3]$ [5][41]

0·331 5,5

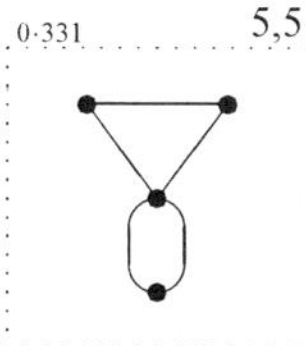

$[42^3]$ [5][32]

5 edges

5 vertices

0·383 1

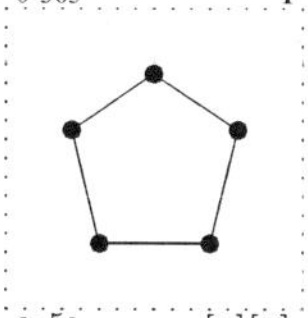

$[2^5]$ [5][5]

Genus 1 – the torus

3 edges

1 vertex

1·4 1

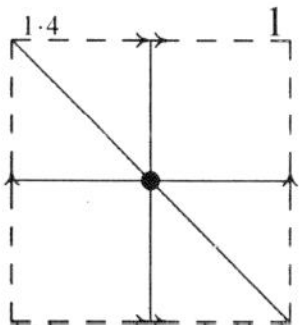

[6] [3][3]

4 edges

1 vertex

1·14 4,4

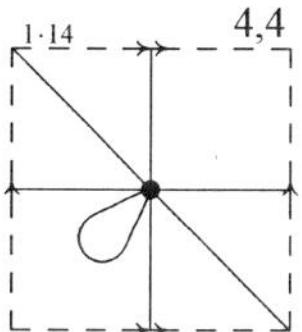

[8] [4][31]

1·16 1,1

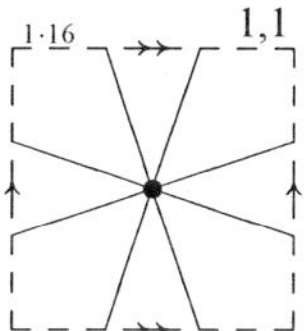

[8] $[4][2^2]$

4 edges

2 vertices

1·34 4

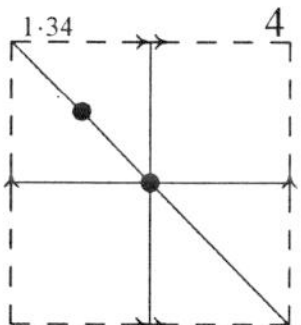

[62] [4][4]

1·41 1

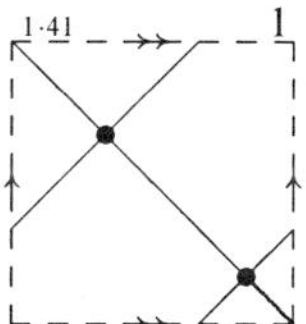

$[4^2]$ [4][4]

5 edges

1 vertex

1·75 5,5

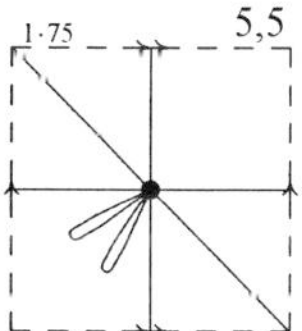

[10] $[5][31^2]$

1·76 5,5

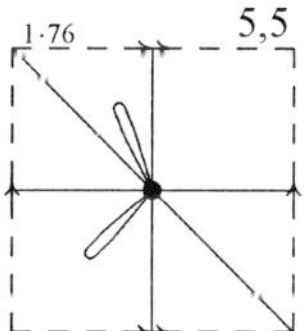

[10] $[5][31^2]$

1·82 5,5

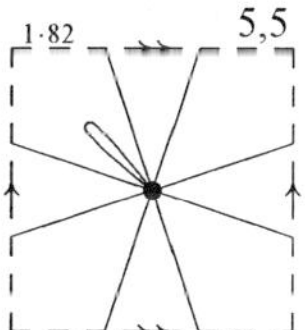

[10] $[5][2^2 1]$

1·85 5,5

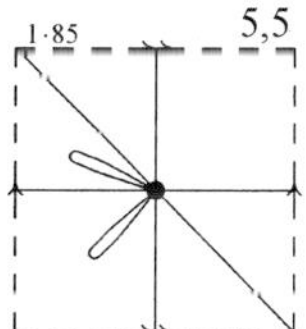

[10] [41][41]

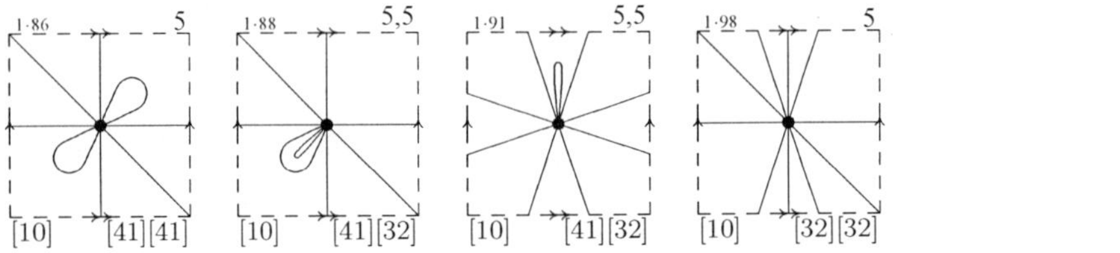

5 edges

2 vertices

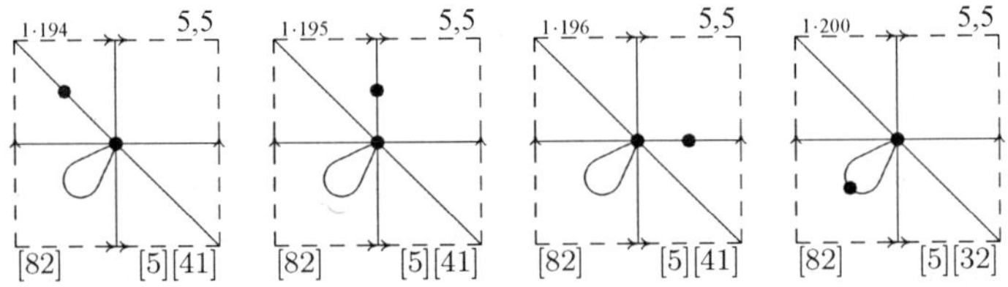

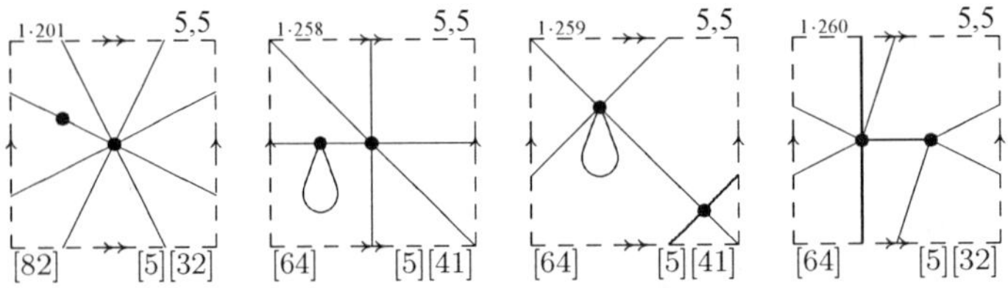

5 edges

3 vertices

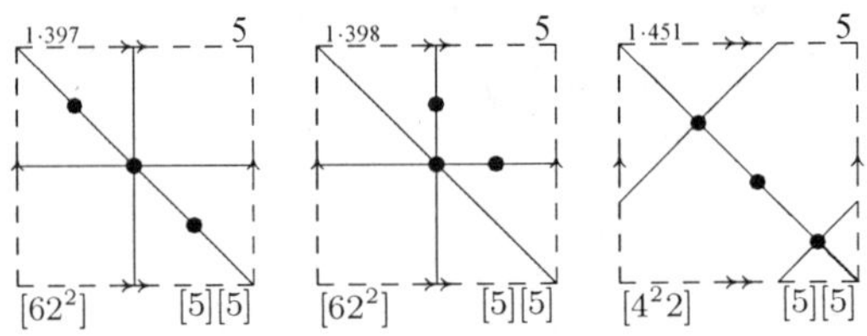

Genus 2 – the double torus

5 edges

1 vertex

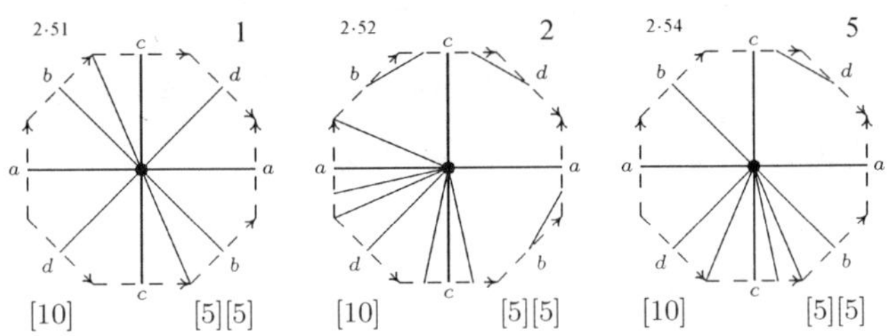

7.3.2 Nonorientable of genus 1, 2 and 3

Genus 1 the projective plane

2 edges

1 vertex

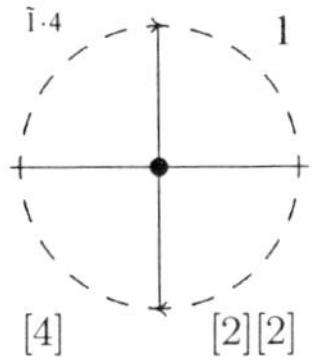

3 edges

1 vertex

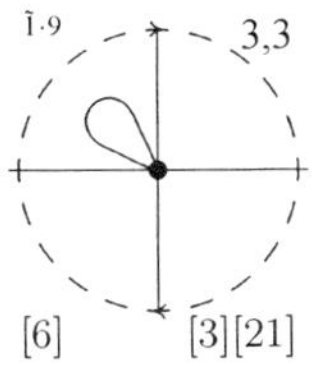

3 edges

2 vertices

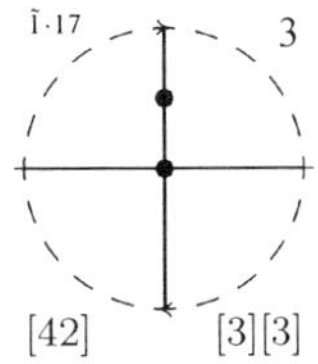

4 edges

1 vertex

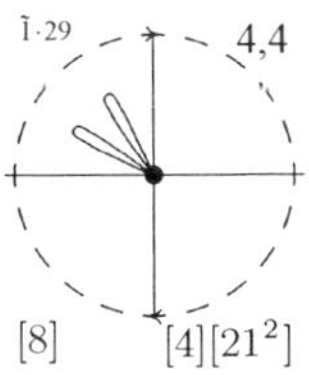

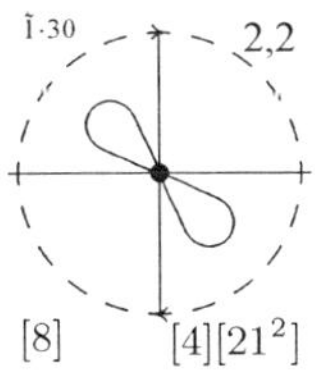

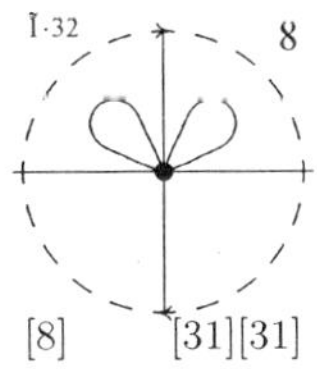

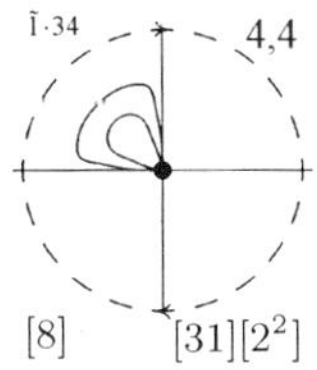

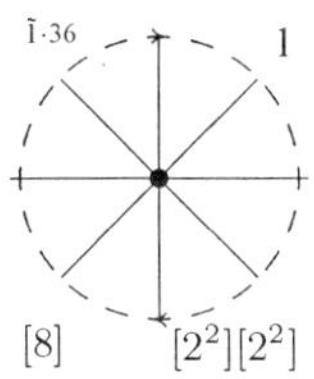

4 edges

2 vertices

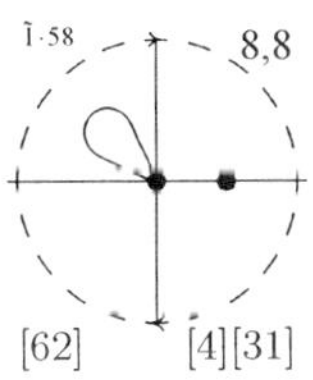

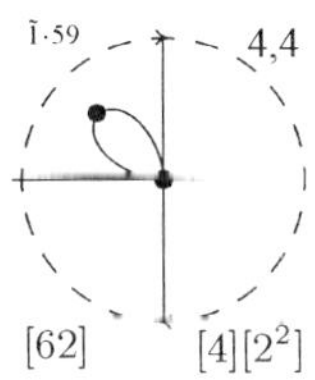

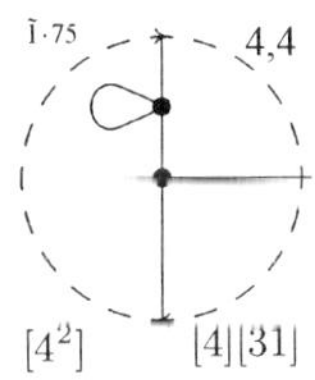

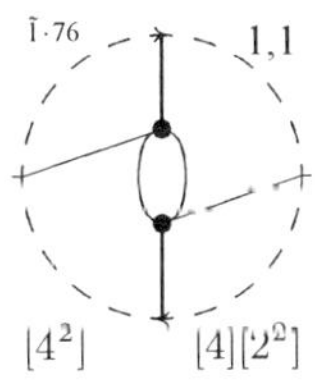

4 edges

3 vertices

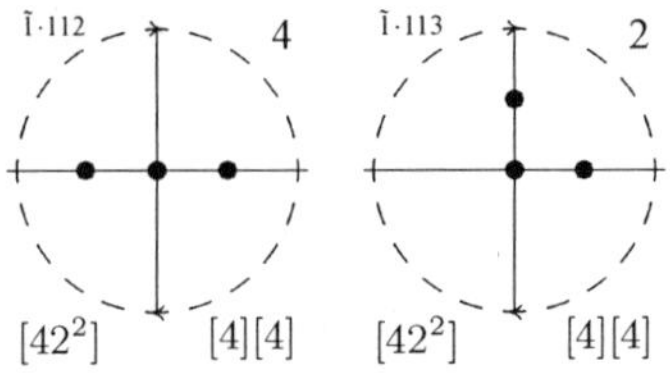

Genus 2 – the Klein bottle

3 edges

1 vertex

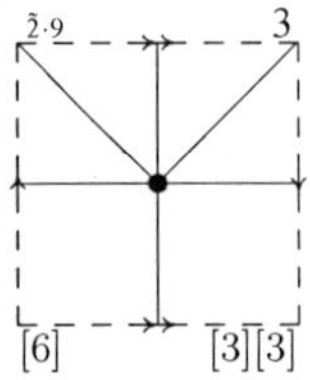

4 edges

1 vertex

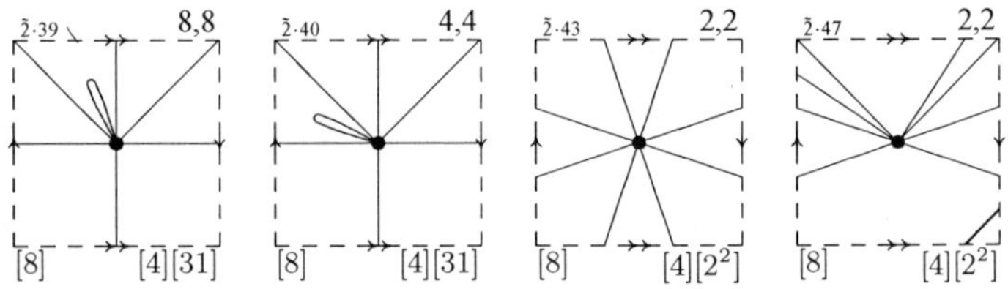

4 edges

2 vertices

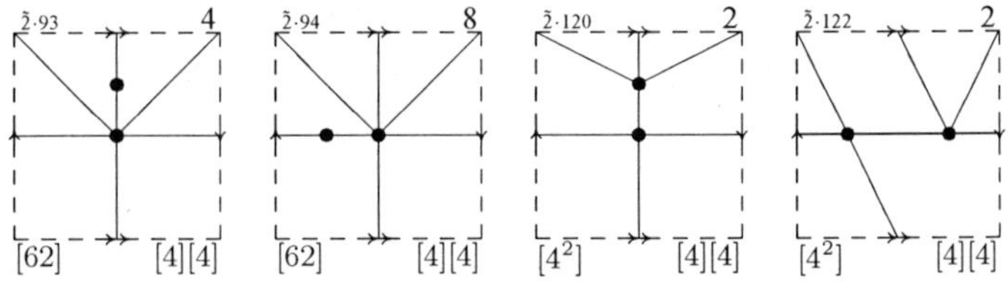

Genus 3 – the crosscapped torus

4 edges

1 vertex

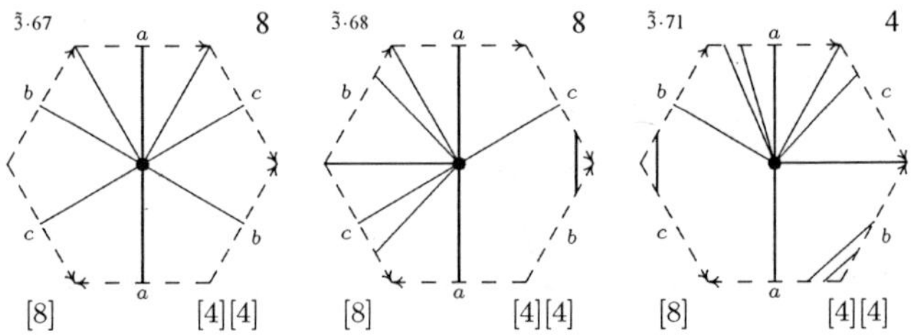

Chapter 8

Associated graphs and their maps

The associated graphs are listed by number of *edges* and *vertices*, where these are listed in decreasing order of subordination. Running footers display this information, and the maps appear in decreasing *lexicographic* order of *vertex partition* and then *face partition*. For each entry in this part of the Atlas we give:

- a *title* giving the number of edges, the number of vertices, and an index of the form $\mathfrak{g}_i$;
- a *list of maps* in orientable surfaces in increasing order of genus. The first map in this list serves two purposes: i) as the *associated graph* of each of the maps in the list, ii) as a *map* of genus 0 of the list (this is feasible since, under the conditions placed on the Atlas each $\mathfrak{g}_i$ is embeddable in the sphere);
- The *genus distribution*, which appears at the end of each list of maps. This information is summarized in Section 12.2.

1 edge
1 vertex
$\mathfrak{g}_1$

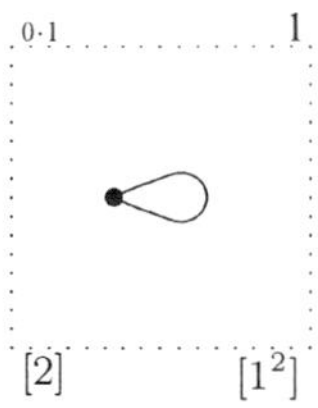

1

1 edge
2 vertices
$\mathfrak{g}_2$

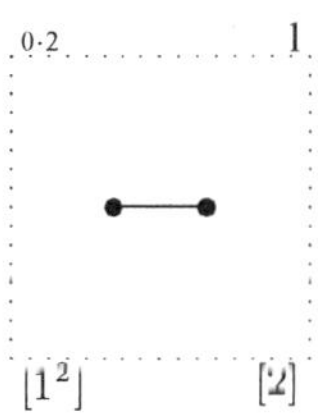

1

2 edges
1 vertex
$\mathfrak{g}_3$

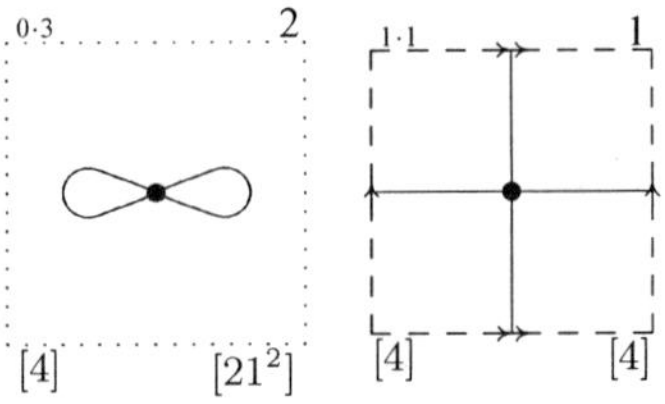

$1 + u$

2 edges
2 vertices
$\mathfrak{g}_4$

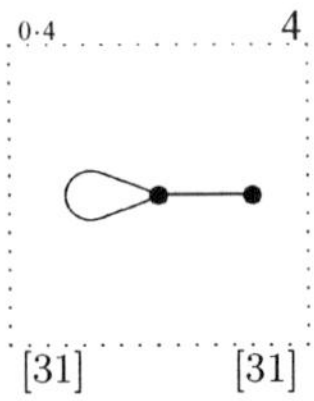

1

2 edges
2 vertices
$\mathfrak{g}_5$

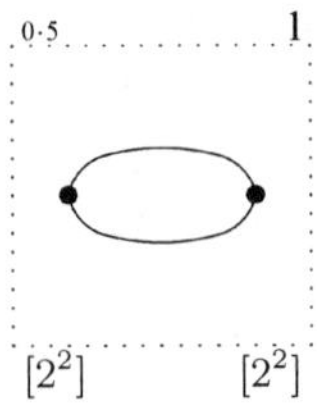

1

2 edges
3 vertices
$\mathfrak{g}_6$

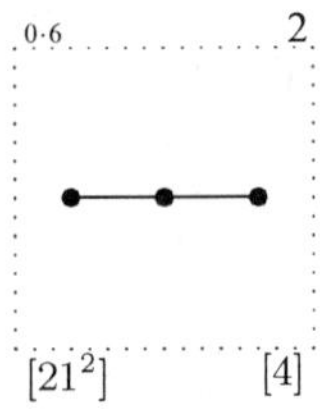

1

3 edges
1 vertex
$\mathfrak{g}_7$

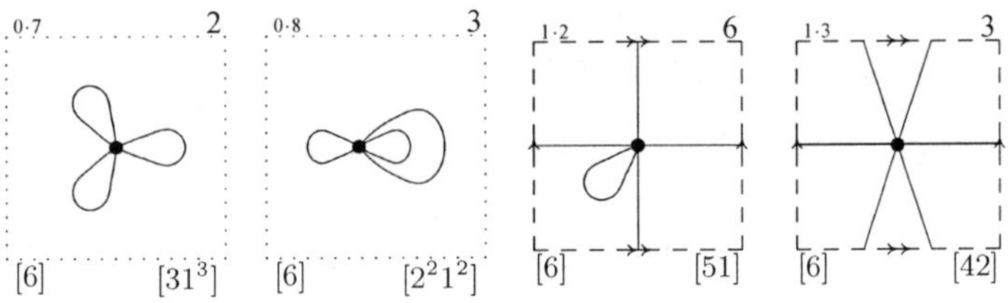

3e 2v

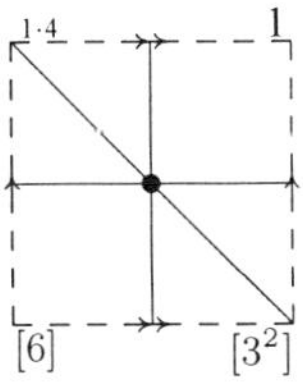

$2 + 3u$

3 edges
2 vertices
$\mathfrak{g}_8$

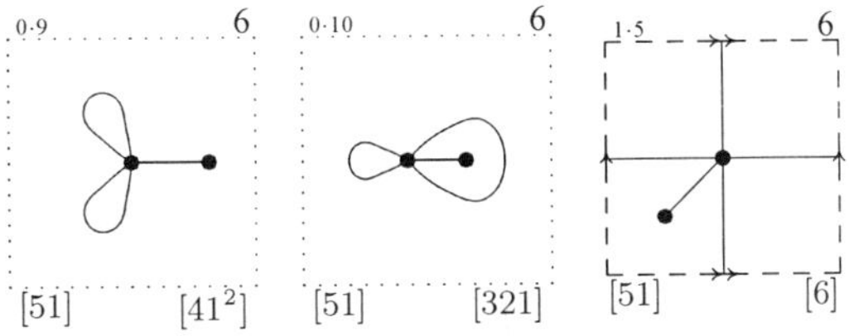

$2 + u$

3 edges
2 vertices
$\mathfrak{g}_9$

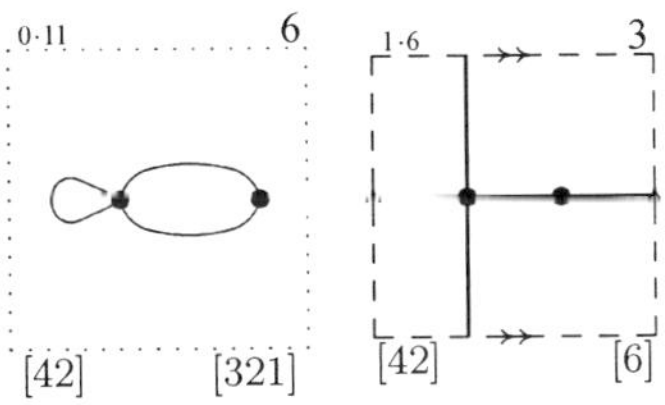

$1 + u$

3 edges
2 vertices
$\mathfrak{g}_{10}$

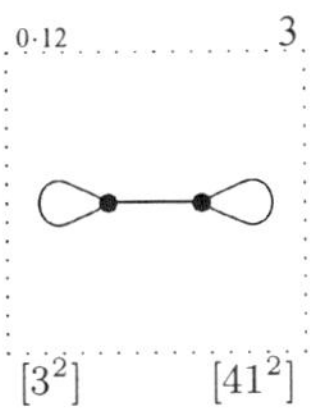

1

3 edges
2 vertices
$\mathfrak{g}_{11}$

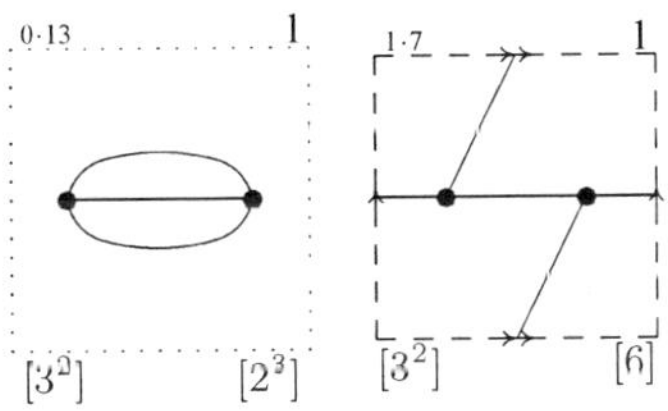

$1 + u$

3e 3v

3 edges
3 vertices
$\mathfrak{g}_{12}$

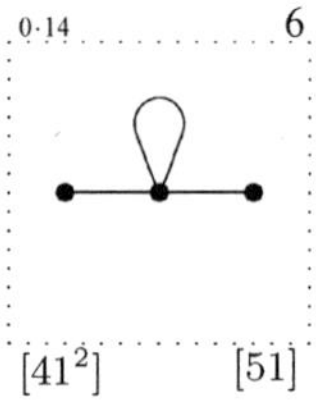

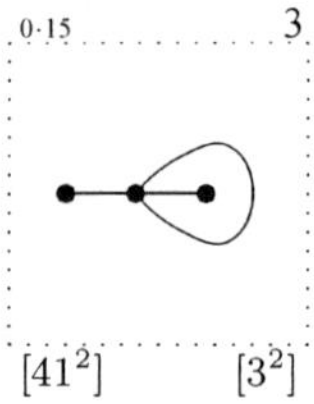

2

3 edges
3 vertices
$\mathfrak{g}_{13}$

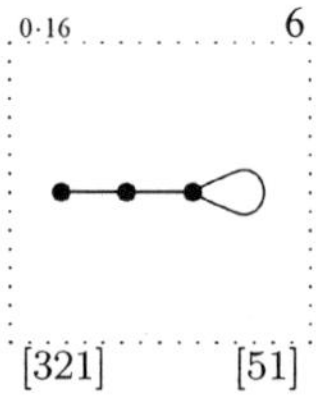

1

3 edges
3 vertices
$\mathfrak{g}_{14}$

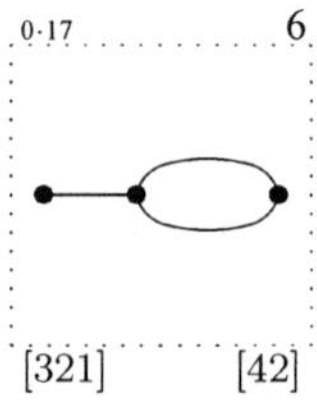

1

3 edges
3 vertices
$\mathfrak{g}_{15}$

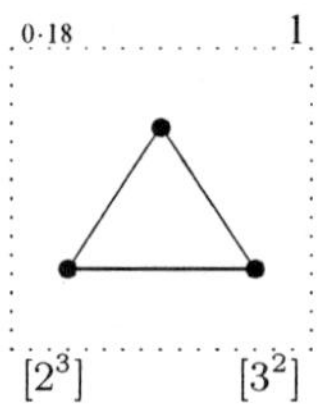

1

3 edges
4 vertices
$\mathfrak{g}_{16}$

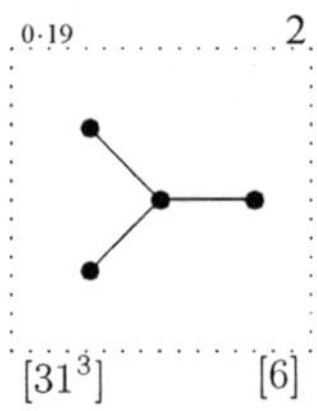

1

4e 1v

3 edges
4 vertices
$\mathfrak{g}_{17}$

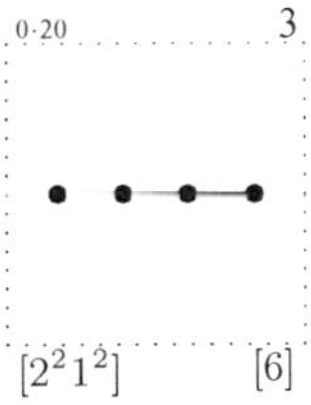

1

4 edges
1 vertex
$\mathfrak{g}_{18}$

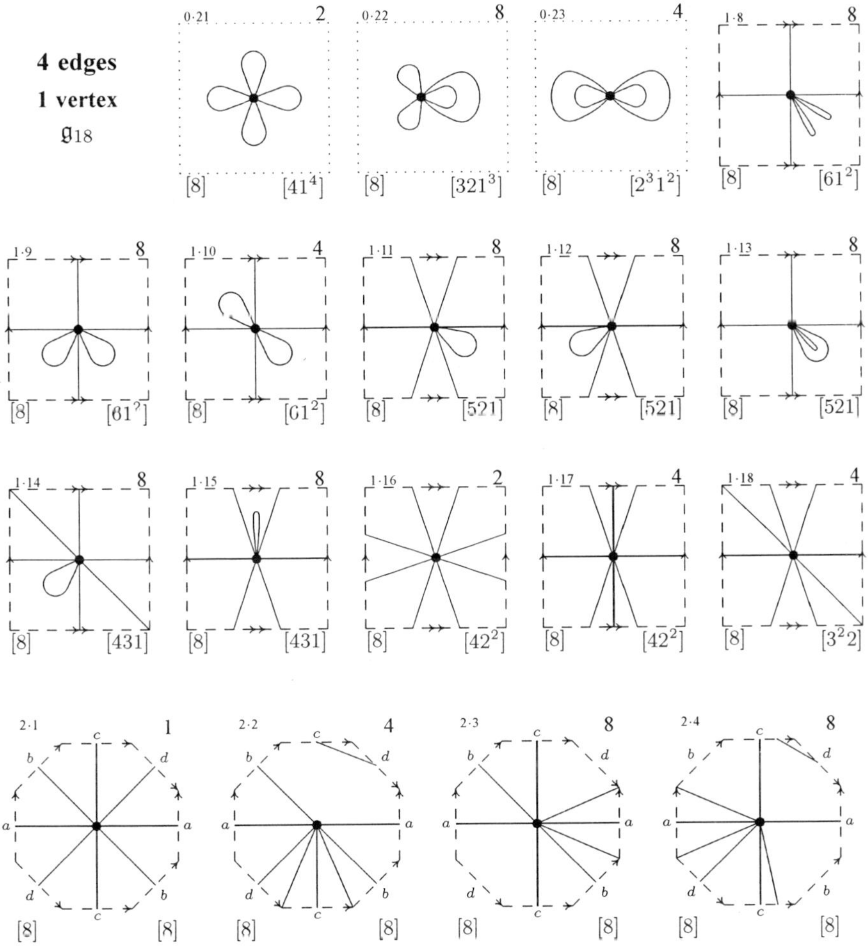

$3 + 11u + 4u^2$

4e 2v

4 edges
2 vertices
$\mathfrak{g}_{19}$

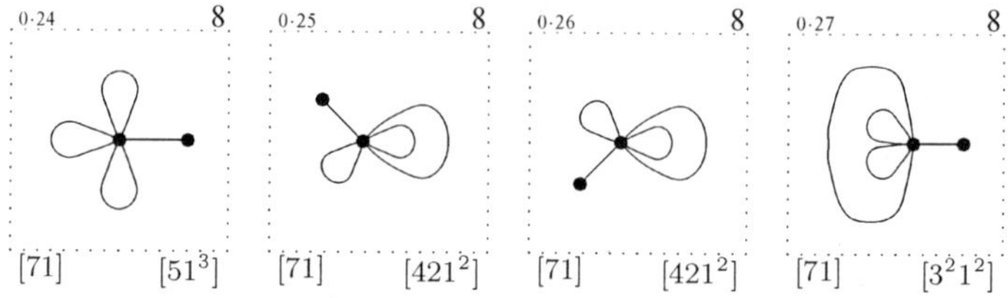

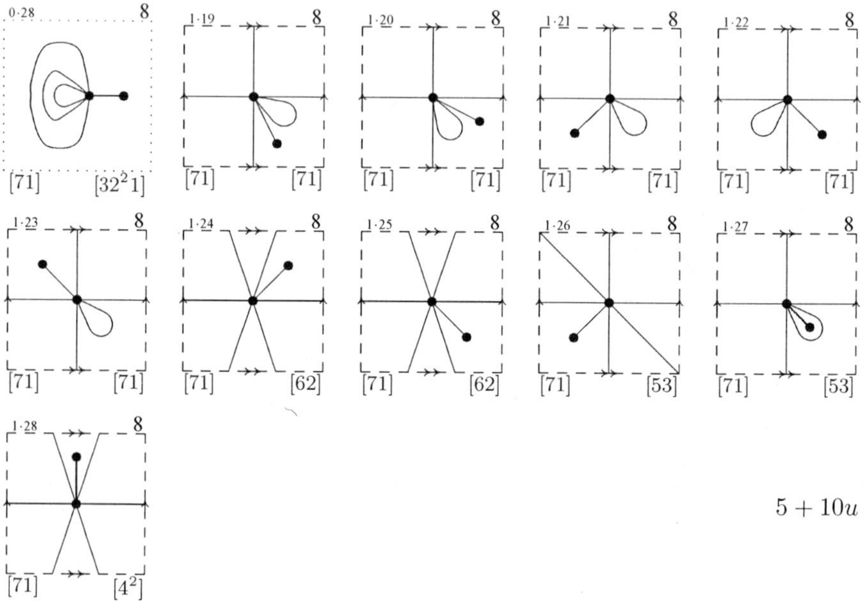

$5 + 10u$

4 edges
2 vertices
$\mathfrak{g}_{20}$

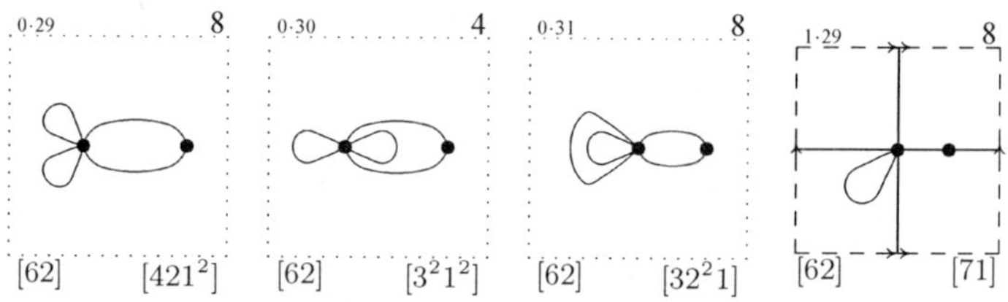

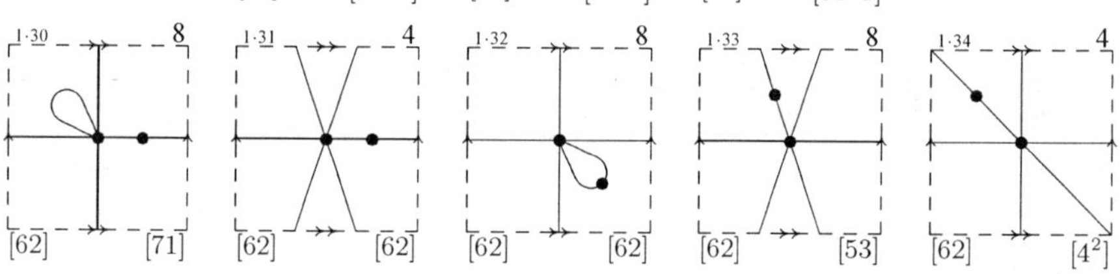

$3 + 6u$

4 edges
2 vertices
$\mathfrak{g}_{21}$

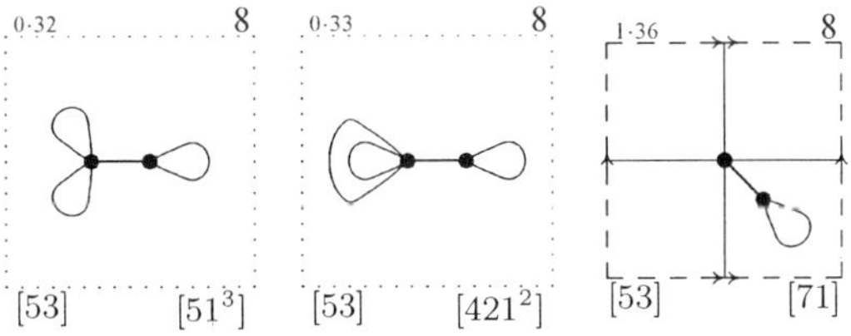

$2 + u$

4 edges
2 vertices
$\mathfrak{g}_{22}$

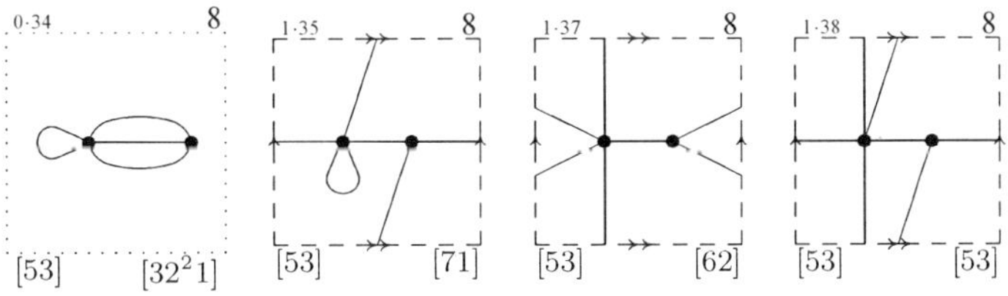

$1 + 3u$

4 edges
2 vertices
$\mathfrak{g}_{23}$

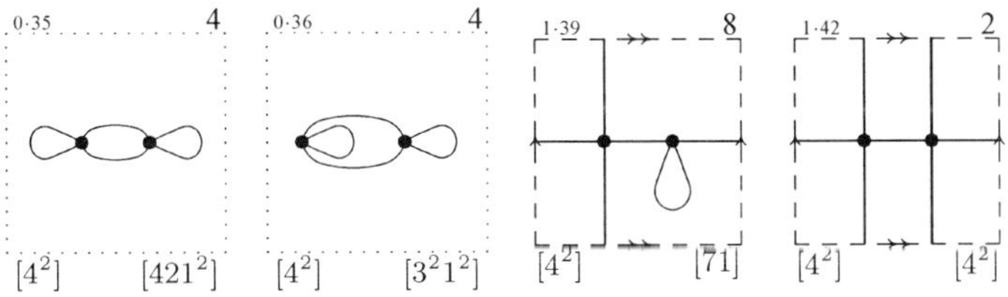

$2 + 2u$

4 edges
2 vertices
$\mathfrak{g}_{24}$

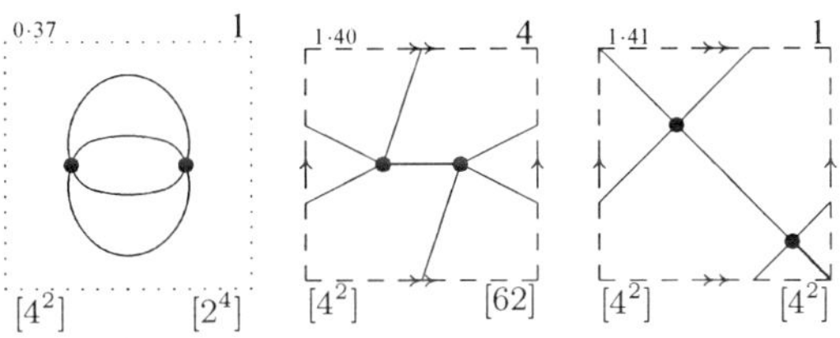

$1 + 2u$

4 edges
3 vertices
$\mathfrak{g}_{25}$

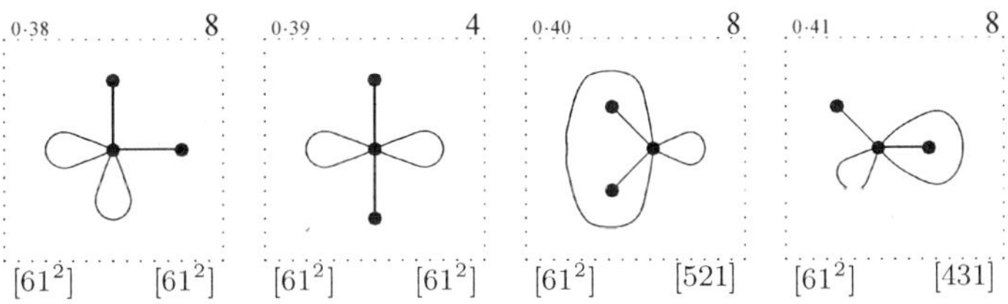

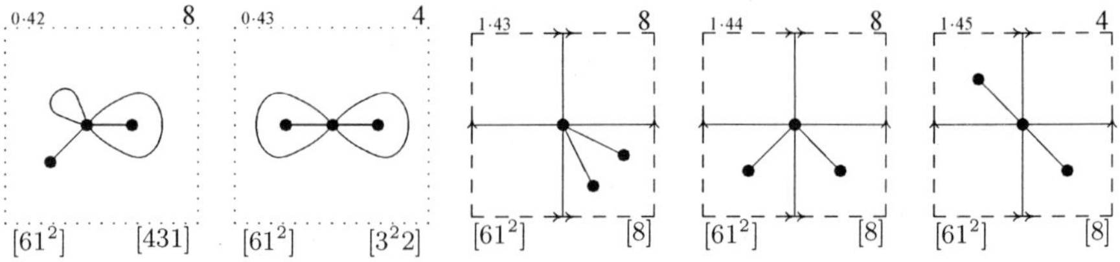

$6 + 3u$

4 edges
3 vertices
$\mathfrak{g}_{26}$

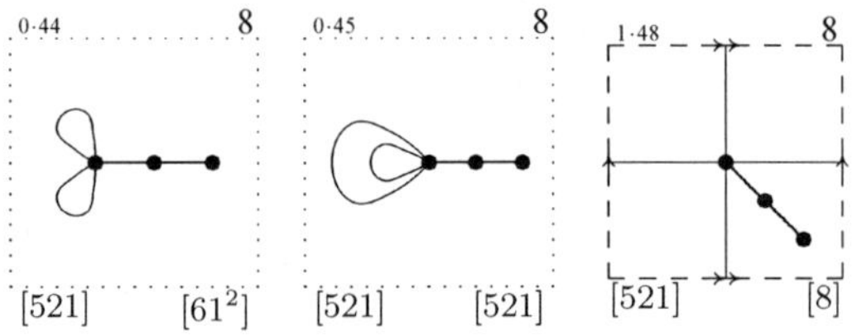

$2 + u$

4 edges
3 vertices
$\mathfrak{g}_{27}$

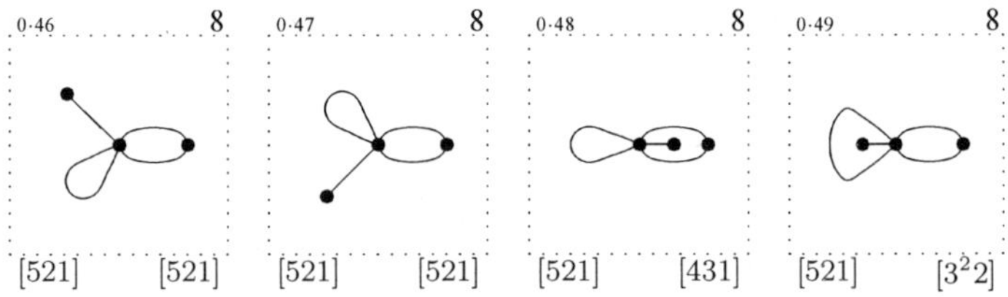

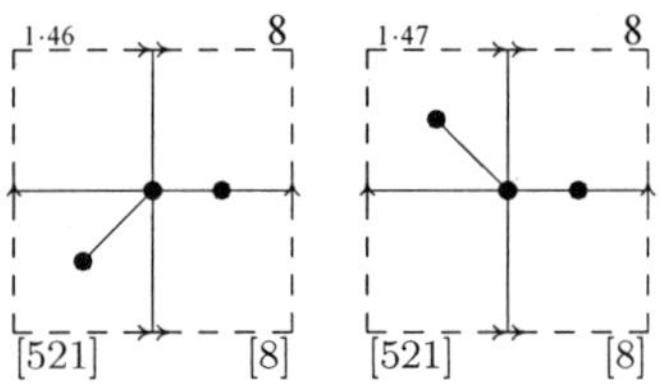

$4 + 2u$

4 edges
3 vertices
$\mathfrak{g}_{28}$

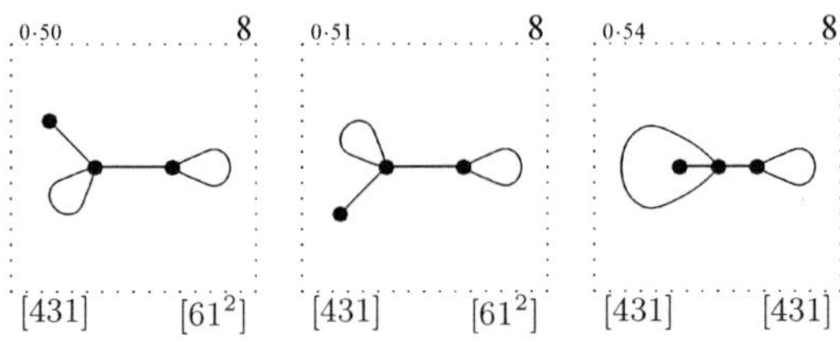

3

4 edges
3 vertices
$\mathfrak{g}_{29}$

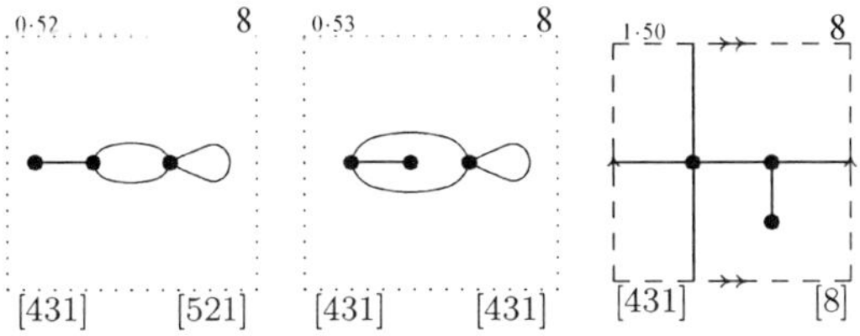

$2 + u$

4 edges
3 vertices
$\mathfrak{g}_{30}$

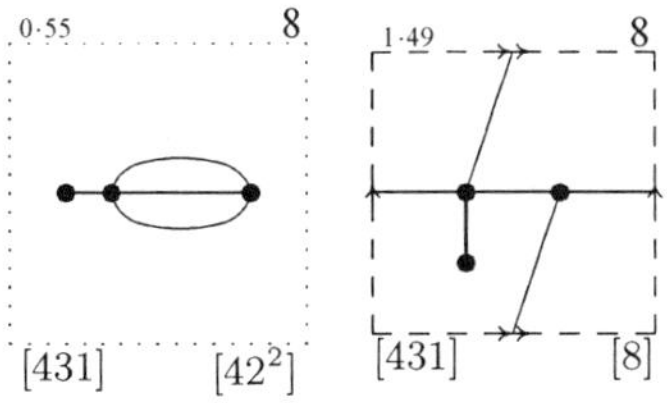

$1 + u$

4 edges
3 vertices
$\mathfrak{g}_{31}$

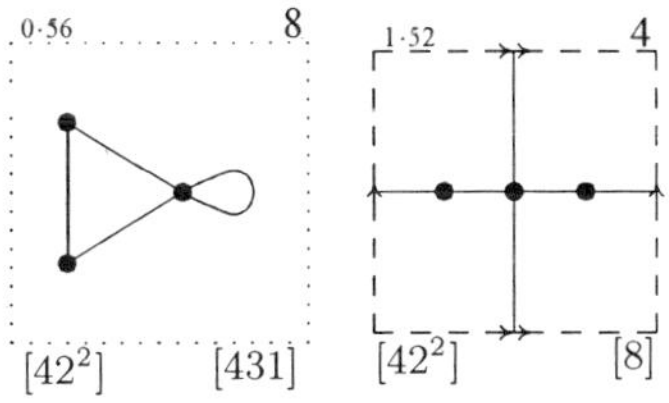

$1 + u$

4 edges
3 vertices
$\mathfrak{g}_{32}$

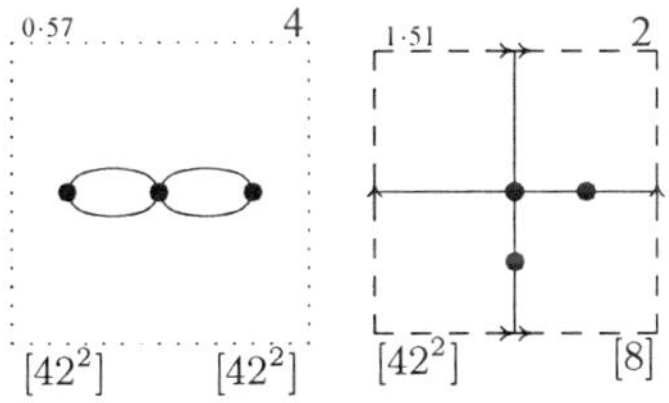

$1 + u$

4 edges
3 vertices
$\mathfrak{g}_{33}$

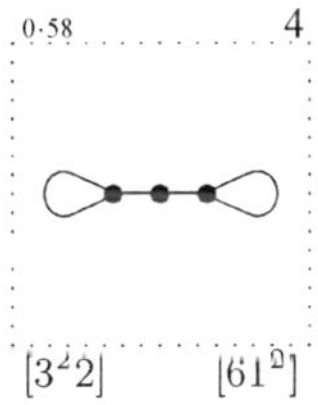

1

4 edges
3 vertices
$\mathfrak{g}_{34}$

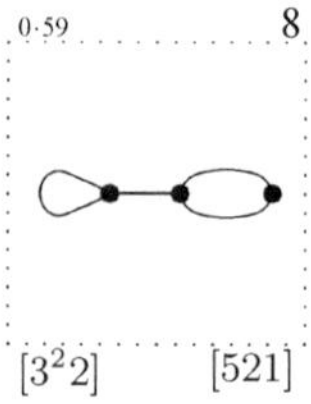

1

4 edges
3 vertices
$\mathfrak{g}_{35}$

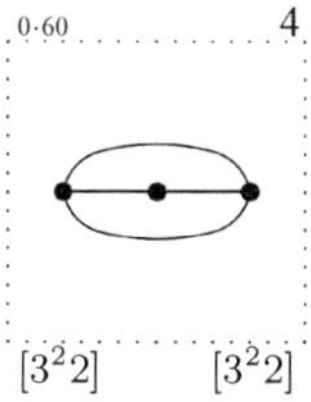

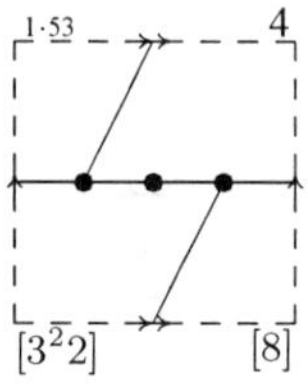

$1 + u$

4 edges
4 vertices
$\mathfrak{g}_{36}$

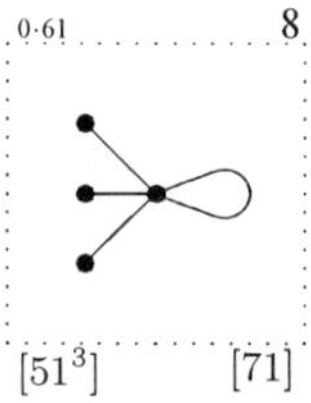

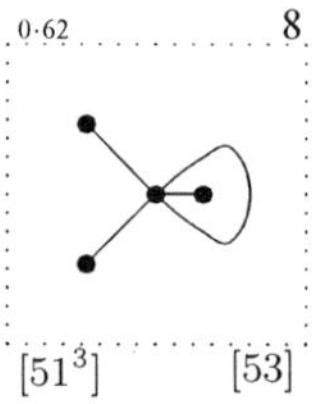

2

4 edges
4 vertices
$\mathfrak{g}_{37}$

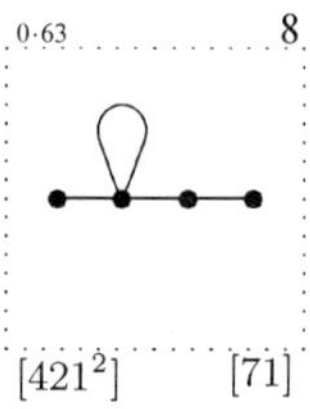

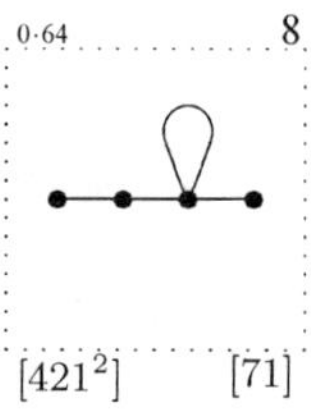

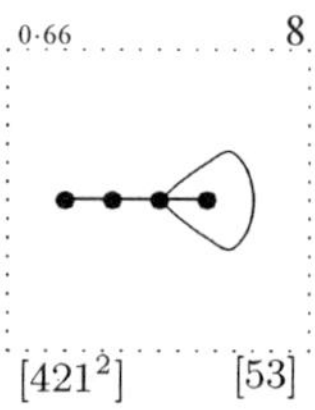

3

4 edges
4 vertices
$\mathfrak{g}_{38}$

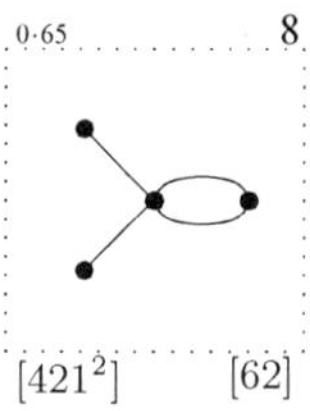

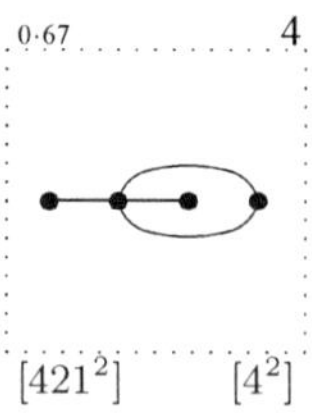

2

4 edges
4 vertices
$\mathfrak{g}_{39}$

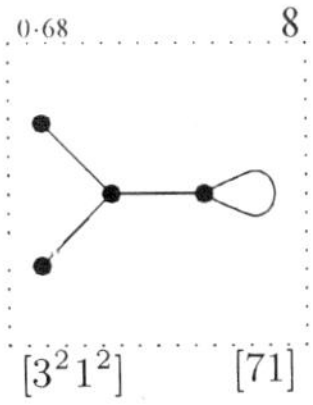

1

4 edges
4 vertices
$\mathfrak{g}_{40}$

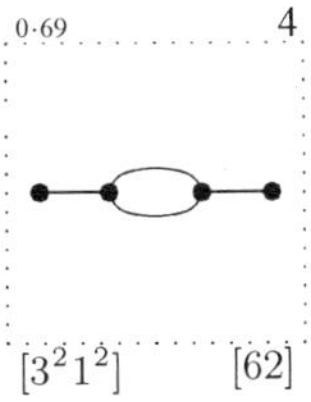

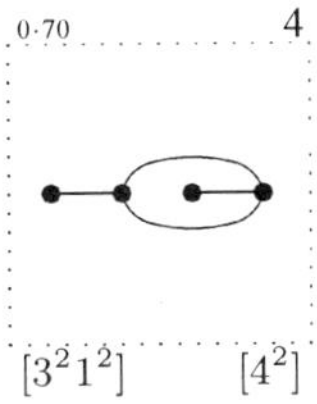

2

4 edges
4 vertices
$\mathfrak{g}_{41}$

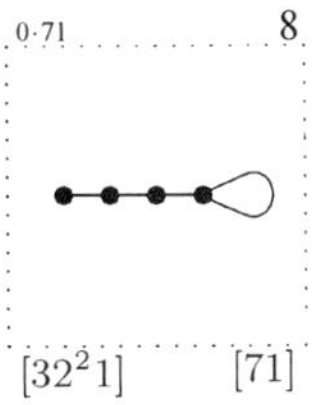

1

4 edges
4 vertices
$\mathfrak{g}_{42}$

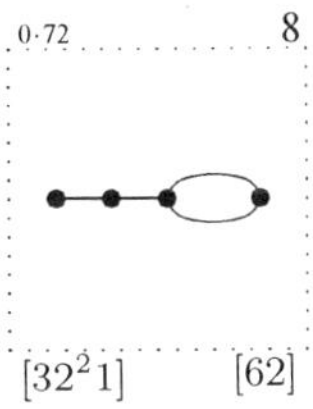

1

4 edges
4 vertices
$\mathfrak{g}_{43}$

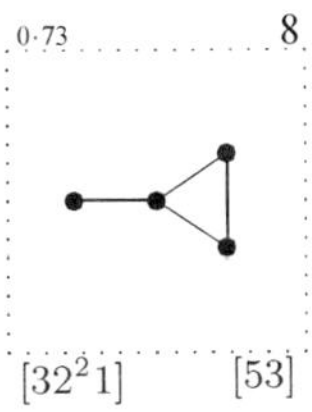

1

4 edges
4 vertices
$\mathfrak{g}_{44}$

0·74 1
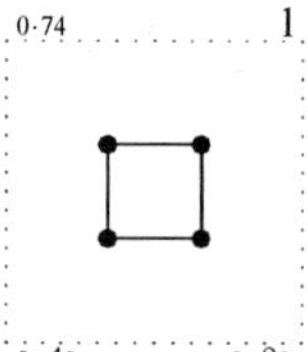
$[2^4]$ $[4^2]$

1

4 edges
5 vertices
$\mathfrak{g}_{45}$

0·75 2
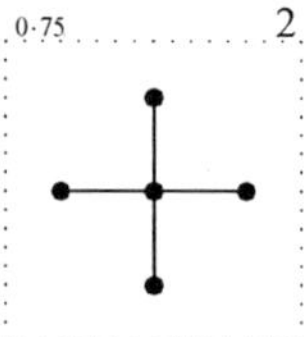
$[41^4]$ $[8]$

1

4 edges
5 vertices
$\mathfrak{g}_{46}$

0·76 8
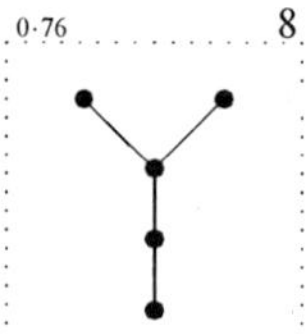
$[321^3]$ $[8]$

1

4 edges
5 vertices
$\mathfrak{g}_{47}$

0·77 4
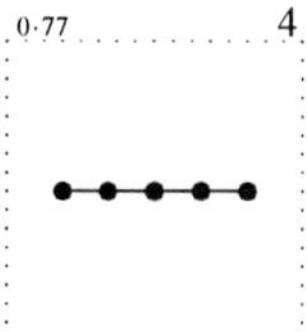
$[2^3 1^2]$ $[8]$

1

5 edges
1 vertex
$\mathfrak{g}_{48}$

0·78 2

$[10]$ $[51^5]$

0·79 10
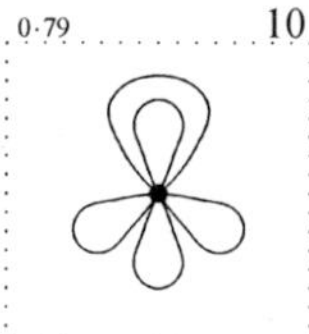
$[10]$ $[421^4]$

0·80 5
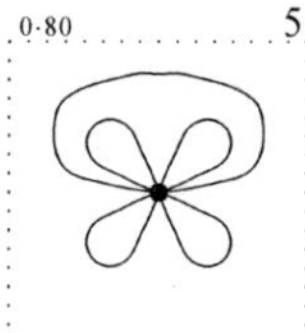
$[10]$ $[3^2 1^4]$

0·81 10
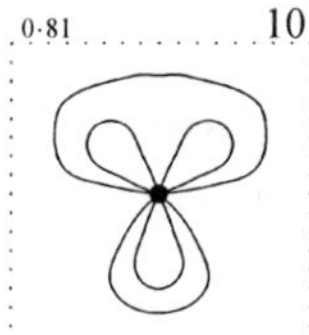
$[10]$ $[32^2 1^3]$

5e 1v

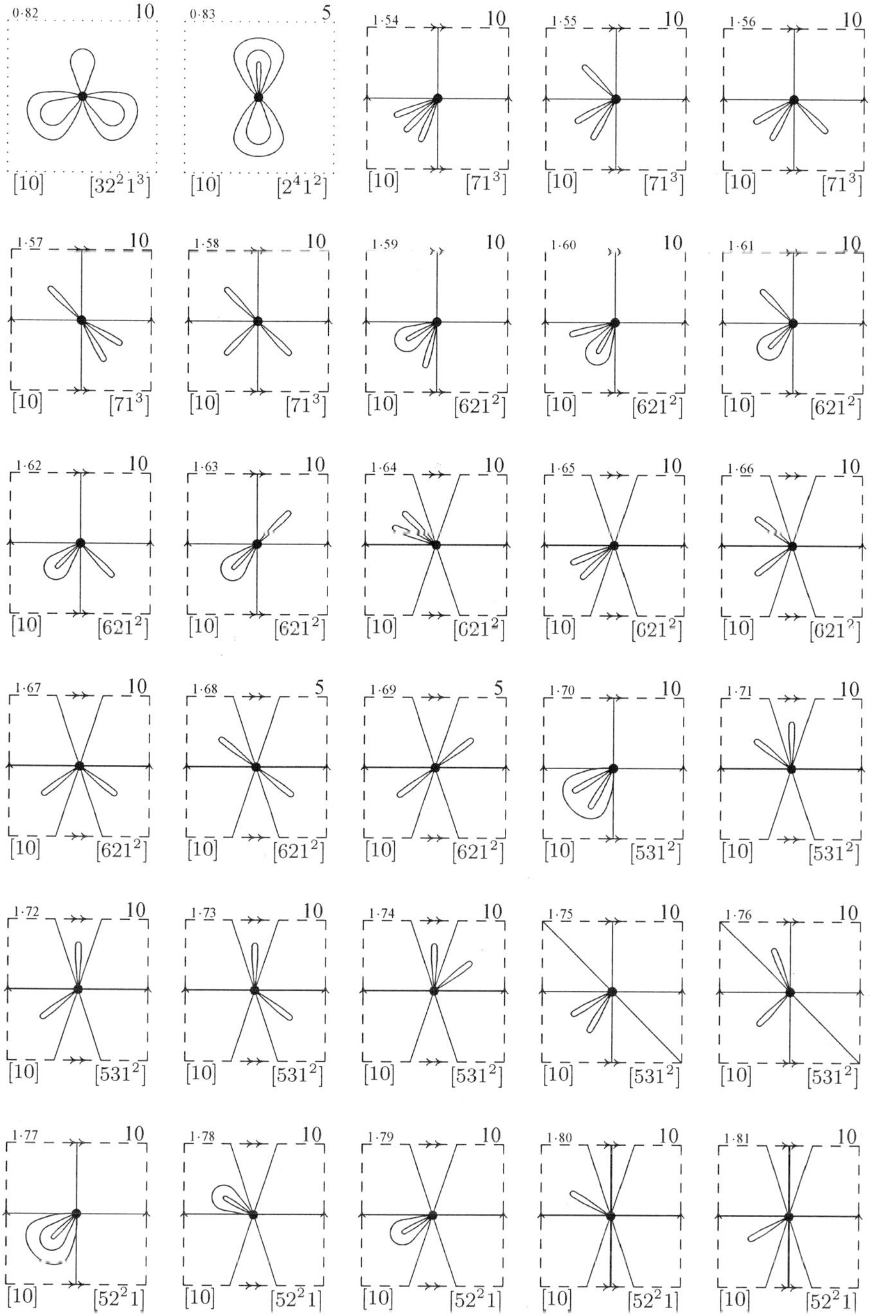
0·82 10 [10] [32²1³]
0·83 5 [10] [2⁴1²]
1·54 10 [10] [71³]
1·55 10 [10] [71³]
1·56 10 [10] [71³]
1·57 10 [10] [71³]
1·58 10 [10] [71³]
1·59 10 [10] [621²]
1·60 10 [10] [621²]
1·61 10 [10] [621²]
1·62 10 [10] [621²]
1·63 10 [10] [621²]
1·64 10 [10] [621²]
1·65 10 [10] [621²]
1·66 10 [10] [621²]
1·67 10 [10] [621²]
1·68 5 [10] [621²]
1·69 5 [10] [621²]
1·70 10 [10] [531²]
1·71 10 [10] [531²]
1·72 10 [10] [531²]
1·73 10 [10] [531²]
1·74 10 [10] [531²]
1·75 10 [10] [531²]
1·76 10 [10] [531²]
1·77 10 [10] [52²1]
1·78 10 [10] [52²1]
1·79 10 [10] [52²1]
1·80 10 [10] [52²1]
1·81 10 [10] [52²1]

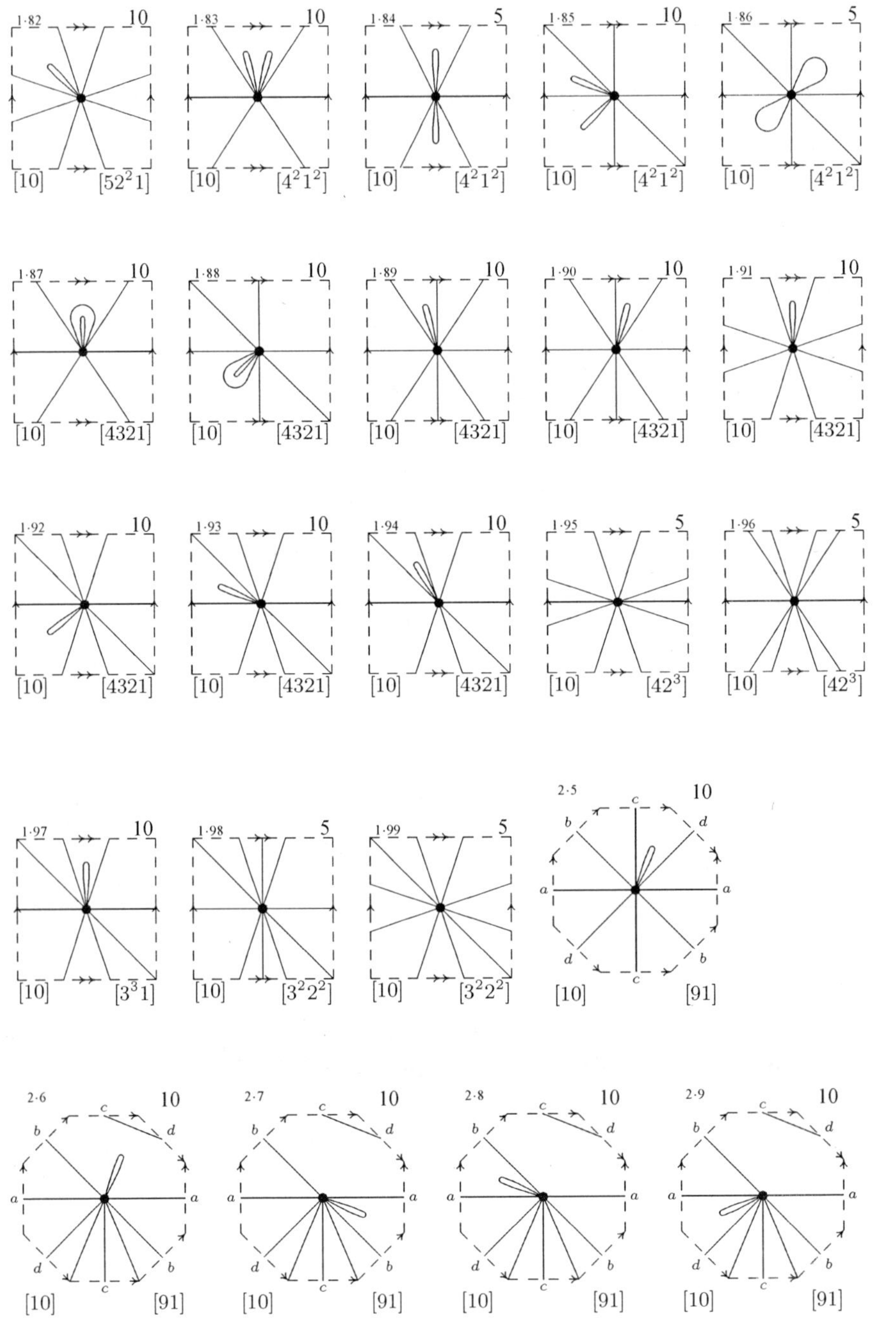
1·82 10 [10] [52²1]
1·83 10 [10] [4²1²]
1·84 5 [10] [4²1²]
1·85 10 [10] [4²1²]
1·86 5 [10] [4²1²]
1·87 10 [10] [4321]
1·88 10 [10] [4321]
1·89 10 [10] [4321]
1·90 10 [10] [4321]
1·91 10 [10] [4321]
1·92 10 [10] [4321]
1·93 10 [10] [4321]
1·94 10 [10] [4321]
1·95 5 [10] [42³]
1·96 5 [10] [42³]
1·97 10 [10] [3³1]
1·98 5 [10] [3²2²]
1·99 5 [10] [3²2²]
2·5 10 [10] [91]
2·6 10 [10] [91]
2·7 10 [10] [91]
2·8 10 [10] [91]
2·9 10 [10] [91]

5e 1v

2·10 10

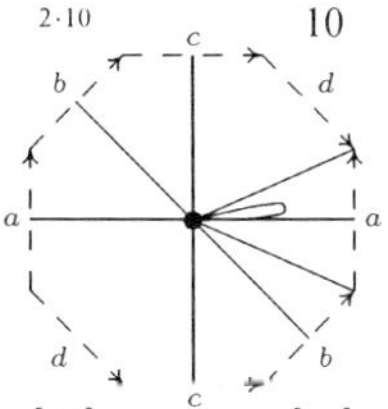

[10] [91]

2·11 10

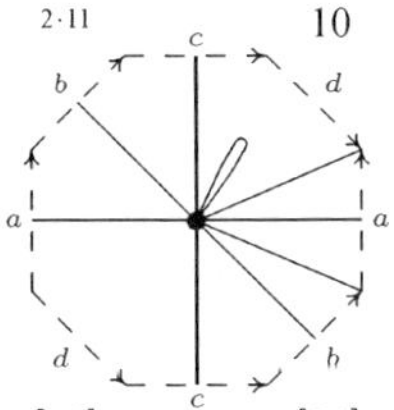

[10] [91]

2·12 10

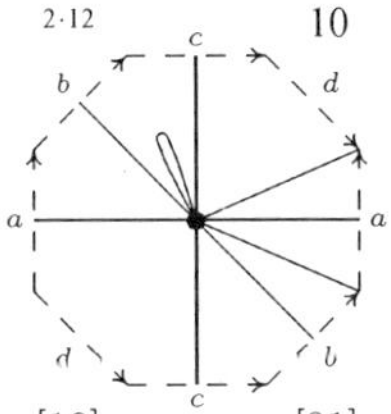

[10] [91]

2·13 10

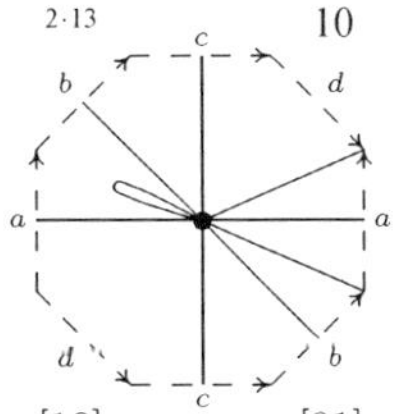

[10] [91]

2·14 10

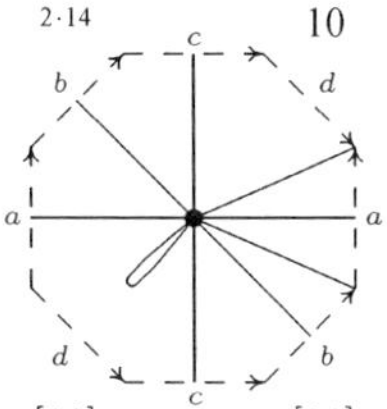

[10] [91]

2·15 10

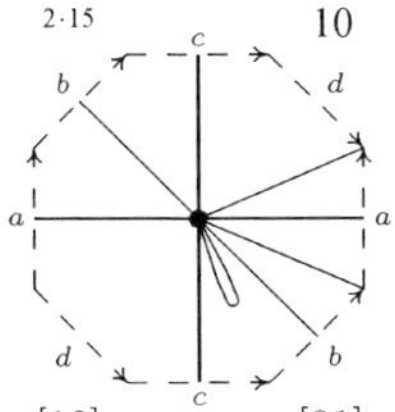

[10] [91]

2·16 10

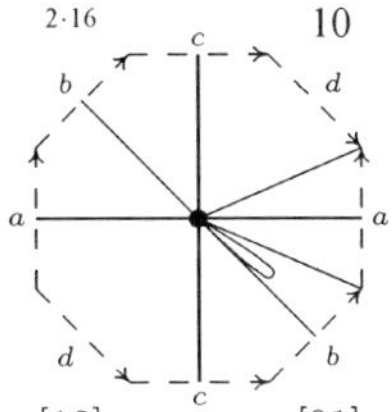

[10] [91]

2·17 10

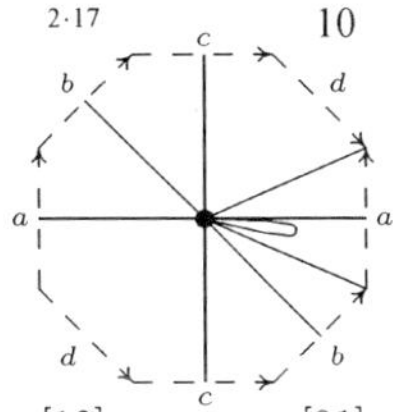

[10] [91]

2·18 10

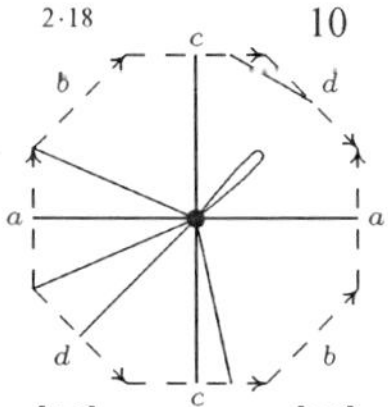

[10] [91]

2·19 10

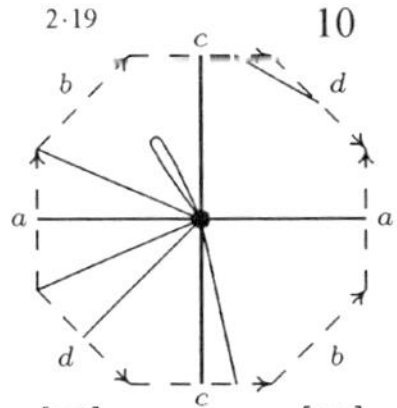

[10] [91]

2·20 10

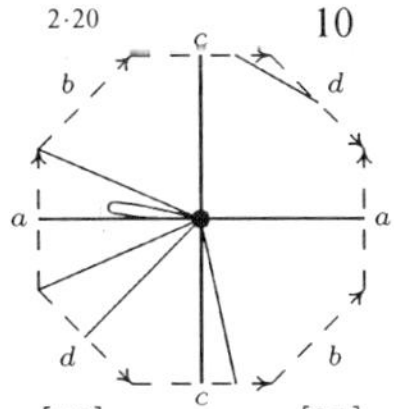

[10] [91]

2·21 10

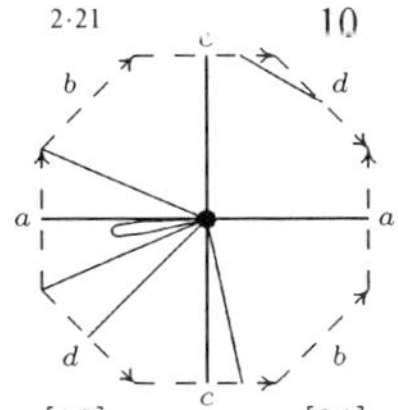

[10] [91]

2·22 10

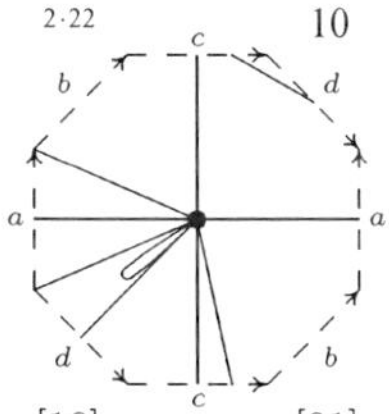

[10] [91]

2·23 10

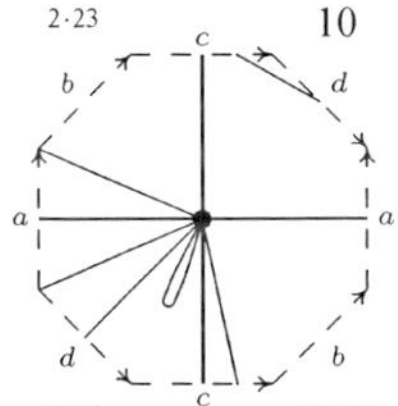

[10] [91]

2·24 10

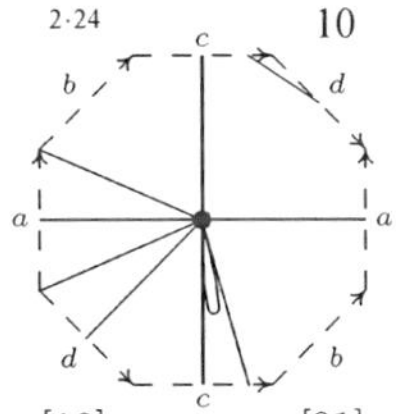

[10] [91]

2·25 10

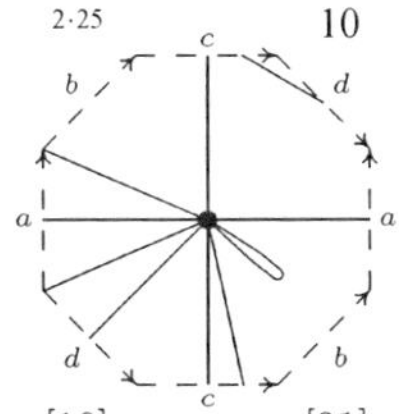

[10] [91]

2·26 5

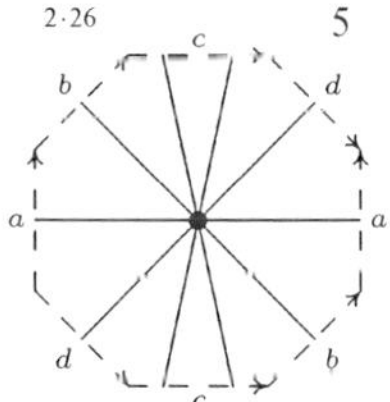

[10] [82]

2·27 10

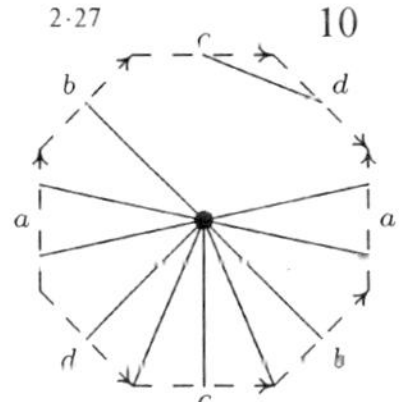

[10] [82]

2·28 10

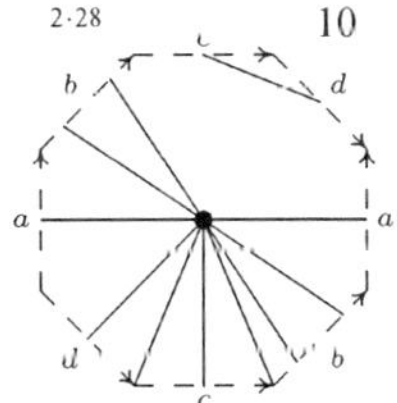

[10] [82]

2·29 10

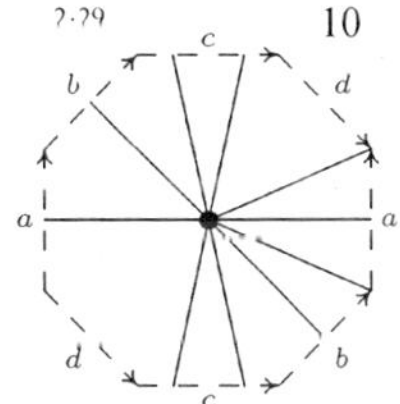

[10] [82]

2·30 10
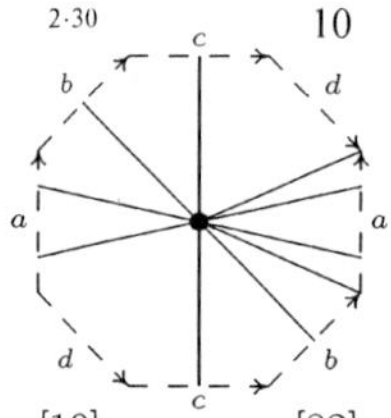
[10] [82]

2·31 10
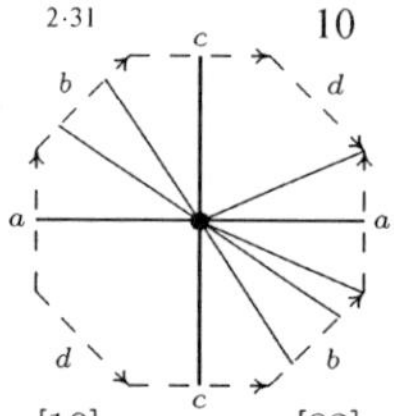
[10] [82]

2·32 10
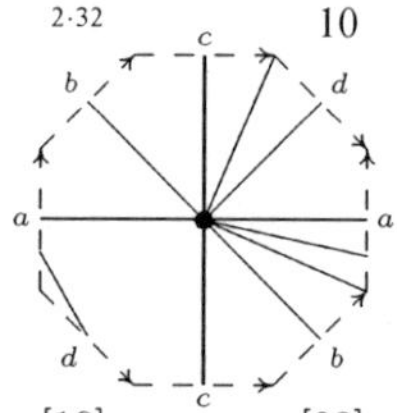
[10] [82]

2·33 10
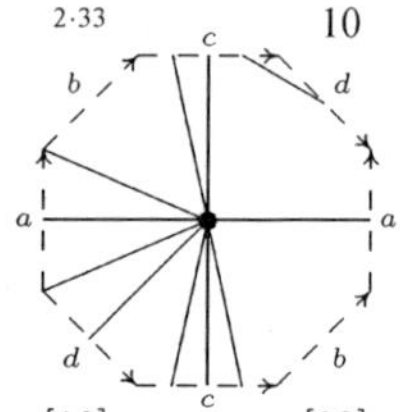
[10] [82]

2·34 10
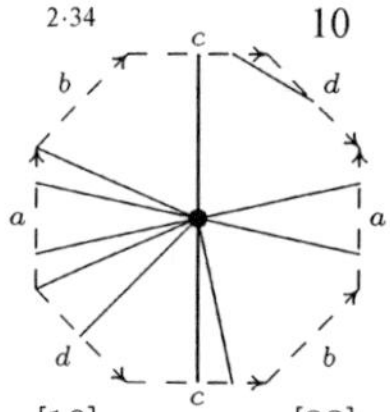
[10] [82]

2·35 10
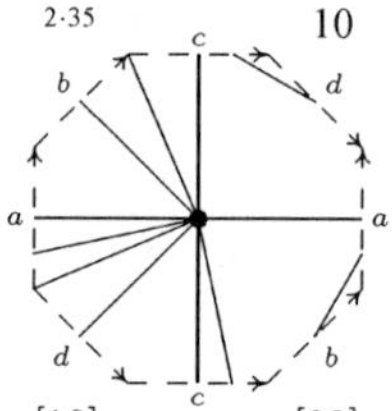
[10] [82]

2·36 10
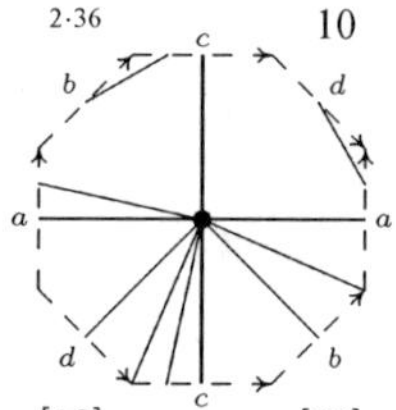
[10] [82]

2·37 10
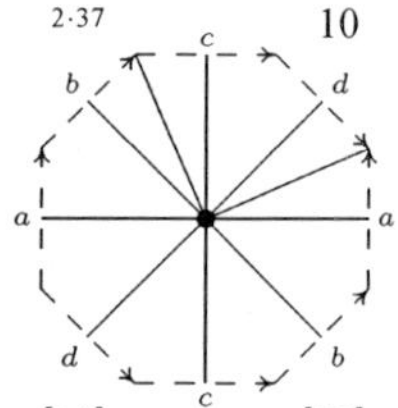
[10] [73]

2·38 10
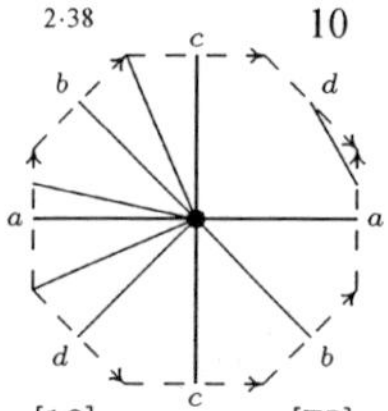
[10] [73]

2·39 10
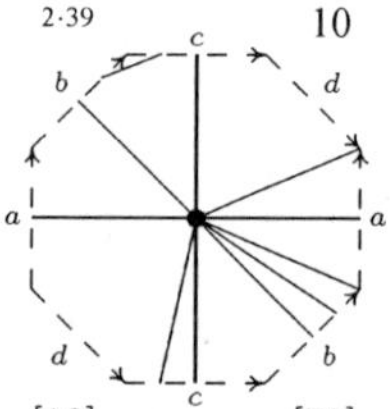
[10] [73]

2·40 10
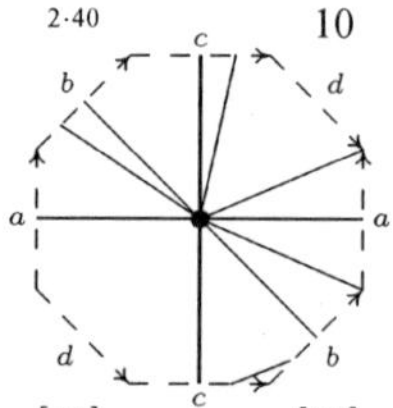
[10] [73]

2·41 10
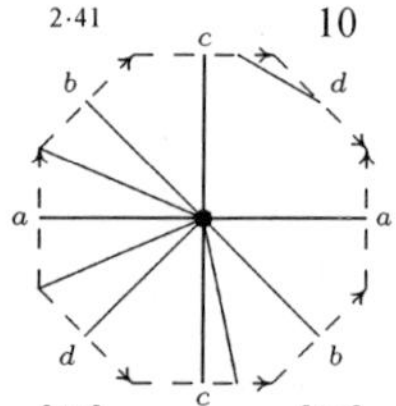
[10] [73]

2·42 10
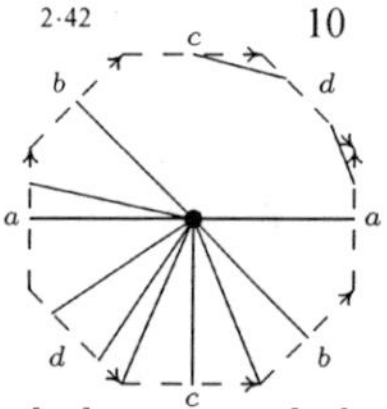
[10] [73]

2·43 10
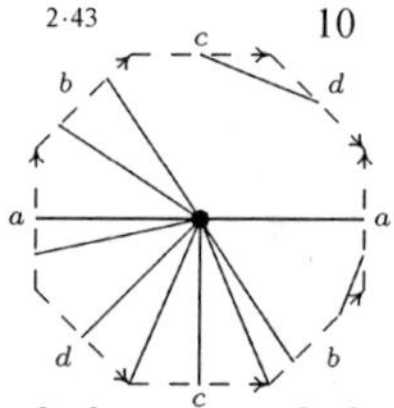
[10] [73]

2·44 5
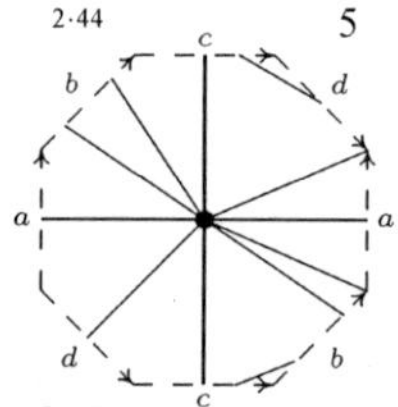
[10] [64]

2·45 10
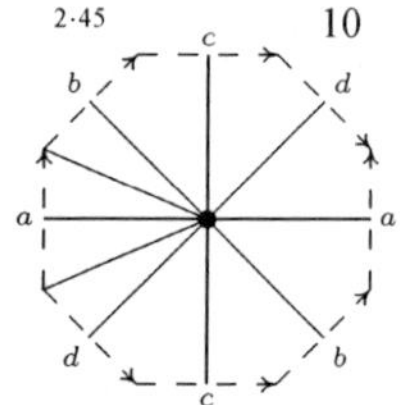
[10] [64]

2·46 10
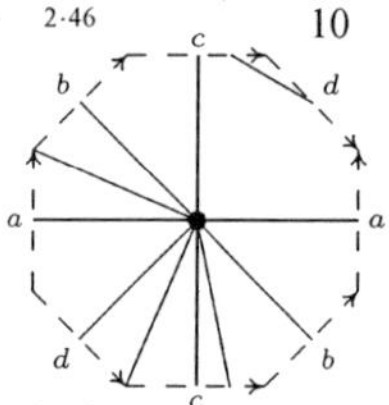
[10] [64]

2·47 10
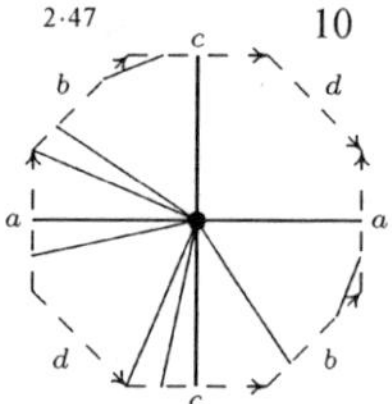
[10] [64]

2·48 10
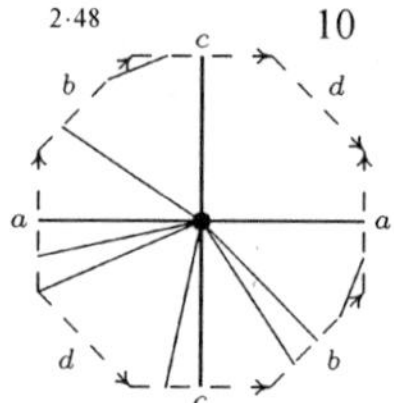
[10] [64]

2·49 10
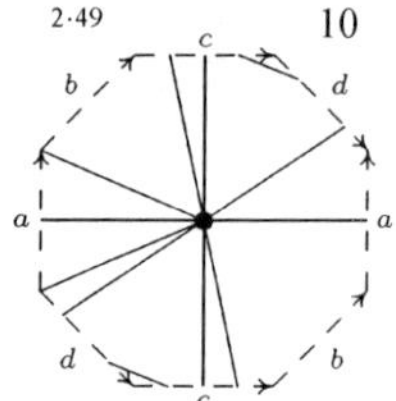
[10] [64]

5e 1v

2·50 10

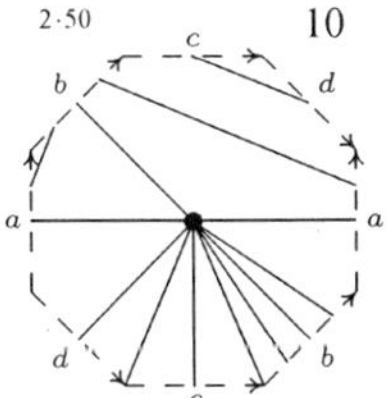

[10] [64]

2·51 1

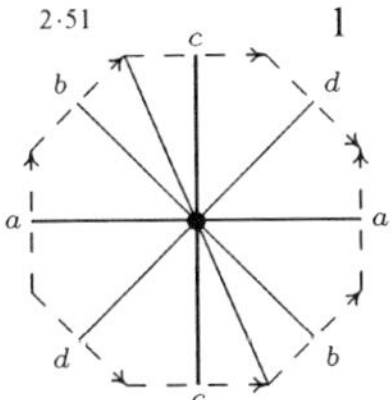

[10] $[5^2]$

2·52 2

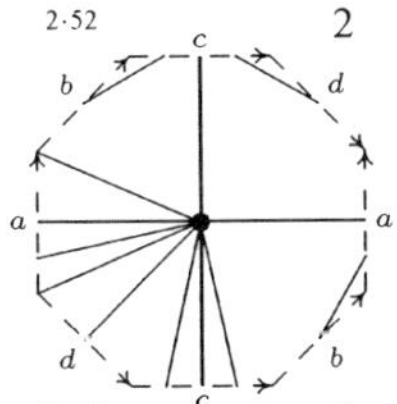

[10] $[5^2]$

2·53 10

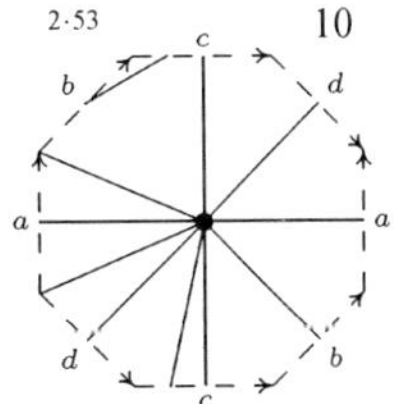

[10] $[5^2]$

2·54 5

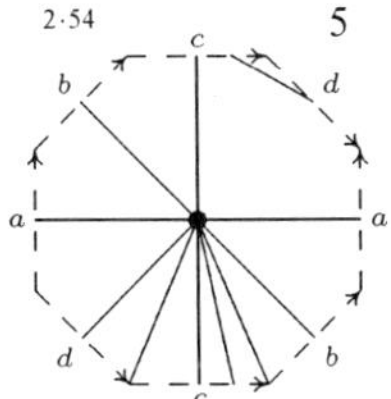

[10] $[5^2]$

2·55 5

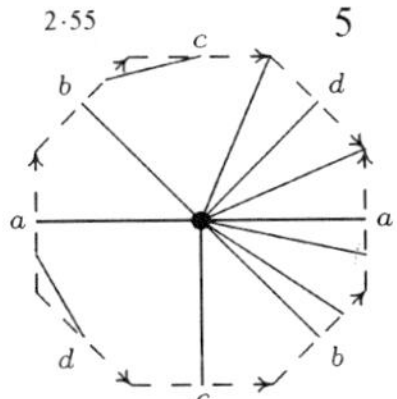

[10] $[5^2]$

2·56 5

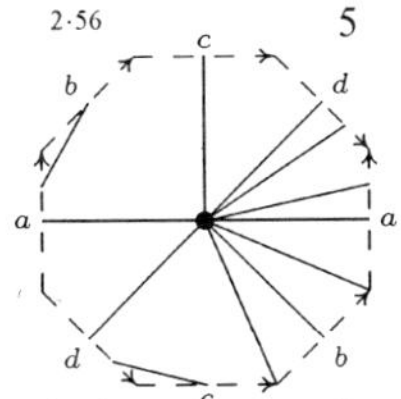

[10] $[5^2]$

2·57 5

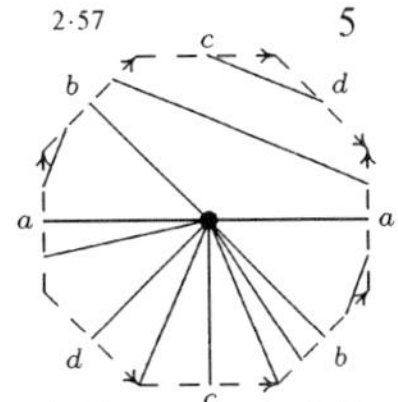

[10] $[5^2]$

$$6 + 46u + 53u^2$$

5 edges

2 vertices

$\mathfrak{g}_{49}$

0·84 10

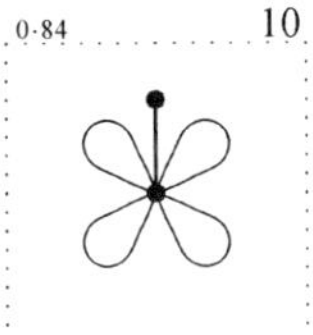

[91] $[61^4]$

0·85 10

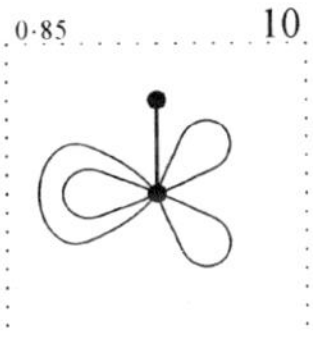

[91] $[521^3]$

0·86 10

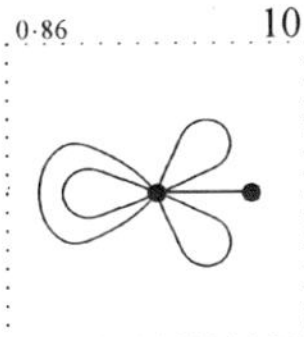

[91] $[521^3]$

0·87 10

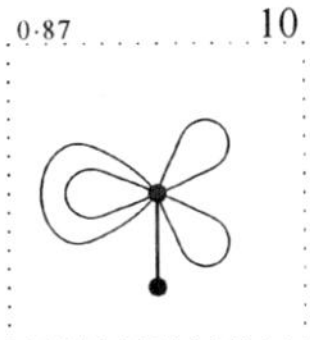

[91] $[521^3]$

0·88 10

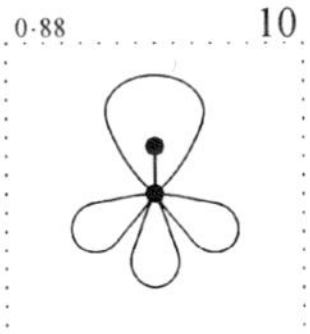

[91] $[431^3]$

0·89 10

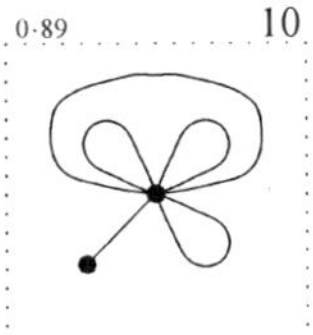

[91] $[431^3]$

0·90 10

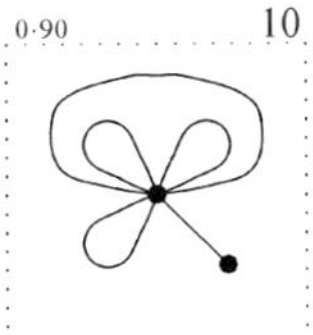

[91] $[431^3]$

0·91 10

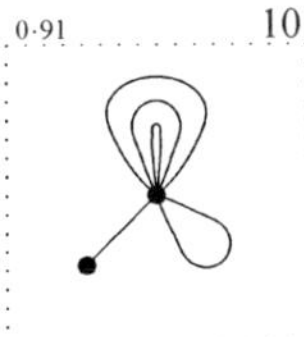

[91] $[42^21^2]$

0·92 10

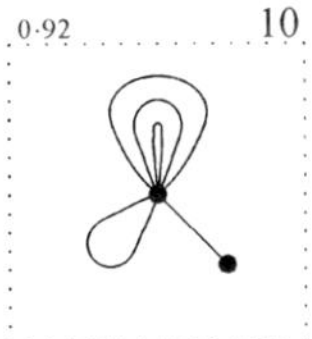

[91] $[42^21^2]$

0·93 10

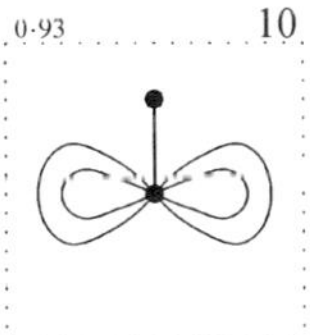

[91] $[42^21^2]$

0·94 10

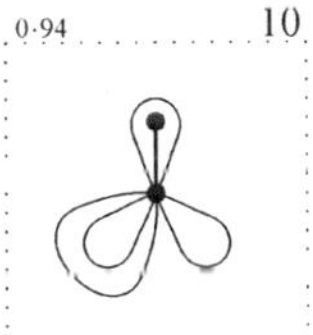

[91] $[3^221^2]$

0·95 10

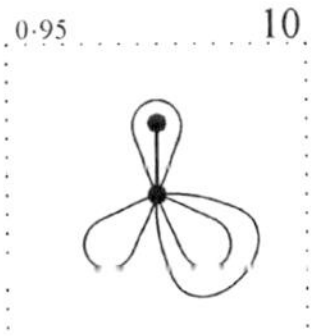

[91] $[3^221^2]$

0·96 10

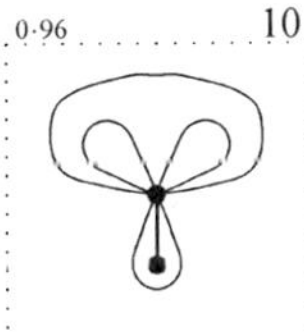

[91] $[3^221^2]$

0·97 10

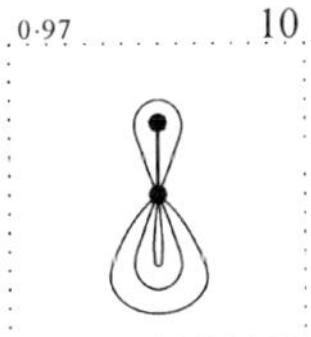

[91] $[32^31]$

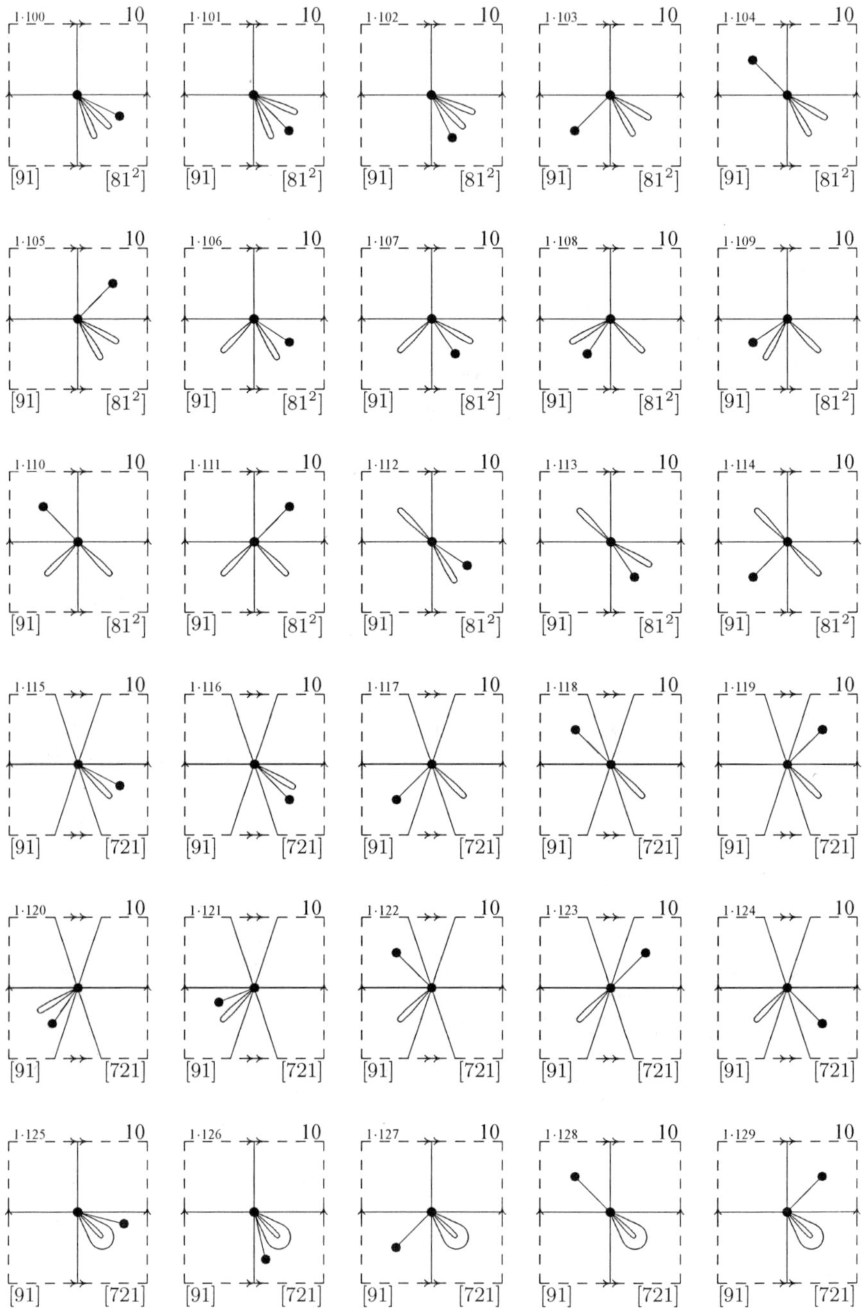
1·100 10 [91] [81²]
1·101 10 [91] [81²]
1·102 10 [91] [81²]
1·103 10 [91] [81²]
1·104 10 [91] [81²]
1·105 10 [91] [81²]
1·106 10 [91] [81²]
1·107 10 [91] [81²]
1·108 10 [91] [81²]
1·109 10 [91] [81²]
1·110 10 [91] [81²]
1·111 10 [91] [81²]
1·112 10 [91] [81²]
1·113 10 [91] [81²]
1·114 10 [91] [81²]
1·115 10 [91] [721]
1·116 10 [91] [721]
1·117 10 [91] [721]
1·118 10 [91] [721]
1·119 10 [91] [721]
1·120 10 [91] [721]
1·121 10 [91] [721]
1·122 10 [91] [721]
1·123 10 [91] [721]
1·124 10 [91] [721]
1·125 10 [91] [721]
1·126 10 [91] [721]
1·127 10 [91] [721]
1·128 10 [91] [721]
1·129 10 [91] [721]

5e 2v

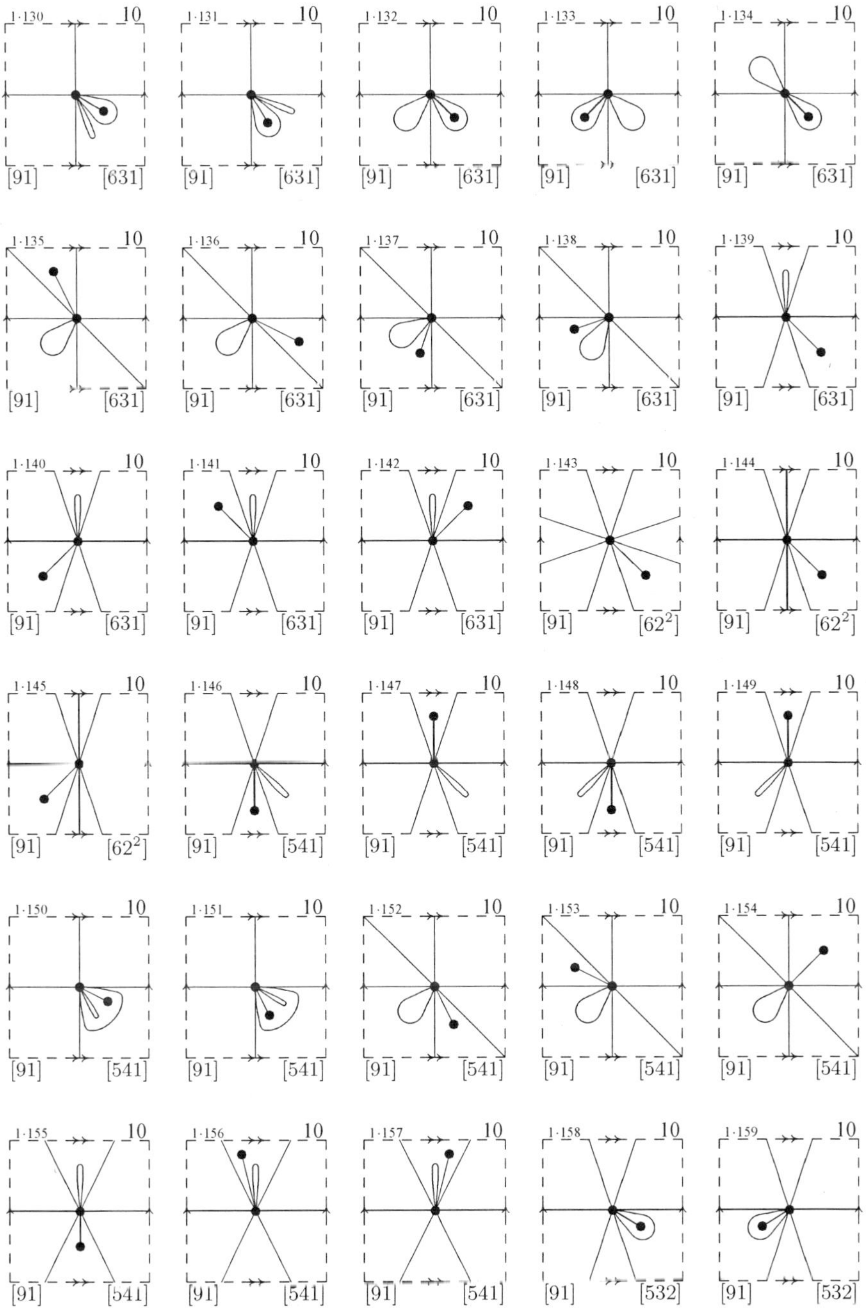
1·130 10 $[9\bar{1}]$ $[63\bar{1}]$
1·131 10 $[9\bar{1}]$ $[63\bar{1}]$
1·132 10 $[9\bar{1}]$ $[63\bar{1}]$
1·133 10 $[9\bar{1}]$ $[63\bar{1}]$
1·134 10 $[9\bar{1}]$ $[63\bar{1}]$
1·135 10 $[91]$ $[63\bar{1}]$
1·136 10 $[9\bar{1}]$ $[63\bar{1}]$
1·137 10 $[9\bar{1}]$ $[63\bar{1}]$
1·138 10 $[9\bar{1}]$ $[63\bar{1}]$
1·139 10 $[9\bar{1}]$ $[63\bar{1}]$
1·140 10 $[9\bar{1}]$ $[63\bar{1}]$
1·141 10 $[9\bar{1}]$ $[63\bar{1}]$
1·142 10 $[9\bar{1}]$ $[63\bar{1}]$
1·143 10 $[9\bar{1}]$ $[62^2]$
1·144 10 $[9\bar{1}]$ $[62^2]$
1·145 10 $[9\bar{1}]$ $[62^2]$
1·146 10 $[9\bar{1}]$ $[54\bar{1}]$
1·147 10 $[9\bar{1}]$ $[54\bar{1}]$
1·148 10 $[9\bar{1}]$ $[54\bar{1}]$
1·149 10 $[9\bar{1}]$ $[54\bar{1}]$
1·150 10 $[9\bar{1}]$ $[54\bar{1}]$
1·151 10 $[9\bar{1}]$ $[54\bar{1}]$
1·152 10 $[9\bar{1}]$ $[54\bar{1}]$
1·153 10 $[9\bar{1}]$ $[54\bar{1}]$
1·154 10 $[9\bar{1}]$ $[54\bar{1}]$
1·155 10 $[9\bar{1}]$ $[54\bar{1}]$
1·156 10 $[9\bar{1}]$ $[54\bar{1}]$
1·157 10 $[9\bar{1}]$ $[54\bar{1}]$
1·158 10 $[91]$ $[532]$
1·159 10 $[9\bar{1}]$ $[532]$

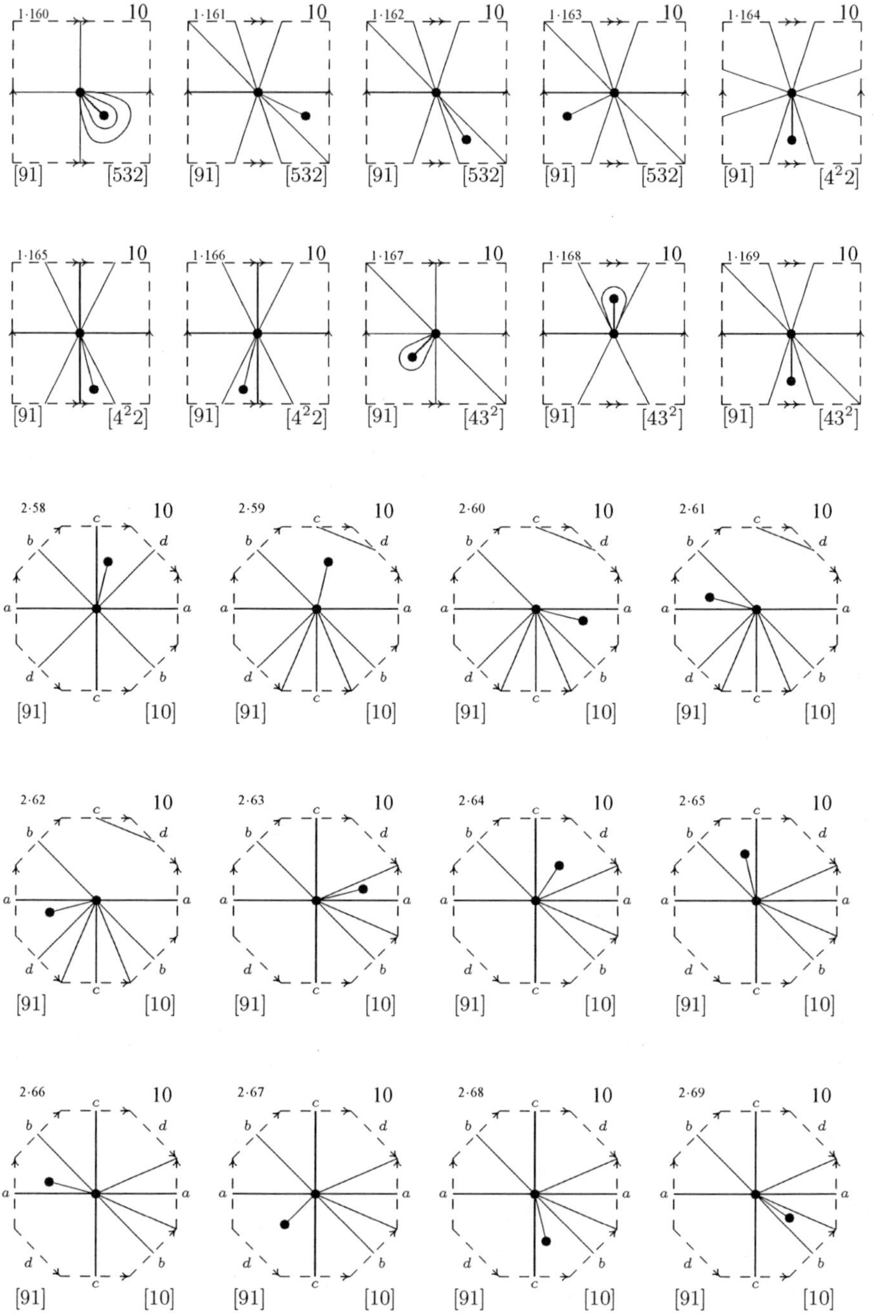
1·160 10 [91] [532]
1·161 10 [91] [532]
1·162 10 [91] [532]
1·163 10 [91] [532]
1·164 10 [91] [4²2]
1·165 10 [91] [4²2]
1·166 10 [91] [4²2]
1·167 10 [91] [43²]
1·168 10 [91] [43²]
1·169 10 [91] [43²]
2·58 10 [91] [10]
2·59 10 [91] [10]
2·60 10 [91] [10]
2·61 10 [91] [10]
2·62 10 [91] [10]
2·63 10 [91] [10]
2·64 10 [91] [10]
2·65 10 [91] [10]
2·66 10 [91] [10]
2·67 10 [91] [10]
2·68 10 [91] [10]
2·69 10 [91] [10]

5e 2v

2·70 10

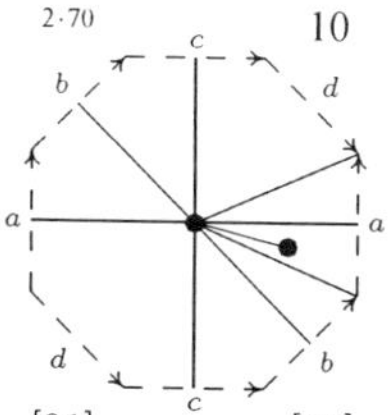

[91] [10]

2·71 10

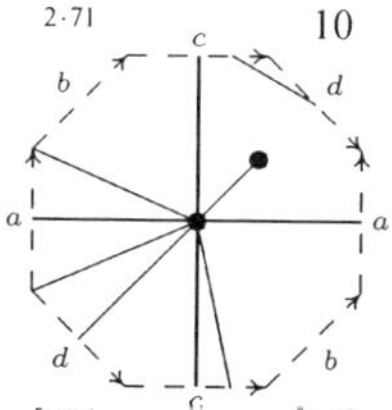

[91] [10]

2·72 10

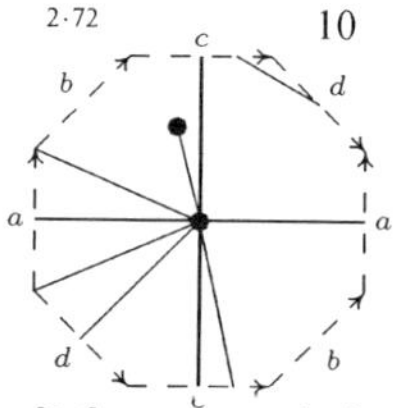

[91] [10]

2·73 10

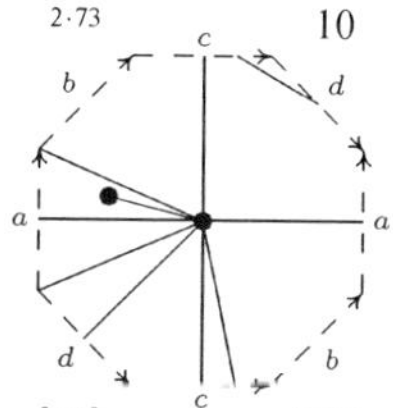

[91] [10]

2·74 10

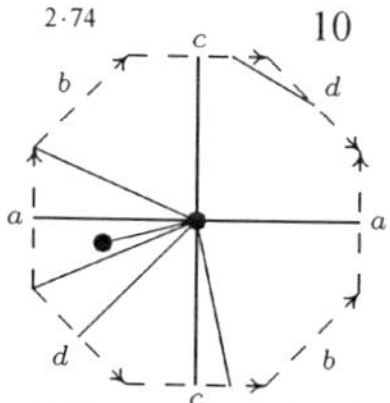

[91] [10]

2·75 10

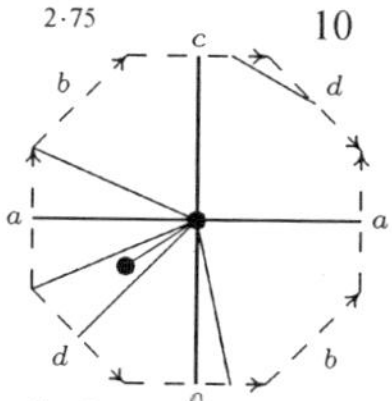

[91] [10]

2·76 10

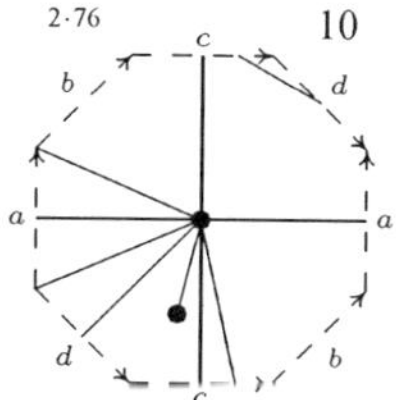

[91] [10]

2·77 10

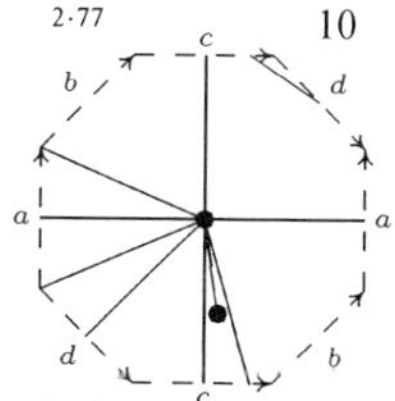

[91] [10]

2·78 10

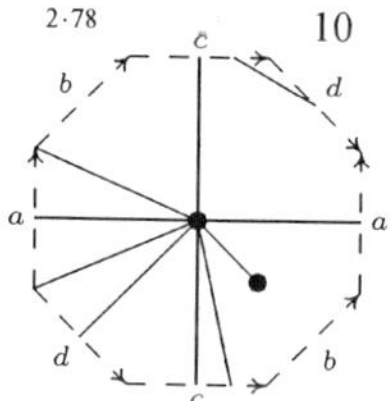

[91] [10]

$$14 + 70u + 21u^2$$

5 edges

2 vertices

$\mathfrak{g}$50

0·98 10

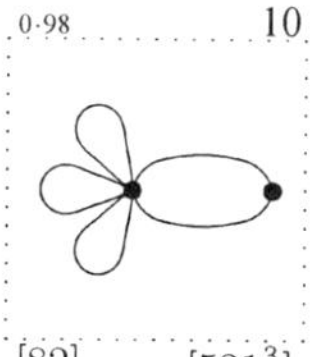

[82] [521^3]

0·99 10

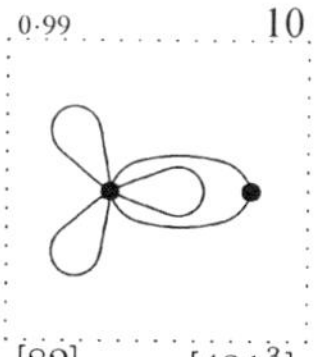

[82] [431^3]

0·100 10

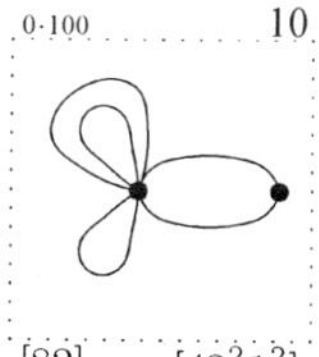

[82] [42^{2}1^2]

0·101 10

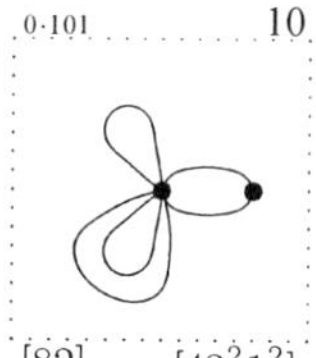

[82] [42^{2}1^2]

0·102 10

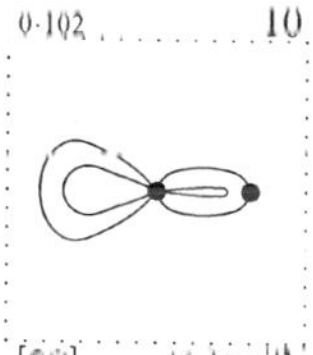

[82] [3^{2}21^2]

0·103 10

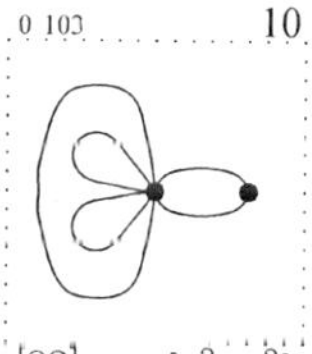

[82] [3^{2}21^2]

0·104 10

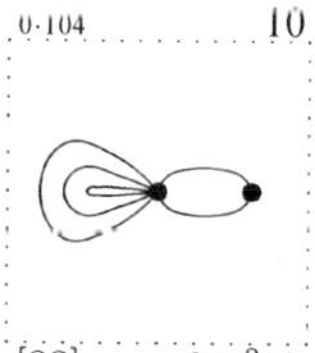

[82] [32^{3}1]

1·170 10

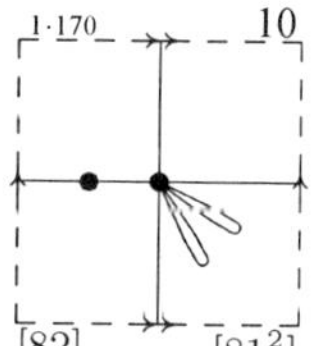

[82] [81^2]

1·171 10

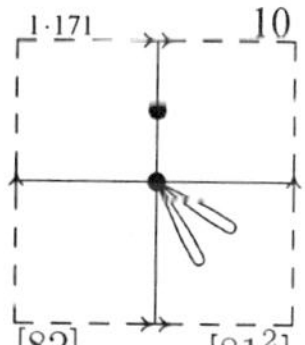

[82] [81^2]

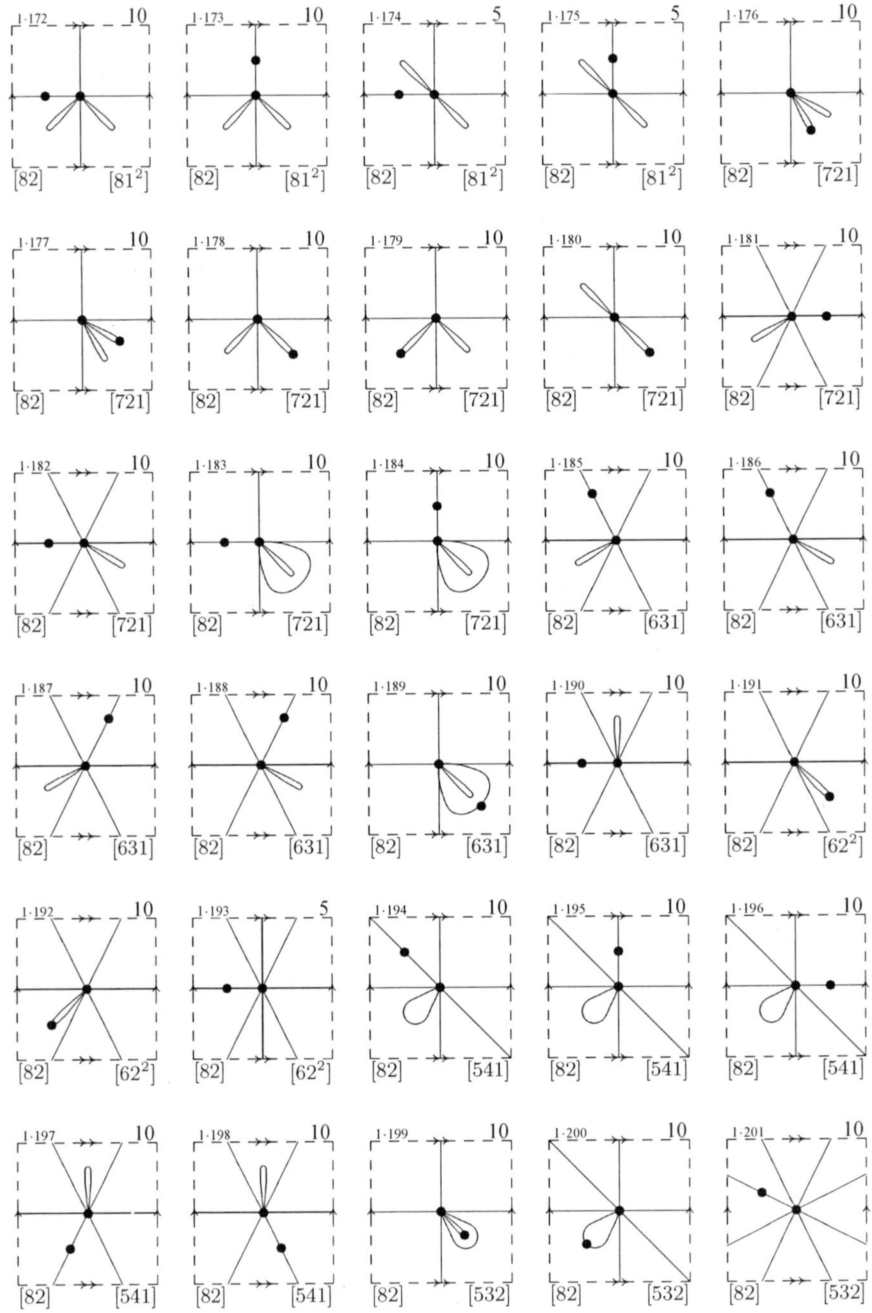
1·172 10 [82] $[81^2]$
1·173 10 [82] $[81^2]$
1·174 5 [82] $[81^2]$
1·175 5 [82] $[81^2]$
1·176 10 [82] [721]
1·177 10 [82] [721]
1·178 10 [82] [721]
1·179 10 [82] [721]
1·180 10 [82] [721]
1·181 10 [82] [721]
1·182 10 [82] [721]
1·183 10 [82] [721]
1·184 10 [82] [721]
1·185 10 [82] [631]
1·186 10 [82] [631]
1·187 10 [82] [631]
1·188 10 [82] [631]
1·189 10 [82] [631]
1·190 10 [82] [631]
1·191 10 [82] $[62^2]$
1·192 10 [82] $[62^2]$
1·193 5 [82] $[62^2]$
1·194 10 [82] [541]
1·195 10 [82] [541]
1·196 10 [82] [541]
1·197 10 [82] [541]
1·198 10 [82] [541]
1·199 10 [82] [532]
1·200 10 [82] [532]
1·201 10 [82] [532]

5e 2v

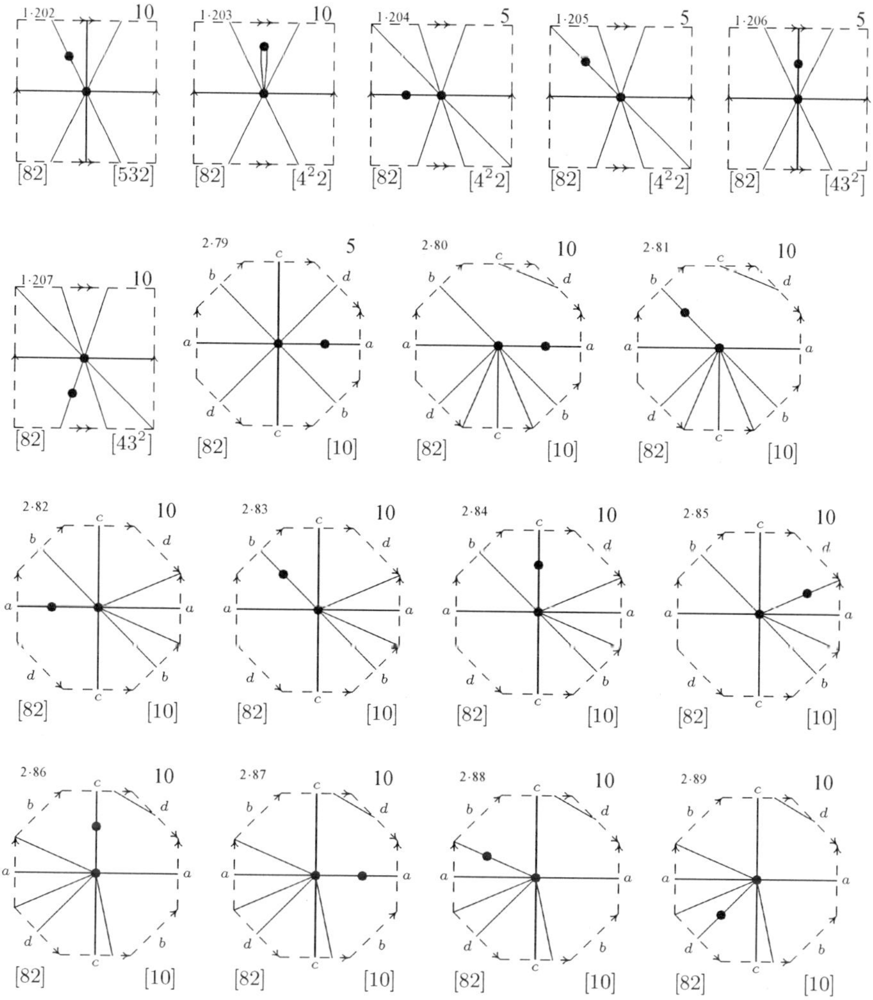

$$7 + 38u + 11u^2$$

5 edges

2 vertices

$\mathfrak{g}_{51}$

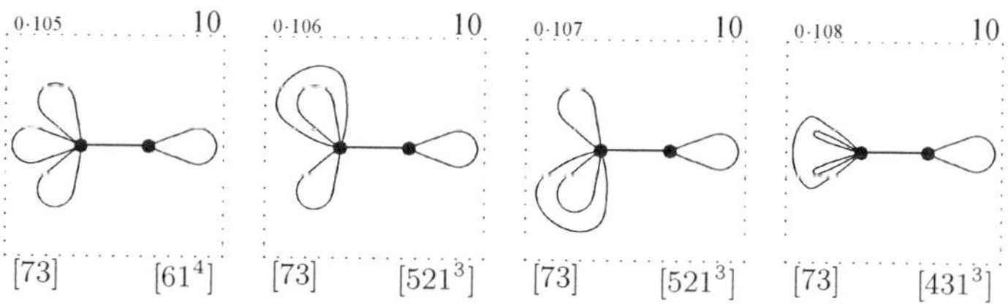

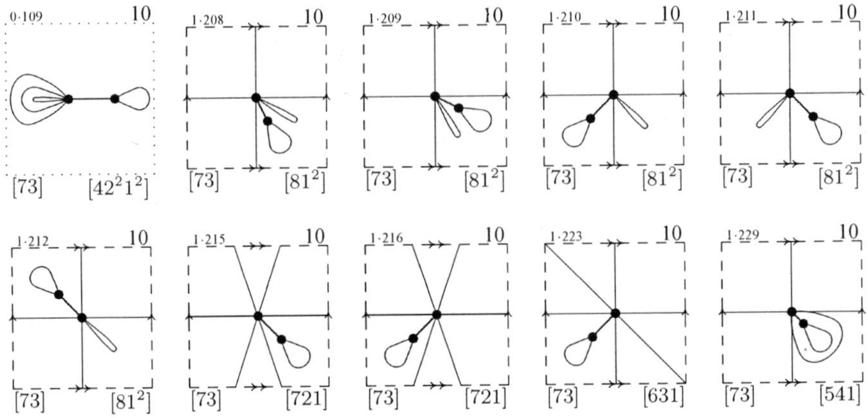

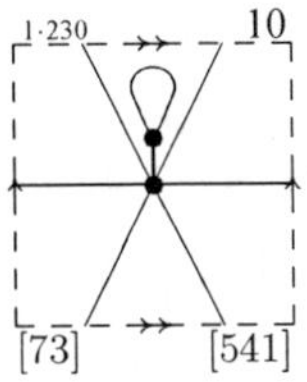

$5 + 10u$

5 edges
2 vertices
$\mathfrak{g}_{52}$

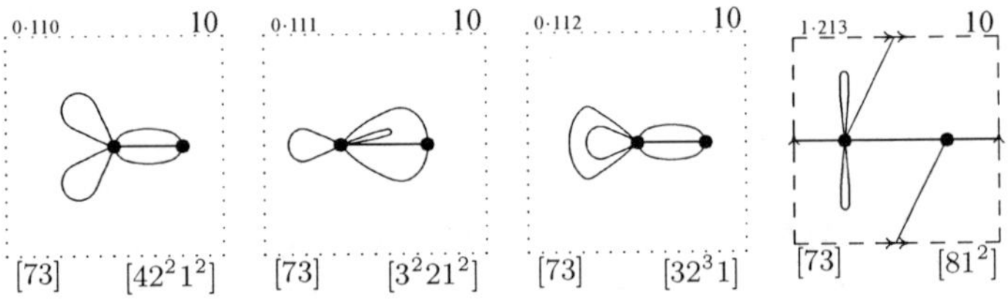

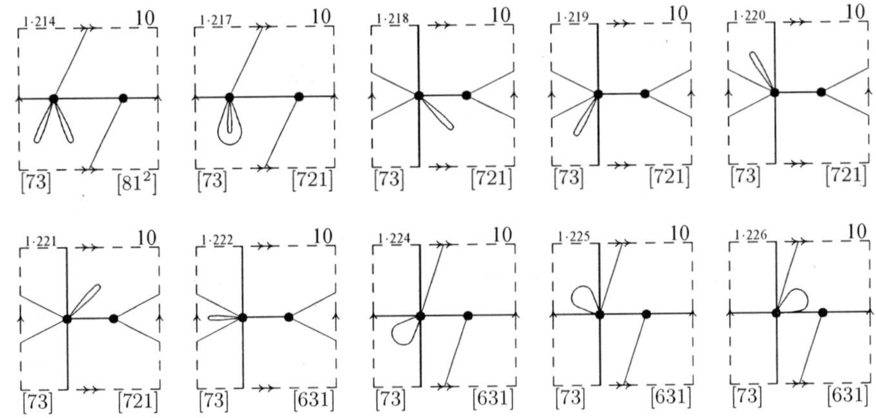

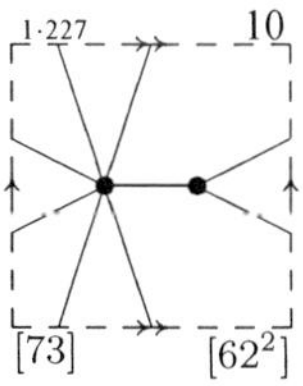

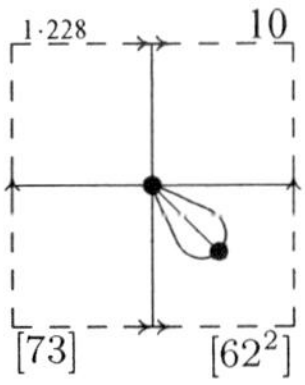

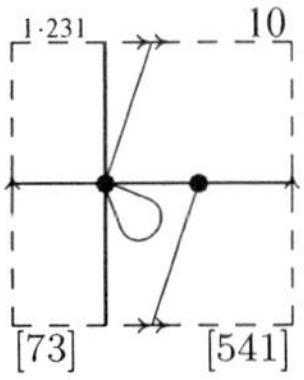

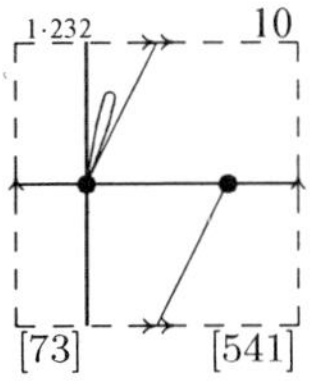

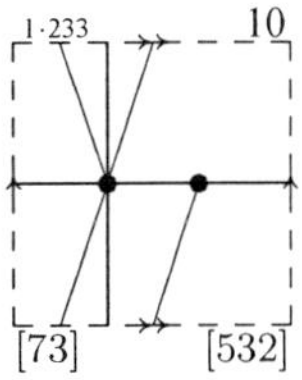

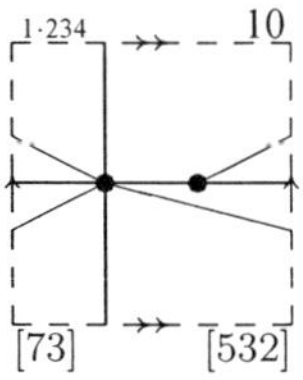

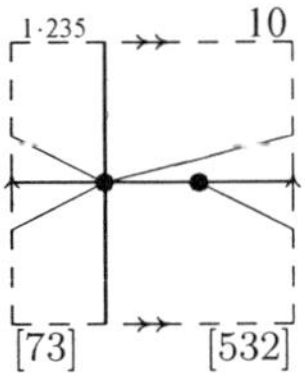

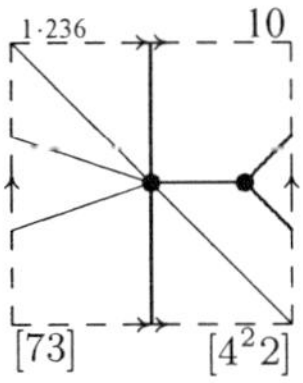

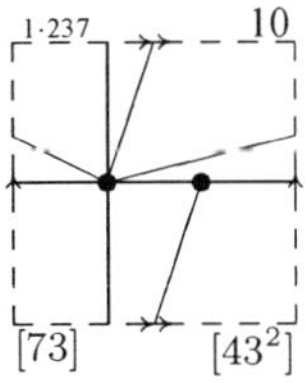

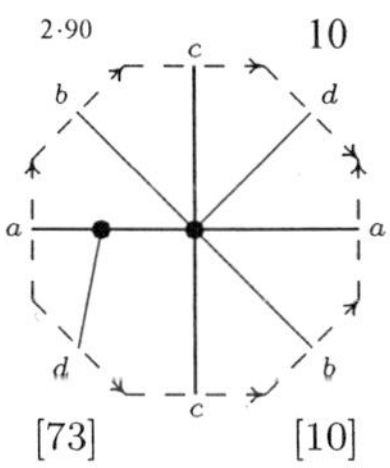

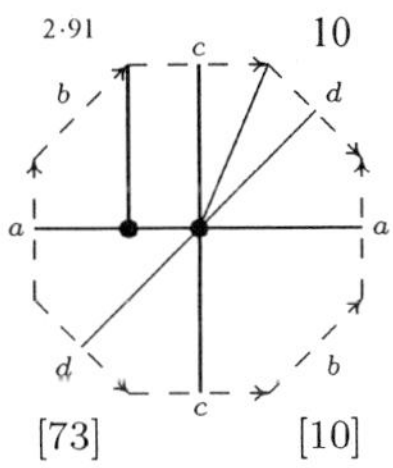

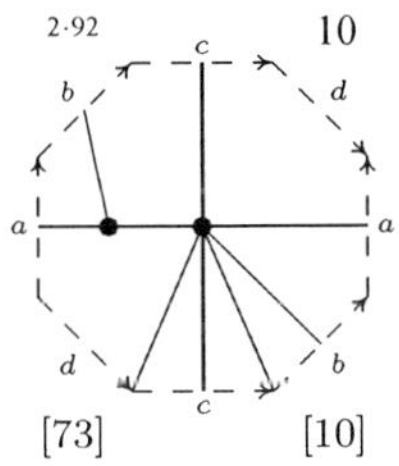

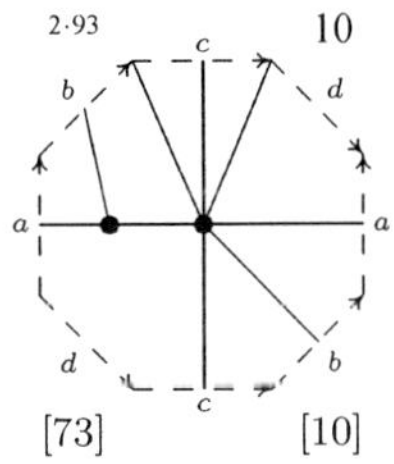

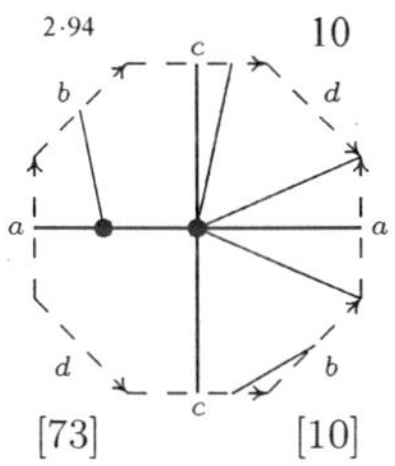

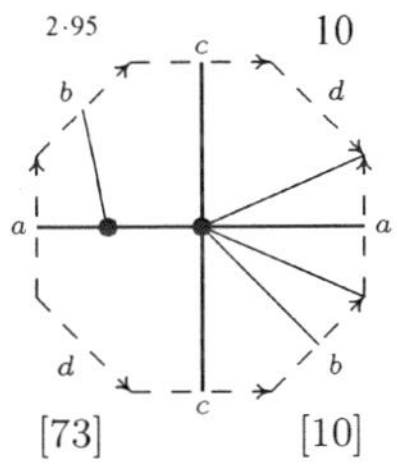

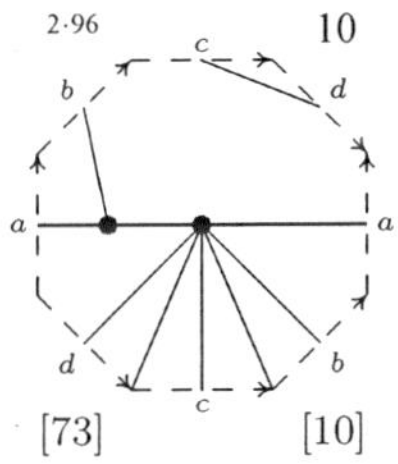

$$3 + 20u + 7u^2$$

5 edges

2 vertices

$\mathfrak{g}_{53}$

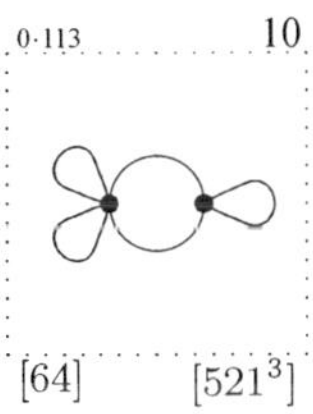

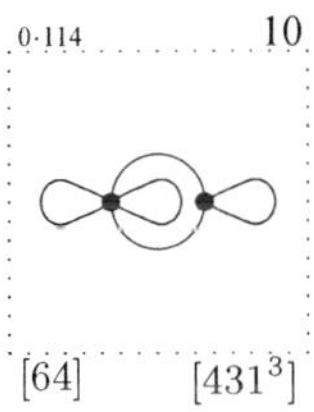

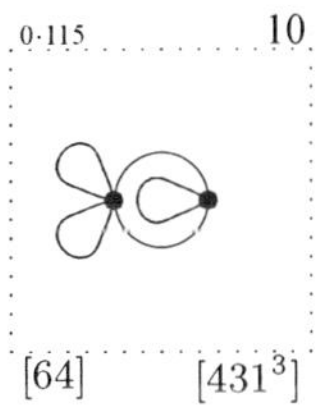

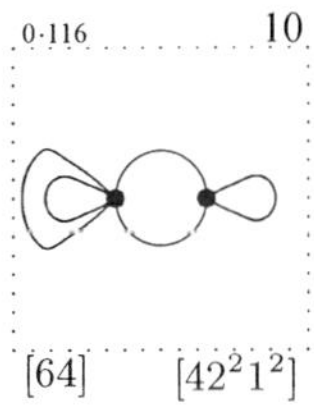

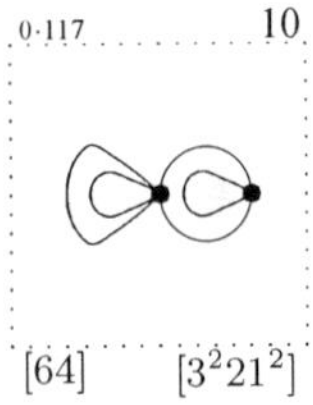

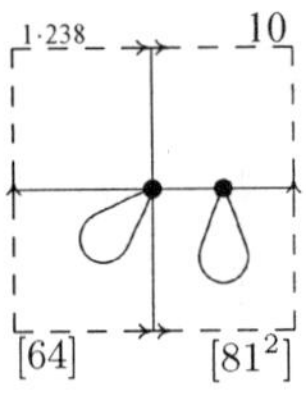

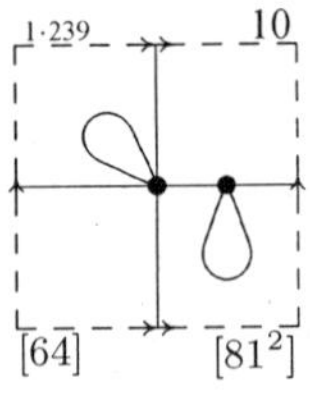

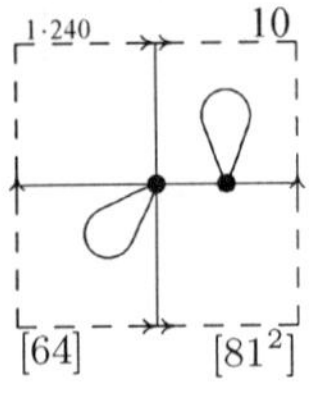

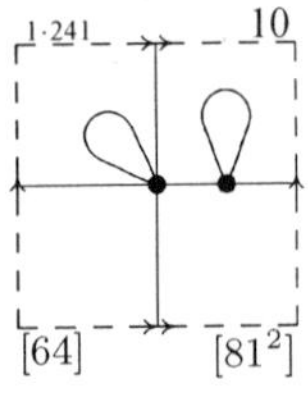

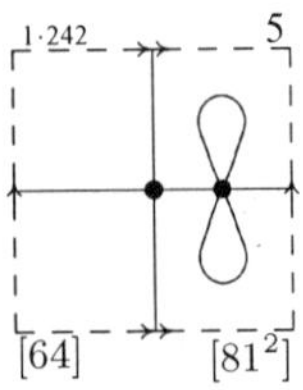

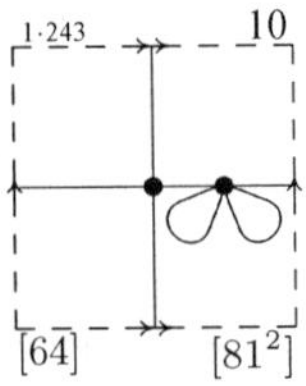

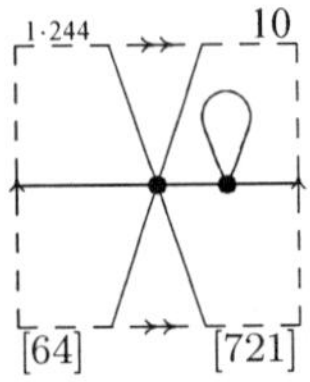

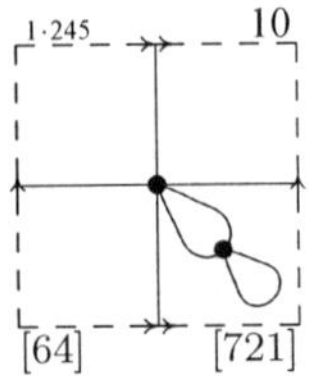

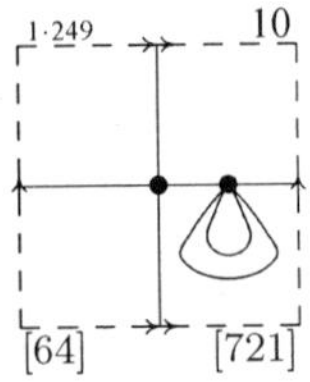

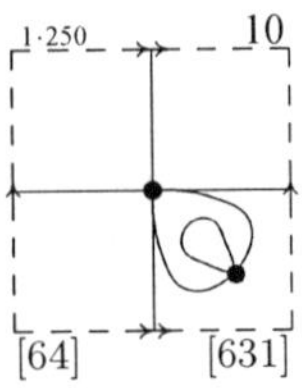

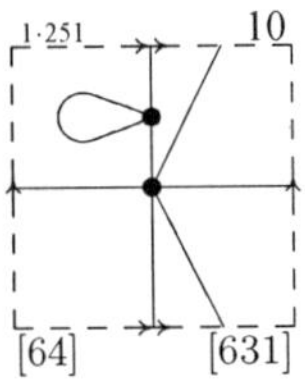

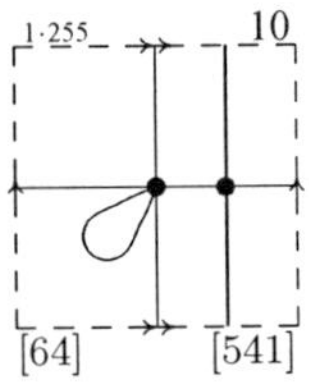

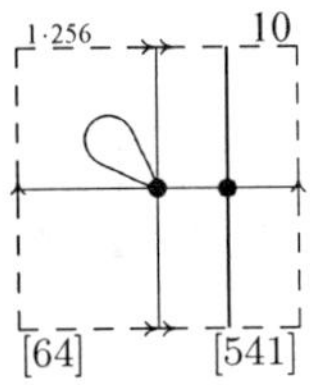

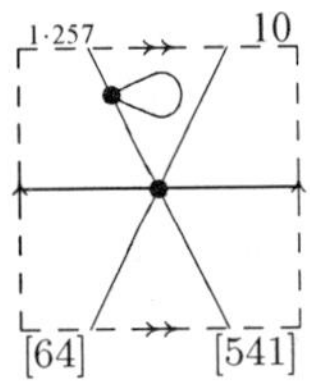

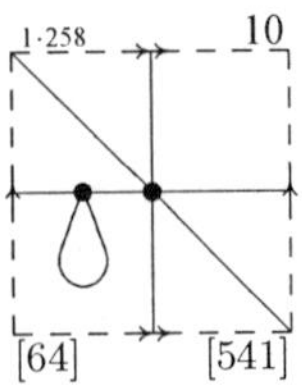

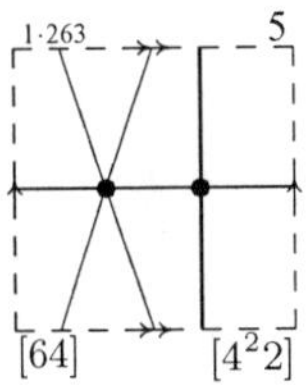

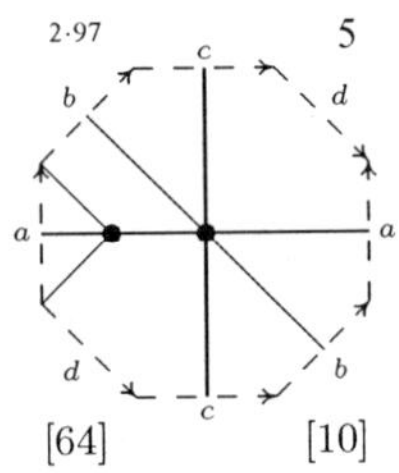

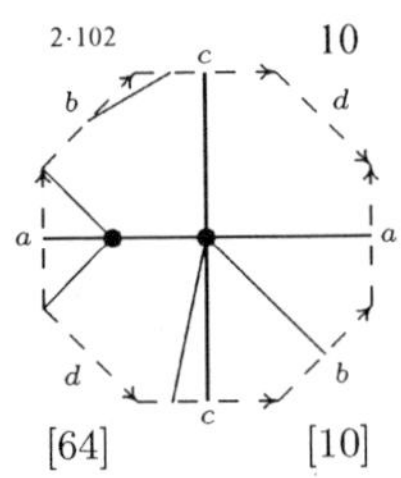

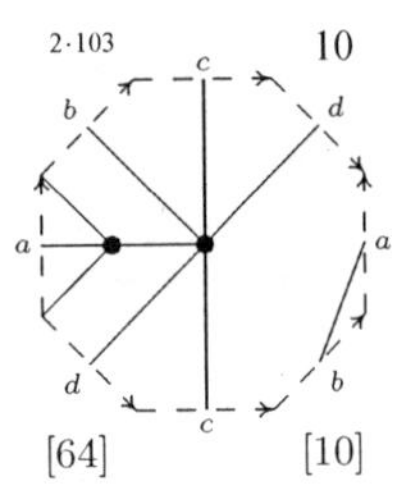

$5 + 16u + 3u^2$

5e 2v

5 edges
2 vertices
$\mathfrak{g}_{54}$

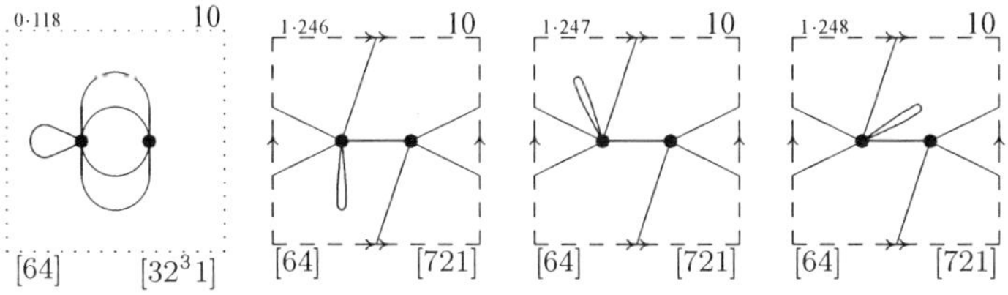

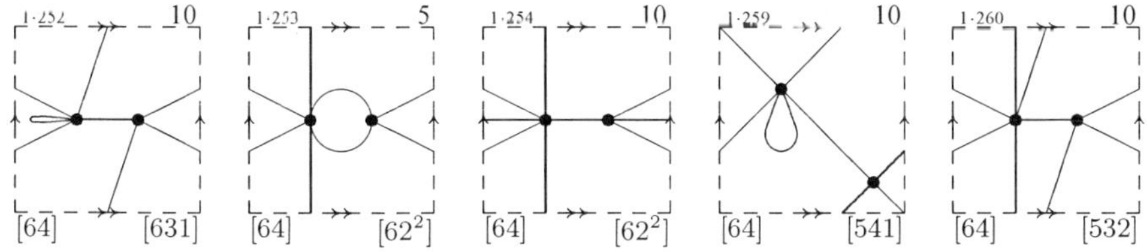

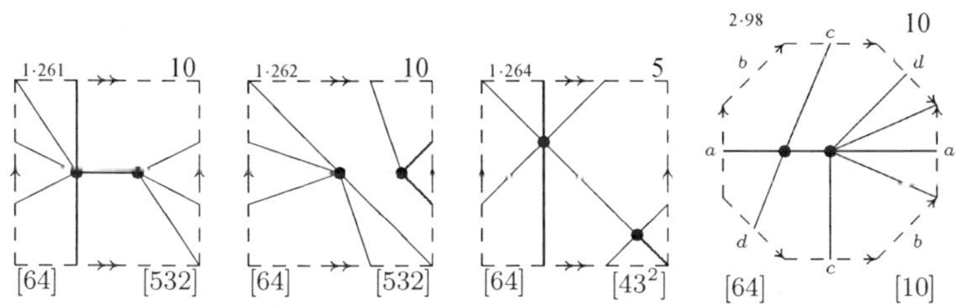

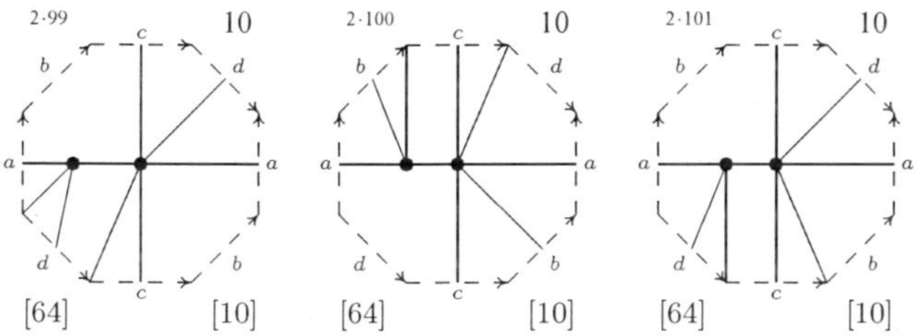

$$1 + 11u + 4u^2$$

5 edges
2 vertices
$\mathfrak{g}_{55}$

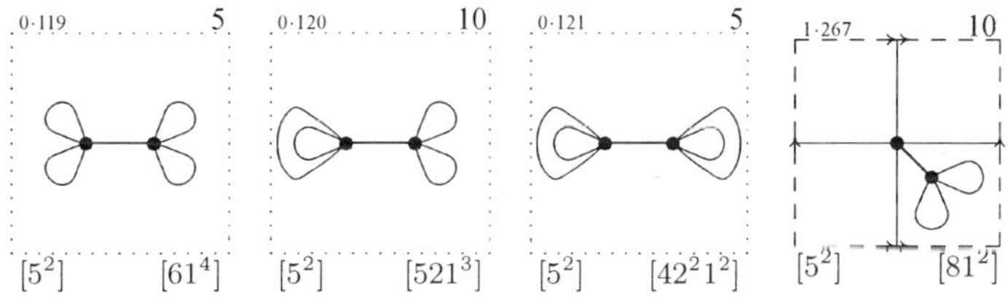

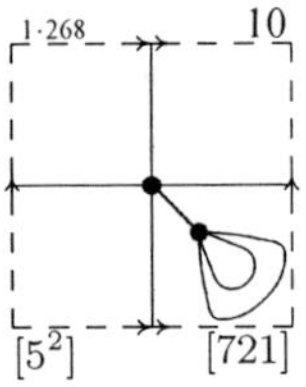

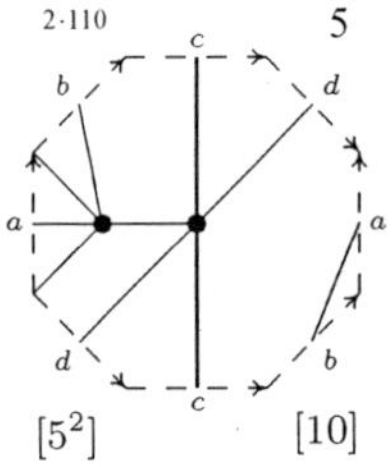

$3 + 2u + u^2$

5 edges

2 vertices

$\mathfrak{g}56$

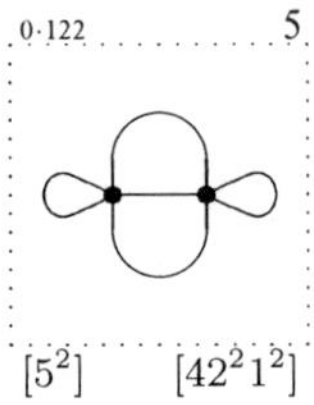

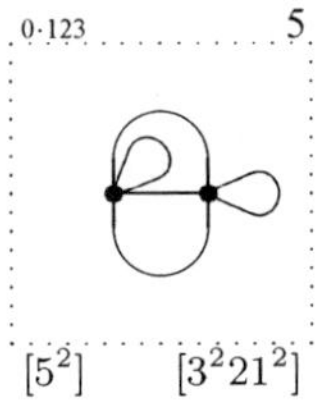

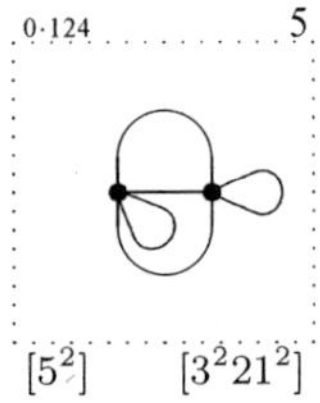

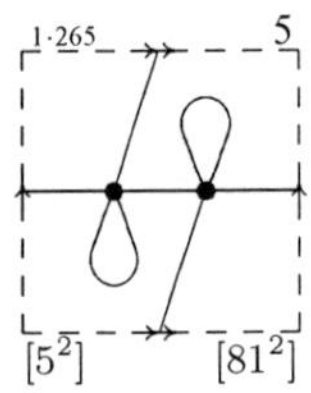

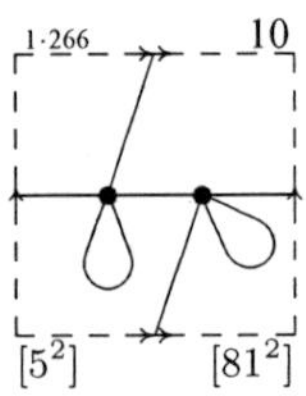

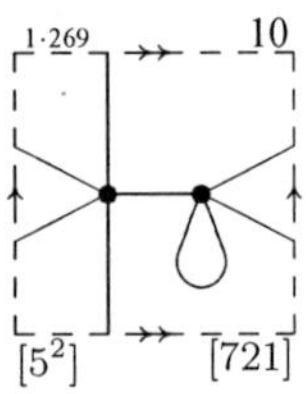

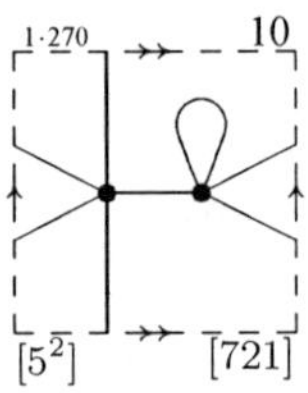

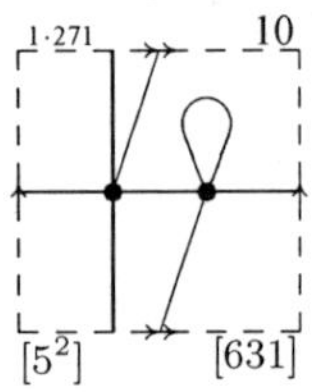

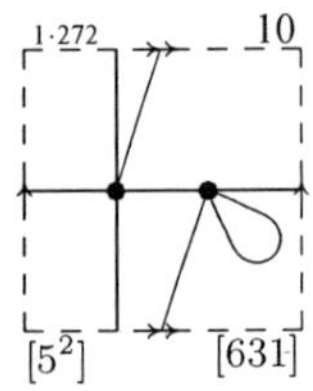

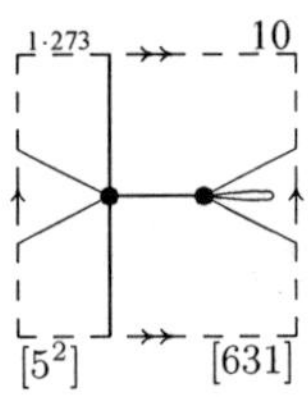

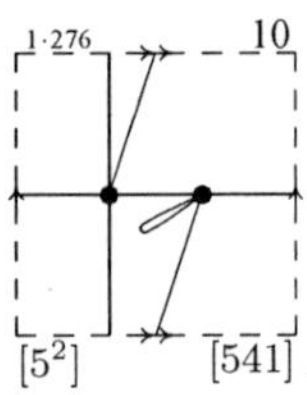

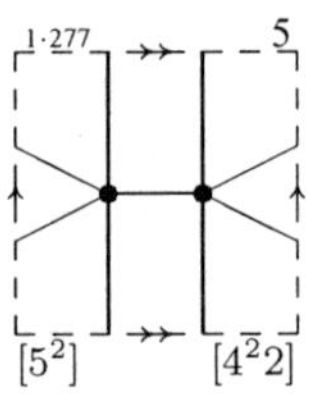

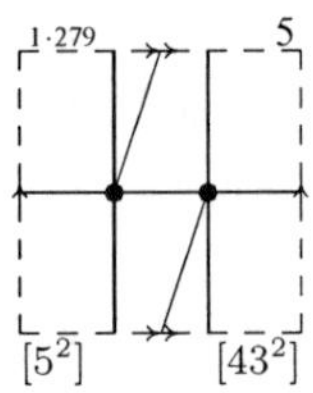

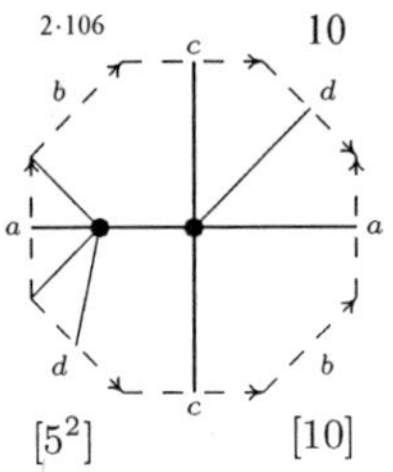

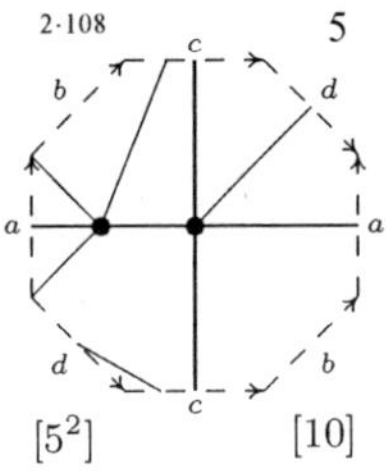

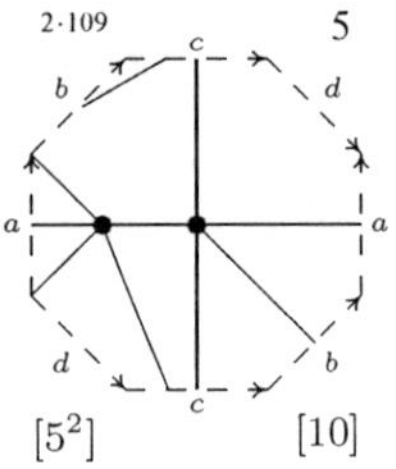

$3 + 10u + 3u^2$

5 edges
2 vertices
$\mathfrak{g}_{57}$

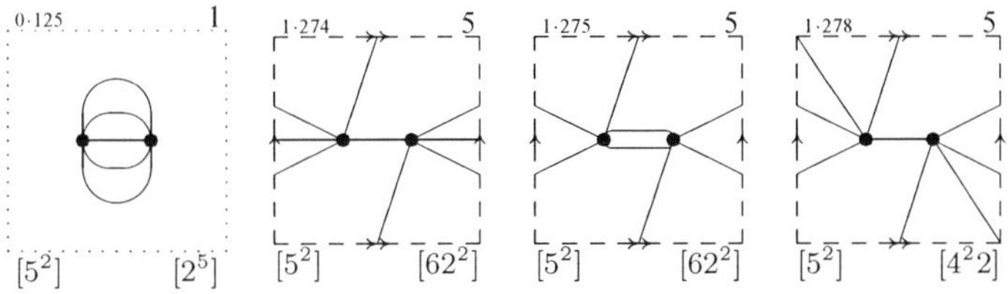

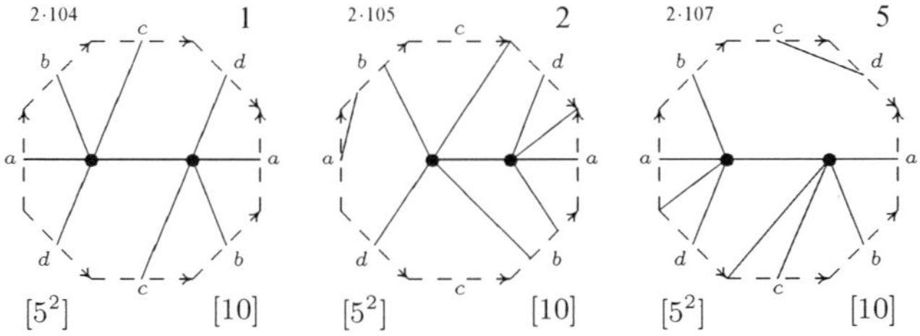

$$1 + 3u + 3u^2$$

5 edges
3 vertices
$\mathfrak{g}_{58}$

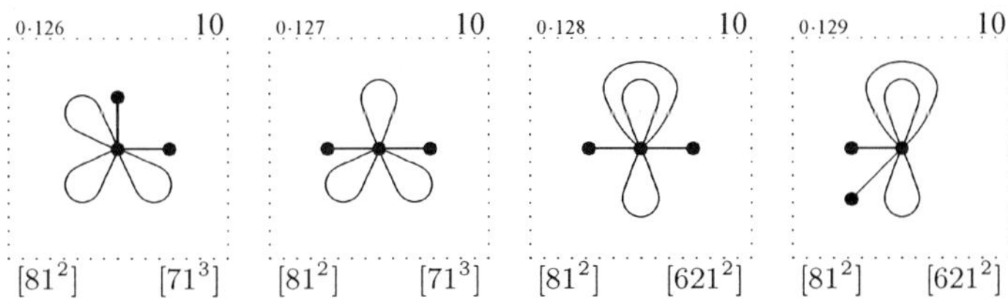

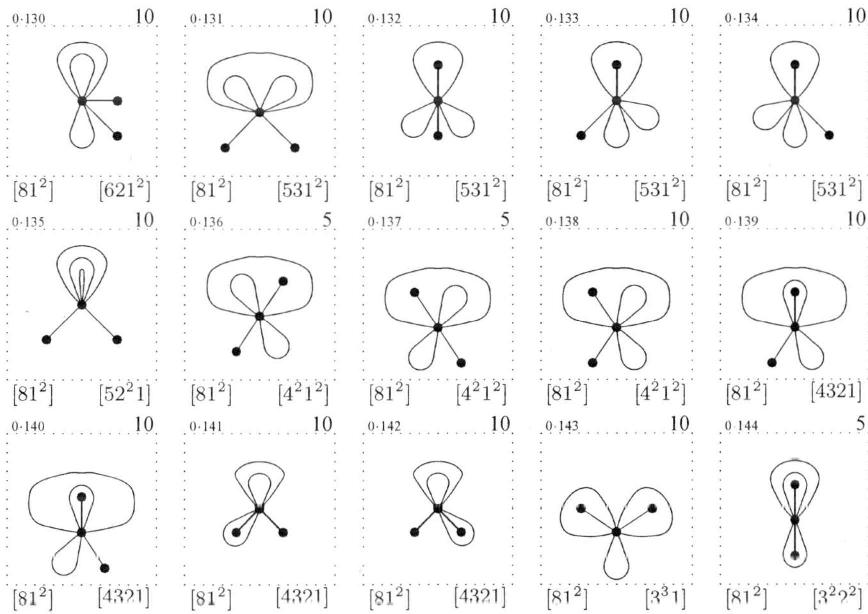

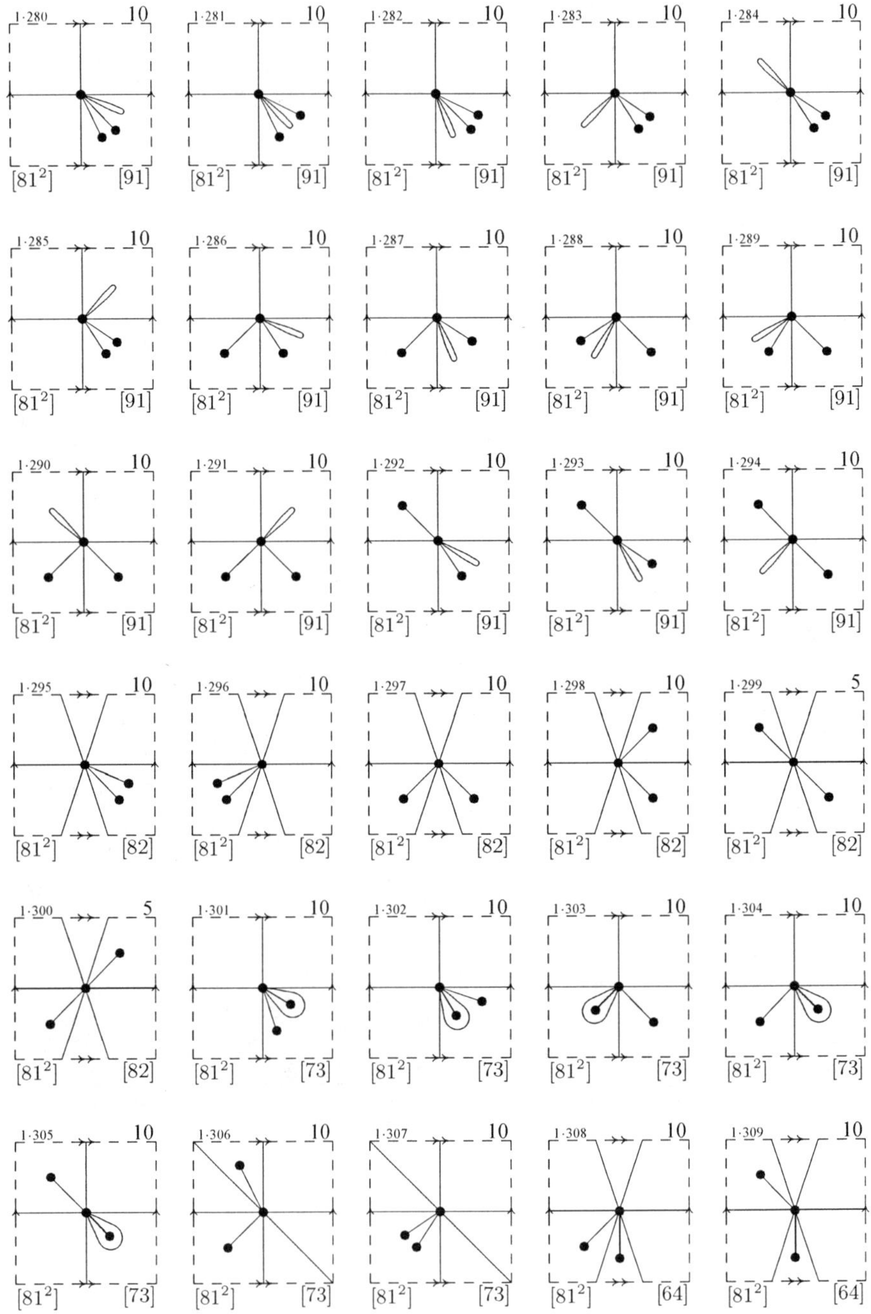
1·280 10 [81²] [91]
1·281 10 [81²] [91]
1·282 10 [81²] [91]
1·283 10 [81²] [91]
1·284 10 [81²] [91]
1·285 10 [81²] [91]
1·286 10 [81²] [91]
1·287 10 [81²] [91]
1·288 10 [81²] [91]
1·289 10 [81²] [91]
1·290 10 [81²] [91]
1·291 10 [81²] [91]
1·292 10 [81²] [91]
1·293 10 [81²] [91]
1·294 10 [81²] [91]
1·295 10 [81²] [82]
1·296 10 [81²] [82]
1·297 10 [81²] [82]
1·298 10 [81²] [82]
1·299 5 [81²] [82]
1·300 5 [81²] [82]
1·301 10 [81²] [73]
1·302 10 [81²] [73]
1·303 10 [81²] [73]
1·304 10 [81²] [73]
1·305 10 [81²] [73]
1·306 10 [81²] [73]
1·307 10 [81²] [73]
1·308 10 [81²] [64]
1·309 10 [81²] [64]

5e 3v

1·310 10
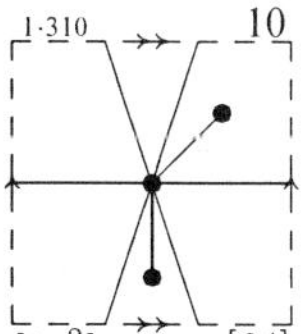
$[81^2]$ [64]

1·311 10
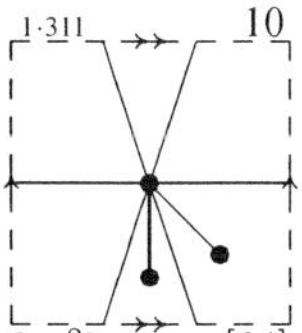
$[81^2]$ [64]

1·312 5
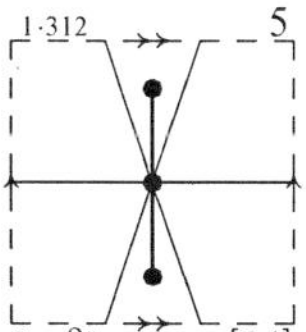
$[81^2]$ [64]

1·313 10
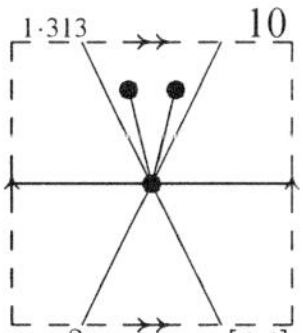
$[81^2]$ [64]

1·314 5
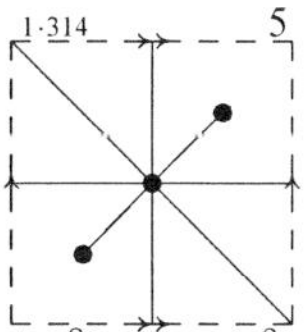
$[81^2]$ $[5^2]$

1·315 10
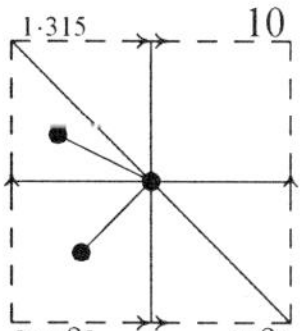
$[81^2]$ $[5^2]$

1·316 10
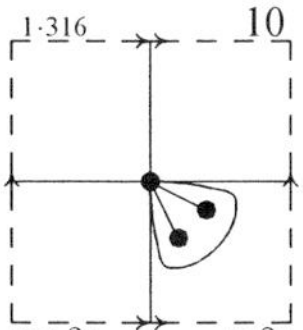
$[81^2]$ $[5^2]$

$19 + 37u$

5 edges
3 vertices
$\mathfrak{g}_{59}$

0·145 10
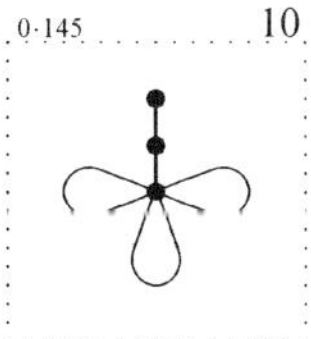
[721] $[71^3]$

0·146 10
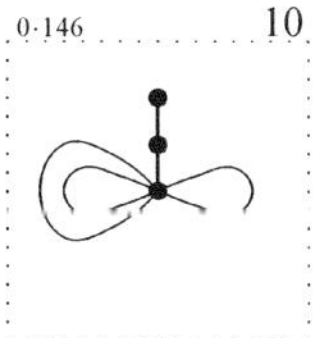
[721] $[621^2]$

0·147 10
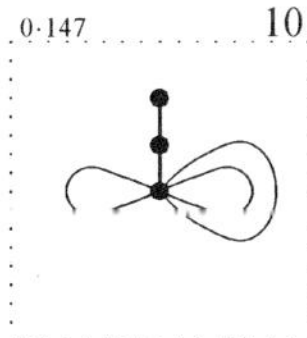
[721] $[621^2]$

0·151 10
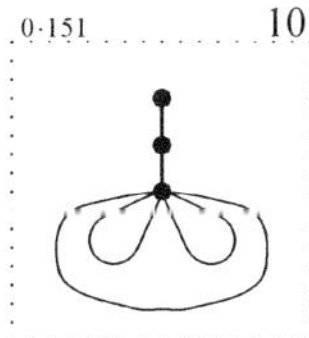
[721] $[531^2]$

0·154 10
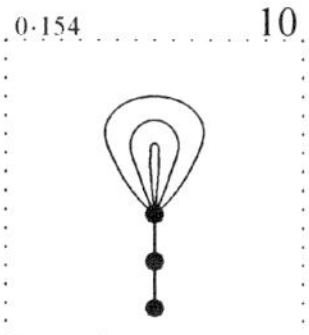
[721] $[52^21]$

1·327 10
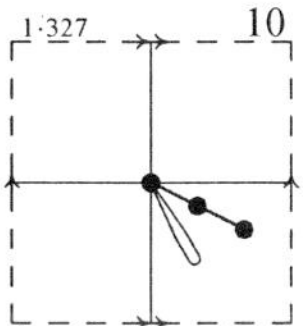
[721] [91]

1·328 10
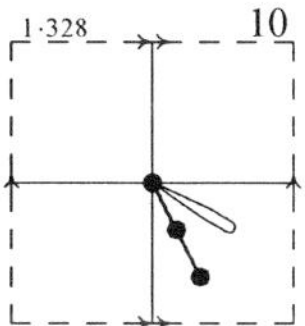
[721] [91]

1·329 10
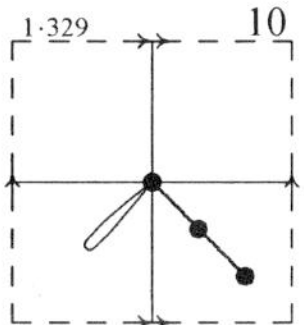
[721] [91]

1·330 10
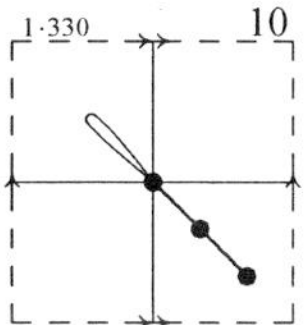
[721] [91]

1·331 10
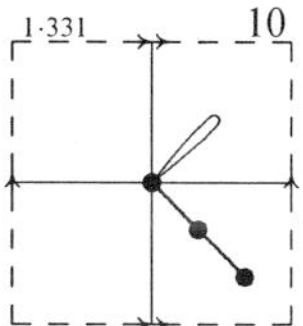
[721] [91]

1·339 10
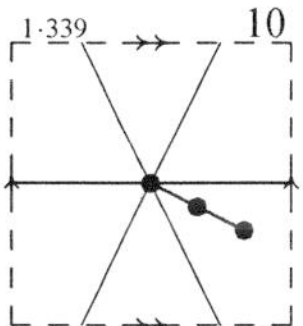
[721] [82]

1·340 10
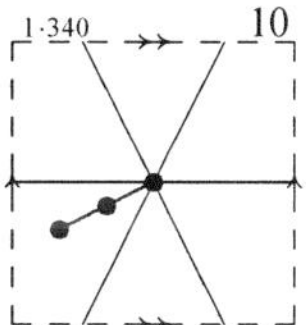
[721] [82]

1·343 10
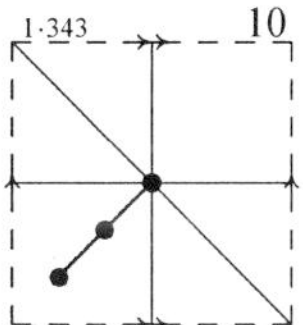
[721] [73]

1·353 10
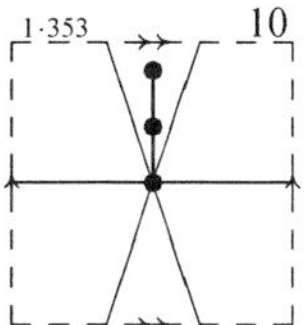
[721] [64]

1·354 10
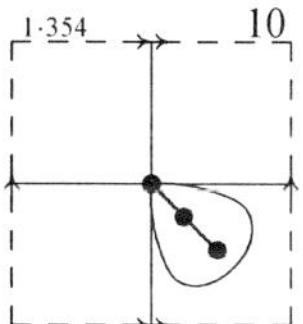
[721] $[5^2]$

$5 + 10u$

5 edges
3 vertices
$\mathfrak{g}_{60}$

0·148 10
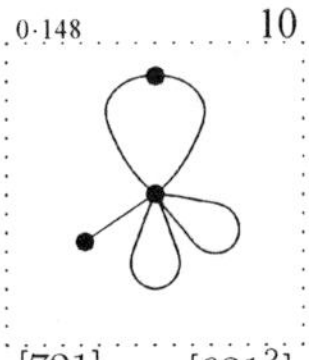
[721] $[621^2]$

0·149 10
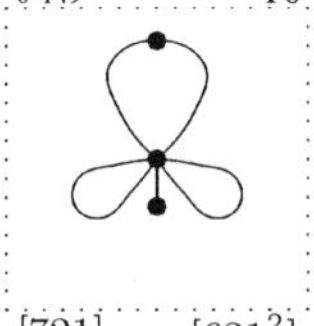
[721] $[621^2]$

0·150 10
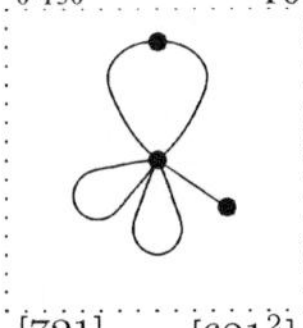
[721] $[621^2]$

0·152 10
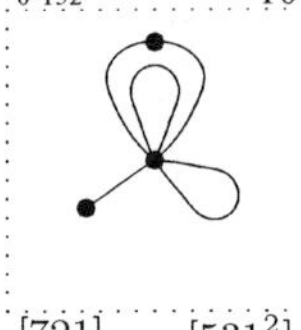
[721] $[531^2]$

0·153 10
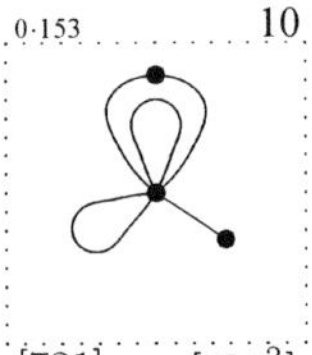
[721] $[531^2]$

0·155 10
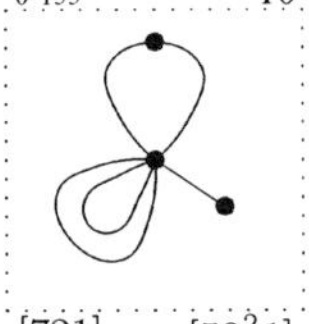
[721] $[52^21]$

0·156 10
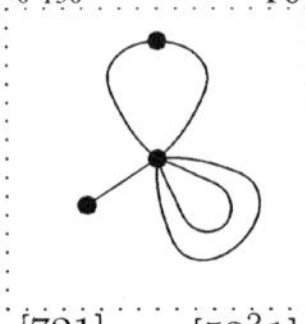
[721] $[52^21]$

0·157 10
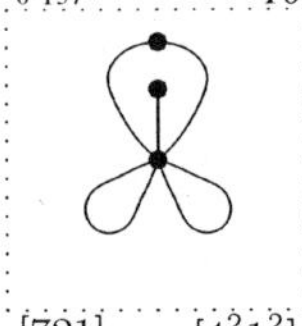
[721] $[4^21^2]$

0·158 10
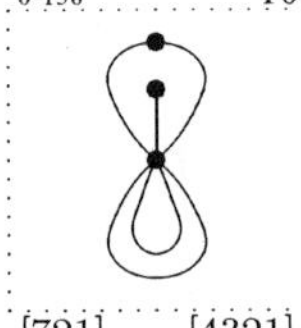
[721] [4321]

0·159 10
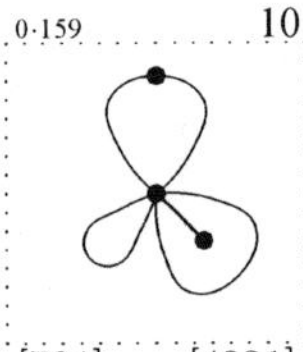
[721] [4321]

0·160 10
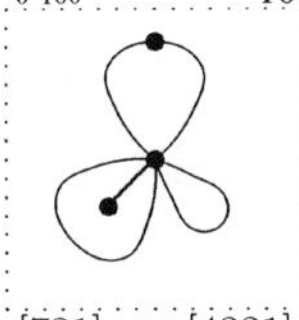
[721] [4321]

0·161 10
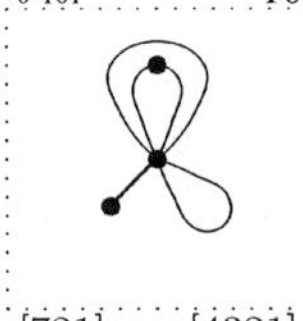
[721] [4321]

0·162 10
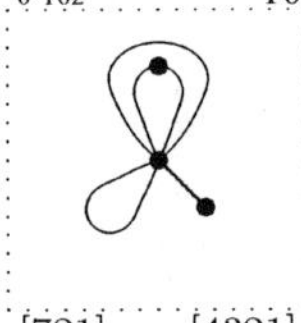
[721] [4321]

0·163 10
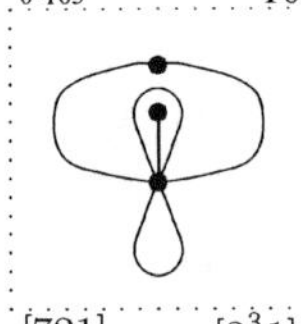
[721] $[3^31]$

0·164 10
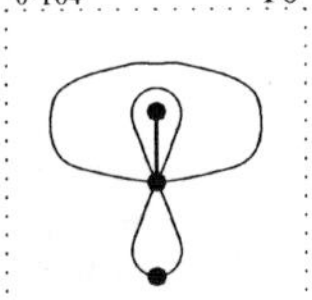
[721] $[3^22^2]$

1·317 10
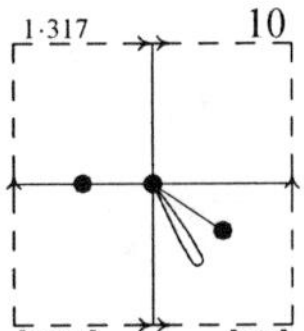
[721] [91]

1·318 10
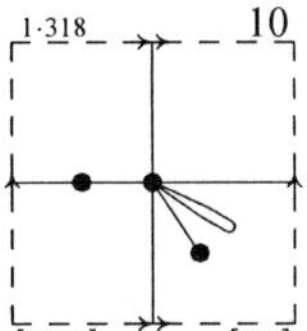
[721] [91]

1·319 10
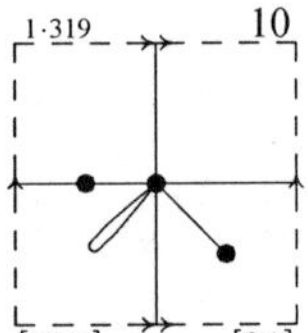
[721] [91]

1·320 10
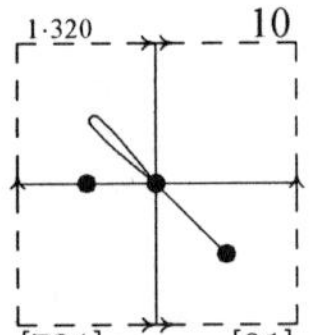
[721] [91]

1·321 10
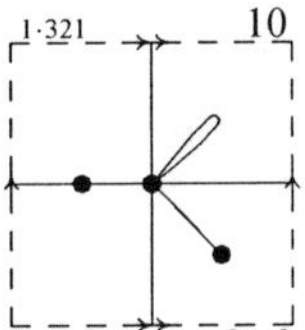
[721] [91]

1·322 10
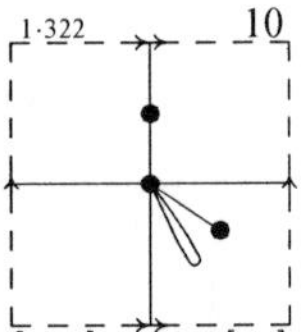
[721] [91]

1·323 10
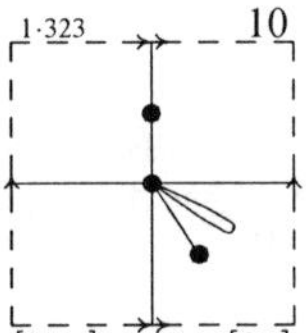
[721] [91]

1·324 10
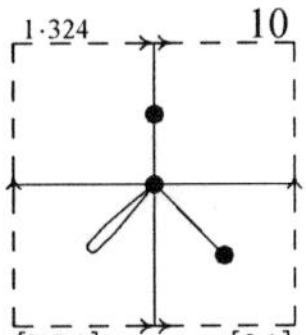
[721] [91]

1·325 10
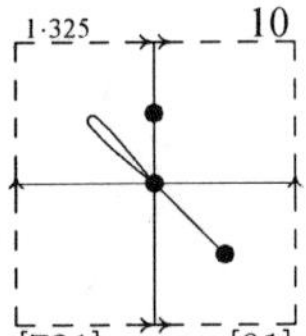
[721] [91]

1·326 10
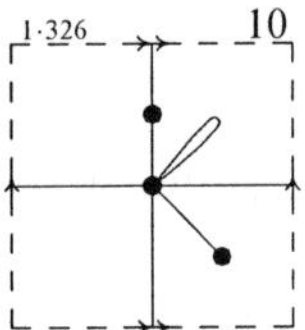
[721] [91]

1·332 10
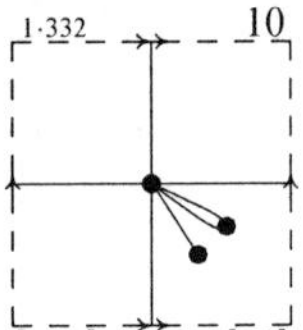
[721] [82]

1·333 10
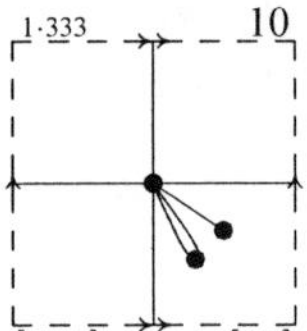
[721] [82]

1·334 10
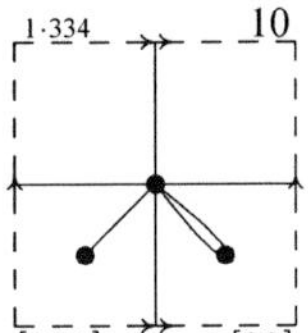
[721] [82]

1·335 10
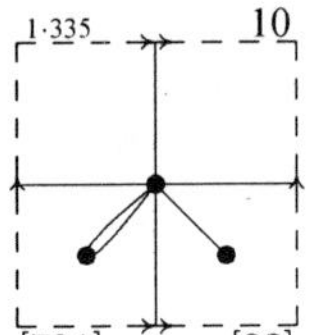
[721] [82]

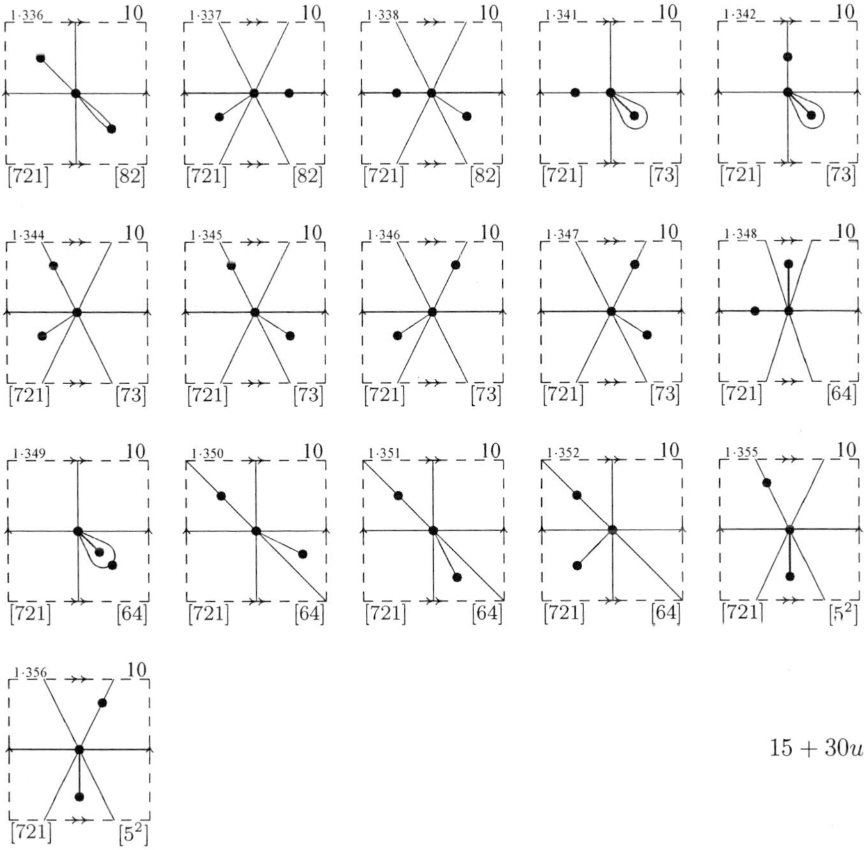

$15 + 30u$

5 edges
3 vertices
$\mathfrak{g}_{61}$

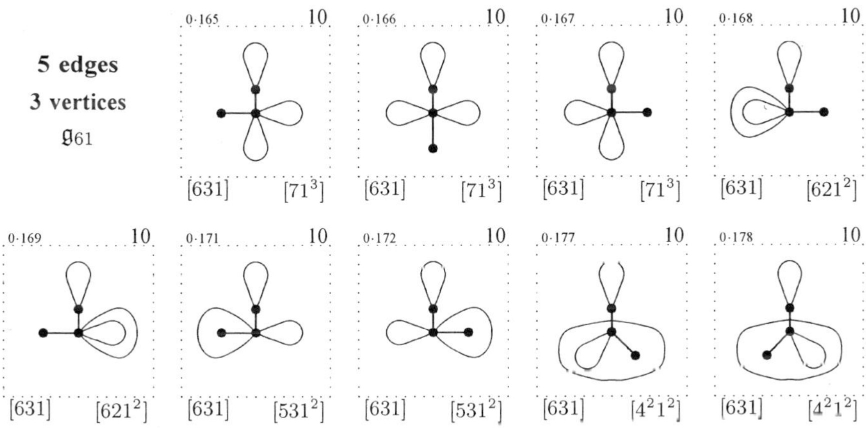

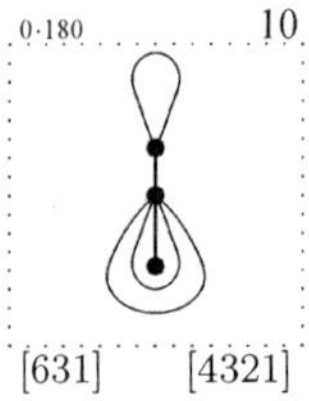

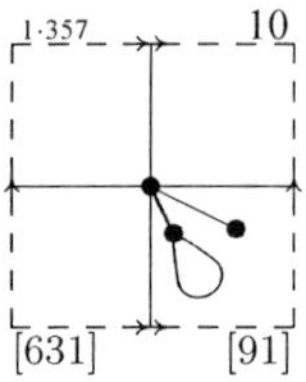

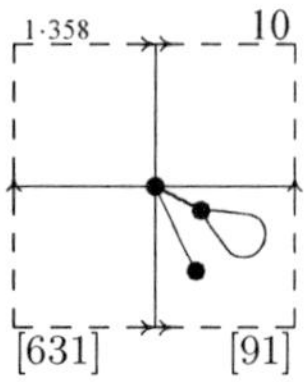

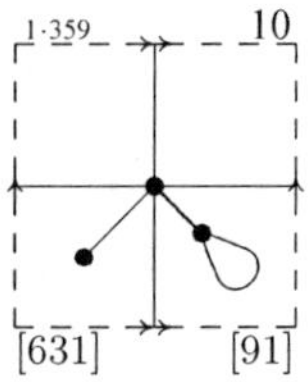

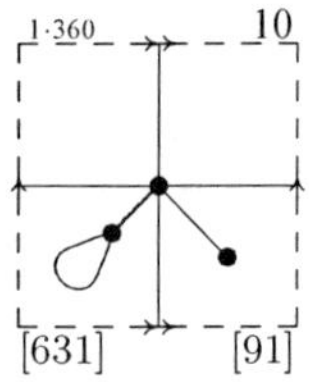

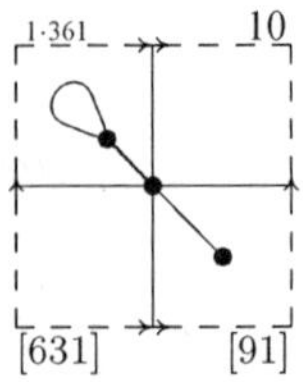

$10 + 5u$

5 edges

3 vertices

$\mathfrak{g}_{62}$

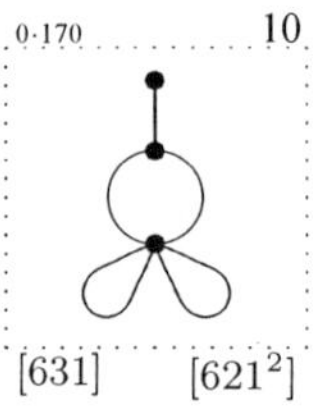

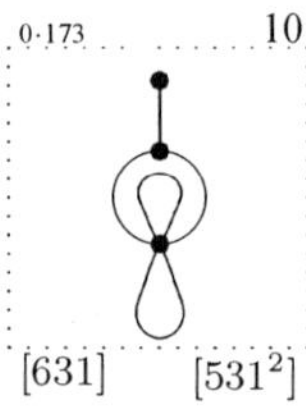

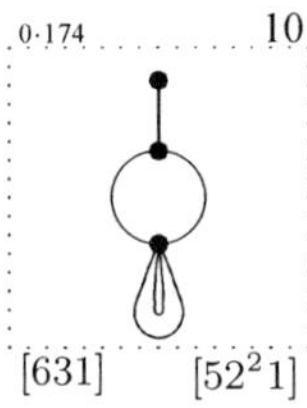

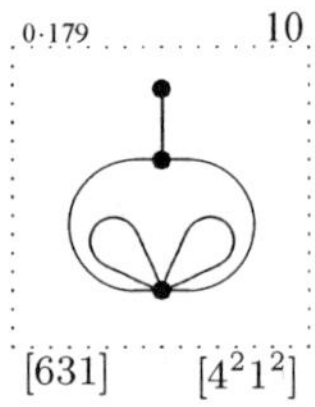

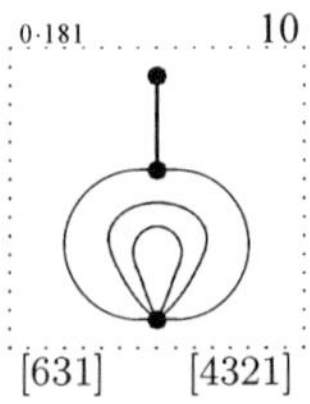

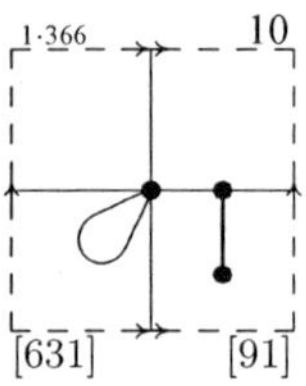

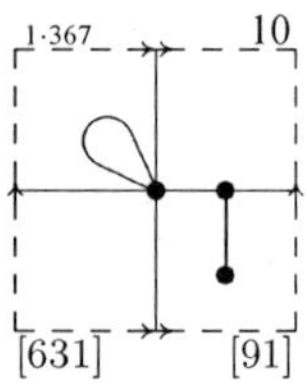

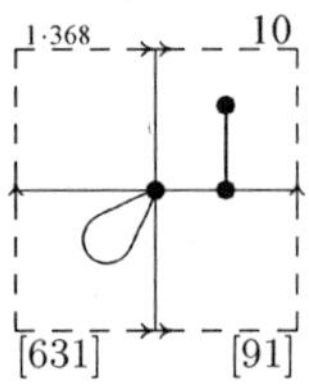

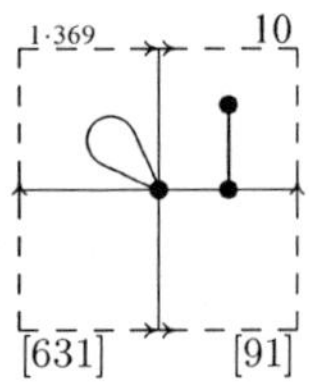

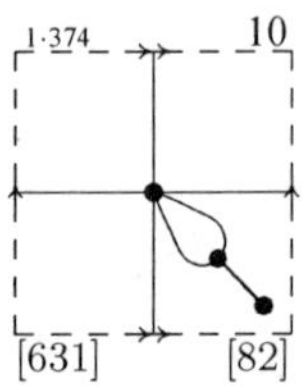

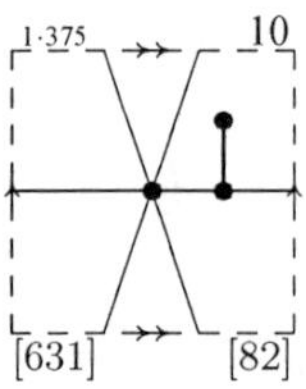

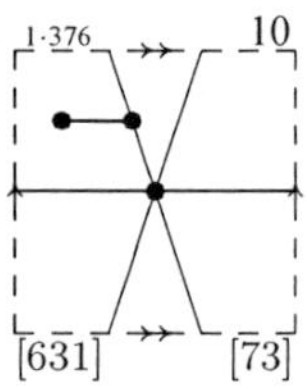

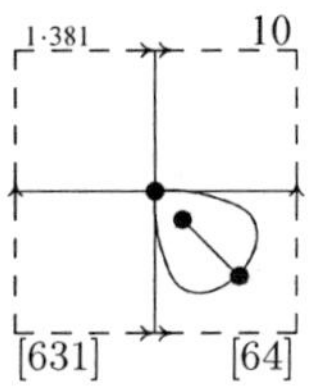

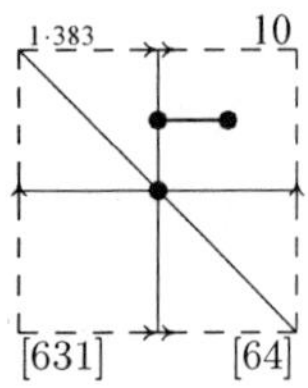

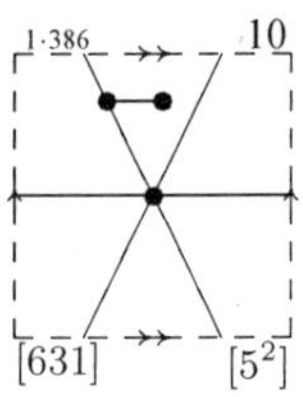

$5 + 10u$

5 edges
3 vertices
$\mathfrak{g}_{63}$

0·175 10
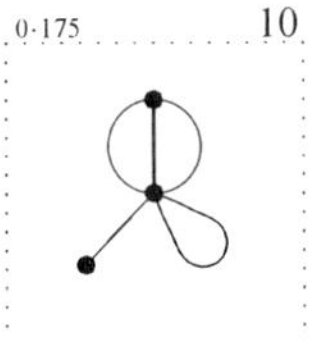
[631] [$52^2 1$]

0·176 10
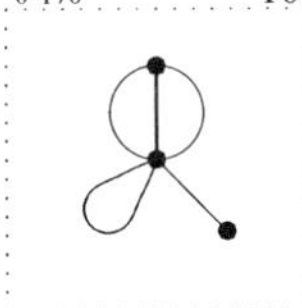
[631] [$52^2 1$]

0·182 10
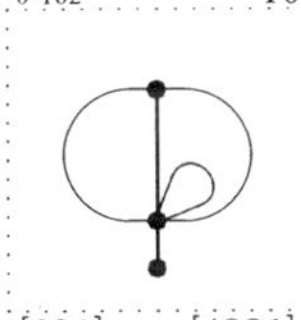
[631] [4321]

0·183 10
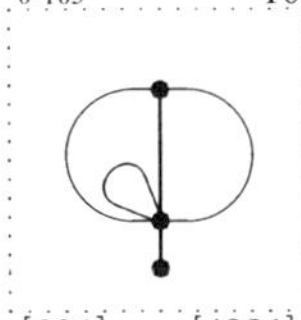
[631] [4321]

0·184 10
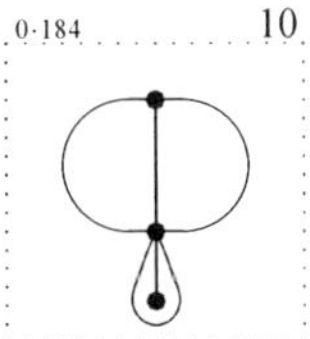
[631] [$3^2 2^2$]

1·362 10
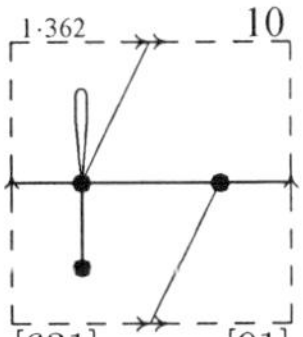
[631] [91]

1·363 10
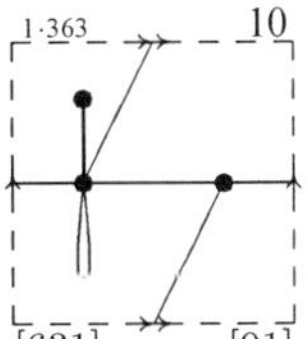
[631] [91]

1·364 10
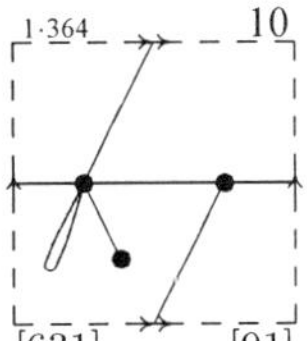
[631] [91]

1·365 10
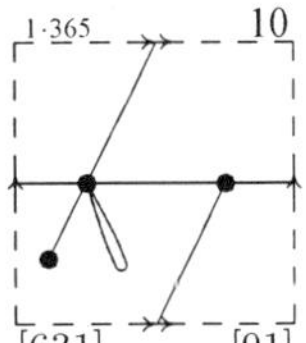
[631] [91]

1·370 10
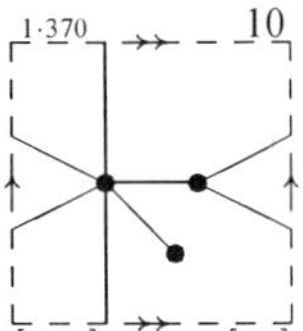
[631] [82]

1·371 10
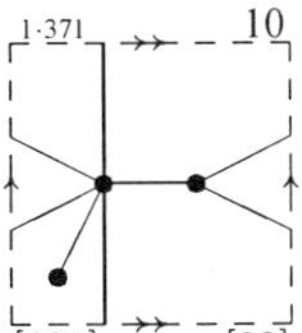
[631] [82]

1·372 10
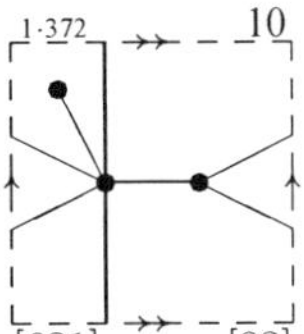
[631] [82]

1·373 10
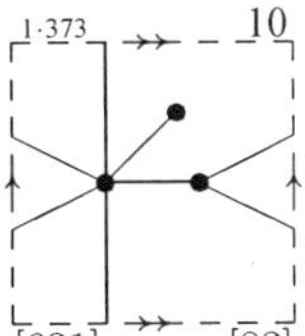
[631] [82]

1·377 10
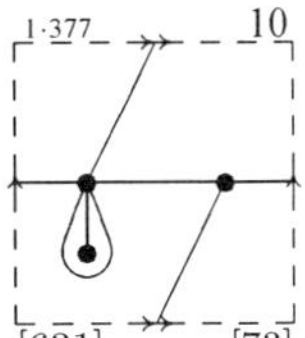
[631] [73]

1·378 10
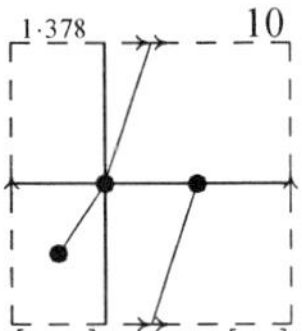
[631] [73]

1·379 10
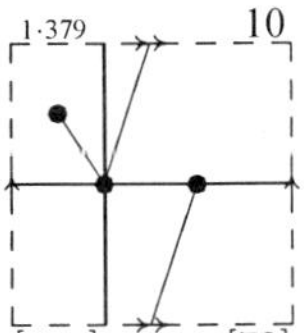
[631] [73]

1·380 10
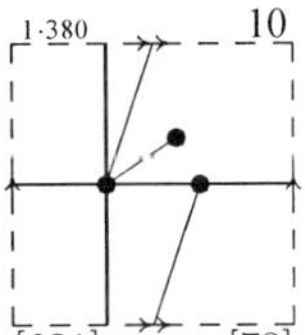
[631] [73]

1·382 10
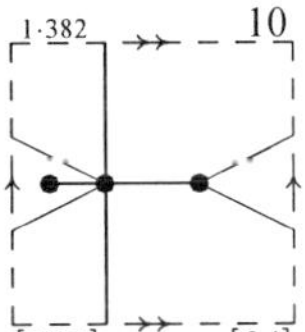
[631] [64]

1·384 10
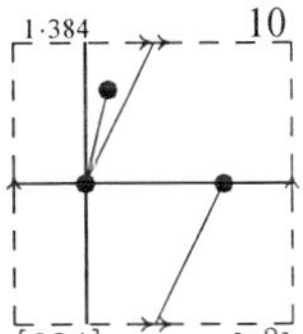
[631] [5^2]

1·385 10
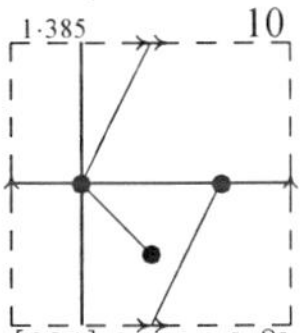
[631] [5^2]

$5 + 15u$

5 edges
3 vertices
$\mathfrak{g}_{64}$

0·185 10
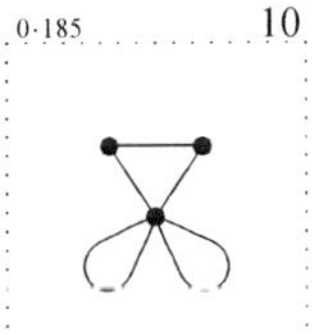
[62^2] [531^2]

0·187 5
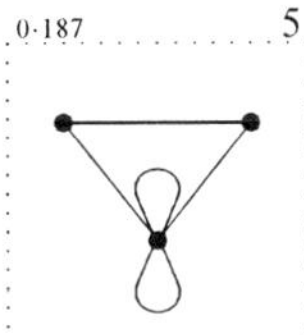
[62^2] [$4^2 1^2$]

0·189 10
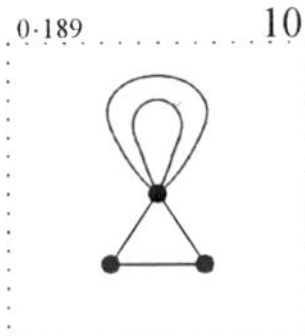
[62^2] [4321]

1·388 10
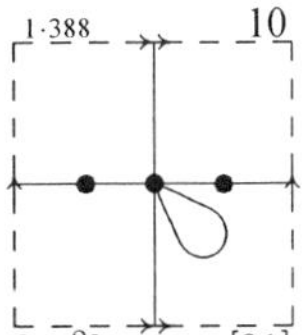
[62^2] [91]

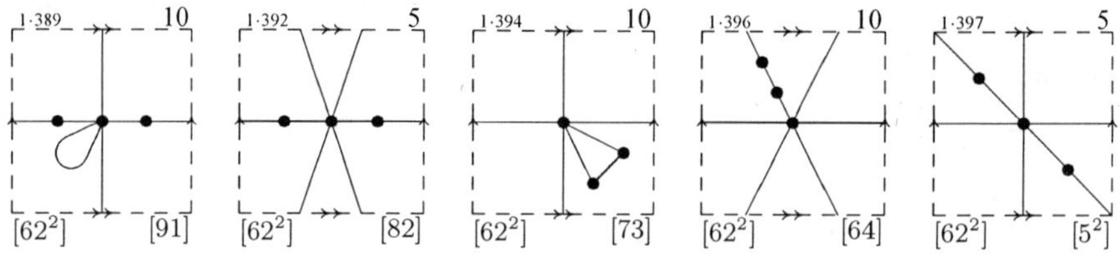

$3 + 6u$

5 edges

3 vertices

$\mathfrak{g}_{65}$

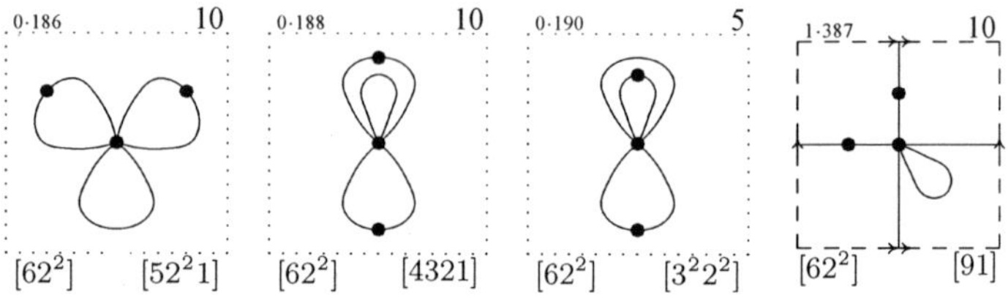

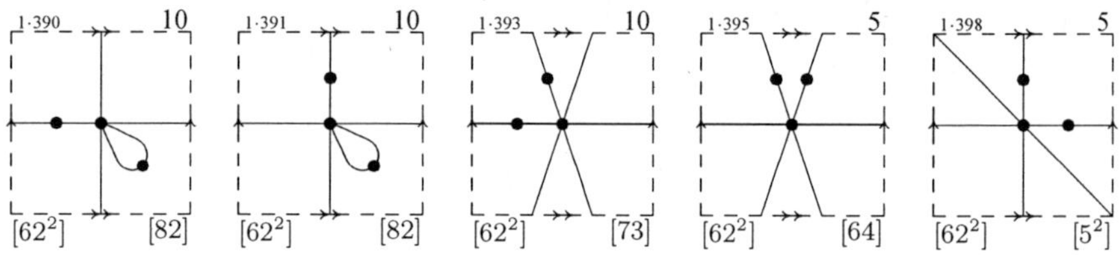

$3 + 6u$

5 edges

3 vertices

$\mathfrak{g}_{66}$

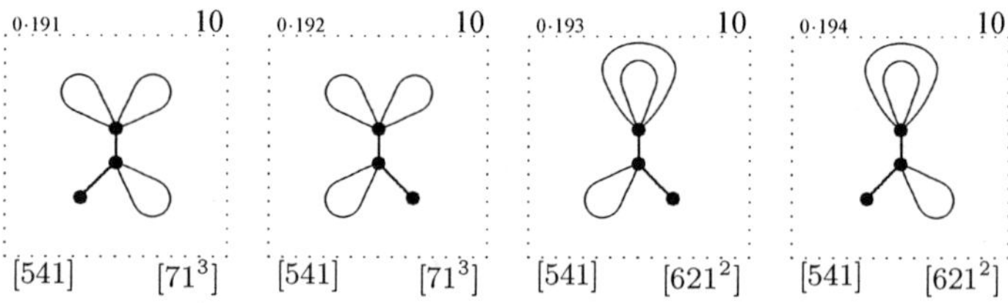

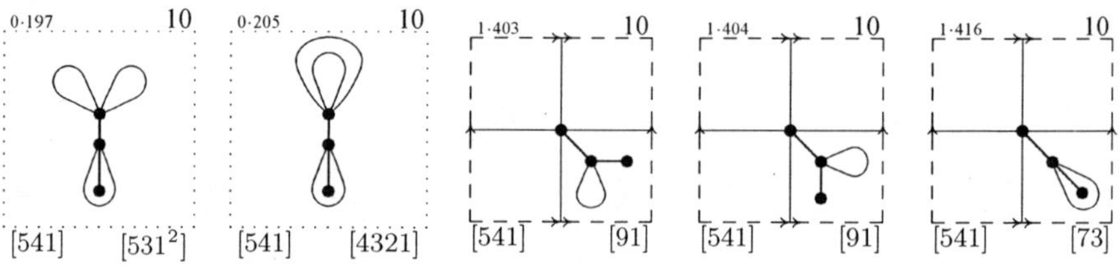

$6 + 3u$

5 edges
3 vertices
$\mathfrak{g}_{67}$

0·195 10
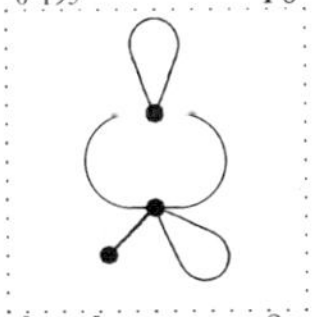
[541] $[621^2]$

0·196 10
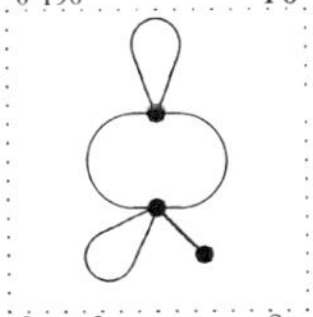
[541] $[621^2]$

0·198 10
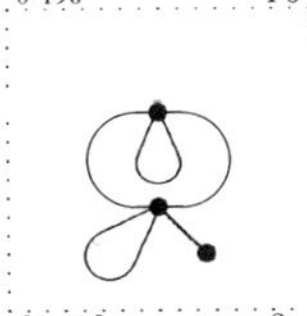
[541] $[531^2]$

0·199 10
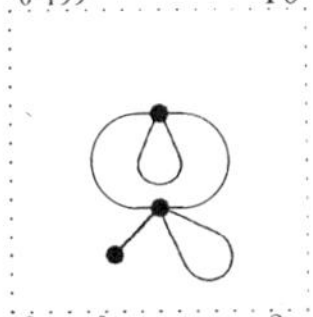
[541] $[531^2]$

0·200 10
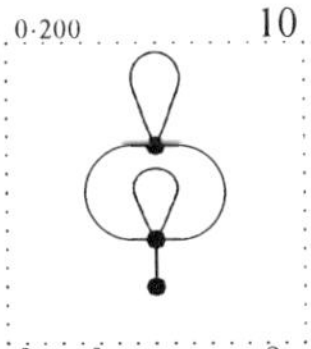
[541] $[531^2]$

0·202 10
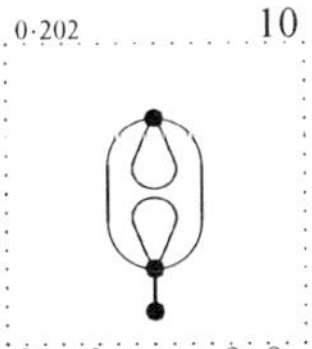
[541] $[4^21^2]$

0·206 10
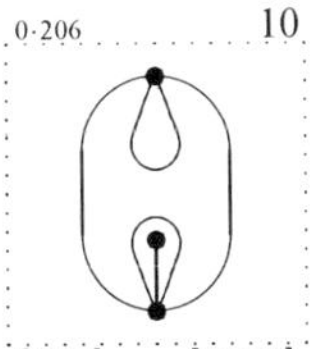
[541] [4321]

0·208 10
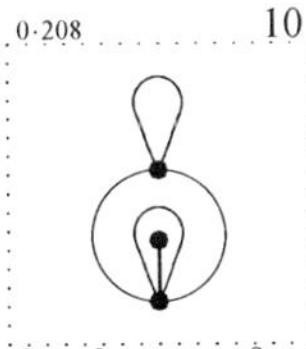
[541] $[3^31]$

1·399 10
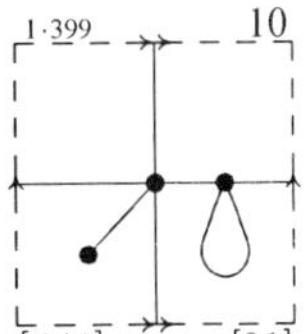
[541] [91]

1·400 10
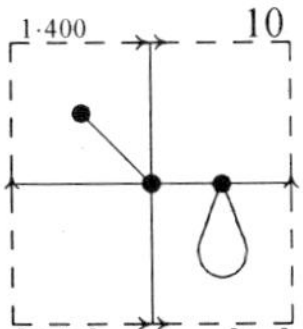
[541] [91]

1·401 10
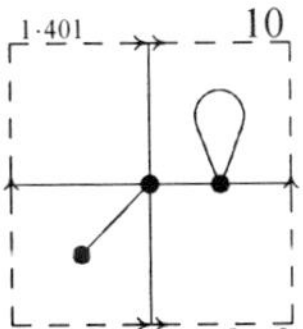
[541] [91]

1·402 10
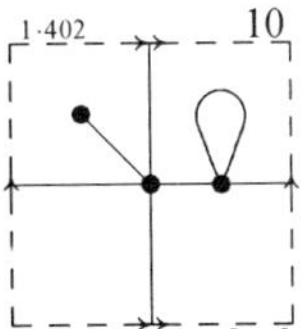
[541] [91]

1·408 10
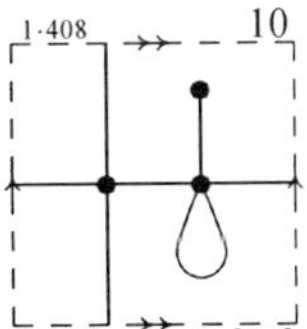
[541] [91]

1·409 10
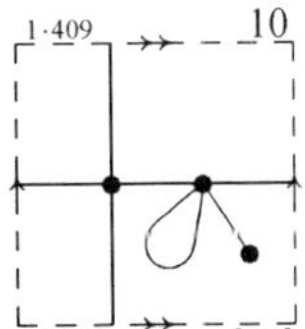
[541] [91]

1·410 10
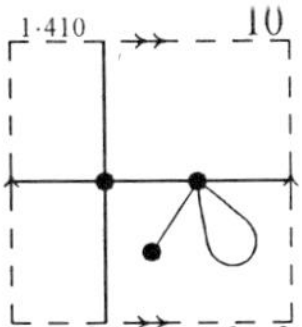
[541] [91]

1·417 10
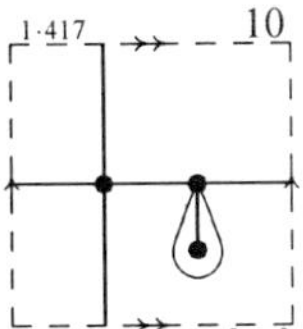
[541] [73]

1·420 10
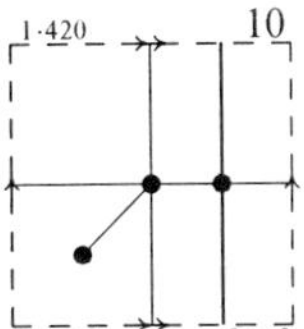
[541] [64]

1·421 10
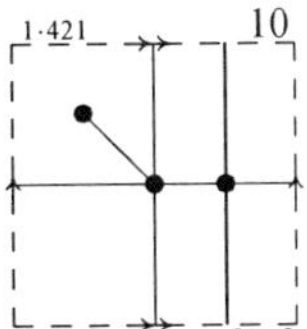
[541] [64]

$8 + 10u$

5 edges
3 vertices
$\mathfrak{g}_{68}$

0·201 10
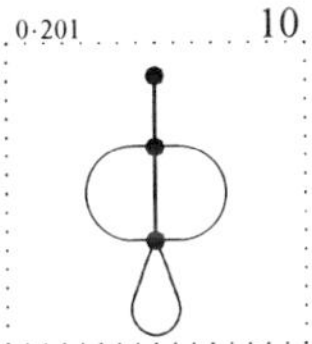
[541] $[52^21]$

0·203 10
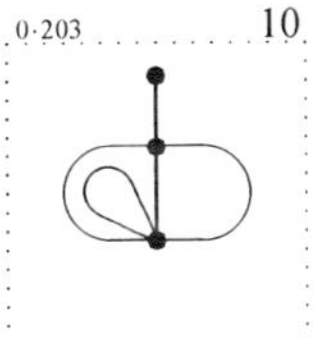
[541] [4321]

0·204 10
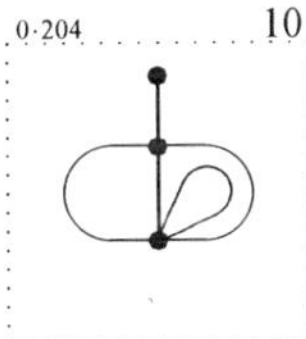
[541] [4321]

1·405 10
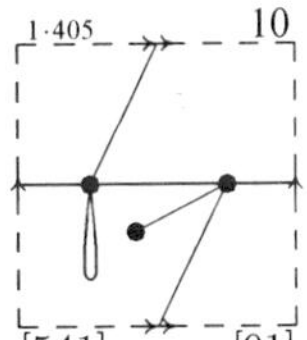
[541] [91]

1·406 10
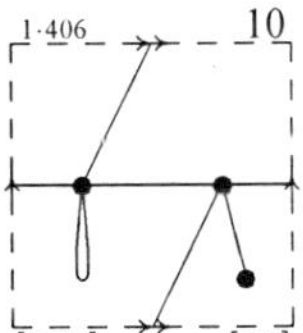
[541] [91]

1·407 10
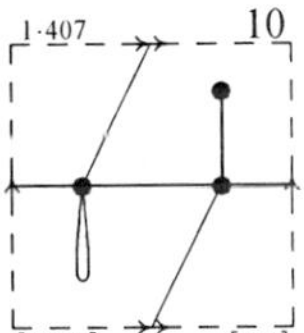
[541] [91]

1·414 10
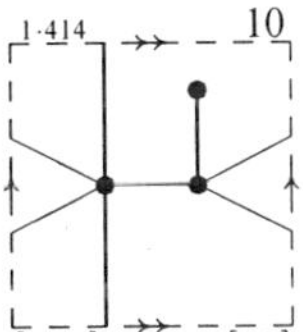
[541] [82]

1·415 10
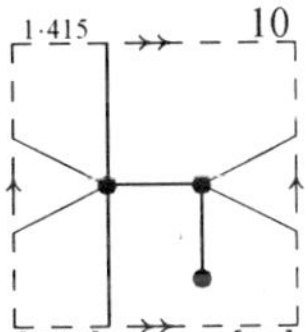
[541] [82]

1·418 10
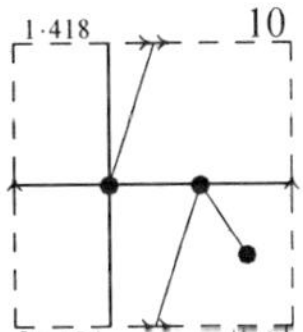
[541] [73]

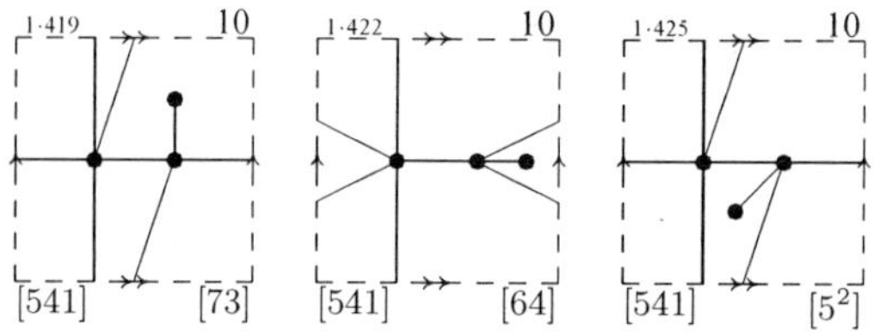

$3 + 9u$

5 edges
3 vertices
$\mathfrak{g}_{69}$

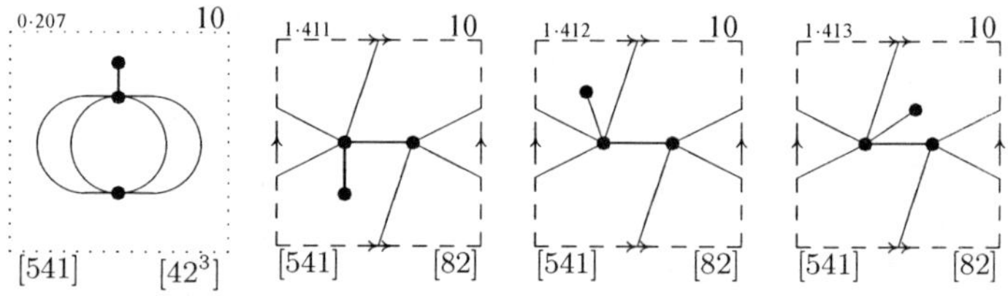

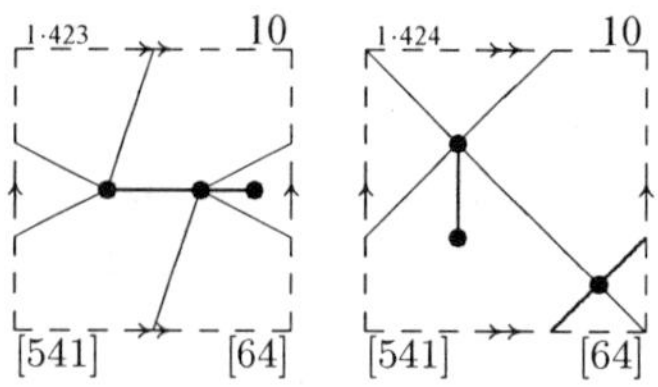

$1 + 5u$

5 edges
3 vertices
$\mathfrak{g}_{70}$

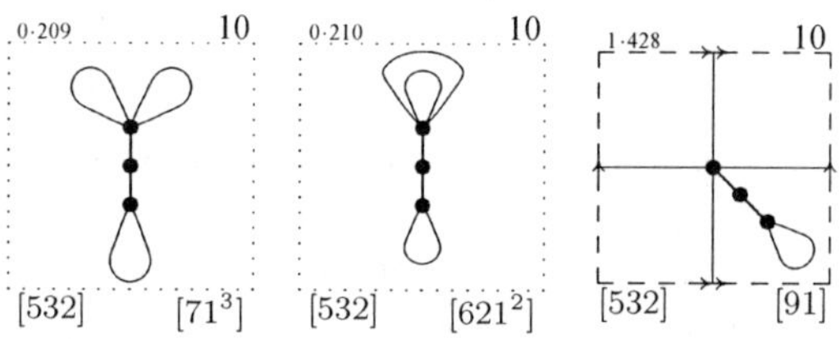

$2 + u$

5 edges
3 vertices
$\mathfrak{g}_{71}$

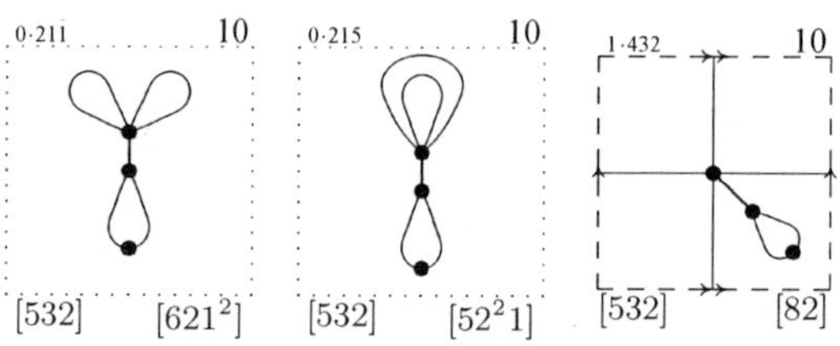

$2 + u$

5 edges
3 vertices
$\mathfrak{g}_{72}$

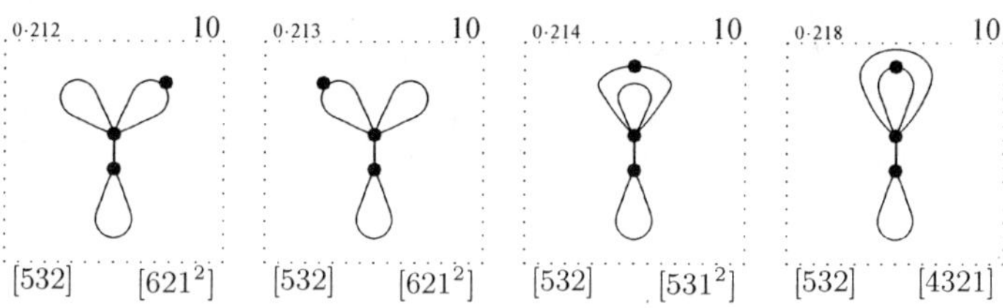

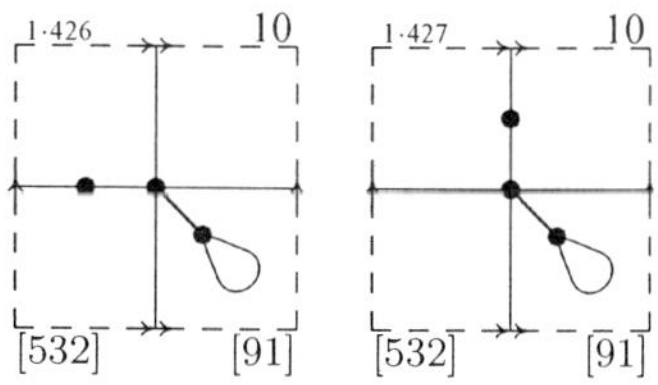

$4 + 2u$

5 edges
3 vertices
$\mathfrak{g}_{73}$

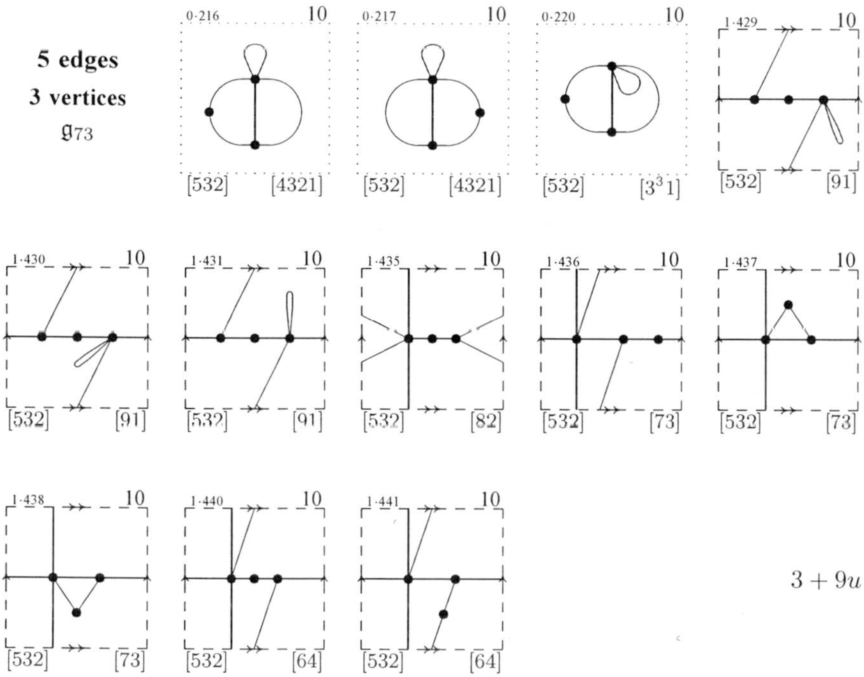

$3 + 9u$

5 edges
3 vertices
$\mathfrak{g}_{74}$

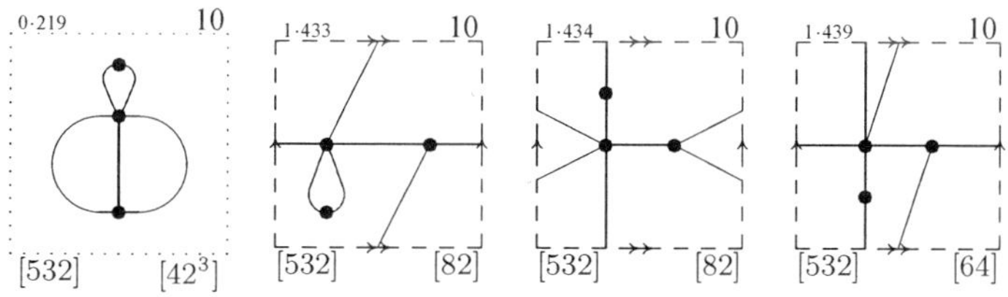

$1 + 3u$

5 edges
3 vertices
$\mathfrak{g}75$

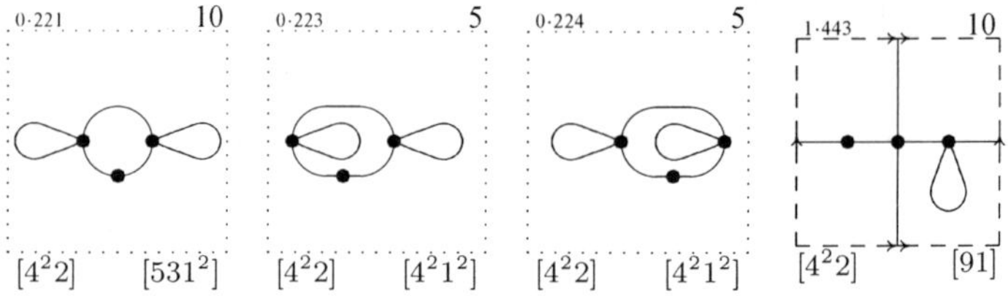

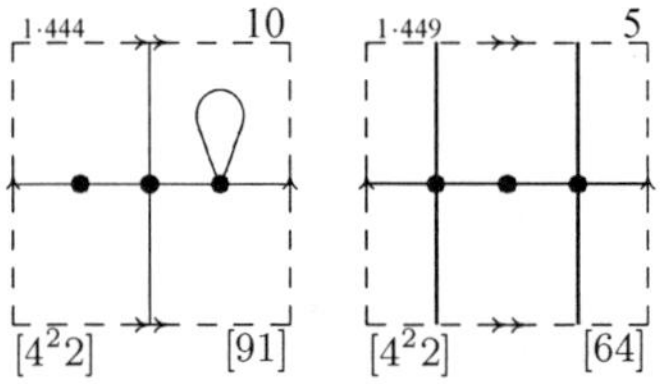

$3 + 3u$

5 edges
3 vertices
$\mathfrak{g}76$

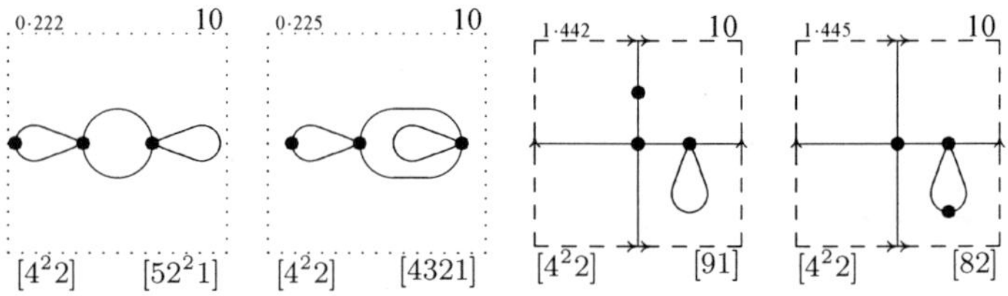

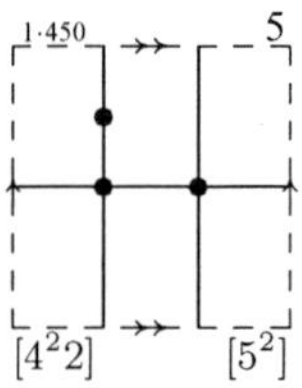

$2 + 3u$

5 edges
3 vertices
$\mathfrak{g}77$

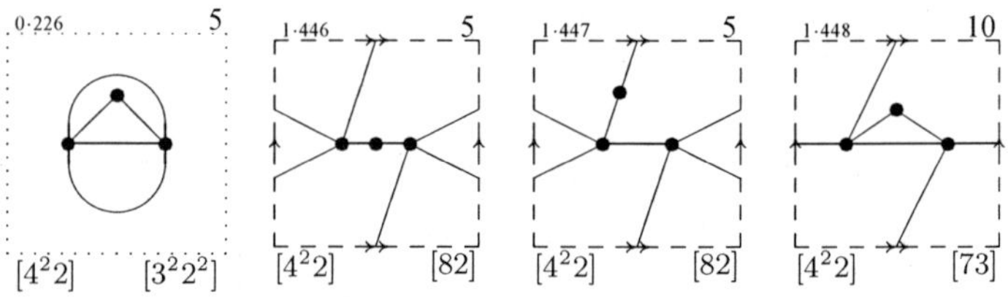

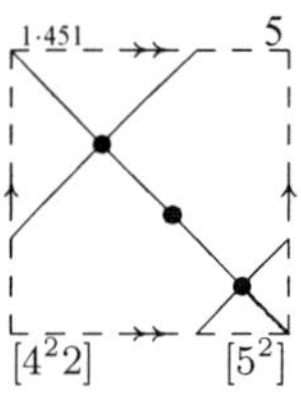

$1 + 4u$

5 edges
3 vertices
$\mathfrak{g}_{78}$

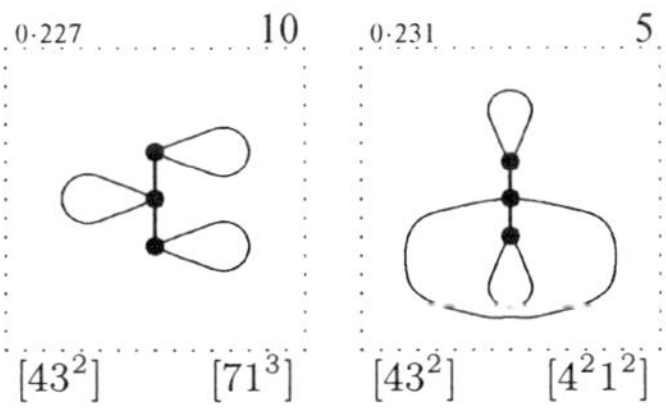

2

5 edges
3 vertices
$\mathfrak{g}_{79}$

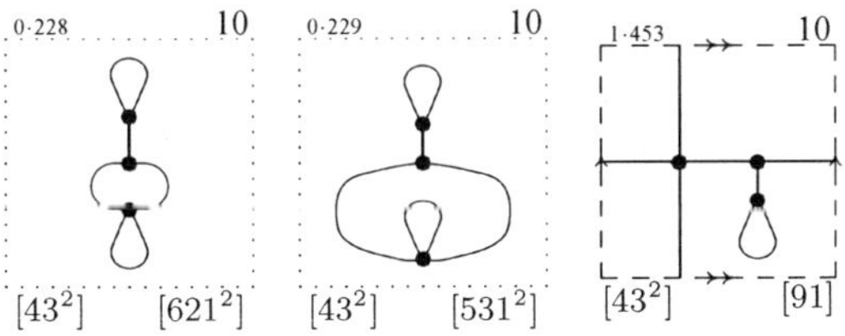

$2 + u$

5 edges
3 vertices
$\mathfrak{g}_{80}$

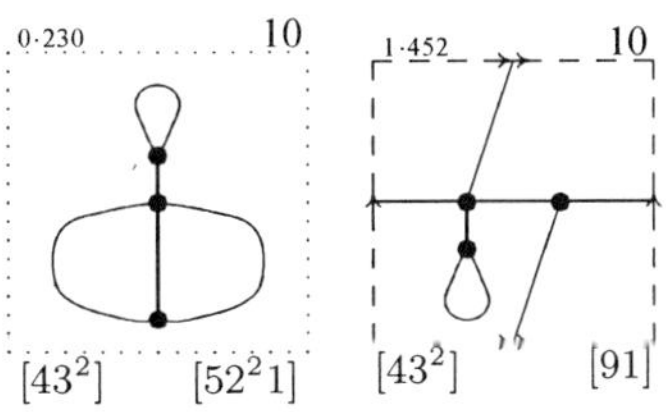

$1 + u$

5 edges
3 vertices
$\mathfrak{g}_{81}$

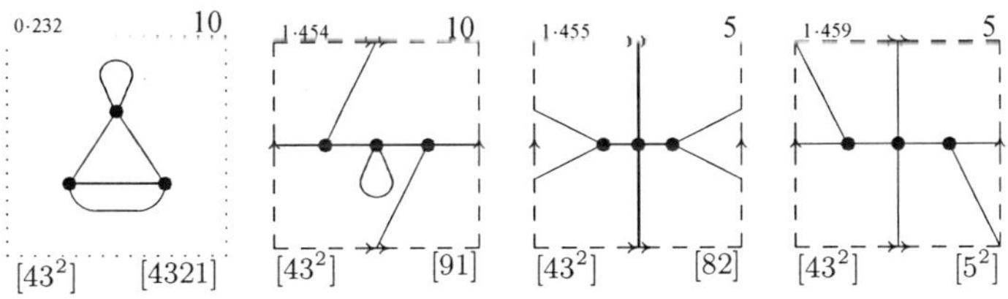

$1 + 3u$

5 edges
3 vertices
$\mathfrak{g}_{82}$

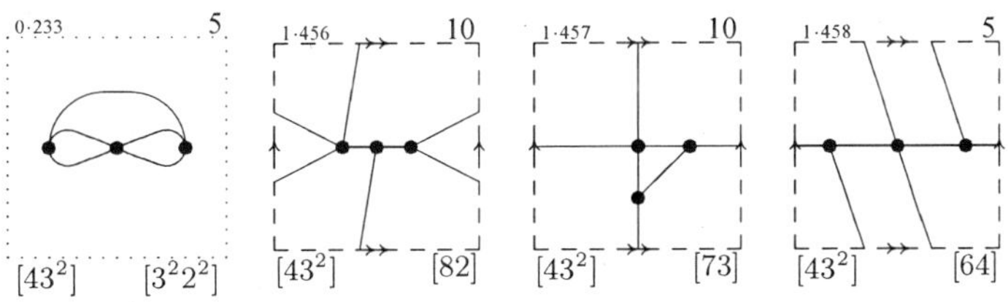

$1 + 3u$

5 edges
4 vertices
$\mathfrak{g}_{83}$

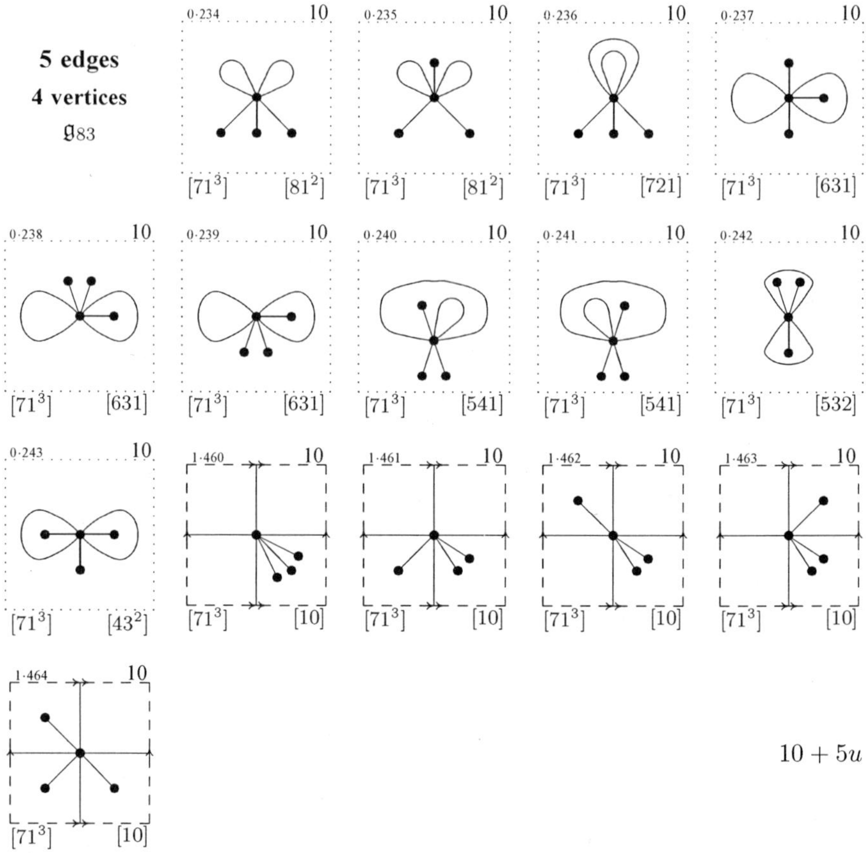

$10 + 5u$

5 edges
4 vertices
$\mathfrak{g}_{84}$

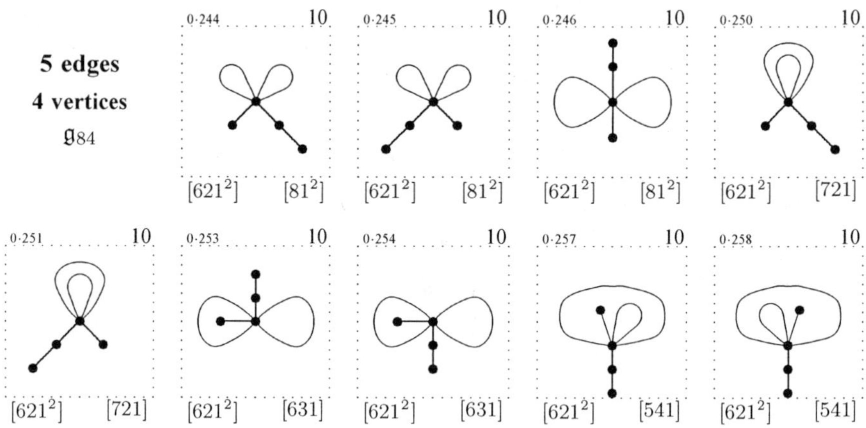

0·262 10
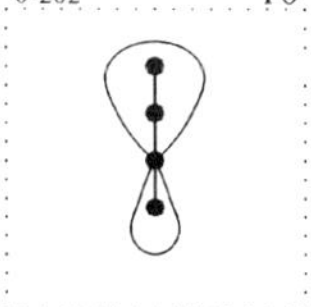
$[621^2]$ $[532]$

1·465 10
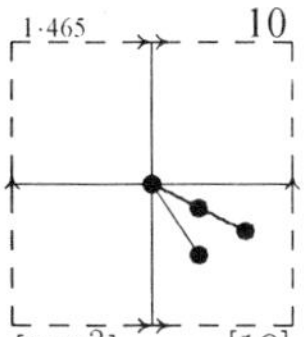
$[621^2]$ $[10]$

1·466 10
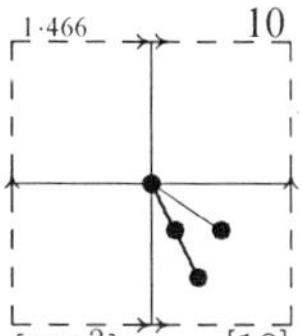
$[621^2]$ $[10]$

1·467 10
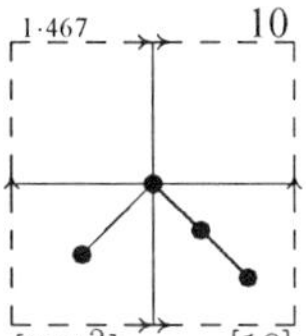
$[621^2]$ $[10]$

1·468 10
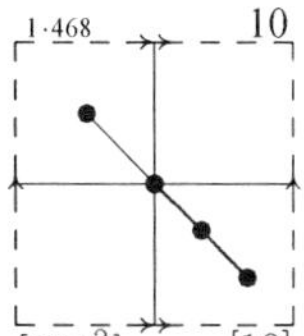
$[621^2]$ $[10]$

1·469 10
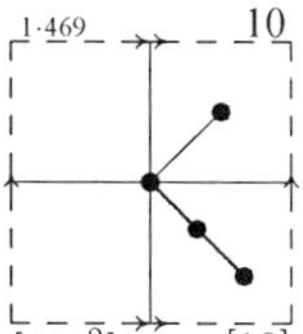
$[621^2]$ $[10]$

$10 + 5u$

5 edges
4 vertices
$\mathfrak{g}_{85}$

0·247 10
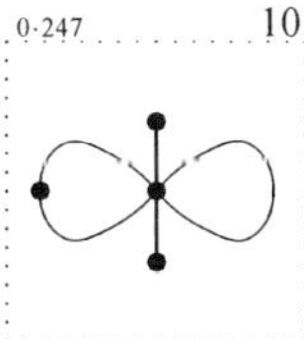
$[621^2]$ $[721]$

0·248 10
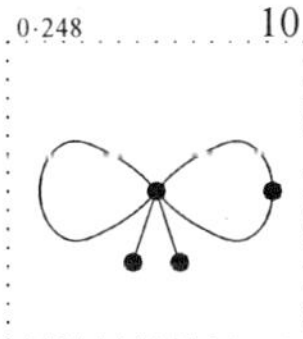
$[621^2]$ $[721]$

0·249 10
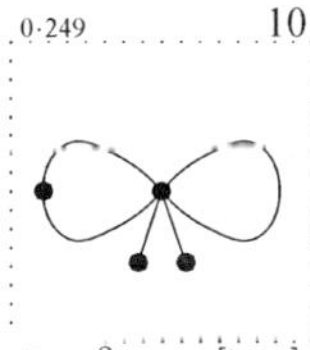
$[621^2]$ $[721]$

0·252 10
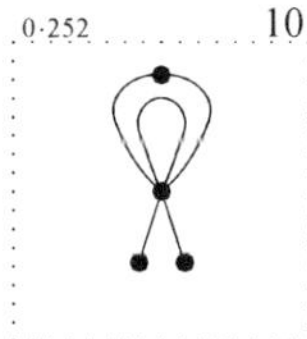
$[621^2]$ $[631]$

0·255 10
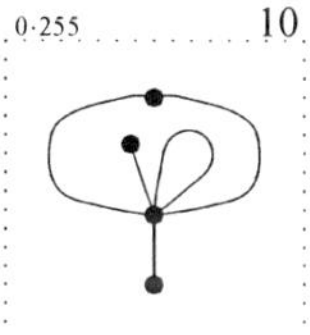
$[621^2]$ $[541]$

0·256 10
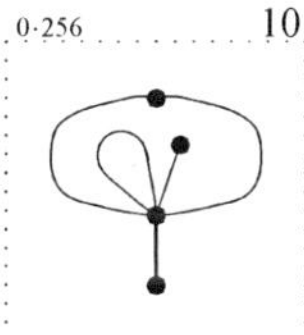
$[621^2]$ $[541]$

0·259 10
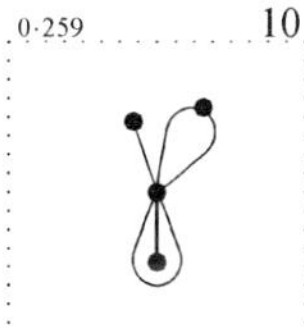
$[621^2]$ $[532]$

0·260 10
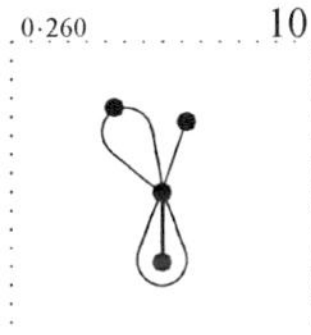
$[621^2]$ $[532]$

0·261 10
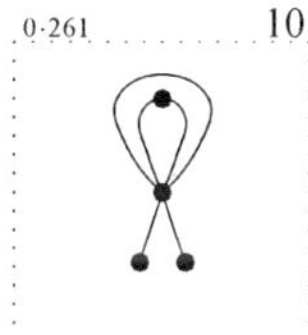
$[621^2]$ $[532]$

0·263 10
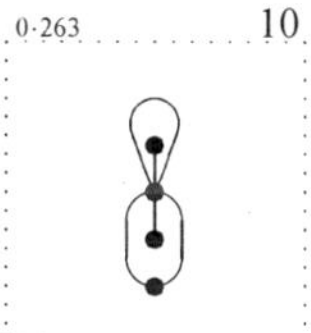
$[621^2]$ $[43^2]$

1·470 10
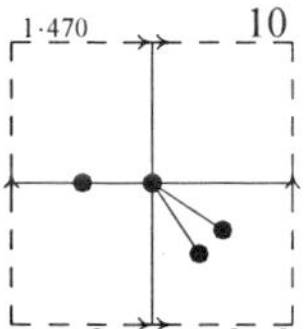
$[621^2]$ $[10]$

1·471 10
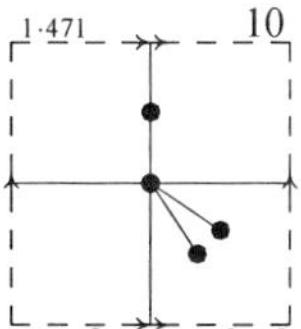
$[621^2]$ $[10]$

1·472 10
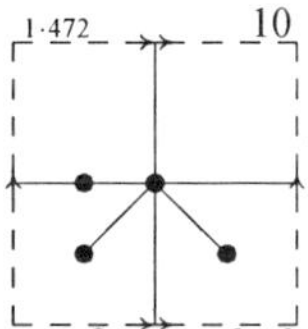
$[621^2]$ $[10]$

1·473 10
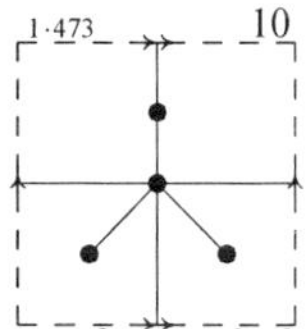
$[621^2]$ $[10]$

1·474 5
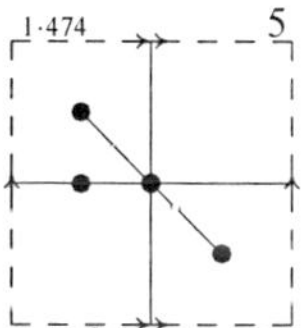
$[621^2]$ $[10]$

1·475 5
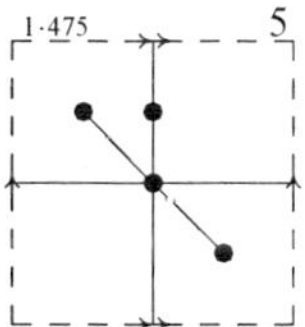
$[621^2]$ $[10]$

$10 + 6u$

5 edges
4 vertices
$\mathfrak{g}_{86}$

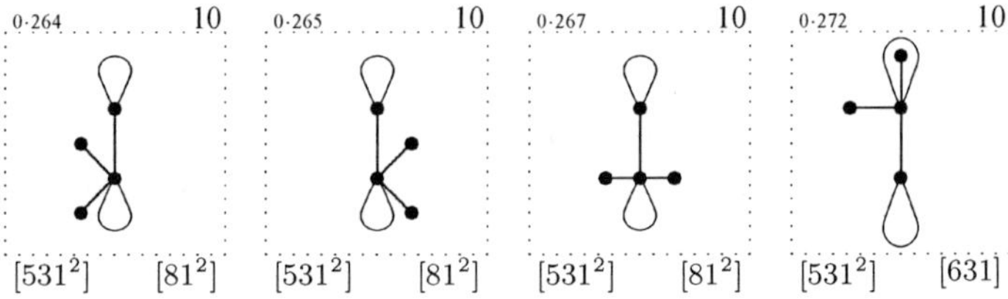

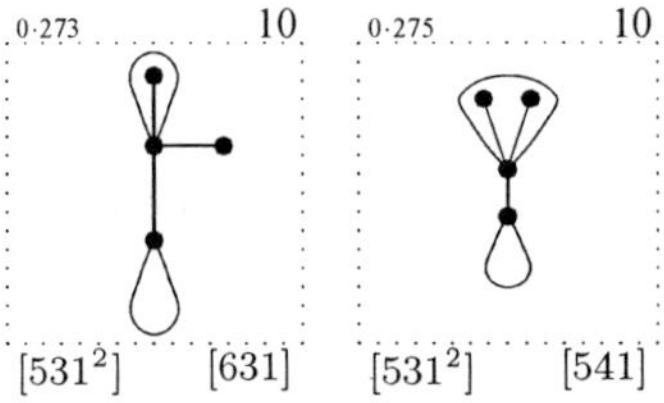

6

5 edges
4 vertices
$\mathfrak{g}_{87}$

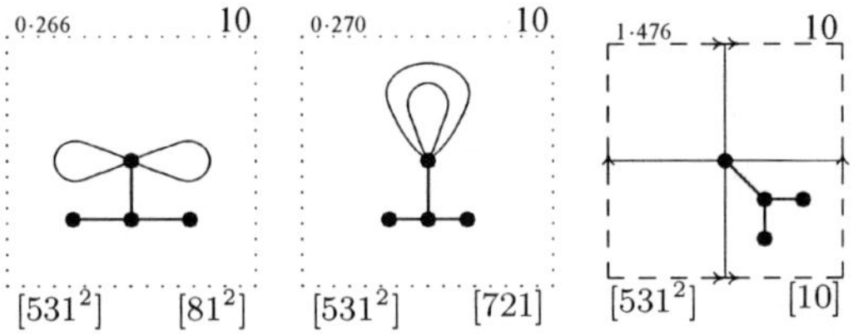

$2+u$

5 edges
4 vertices
$\mathfrak{g}_{88}$

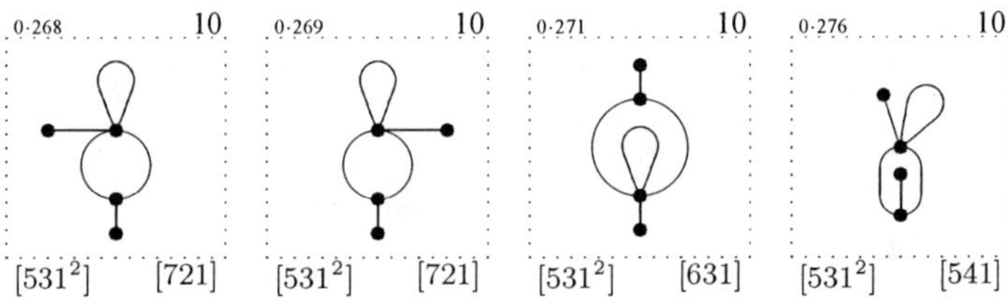

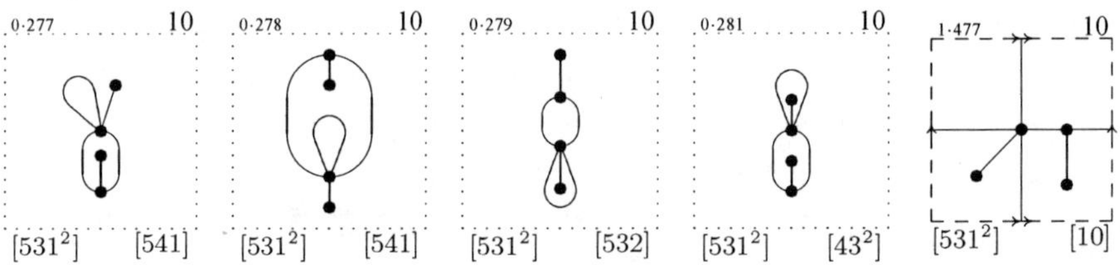

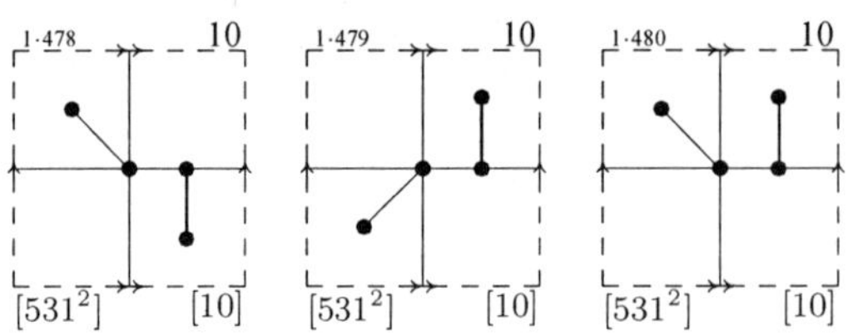

$8+4u$

5 edges
4 vertices
$\mathfrak{g}_{89}$

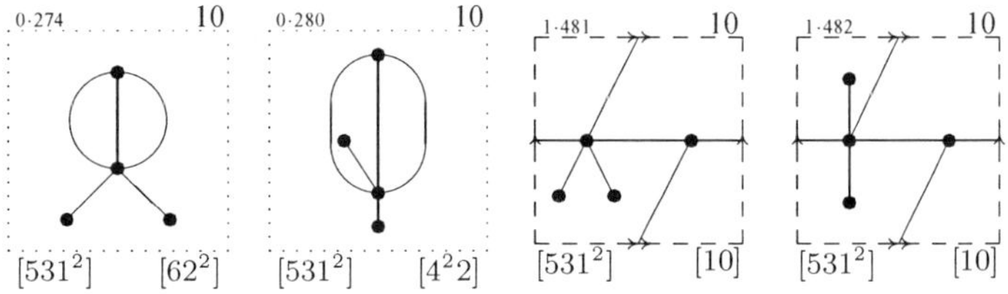

$2 + 2u$

5 edges
4 vertices
$\mathfrak{g}_{90}$

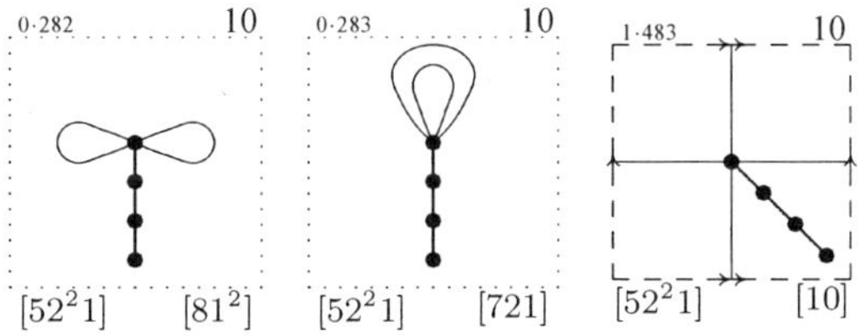

$2 + u$

5 edges
4 vertices
$\mathfrak{g}_{91}$

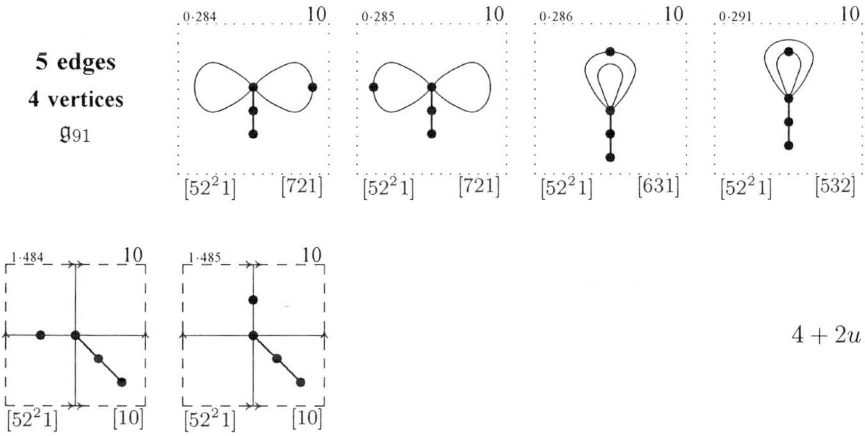

$4 + 2u$

5 edges
4 vertices
$\mathfrak{g}_{92}$

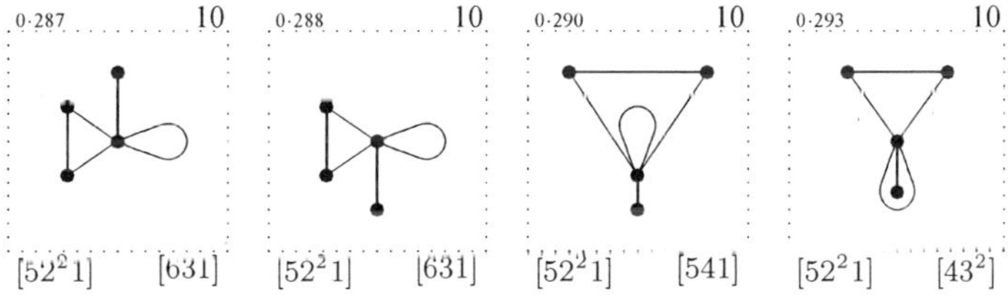

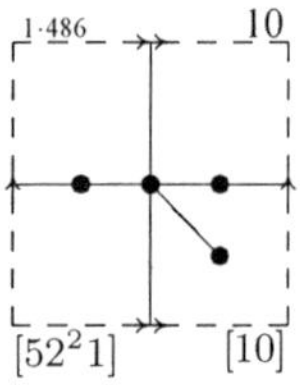

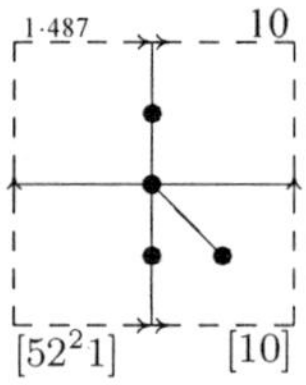

$4 + 2u$

5 edges
4 vertices
$\mathfrak{g}93$

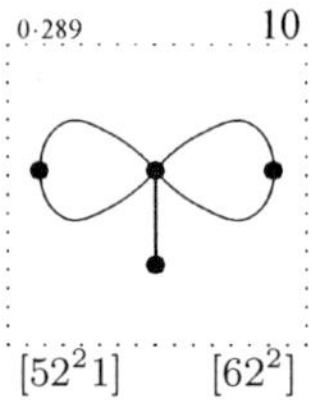

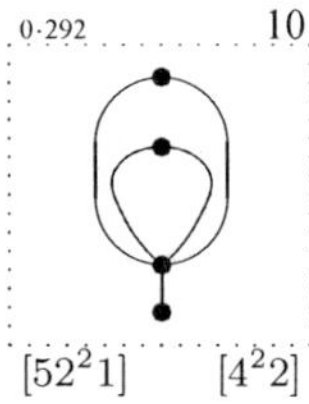

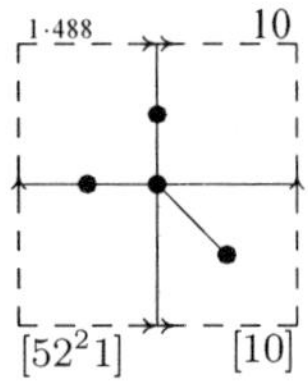

$2 + u$

5 edges
4 vertices
$\mathfrak{g}94$

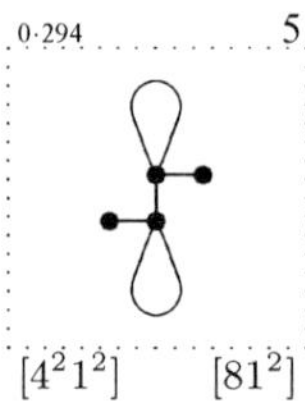

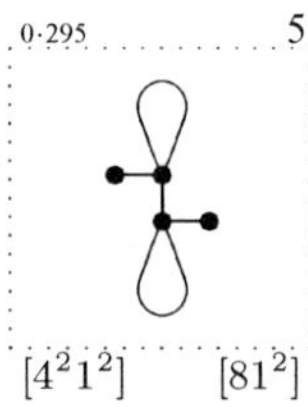

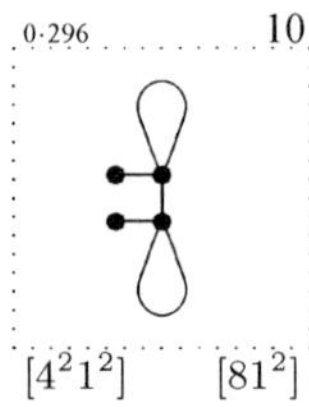

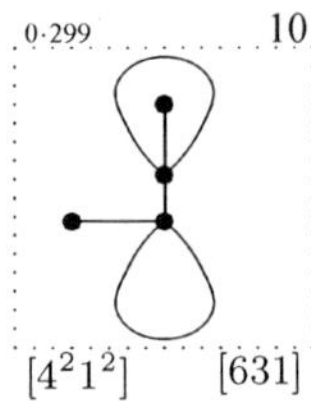

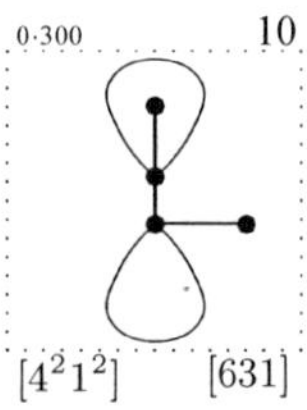

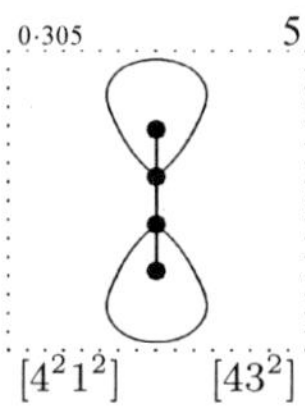

6

5 edges
4 vertices
$\mathfrak{g}95$

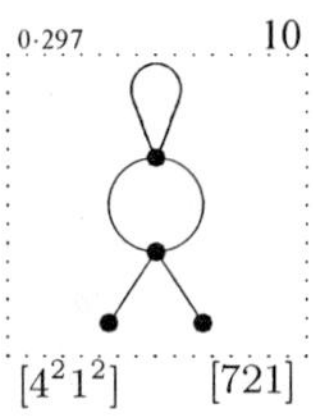

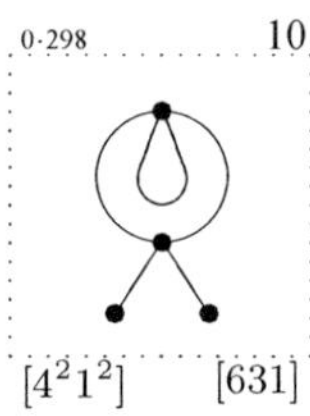

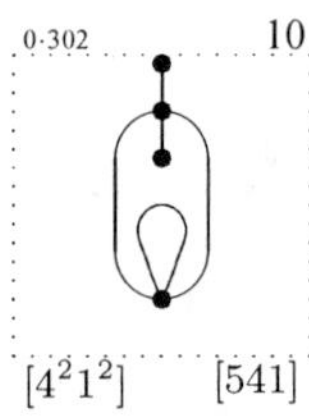

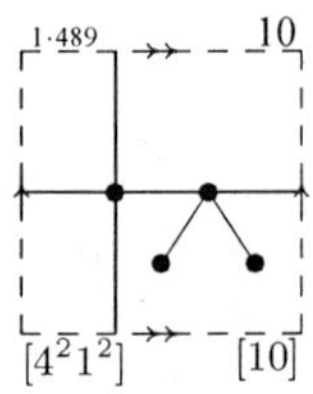

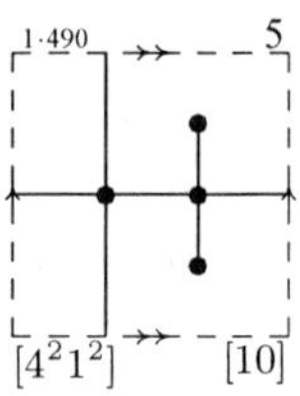

$3 + 2u$

5 edges
4 vertices
$\mathfrak{g}_{96}$

0·301 5
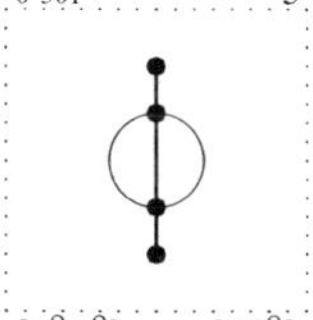
$[4^2 1^2]$ $[62^2]$

0·303 5
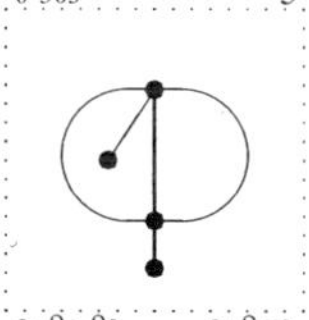
$[4^2 1^2]$ $[4^2 2]$

0·304 5
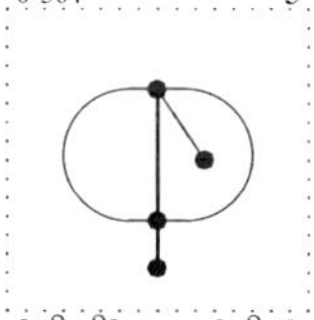
$[4^2 1^2]$ $[4^2 2]$

1·491 10
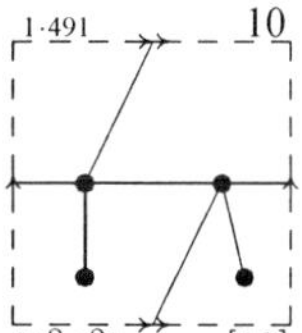
$[4^2 1^2]$ $[10]$

1·492 5
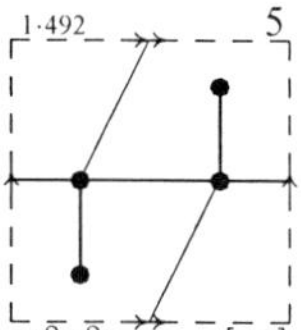
$[4^2 1^2]$ $[10]$

$3 + 2u$

5 edges
4 vertices
$\mathfrak{g}_{97}$

0·306 10
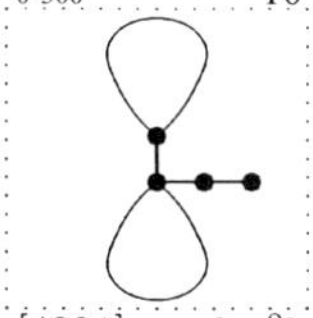
$[4321]$ $[81^2]$

0·307 10
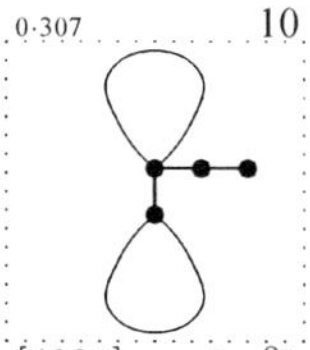
$[4321]$ $[81^2]$

0·323 10
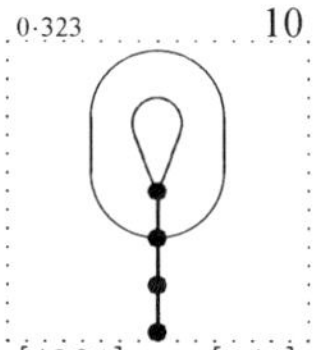
$[4321]$ $[541]$

3

5 edges
4 vertices
$\mathfrak{g}_{98}$

0·308 10
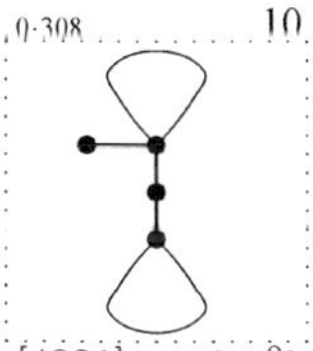
$[4321]$ $[81^2]$

0·309 10
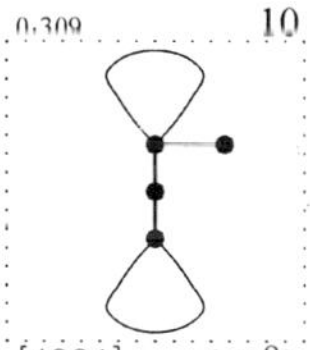
$[4321]$ $[81^2]$

0·317 10
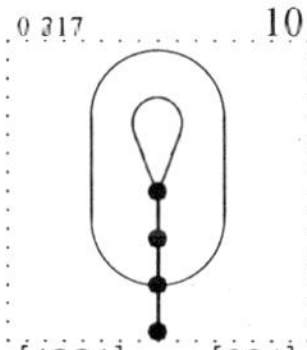
$[4321]$ $[631]$

3

5 edges
4 vertices
$\mathfrak{g}_{99}$

0·310 10
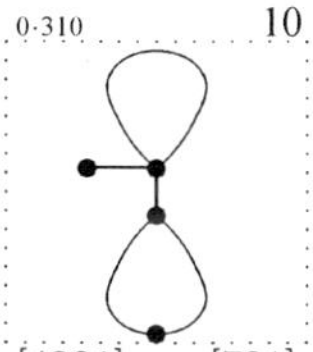
$[4321]$ $[721]$

0·311 10
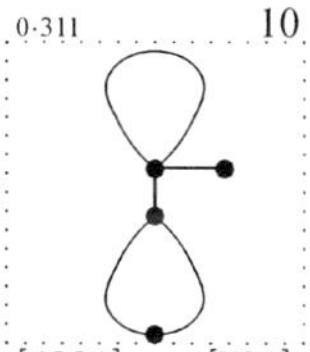
$[4321]$ $[721]$

0·327 10
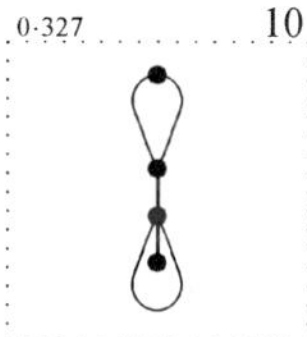
$[4321]$ $[532]$

3

5 edges
4 vertices
$\mathfrak{g}_{100}$

0·312 10
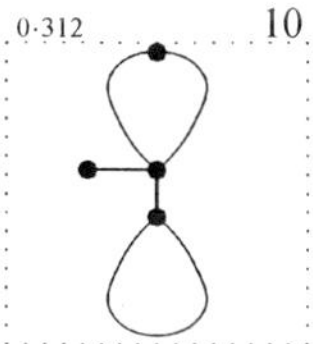
$[4321]$ $[721]$

0·313 10
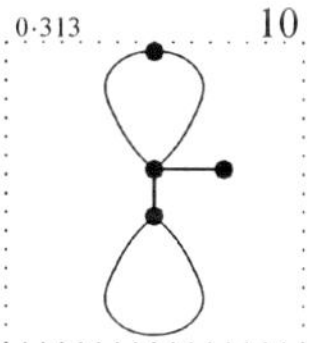
$[4321]$ $[721]$

0·324 10
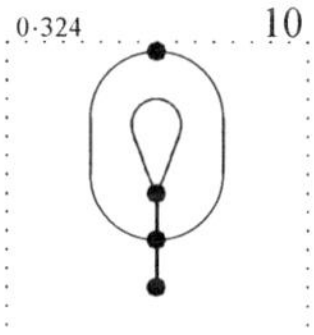
$[4321]$ $[541]$

3

5 edges
4 vertices
$\mathfrak{g}_{101}$

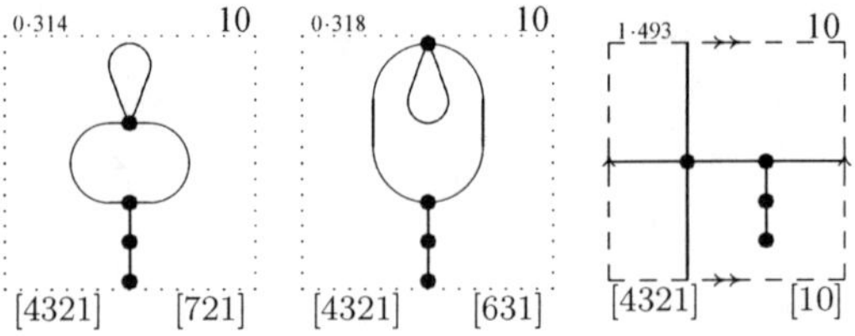

$2+u$

5 edges
4 vertices
$\mathfrak{g}_{102}$

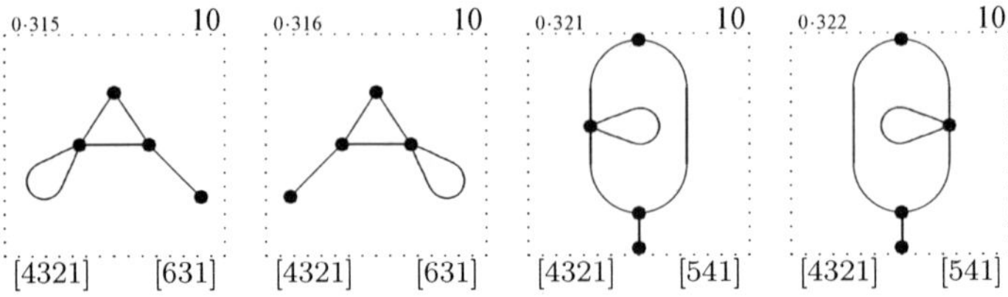

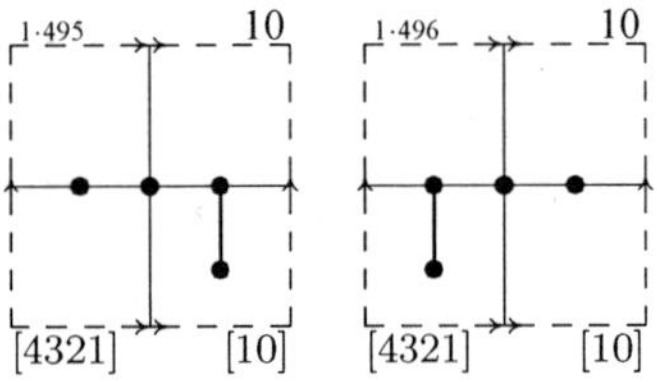

$4+2u$

5 edges
4 vertices
$\mathfrak{g}_{103}$

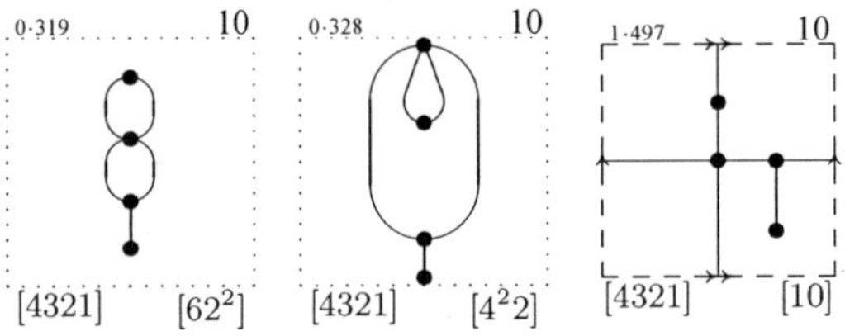

$2+u$

5 edges
4 vertices
$\mathfrak{g}_{104}$

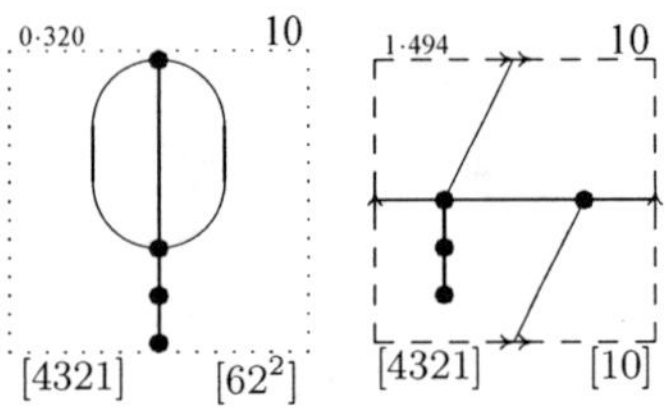

$1+u$

5 edges
4 vertices
$\mathfrak{g}_{105}$

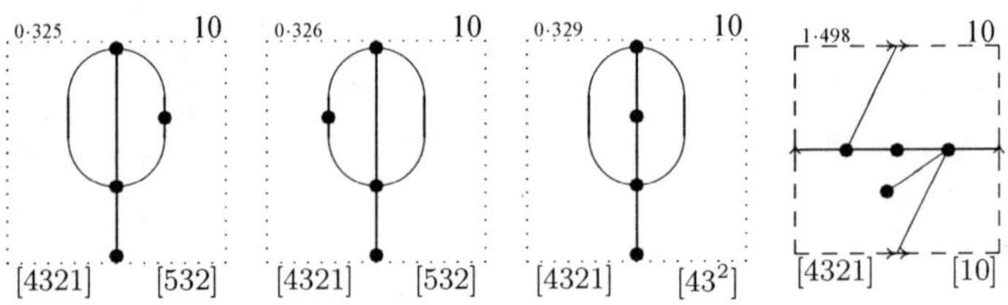

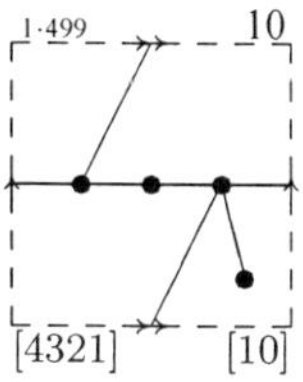

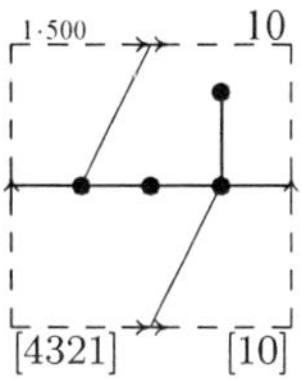

$3 + 3u$

5 edges

4 vertices

$\mathfrak{g}_{106}$

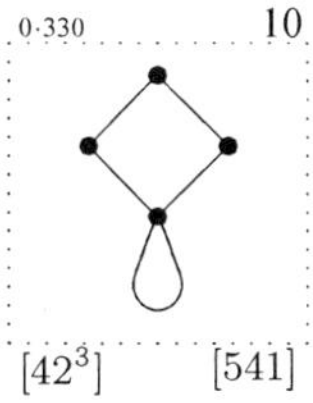

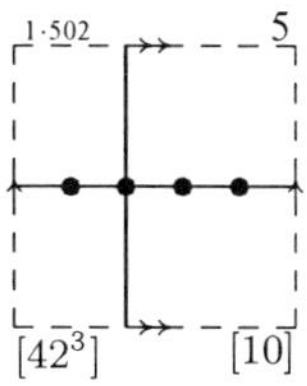

$1 + u$

5 edges

4 vertices

$\mathfrak{g}_{107}$

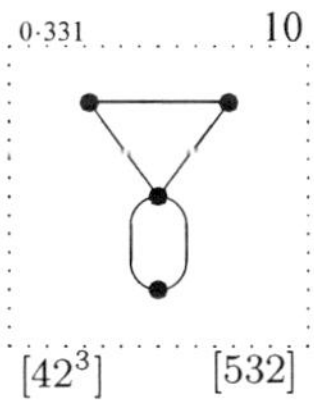

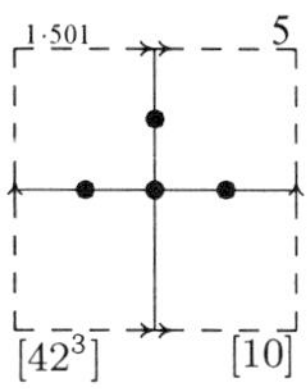

$1 + u$

5 edges

4 vertices

$\mathfrak{g}_{108}$

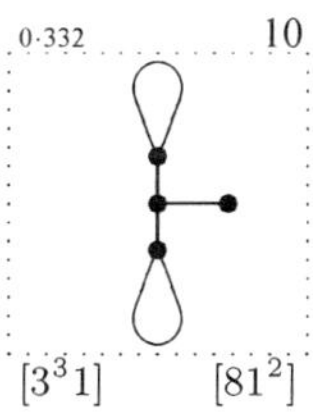

1

5 edges

4 vertices

$\mathfrak{g}_{109}$

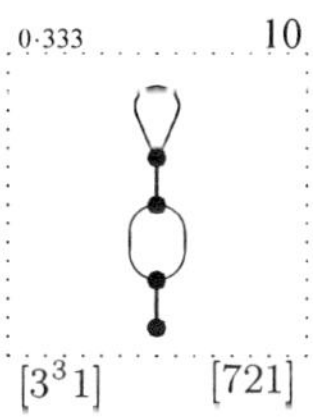

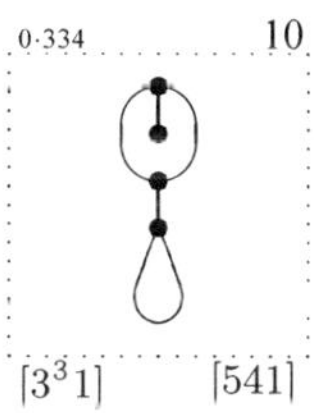

2

5 edges
4 vertices
$\mathfrak{g}_{110}$

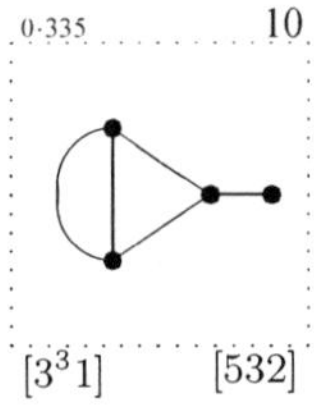

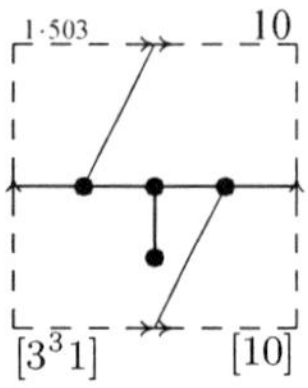

$1 + u$

5 edges
4 vertices
$\mathfrak{g}_{111}$

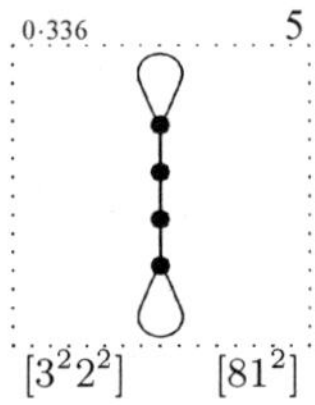

1

5 edges
4 vertices
$\mathfrak{g}_{112}$

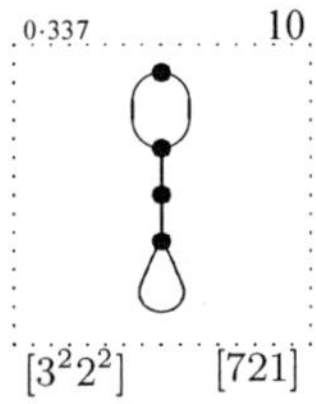

1

5 edges
4 vertices
$\mathfrak{g}_{113}$

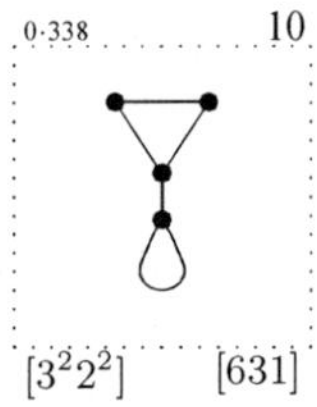

1

5 edges
4 vertices
$\mathfrak{g}_{114}$

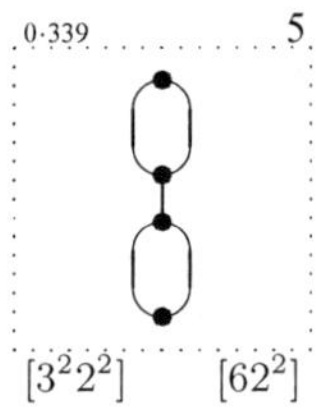

1

5 edges
4 vertices
$\mathfrak{g}_{115}$

0·340 5
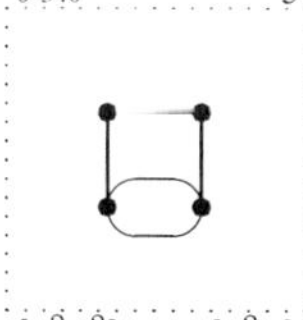
$[3^2 2^2]$ $[4^2 2]$

1·504 5
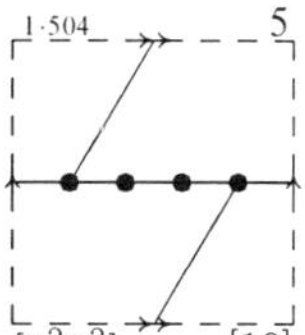
$[3^2 2^2]$ $[10]$

$1 + u$

5 edges
4 vertices
$\mathfrak{g}_{116}$

0·341 5
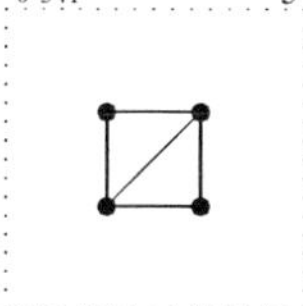
$[3^2 2^2]$ $[43^2]$

1·505 5
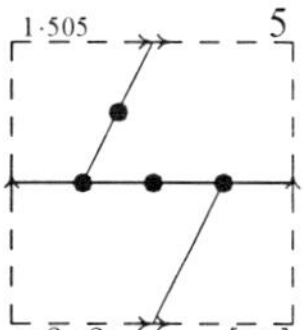
$[3^2 2^2]$ $[10]$

$1 + u$

5 edges
5 vertices
$\mathfrak{g}_{117}$

0·342 10
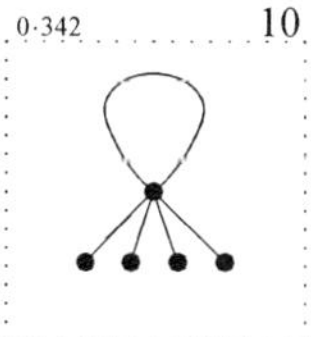
$[61^4]$ $[91]$

0·343 10
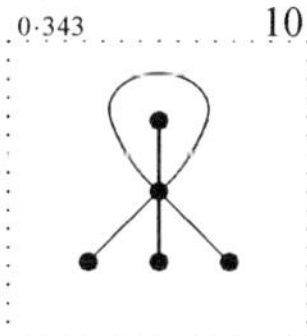
$[61^4]$ $[73]$

0·344 5
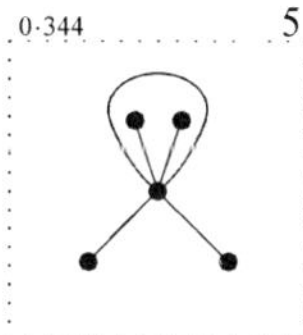
$[61^4]$ $[5^2]$

3

5 edges
5 vertices
$\mathfrak{g}_{118}$

0·345 10
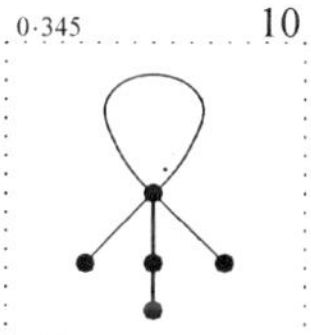
$[521^3]$ $[91]$

0·346 10
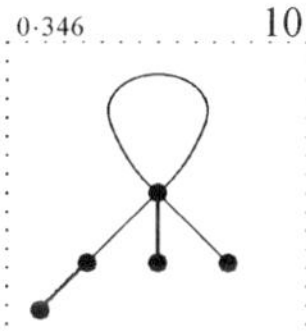
$[521^3]$ $[91]$

0·347 10
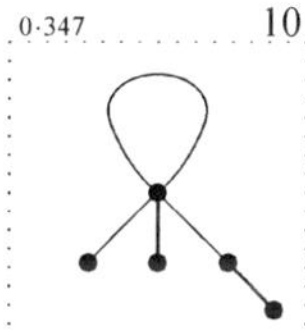
$[521^3]$ $[91]$

0·349 10
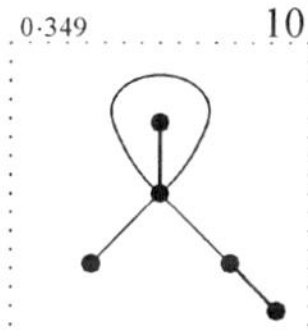
$[521^3]$ $[73]$

0·350 10
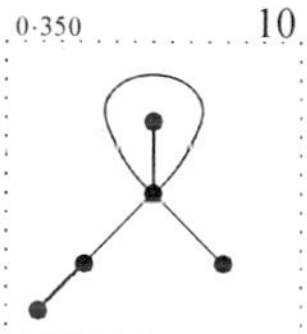
$[521^3]$ $[73]$

0·352 10
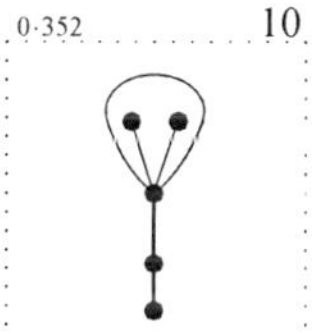
$[521^3]$ $[5^2]$

6

5 edges
5 vertices
$\mathfrak{g}_{119}$

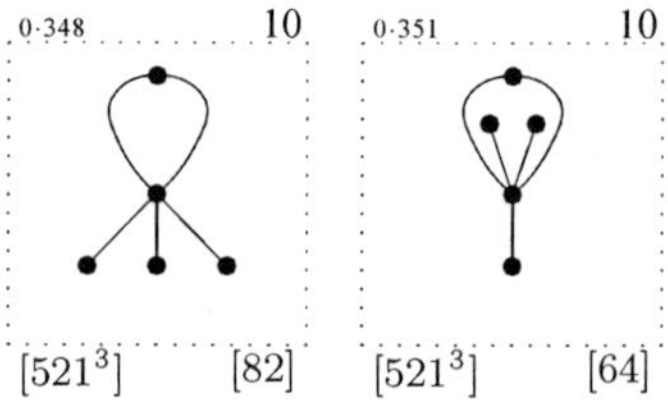

2

5 edges
5 vertices
$\mathfrak{g}_{120}$

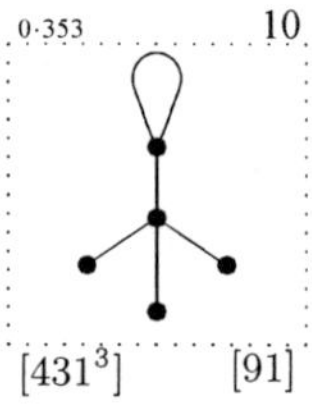

1

5 edges
5 vertices
$\mathfrak{g}_{121}$

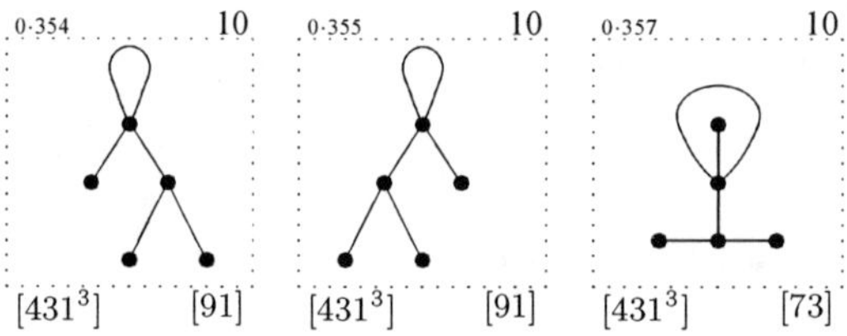

3

5 edges
5 vertices
$\mathfrak{g}_{122}$

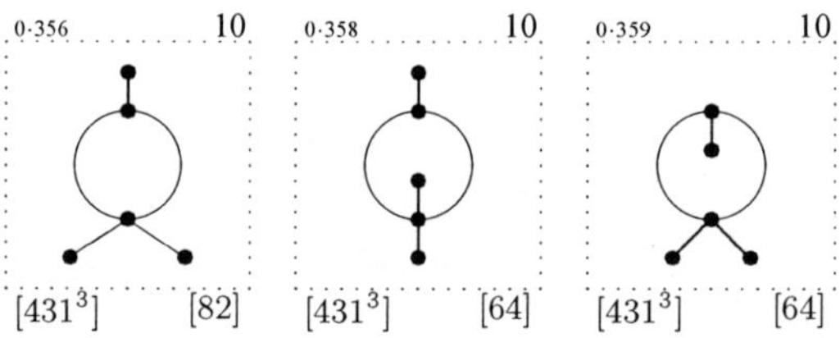

3

5 edges
5 vertices
$\mathfrak{g}_{123}$

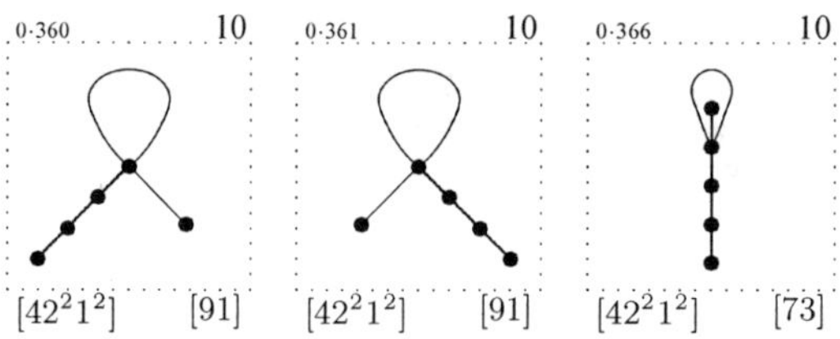

3

5 edges
5 vertices
$\mathfrak{g}_{124}$

0·362 10
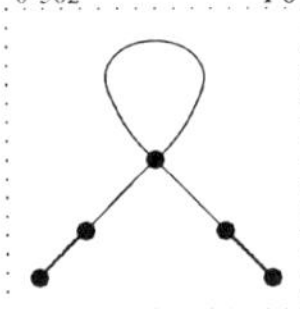
$[42^2 1^2]$ [91]

0·368 5
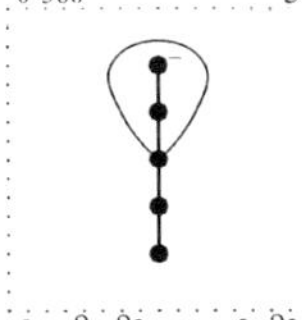
$[42^2 1^2]$ $[5^2]$

2

5 edges
5 vertices
$\mathfrak{g}_{125}$

0·363 10
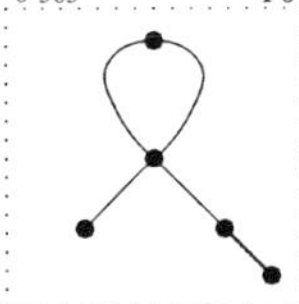
$[42^2 1^2]$ [82]

0·364 10
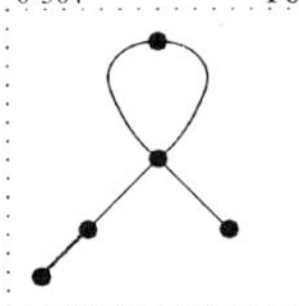
$[42^2 1^2]$ [82]

0·367 10
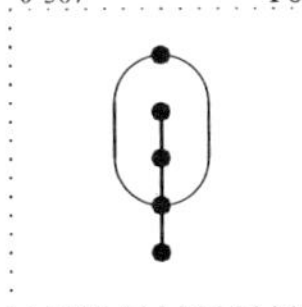
$[42^2 1^2]$ [64]

3

5 edges
5 vertices
$\mathfrak{g}_{126}$

0·365 10
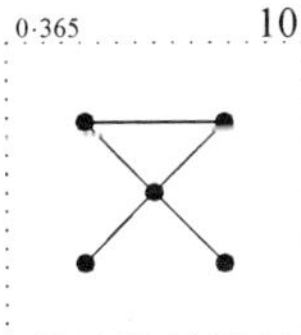
$[42^2 1^2]$ [73]

0·369 5
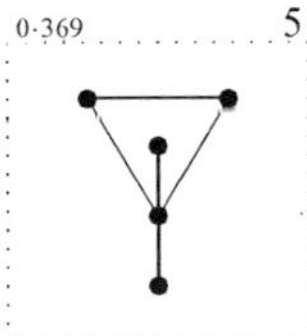
$[42^2 1^2]$ $[5^2]$

2

5 edges
5 vertices
$\mathfrak{g}_{127}$

0·370 10
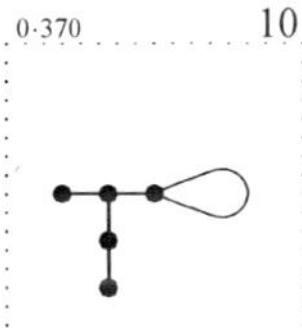
$[3^2 21^2]$ [91]

0·371 10
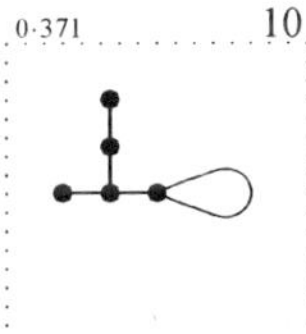
$[3^2 21^2]$ [91]

2

5 edges
5 vertices
$\mathfrak{g}_{128}$

0·372 10
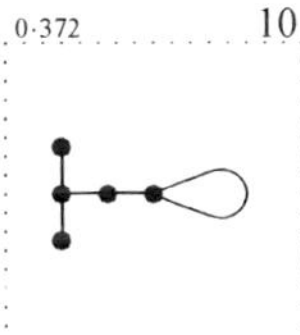
$[3^2 21^2]$ [91]

1

5 edges
5 vertices
$\mathfrak{g}_{129}$

0·373 10
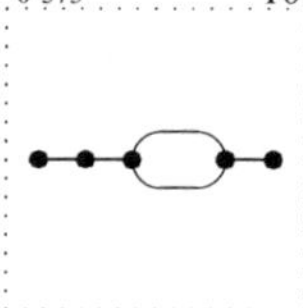
$[3^2 21^2]$ [82]

0·376 10
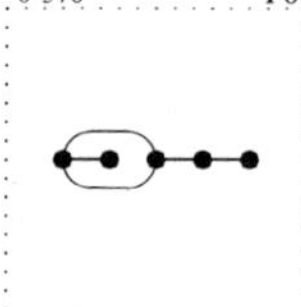
$[3^2 21^2]$ [64]

2

5 edges
5 vertices
$\mathfrak{g}_{130}$

0·374 10
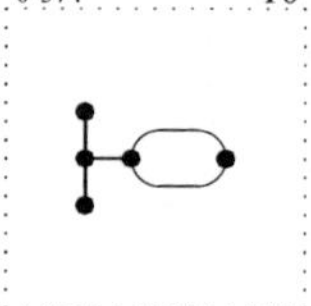
$[3^2 21^2]$ [82]

1

5 edges
5 vertices
$\mathfrak{g}_{131}$

0·375 10
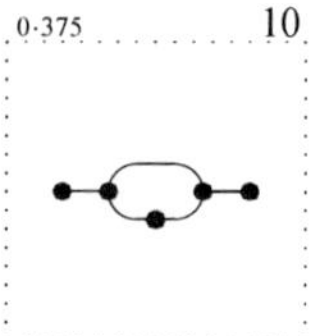
$[3^2 21^2]$ [73]

0·377 5
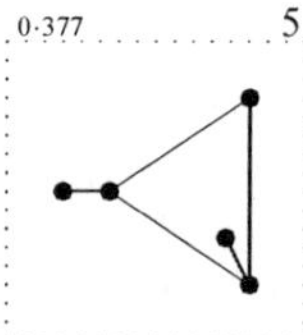
$[3^2 21^2]$ $[5^2]$

0·378 5
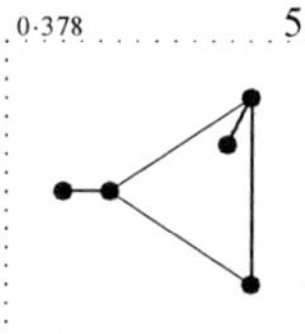
$[3^2 21^2]$ $[5^2]$

3

5 edges
5 vertices
$\mathfrak{g}_{132}$

0·379 10
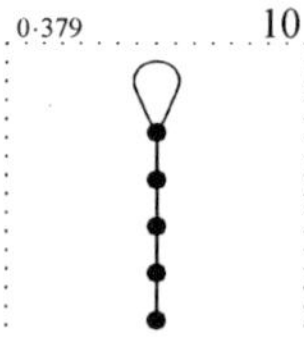
$[32^3 1]$ [91]

1

5 edges
5 vertices
$\mathfrak{g}_{133}$

0·380 10
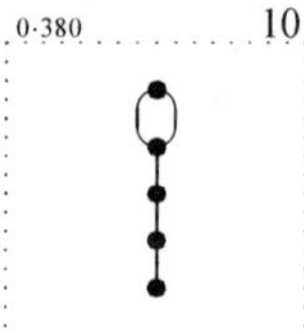
$[32^3 1]$ [82]

1

5 edges
5 vertices
$\mathfrak{g}_{134}$

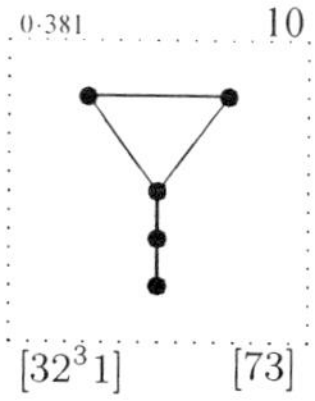

1

5 edges
5 vertices
$\mathfrak{g}_{135}$

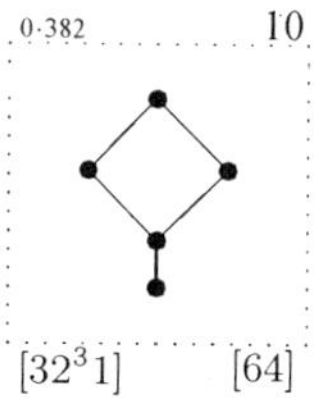

1

5 edges
5 vertices
$\mathfrak{g}_{136}$

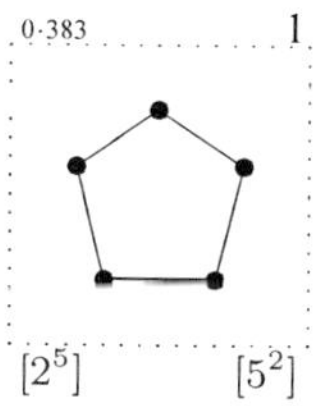

1

5 edges
6 vertices
$\mathfrak{g}_{137}$

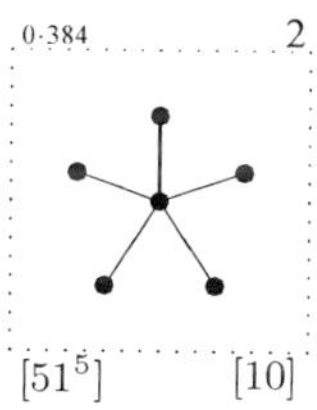

1

5 edges
6 vertices
$\mathfrak{g}_{138}$

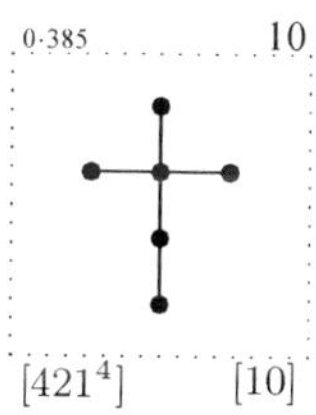

1

5 edges
6 vertices
$\mathfrak{g}_{139}$

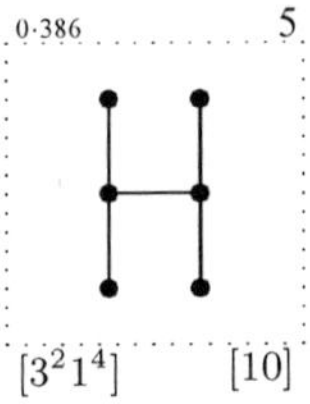

1

5 edges
6 vertices
$\mathfrak{g}_{140}$

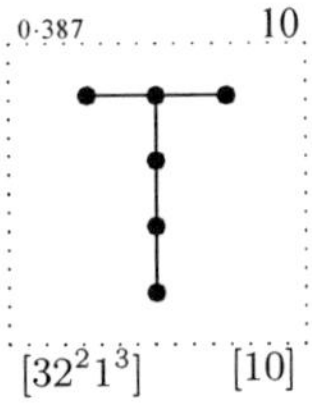

1

5 edges
6 vertices
$\mathfrak{g}_{141}$

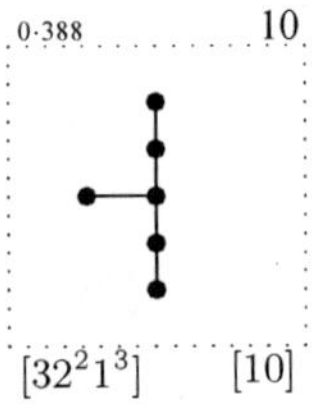

1

5 edges
6 vertices
$\mathfrak{g}_{142}$

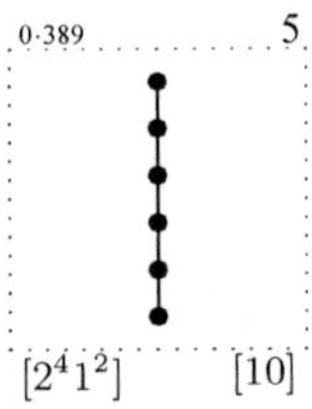

1

Part III

Tables

Chapter 9

Numbers of rooted maps

The following tables give the number of rooted maps in a given surface for a given number of edges for all proper pairs of vertex and face partitions. In each table the number of edges appears in bold face in the top left-hand corner. Below this appears the subtitle $a : b$ specifying the number of vertices and faces a, b in some order. The tables are listed only for $a \geq b$ since, by duality, the table for $b : a$ is the same. The rows and columns of each table are indexed by vertex partition and face partition.

Section 9.1 lists the tables for maps with at most 6 edges in orientable surfaces in increasing order of genus and number of edges.

Section 9.2 lists the tables for maps with at most 6 edges in nonorientable surfaces in increasing order of genus and number of edges.

9.1 Orientable: by vertex and face partition

Genus 0: 1 edge

1	2
2 : 1	
[1^2]	1

Genus 0: 2 edges

2	3	2^2
2 : 2	1	
[3 1]	4	·
[2^2]	·	1

2	4
3 : 1	
[2 1^2]	2

Genus 0: 3 edges

3 3 : 2	5 1	4 2	3^2
$[4\,1^2]$	6	·	3
$[3\,2\,1]$	6	6	·
$[2^3]$	·	·	1

3 4 : 1	6
$[3\,1^3]$	2
$[2^2\,1^2]$	3

Genus 0: 4 edges

4 3 : 3	6 1^2	5 2 1	4 3 1	4 2^2	3^2 2
$[6\,1^2]$	12	8	16	·	4
$[5\,2\,1]$	8	24	8	·	8
$[4\,3\,1]$	16	8	16	8	·
$[4\,2^2]$	·	·	8	4	·
$[3^2\,2]$	4	8	·	·	4

4 4 : 2	7 1	6 2	5 3	4^2
$[5\,1^3]$	8	·	8	·
$[4\,2\,1^2]$	16	8	8	4
$[3^2\,1^2]$	8	4	·	4
$[3\,2^2\,1]$	8	8	8	·
$[2^4]$	·	·	·	1

4 5 : 1	8
$[4\,1^4]$	2
$[3\,2\,1^3]$	8
$[2^3\,1^2]$	4

Genus 0: 5 edges

5 4 : 3	8 1^2	7 2 1	6 3 1	6 2^2	5 4 1	5 3 2	4^2 2	4 3^2
$[7\,1^3]$	20	10	30	·	20	10	·	10
$[6\,2\,1^2]$	30	50	30	·	40	40	·	10
$[5\,3\,1^2]$	40	30	30	10	40	10	10	10
$[5\,2^2\,1]$	10	30	30	10	10	10	10	10
$[4^2\,1^2]$	20	10	30	5	10	·	10	5
$[4\,3\,2\,1]$	40	50	40	20	40	30	10	10
$[4\,2^3]$	·	·	·	·	10	10	·	·
$[3^3\,1]$	10	10	·	·	10	10	·	·
$[3^2\,2^2]$	5	10	10	5	·	·	5	5

5 5 : 2	9 1	8 2	7 3	6 4	5^2
$[6\,1^4]$	10	·	10	·	5
$[5\,2\,1^3]$	30	10	20	10	10
$[4\,3\,1^3]$	30	10	10	20	·
$[4\,2^2\,1^2]$	30	20	20	10	10
$[3^2\,2\,1^2]$	30	20	10	10	10
$[3\,2^3\,1]$	10	10	10	10	·
$[2^5]$	·	·	·	·	1

5 6 : 1	10
$[5\,1^5]$	2
$[4\,2\,1^4]$	10
$[3^2\,1^4]$	5
$[3\,2^2\,1^3]$	20
$[2^4\,1^2]$	5

Genus 0: 6 edges

6 4 : 4	$9\,1^3$	$8\,2\,1^2$	$7\,3\,1^2$	$7\,2^2\,1$	$6\,4\,1^2$	$6\,3\,2\,1$	$6\,2^3$	$5^2\,1^2$	$5\,4\,2\,1$	$5\,3^2\,1$	$5\,3\,2^2$	$4^2\,3\,1$	$4^2\,2^2$	$4\,3^2\,2$	3^4
$[9\,1^3]$	40	48	84	12	72	72	·	36	48	60	12	48	·	24	4
$[8\,2\,1^2]$	48	144	96	60	96	168	·	60	144	84	48	48	·	48	6
$[7\,3\,1^2]$	84	96	108	72	144	84	·	48	120	36	48	84	·	36	·
$[7\,2^2\,1]$	12	60	72	72	36	84	·	12	84	36	48	48	·	36	·
$[6\,4\,1^2]$	72	96	144	36	108	108	12	30	96	72	12	60	18	36	·
$[6\,3\,2\,1]$	72	168	84	84	108	216	24	72	144	60	36	48	36	36	12
$[6\,2^3]$	·	·	·	·	12	24	4	6	24	12	·	·	6	12	·
$[5^2\,1^2]$	36	60	48	12	30	72	6	42	36	36	·	24	18	6	6
$[5\,4\,2\,1]$	48	144	120	84	96	144	24	36	132	84	36	60	24	48	·
$[5\,3^2\,1]$	60	84	36	36	72	60	12	36	84	36	12	12	12	24	·
$[5\,3\,2^2]$	12	48	48	48	12	36	·	·	36	12	48	48	·	12	·
$[4^2\,3\,1]$	48	48	84	48	60	48	·	24	60	12	48	48	·	12	·
$[4^2\,2^2]$	·	·	·	·	18	36	6	18	24	12	·	·	12	6	3
$[4\,3^2\,2]$	24	48	36	36	36	36	12	6	48	24	12	12	6	24	·
$[3^4]$	4	6	·	·	·	12	·	6	·	·	·	·	3	·	1

6	10	9	8	8	7	7	6	6	6	5^2	5	4^3
5 : 3	1^2	2	3	2^2	4	3	5	4	3^2	2	4	
		1	1		1	2	1	2			3	
$[8\,1^4]$	30	12	48	·	24	12	36	·	18	6	24	·
$[7\,2\,1^3]$	72	84	84	·	84	72	84	·	24	36	60	·
$[6\,3\,1^3]$	84	60	72	12	96	36	60	24	12	12	48	4
$[6\,2^2\,1^2]$	54	108	84	18	60	60	72	36	36	30	36	6
$[5\,4\,1^3]$	84	48	108	12	48	12	72	48	36	·	24	12
$[5\,3\,2\,1^2]$	144	180	132	48	108	72	156	84	48	36	60	12
$[5\,2^3\,1]$	12	36	36	12	36	36	12	12	·	12	36	·
$[4^2\,2\,1^2]$	72	72	96	24	60	36	60	48	12	12	48	·
$[4\,3^2\,1^2]$	90	72	48	18	96	48	48	12	·	30	36	6
$[4\,3\,2^2\,1]$	72	108	96	48	84	72	72	48	36	36	36	12
$[4\,2^4]$	·	·	·	·	·	·	12	12	6	·	·	·
$[3^3\,2\,1]$	36	48	24	12	12	12	36	36	12	·	12	·
$[3^2\,2^3]$	6	12	12	6	12	12	·	·	·	6	12	2

6	11	10	9	8	7	6^2
6 : 2	1	2	3	4	5	
$[7\,1^5]$	12	·	12	·	12	·
$[6\,2\,1^4]$	48	12	36	12	36	6
$[5\,3\,1^4]$	48	12	24	24	12	12
$[5\,2^2\,1^3]$	72	36	48	24	48	12
$[4^2\,1^4]$	24	6	12	12	·	9
$[4\,3\,2\,1^3]$	144	72	60	60	60	24
$[4\,2^3\,1^2]$	48	36	36	24	24	12
$[3^3\,1^3]$	24	12	4	12	12	·
$[3^2\,2^2\,1^2]$	72	54	36	30	24	24
$[3\,2^4\,1]$	12	12	12	12	12	·
$[2^6]$	·	·	·	·	·	1

6	12
7 : 1	
$[6\,1^6]$	2
$[5\,2\,1^5]$	12
$[4\,3\,1^5]$	12
$[4\,2^2\,1^4]$	30
$[3^2\,2\,1^4]$	30
$[3\,2^3\,1^3]$	40
$[2^5\,1^2]$	6

Genus 1: 2 edges

2	4
1 : 1	
$[4]$	1

Genus 1: 3 edges

3 2 : 1	6
[5 1]	6
[4 2]	3
[3^2]	1

Genus 1: 4 edges

4 2 : 2	7 1	6 2	5 3	4^2
[7 1]	40	16	16	8
[6 2]	16	12	8	4
[5 3]	16	8	8	·
[4^2]	8	4	·	3

4 3 : 1	8
[6 1^2]	20
[5 2 1]	24
[4 3 1]	16
[4 2^2]	6
[3^2 2]	4

Genus 1: 5 edges

5 3 : 2	9 1	8 2	7 3	6 4	5^2
[8 1^2]	150	50	70	55	25
[7 2 1]	150	90	70	60	30
[6 3 1]	130	60	50	30	30
[6 2^2]	30	25	20	15	10
[5 4 1]	120	50	40	50	10
[5 3 2]	60	40	30	30	·
[4^2 2]	30	20	10	5	10
[4 3^2]	30	15	10	5	5

5 4 : 1	10
[7 1^3]	50
[6 2 1^2]	100
[5 3 1^2]	70
[5 2^2 1]	60
[4^2 1^2]	30
[4 3 2 1]	80
[4 2^3]	10
[3^3 1]	10
[3^2 2^2]	10

Genus 1: 6 edges

6	10	9	8	8	7	7	6	6	6	5^2	5	4^3
3 : 3	1^2	2	3	2^2	4	3	5	4	3^2	2	4	
		1	1		1	2	1	2			3	
$[10\,1^2]$	630	540	600	90	504	252	456	198	144	90	240	36
$[9\,2\,1]$	540	780	564	180	468	348	468	288	144	144	240	36
$[8\,3\,1]$	600	564	432	144	408	228	396	192	96	108	168	24
$[8\,2^2]$	90	180	144	60	108	108	108	90	36	42	72	12
$[7\,4\,1]$	504	468	408	108	408	228	288	120	48	108	168	24
$[7\,3\,2]$	252	348	228	108	228	168	132	96	48	72	108	12
$[6\,5\,1]$	456	468	396	108	288	132	312	204	96	36	120	24
$[6\,4\,2]$	198	288	192	90	120	96	204	168	60	18	60	6
$[6\,3^2]$	144	144	96	36	48	48	96	60	36	12	12	8
$[5^2\,2]$	90	144	108	42	108	72	36	18	12	42	36	12
$[5\,4\,3]$	240	240	168	72	168	108	120	60	12	36	72	·
$[4^3]$	36	36	24	12	24	12	24	6	8	12	·	4

6	11	10	9	8	7	6^2
4 : 2	1	2	3	4	5	
$[9\,1^3]$	420	120	220	144	144	72
$[8\,2\,1^2]$	720	360	348	264	276	132
$[7\,3\,1^2]$	600	252	228	192	228	60
$[7\,2^2\,1]$	360	252	204	156	168	60
$[6\,4\,1^2]$	528	198	216	186	144	78
$[6\,3\,2\,1]$	624	372	276	216	180	132
$[6\,2^3]$	48	42	36	30	24	20
$[5^2\,1^2]$	252	90	108	96	48	54
$[5\,4\,2\,1]$	576	324	240	204	192	84
$[5\,3^2\,1]$	276	132	96	84	60	48
$[5\,3\,2^2]$	144	108	84	72	72	·
$[4^2\,3\,1]$	264	120	84	72	96	12
$[4^2\,2^2]$	72	54	36	24	12	27
$[4\,3^2\,2]$	144	90	60	42	36	18
$[3^4]$	12	6	4	3	·	3

6	12
5 : 1	
$[8\,1^4]$	105
$[7\,2\,1^3]$	300
$[6\,3\,1^3]$	220
$[6\,2^2\,1^2]$	300
$[5\,4\,1^3]$	180
$[5\,3\,2\,1^2]$	420
$[5\,2^3\,1]$	120
$[4^2\,2\,1^2]$	180
$[4\,3^2\,1^2]$	150
$[4\,3\,2^2\,1]$	240
$[4\,2^4]$	15
$[3^3\,2\,1]$	60
$[3^2\,2^3]$	20

Genus 2: 4 edges

4 1 : 1	8
[8]	21

Genus 2: 5 edges

5 2 : 1	10
[9 1]	210
[8 2]	105
[7 3]	70
[6 4]	65
[5^2]	33

Genus 2: 6 edges

6 2 : 2	11 1	10 2	9 3	8 4	7 5	6^2
[11 1]	2268	1008	840	720	648	312
[10 2]	1008	630	420	360	324	156
[9 3]	840	420	300	216	216	108
[8 4]	720	360	216	204	192	93
[7 5]	648	324	216	192	204	60
[6^2]	312	156	108	93	60	66

6 3 : 1	12
[10 1^2]	1134
[9 2 1]	1260
[8 3 1]	840
[8 2^2]	315
[7 4 1]	720
[7 3 2]	420
[6 5 1]	708
[6 4 2]	390
[6 3^2]	150
[5^2 2]	198
[5 4 3]	288
[4^3]	45

Genus 3: 6 edges

6 1 : 1	12
[12]	1485

9.2 Nonorientable: by vertex and face partition

Genus $\tilde{1}$: 1 edge

1	2
1 : 1	
[2]	1

Genus $\tilde{1}$: 2 edges

2	4
2 : 1	
[3 1]	4
$[2^2]$	1

Genus $\tilde{1}$: 3 edges

3	5	4	3^2
2 : 2	1	2	
[5 1]	18	6	6
[4 2]	6	6	3
$[3^2]$	6	3	·

3	6
3 : 1	
$[4\,1^2]$	9
[3 2 1]	12
$[2^3]$	1

Genus $\tilde{1}$: 4 edges

4	7	6	5	4^2
3 : 2	1	2	3	
$[6\,1^2]$	48	12	32	8
[5 2 1]	48	32	32	8
[4 3 1]	56	24	16	16
$[4\,2^2]$	8	8	8	6
$[3^2\,2]$	16	12	8	·

4	8
4 : 1	
$[5\,1^3]$	16
$[4\,2\,1^2]$	36
$[3^2\,1^2]$	16
$[3\,2^2\,1]$	24
$[2^4]$	1

Genus $\tilde{1}$: 5 edges

5 3 : 3	$8\ 1^2$	$7\ 2\ 1$	$6\ 3\ 1$	$6\ 2^2$	$5\ 4\ 1$	$5\ 3\ 2$	$4^2\ 2$	$4\ 3^2$
$[8\ 1^2]$	150	120	180	15	150	80	20	55
$[7\ 2\ 1]$	120	210	160	40	140	130	30	50
$[6\ 3\ 1]$	180	160	120	40	170	90	30	30
$[6\ 2^2]$	15	40	40	15	40	40	15	15
$[5\ 4\ 1]$	150	140	170	40	110	40	50	40
$[5\ 3\ 2]$	80	130	90	40	40	40	30	30
$[4^2\ 2]$	20	30	30	15	50	30	10	5
$[4\ 3^2]$	55	50	30	15	40	30	5	5

5 4 : 2	$9\ 1$	$8\ 2$	$7\ 3$	$6\ 4$	5^2
$[7\ 1^3]$	100	20	70	30	30
$[6\ 2\ 1^2]$	180	90	110	60	60
$[5\ 3\ 1^2]$	180	70	70	80	30
$[5\ 2^2\ 1]$	90	70	70	50	20
$[4^2\ 1^2]$	90	30	40	45	5
$[4\ 3\ 2\ 1]$	210	130	90	80	50
$[4\ 2^3]$	10	10	10	10	10
$[3^3\ 1]$	40	20	10	10	10
$[3^2\ 2^2]$	30	25	20	15	·

5 5 : 1	10
$[6\ 1^4]$	25
$[5\ 2\ 1^3]$	80
$[4\ 3\ 1^3]$	70
$[4\ 2^2\ 1^2]$	90
$[3^2\ 2\ 1^2]$	80
$[3\ 2^3\ 1]$	40
$[2^5]$	1

Genus $\tilde{1}$: 6 edges

6 4 : 3	$10\ 1^2$	$9\ 2\ 1$	$8\ 3\ 1$	$8\ 2^2$	$7\ 4\ 1$	$7\ 3\ 2$	$6\ 5\ 1$	$6\ 4\ 2$	$6\ 3^2$	$5^2\ 2$	$5\ 4\ 3$	4^3
$[9\,1^3]$	360	240	456	24	324	168	360	72	144	72	228	16
$[8\,2\,1^2]$	540	720	648	108	540	444	612	204	204	216	360	24
$[7\,3\,1^2]$	648	540	516	120	636	300	492	180	96	144	300	36
$[7\,2^2\,1]$	216	432	372	120	312	300	252	156	108	132	216	24
$[6\,4\,1^2]$	594	468	648	102	492	216	420	252	144	66	252	42
$[6\,3\,2\,1]$	648	864	600	228	492	396	636	372	216	180	240	48
$[6\,2^3]$	18	48	48	18	48	48	60	60	24	18	48	2
$[5^2\,1^2]$	288	252	300	48	168	84	252	138	108	24	84	30
$[5\,4\,2\,1]$	540	732	660	216	504	336	480	348	144	132	312	36
$[5\,3^2\,1]$	408	384	240	96	276	144	288	156	60	72	120	12
$[5\,3\,2^2]$	144	264	216	96	168	168	72	60	36	72	120	24
$[4^2\,3\,1]$	360	300	288	96	324	168	216	84	24	96	132	24
$[4^2\,2^2]$	36	60	60	30	60	48	108	78	36	24	24	6
$[4\,3^2\,2]$	198	252	168	90	156	108	132	108	48	42	72	6
$[3^4]$	24	24	12	3	·	12	24	12	6	·	·	1

6 5 : 2	$11\ 1$	$10\ 2$	$9\ 3$	$8\ 4$	$7\ 5$	6^2
$[8\,1^4]$	180	30	132	48	108	27
$[7\,2\,1^3]$	480	192	300	156	300	72
$[6\,3\,1^3]$	456	156	204	192	180	72
$[6\,2^2\,1^2]$	432	270	288	174	240	96
$[5\,4\,1^3]$	444	132	228	180	96	108
$[5\,3\,2\,1^2]$	864	468	396	324	348	180
$[5\,2^3\,1]$	144	120	120	96	96	24
$[4^2\,2\,1^2]$	432	216	204	180	156	72
$[4\,3^2\,1^2]$	444	198	132	174	168	42
$[4\,3\,2^2\,1]$	504	360	276	216	216	108
$[4\,2^4]$	12	12	12	12	12	15
$[3^3\,2\,1]$	192	120	72	60	48	48
$[3^2\,2^3]$	48	42	36	30	24	·

6 6 : 1	12
$[7\,1^5]$	36
$[6\,2\,1^4]$	150
$[5\,3\,1^4]$	132
$[5\,2^2\,1^3]$	240
$[4^2\,1^4]$	63
$[4\,3\,2\,1^3]$	420
$[4\,2^3\,1^2]$	180
$[3^3\,1^3]$	64
$[3^2\,2^2\,1^2]$	240
$[3\,2^4\,1]$	60
$[2^6]$	1

Genus $\tilde{2}$: 2 edges

2	4
1 : 1	
[4]	4

Genus $\tilde{2}$: 3 edges

3	6
2 : 1	
[5 1]	24
[4 2]	12
[3^2]	6

Genus $\tilde{2}$: 4 edges

4	7	6	5	4^2
2 : 2	1	2	3	
[7 1]	160	64	80	32
[6 2]	64	48	40	16
[5 3]	80	40	32	8
[4^2]	32	16	8	12

4	8
3 : 1	
[6 1^2]	80
[5 2 1]	96
[4 3 1]	80
[4 2^2]	24
[3^2 2]	24

Genus $\tilde{2}$: 5 edges

5	9	8	7	6	5^2
3 : 2	1	2	3	4	
[8 1^2]	600	200	320	220	130
[7 2 1]	600	360	330	240	150
[6 3 1]	600	270	230	190	130
[6 2^2]	120	100	90	70	40
[5 4 1]	540	230	220	230	60
[5 3 2]	300	200	150	120	30
[4^2 2]	120	80	50	50	40
[4 3^2]	160	80	50	40	30

5	10
4 : 1	
[7 1^3]	200
[6 2 1^2]	400
[5 3 1^2]	320
[5 2^2 1]	240
[4^2 1^2]	150
[4 3 2 1]	400
[4 2^3]	40
[3^3 1]	60
[3^2 2^2]	60

Genus $\tilde{2}$: 6 edges

6	10	9	8	8	7	7	6	6	6	5^2	5	4^3
3 : 3	1^2	2	3	2^2	4	3	5	4	3^2	2	4	
		1	1		1	2	1	2			3	
$[10\,1^2]$	2520	2160	2640	360	2160	1152	2112	792	708	468	1200	144
$[9\,2\,1]$	2160	3120	2496	720	2052	1632	2124	1152	708	720	1212	144
$[8\,3\,1]$	2640	2496	1992	612	2040	1128	1944	876	444	552	900	120
$[8\,2^2]$	360	720	612	240	504	492	468	372	180	204	360	48
$[7\,4\,1]$	2160	2052	2040	504	1884	984	1416	732	336	420	876	132
$[7\,3\,2]$	1152	1632	1128	492	984	792	768	516	276	336	492	72
$[6\,5\,1]$	2112	2124	1944	468	1416	768	1548	936	528	252	660	132
$[6\,4\,2]$	792	1152	876	372	732	516	936	696	276	192	372	48
$[6\,3^2]$	708	708	444	180	336	276	528	276	144	96	120	24
$[5^2\,2]$	468	720	552	204	420	336	252	192	96	144	228	48
$[5\,4\,3]$	1200	1212	900	360	876	492	660	372	120	228	384	36
$[4^3]$	144	144	120	48	132	72	132	48	24	48	36	16

6	11	10	9	8	7	6^2
4 : 2	1	2	3	4	5	
$[9\,1^3]$	1680	480	960	576	720	288
$[8\,2\,1^2]$	2880	1440	1560	1056	1356	528
$[7\,3\,1^2]$	2640	1080	1068	948	1044	348
$[7\,2^2\,1]$	1440	1008	912	660	744	276
$[6\,4\,1^2]$	2400	900	1080	912	708	396
$[6\,3\,2\,1]$	2880	1692	1320	1008	1032	588
$[6\,2^3]$	192	168	156	132	120	72
$[5^2\,1^2]$	1152	432	540	420	276	252
$[5\,4\,2\,1]$	2592	1476	1212	1032	924	444
$[5\,3^2\,1]$	1416	672	480	444	408	228
$[5\,3\,2^2]$	720	540	432	336	312	60
$[4^2\,3\,1]$	1320	600	444	456	468	120
$[4^2\,2^2]$	288	216	156	132	120	108
$[4\,3^2\,2]$	768	480	312	252	228	120
$[3^4]$	72	36	24	12	12	18

6	12
5 : 1	
$[8\,1^4]$	420
$[7\,2\,1^3]$	1200
$[6\,3\,1^3]$	960
$[6\,2^2\,1^2]$	1200
$[5\,4\,1^3]$	864
$[5\,3\,2\,1^2]$	1920
$[5\,2^3\,1]$	480
$[4^2\,2\,1^2]$	900
$[4\,3^2\,1^2]$	804
$[4\,3\,2^2\,1]$	1200
$[4\,2^4]$	60
$[3^3\,2\,1]$	360
$[3^2\,2^3]$	120

Genus $\tilde{3}$: 3 edges

3 1 : 1	6
[6]	41

Genus $\tilde{3}$: 4 edges

4 2 : 1	8
[7 1]	328
[6 2]	164
[5 3]	136
[4^2]	62

Genus $\tilde{3}$: 5 edges

5	9	8	7	6	5^2
2 : 2	1	2	3	4	
[9 1]	2870	1230	1260	1030	510
[8 2]	1230	820	630	515	255
[7 3]	1260	630	470	350	210
[6 4]	1030	515	350	355	195
[5^2]	510	255	210	195	40

5 3 : 1	10
[8 1^2]	1435
[7 2 1]	1640
[6 3 1]	1260
[6 2^2]	410
[5 4 1]	1130
[5 3 2]	680
[4^2 2]	310
[4 3^2]	285

Genus $\tilde{3}$: 6 edges

6 3 : 2	11 1	10 2	9 3	8 4	7 5	6^2
$[10\,1^2]$	13776	5166	6504	4812	4776	2226
$[9\,2\,1]$	13776	7872	6768	5184	5328	2472
$[8\,3\,1]$	12516	5748	4728	3936	4128	1752
$[8\,2^2]$	2952	2214	1800	1416	1356	612
$[7\,4\,1]$	10716	4740	4176	3732	3432	1320
$[7\,3\,2]$	6048	3780	2724	2124	2028	816
$[6\,5\,1]$	9960	4428	4128	3408	2592	1740
$[6\,4\,2]$	4944	3090	2136	1818	1584	1098
$[6\,3^2]$	2880	1440	1020	702	696	498
$[5^2\,2]$	2448	1530	1176	984	864	258
$[5\,4\,3]$	5040	2520	1728	1524	1476	564
$[4^3]$	744	372	224	240	252	92

6 4 : 1	12
$[9\,1^3]$	4592
$[8\,2\,1^2]$	8610
$[7\,3\,1^2]$	6504
$[7\,2^2\,1]$	4920
$[6\,4\,1^2]$	5658
$[6\,3\,2\,1]$	7560
$[6\,2^3]$	820
$[5^2\,1^2]$	2712
$[5\,4\,2\,1]$	6780
$[5\,3^2\,1]$	3000
$[5\,3\,2^2]$	2040
$[4^2\,3\,1]$	2796
$[4^2\,2^2]$	930
$[4\,3^2\,2]$	1710
$[3^4]$	128

Genus $\tilde{4}$: 4 edges

4 1 : 1	8
$[8]$	488

Genus $\tilde{4}$: 5 edges

5 2 : 1	10
$[9\,1]$	4880
$[8\,2]$	2440
$[7\,3]$	1900
$[6\,4]$	1640
$[5^2]$	800

Genus $\tilde{4}$: 6 edges

6 2 : 2	11 1	10 2	9 3	8 4	7 5	6^2
[11 1]	52704	23424	21816	17664	16440	7872
[10 2]	23424	14640	10908	8832	8220	3936
[9 3]	21816	10908	7968	6168	6264	2940
[8 4]	17664	8832	6168	5676	5328	2424
[7 5]	16440	8220	6264	5328	4752	1932
[6^2]	7872	3936	2940	2424	1932	1476

6 3 : 1	12
[10 1^2]	26352
[9 2 1]	29280
[8 3 1]	21816
[8 2^2]	7320
[7 4 1]	18648
[7 3 2]	11400
[6 5 1]	17472
[6 4 2]	9840
[6 3^2]	4476
[5^2 2]	4800
[5 4 3]	8040
[4^3]	1236

Genus $\tilde{5}$: 5 edges

5 1 : 1	10
[10]	8229

Genus $\tilde{5}$: 6 edges

6 2 : 1	12
[11 1]	98748
[10 2]	49374
[9 3]	36988
[8 4]	30990
[7 5]	28572
[6^2]	13807

Genus $\tilde{6}$: 6 edges

6 1 : 1	12
[12]	164892

9.3 Summarized by edges and vertices

Section 9.3.1 gives tables of the numbers of rooted maps with at most 11 edges in orientable surfaces and at most 8 edges in nonorientable surfaces. The tables are indexed by the number of edges.

Section 9.3.2 gives the tables that refine this information by indexing, in addition, with respect to the number of vertices.

9.3.1 All maps by number of edges

Orientable

	Genus					
e	0	1	2	3	4	5
1	2	·	·	·	·	·
2	9	1	·	·	·	·
3	54	20	·	·	·	·
4	378	307	21	·	·	·
5	2916	4280	966	·	·	·
6	24057	56914	27954	1485	·	
7	208494	736568	650076	113256	·	·
8	1876446	9370183	13271982	5008230	225225	·
9	17399772	117822512	248371380	167808024	24635754	·
10	165297834	1469283166	4366441128	4721384790	1495900107	59520825
11	1602117468	18210135416	73231116024	117593590752	66519597474	8608033980

Nonorientable

	Genus							
e	1	2	3	4	5	6	7	8
1	1	·	·	·	·	·	·	·
2	10	4	·	·	·	·	·	·
3	98	84	41	·	·	·	·	·
4	982	1340	1380	488	·	·	·	·
5	10062	19280	31225	23320	8229	·	·	·
6	105024	263284	592824	696912	516958	164892	·	·
7	1112757	3486224	10185056	16662492	19381145	12980716	4016613	·
8	11934910	45247084	164037704	348539072	562395292	590136856	382630152	112818960

9.3.2 All maps by numbers of edges and vertices

Orientable

e	v	Genus 0	1	2	3
1	1	1	·	·	·
	2	1	·	·	·
2	1	2	1	·	·
	2	5	·	·	·
	3	2	·	·	·
3	1	5	10	·	·
	2	22	10	·	·
	3	22	·	·	·
	4	5	·	·	·
4	1	14	70	21	·
	2	93	167	·	·
	3	164	70	·	·
	4	93	·	·	·
	5	14	·	·	·
5	1	42	420	483	·
	2	386	1720	483	·
	3	1030	1720	·	·
	4	1030	420	·	·
	5	386	·	·	·
	6	42	·	·	·
6	1	132	2310	6468	1485
	2	1586	14065	15018	·
	3	5868	24164	6468	·
	4	8885	14065	·	·
	5	5868	2310	·	·
	6	1586	·	·	·
	7	132	·	·	·
7	1	429	12012	66066	56628
	2	6476	100156	258972	56628
	3	31388	256116	258972	·
	4	65954	256116	66066	·
	5	65954	100156	·	·
	6	31388	12012	·	·
	7	6476	·	·	·
	8	429	·	·	·

		Genus					
e	v	0	1	2	3	4	5
8	1	1430	60060	570570	1169740	225225	.
	2	26333	649950	3288327	2668750	.	.
	3	160648	2278660	5554188	1169740	.	.
	4	442610	3392843	3288327	.	.	.
	5	614404	2278660	570570	.	.	.
	6	442610	649950	.	.	.	.
	7	160648	60060	.	.	.	.
	8	26333	.	.	.	.	.
	9	1430	.	.	.	.	.
9	1	4862	291720	4390386	17454580	12317877	.
	2	106762	3944928	34374186	66449432	12317877	.
	3	795846	17970784	85421118	66449432	.	.
	4	2762412	36703824	85421118	17454580	.	.
	5	5030004	36703824	34374186	.	.	.
	6	5030004	17970784	4390386	.	.	.
	7	2762412	3944928	.	.	.	.
	8	795846	291720	.	.	.	.
	9	106762	.	.	.	.	.
	10	4862	.	.	.	.	.
10	1	16796	1385670	31039008	211083730	351683046	59520825
	2	431910	22764165	313530000	1171704435	792534015	.
	3	3845020	129726760	1059255456	1955808460	351683046	.
	4	16322085	344468530	1558792200	1171704435	.	.
	5	37460376	472592916	1059255456	211083730	.	.
	6	49145460	344468530	313530000	.	.	.
	7	37460376	129726760	31039008	.	.	.
	8	16322085	22764165	.	.	.	.
	9	3845020	1385670	.	.	.	.
	10	431910	.	.	.	.	.
	11	16796	.	.	.	.	.
11	1	58786	6466460	205633428	2198596400	7034538511	4304016990
	2	1744436	126264820	2583699888	16476937840	26225260226	4304016990
	3	18211380	875029804	11270290416	40121261136	26225260226	.
	4	92400330	2908358552	22555934280	40121261136	7034538511	.
	5	259477218	5188948072	22555934280	16476937840	.	.
	6	429166584	5188948072	11270290416	2198596400	.	.
	7	429166584	2908358552	2583699888	.	.	.
	8	259477218	875029804	205633428	.	.	.
	9	92400330	126264820	.	.	.	.
	10	18211380	6466460	.	.	.	.
	11	1744436	.	.	.	.	.
	12	58786	.	.	.	.	.

Nonorientable

e	v	Genus 1	2	3	4	5	6	7	8
1	1	1	.	.	.	.	.	.	.
2	1	5	4	.	.	.	.	.	.
	2	5	.	.	.	.	.	.	.
3	1	22	42	41	.	.	.	.	.
	2	54	42	.	.	.	.	.	.
	3	22	.	.	.	.	.	.	.
4	1	93	304	690	488	.	.	.	.
	2	398	732	690	.	.	.	.	.
	3	398	304	.	.	.	.	.	.
	4	93	.	.	.	.	.	.	.
5	1	386	1870	7150	11660	8229	.	.	.
	2	2480	7770	16925	11660	.	.	.	.
	3	4330	7770	7150	.	.	.	.	.
	4	2480	1870	.	.	.	.	.	.
	5	386	.	.	.	.	.	.	.
6	1	1586	10488	58760	160680	258479	164892	.	.
	2	14046	65010	237652	375552	258479	.	.	.
	3	36880	112288	237652	160680	.	.	.	.
	4	36880	65010	58760	.	.	.	.	.
	5	14046	10488	.	.	.	.	.	.
	6	1586	.	.	.	.	.	.	.
7	1	6476	55412	420182	1678880	4494224	6490358	4016613	.
	2	74788	471450	2518957	6652366	10392697	6490358	.	.
	3	271138	1216250	4306778	6652366	4494224	.	.	.
	4	407953	1216250	2518957	1678880	.	.	.	.
	5	271138	471450	420182	.	.	.	.	.
	6	74788	55412	.	.	.	.	.	.
	7	6476	.	.	.	.	.	.	.
8	1	26333	280768	2736524	14771680	57462482	137479552	191315076	112818960
	2	381356	3105648	22417804	86303920	223735164	315177752	191315076	.
	3	1805748	11011008	56864524	146387872	223735164	137479552	.	.
	4	3754018	16452236	56864524	86303920	57462482	.	.	.
	5	3754018	11011008	22417804	14771680	.	.	.	.
	6	1805748	3105648	2736524	.	.	.	.	.
	7	381356	280768	.	.	.	.	.	.
	8	26333	.	.	.	.	.	.	.

Chapter 10

Numbers of unrooted maps

The following tables give the number of unrooted maps in a given surface for a given number of edges for all proper pairs of vertex and face partitions. In each table the number of edges appears in bold face in the top left-hand corner. Below this appears the subtitle $a : b$ specifying the number of vertices and faces a, b in some order. The tables are listed only for $a \geq b$ since, by duality, the table for $b : a$ is the same. The rows and columns of each table are indexed by vertex partition and face partition.

Section 10.1 lists the tables for maps with at most 5 edges in orientable surfaces in increasing order of genus and number of edges.

Section 10.2 lists the tables for maps with at most 5 edges in nonorientable surfaces in increasing order of genus and number of edges.

10.1 Orientable: by vertex and face partition

Genus 0: 1 edge

1 2 : 1	2
$[1^2]$	1

Genus 0: 2 edges

2 2 : 2	3 1	2^2
$[3\,1]$	1	·
$[2^2]$		1

2 3 : 1	4
$[2\,1^2]$	1

Genus 0: 3 edges

3	5	4	3^2
3 : 2	1	2	
$[4\,1^2]$	1	·	1
$[3\,2\,1]$	1	1	·
$[2^3]$	·	·	1

3	6
4 : 1	
$[3\,1^3]$	1
$[2^2\,1^2]$	1

Genus 0: 4 edges

4	6	5	4	4	3^2
3 : 3	1^2	2	3	2^2	2
		1	1		
$[6\,1^2]$	2	1	2	·	1
$[5\,2\,1]$	1	3	1	·	1
$[4\,3\,1]$	2	1	2	1	·
$[4\,2^2]$	·	·	1	1	·
$[3^2\,2]$	1	1	·	·	1

4	7	6	5	4^2
4 : 2	1	2	3	
$[5\,1^3]$	1	·	1	·
$[4\,2\,1^2]$	2	1	1	1
$[3^2\,1^2]$	1	1	·	1
$[3\,2^2\,1]$	1	1	1	·
$[2^4]$	·	·	·	1

4	8
5 : 1	
$[4\,1^4]$	1
$[3\,2\,1^3]$	1
$[2^3\,1^2]$	1

Genus 0: 5 edges

5	8	7	6	6	5	5	4^2	4
4 : 3	1^2	2	3	2^2	4	3	2	3^2
		1	1		1	2		
$[7\,1^3]$	2	1	3	·	2	1	·	1
$[6\,2\,1^2]$	3	5	3	·	4	4	·	1
$[5\,3\,1^2]$	4	3	3	1	4	1	1	1
$[5\,2^2\,1]$	1	3	3	1	1	1	1	1
$[4^2\,1^2]$	3	1	3	1	1	·	2	1
$[4\,3\,2\,1]$	4	5	4	2	4	3	1	1
$[4\,2^3]$	·	·	·	·	1	1	·	·
$[3^3\,1]$	1	1	·	·	1	1	·	·
$[3^2\,2^2]$	1	1	1	1	·	·	1	1

5	9	8	7	6	5^2
5 : 2	1	2	3	4	
$[6\,1^4]$	1	·	1	·	1
$[5\,2\,1^3]$	3	1	2	1	1
$[4\,3\,1^3]$	3	1	1	2	·
$[4\,2^2\,1^2]$	3	2	2	1	2
$[3^2\,2\,1^2]$	3	2	1	1	2
$[3\,2^3\,1]$	1	1	1	1	·
$[2^5]$	·	·	·	·	1

5	10
6 : 1	
$[5\,1^5]$	1
$[4\,2\,1^4]$	1
$[3^2\,1^4]$	1
$[3\,2^2\,1^3]$	2
$[2^4\,1^2]$	1

Genus 1: 2 edges

2	4
1 : 1	
[4]	1

Genus 1: 3 edges

3	6
2 : 1	
[5 1]	1
[4 2]	1
$[3^2]$	1

Genus 1: 4 edges

4	7	6	5	4^2
2 : 2	1	2	3	
[7 1]	5	2	2	1
[6 2]	2	2	1	1
[5 3]	2	1	1	·
$[4^2]$	1	1	·	2

4	8
3 : 1	
$[6\,1^2]$	3
[5 2 1]	3
[4 3 1]	2
$[4\,2^2]$	2
$[3^2\,2]$	1

Genus 1: 5 edges

5	9	8	7	6	5^2
3 : 2	1	2	3	4	
$[8\,1^2]$	15	6	7	6	3
[7 2 1]	15	9	7	6	3
[6 3 1]	13	6	5	3	3
$[6\,2^2]$	3	3	2	2	2
[5 4 1]	12	5	4	5	1
[5 3 2]	6	4	3	3	·
$[4^2\,2]$	3	3	1	1	2
$[4\,3^2]$	3	2	1	1	1

5	10
4 : 1	
$[7\,1^3]$	5
$[6\,2\,1^2]$	11
$[5\,3\,1^2]$	7
$[5\,2^2\,1]$	6
$[4^2\,1^2]$	4
[4 3 2 1]	8
$[4\,2^3]$	2
$[3^3\,1]$	1
$[3^2\,2^2]$	2

Genus 2: 4 edges

4 1 : 1	8
[8]	4

Genus 2: 5 edges

5 2 : 1	10
[9 1]	21
[8 2]	11
[7 3]	7
[6 4]	7
$[5^2]$	7

10.2 Nonorientable: by vertex and face partition

Genus $\tilde{1}$: 1 edge

1 1 : 1	2
[2]	1

Genus $\tilde{1}$: 2 edges

2 2 : 1	4
[3 1]	1
$[2^2]$	1

Genus $\tilde{1}$: 3 edges

3	5	4	3^2
2 : 2	1	2	
$[5\,1]$	2	1	1
$[4\,2]$	1	1	1
$[3^2]$	1	1	·

3	6
3 : 1	
$[4\,1^2]$	2
$[3\,2\,1]$	2
$[2^3]$	1

Genus $\tilde{1}$: 4 edges

4	7	6	5	4^2
3 : 2	1	2	3	
$[6\,1^2]$	4	2	3	1
$[5\,2\,1]$	4	3	3	1
$[4\,3\,1]$	4	2	2	2
$[4\,2^2]$	1	1	1	2
$[3^2\,2]$	2	2	1	·

4	8
4 : 1	
$[5\,1^3]$	2
$[4\,2\,1^2]$	4
$[3^2\,1^2]$	2
$[3\,2^2\,1]$	3
$[2^4]$	1

Genus $\tilde{1}$: 5 edges

5	8	7	6	6	5	5	4^2	4
3 : 3	1^2	2	3	2^2	4	3	2	3^2
		1	1		1	2		
$[8\,1^2]$	11	8	10	2	8	6	2	4
$[7\,2\,1]$	8	12	9	3	8	9	2	4
$[6\,3\,1]$	10	9	7	3	10	6	2	3
$[6\,2^2]$	2	3	3	2	3	3	2	2
$[5\,4\,1]$	8	8	10	3	8	3	3	4
$[5\,3\,2]$	6	9	6	3	3	3	3	2
$[4^2\,2]$	2	2	2	2	3	3	1	1
$[4\,3^2]$	4	4	3	2	4	2	1	1

5	9	8	7	6	5^2
4 : 2	1	2	3	4	
$[7\,1^3]$	6	2	5	2	2
$[6\,2\,1^2]$	10	6	7	4	5
$[5\,3\,1^2]$	11	5	4	6	2
$[5\,2^2\,1]$	6	5	5	4	2
$[4^2\,1^2]$	5	3	3	4	1
$[4\,3\,2\,1]$	12	8	6	7	3
$[4\,2^3]$	1	1	1	1	2
$[3^3\,1]$	2	2	1	1	1
$[3^2\,2^2]$	3	3	2	2	·

5	10
5 : 1	
$[6\,1^4]$	3
$[5\,2\,1^3]$	6
$[4\,3\,1^3]$	5
$[4\,2^2\,1^2]$	8
$[3^2\,2\,1^2]$	6
$[3\,2^3\,1]$	4
$[2^5]$	1

Genus $\tilde{2}$: 2 edges

2	4
1 : 1	
[4]	2

Genus $\tilde{2}$: 3 edges

3	6
2 : 1	
[5 1]	3
[4 2]	3
$[3^2]$	2

Genus $\tilde{2}$: 4 edges

4	7	6	5	4^2
2 : 2	1	2	3	
[7 1]	11	5	7	3
[6 2]	5	6	4	3
[5 3]	7	4	3	1
$[4^2]$	3	3	1	5

4	8
3 : 1	
$[6\,1^2]$	9
[5 2 1]	8
[4 3 1]	7
$[4\,2^2]$	5
$[3^2\,2]$	4

Genus $\tilde{2}$: 5 edges

5	9	8	7	6	5^2
3 : 2	1	2	3	4	
$[8\,1^2]$	33	15	18	14	11
[7 2 1]	32	31	19	16	9
[6 3 1]	31	16	14	13	8
$[6\,2^2]$	8	8	7	7	6
[5 4 1]	29	13	14	16	4
[5 3 2]	19	15	9	8	2
$[4^2\,2]$	7	7	4	5	6
$[4\,3^2]$	11	7	4	5	4

5	10
4 : 1	
$[7\,1^3]$	13
$[6\,2\,1^2]$	26
$[5\,3\,1^2]$	19
$[5\,2^2\,1]$	16
$[4^2\,1^2]$	14
[4 3 2 1]	25
$[4\,2^3]$	6
$[3^3\,1]$	4
$[3^2\,2^2]$	8

Genus $\tilde{3}$: 3 edges

3	6
1 : 1	
[6]	8

Genus $\tilde{3}$: 4 edges

4	8
2 : 1	
[7 1]	24
[6 2]	18
[5 3]	12
[4^2]	10

Genus $\tilde{3}$: 5 edges

5	9	8	7	6	5^2
2 : 2	1	2	3	4	
[9 1]	147	66	70	59	29
[8 2]	66	55	38	36	20
[7 3]	70	38	28	23	13
[6 4]	59	36	23	31	17
[5^2]	29	20	13	17	3

5	10
3 : 1	
[8 1^2]	88
[7 2 1]	90
[6 3 1]	70
[6 2^2]	36
[5 4 1]	64
[5 3 2]	42
[4^2 2]	28
[4 3^2]	24

Genus $\tilde{4}$: 4 edges

4	8
1 : 1	
[8]	47

Genus $\tilde{4}$: 5 edges

5 2 : 1	10
[9 1]	258
[8 2]	149
[7 3]	109
[6 4]	111
$[5^2]$	59

Genus $\tilde{5}$: 5 edges

5 1 : 1	10
[10]	473

Chapter 11

Nonrealizable pairs of partitions

Listed below are pairs of partitions of at most 8 that satisfy the Euler-Poincaré Formula, but are not the vertex partition and face partition of any map on the appropriate surface. The entries are indexed by genus and number of edges, and are listed in reverse lexicographical order. They are to be read across the rows of the tables.

11.1 For orientable surfaces

Genus 0: 2 edges

$[3\,1]$ $[2^2]$

Genus 0: 3 edges

$[5\,1]$ $[2^3]$	$[4\,2]$ $[4\,1^2]$	$[4\,2]$ $[2^3]$	$[3^2]$ $[3\,2\,1]$

Genus 0: 4 edges

$[7\,1]$ $[2^4]$	$[6\,2]$ $[5\,1^3]$	$[6\,2]$ $[2^4]$	$[6\,1^2]$ $[4\,2^2]$
$[5\,3]$ $[3^2\,1^2]$	$[5\,3]$ $[2^4]$	$[5\,2\,1]$ $[4\,2^2]$	$[4^2]$ $[5\,1^3]$
$[4^2]$ $[3\,2^2\,1]$	$[4\,3\,1]$ $[3^2\,2]$	$[4\,2^2]$ $[3^2\,2]$	

Genus 0: 5 edges

$[9\,1]\ [2^5]$	$[8\,2]\ [6\,1^4]$	$[8\,2]\ [2^5]$	$[8\,1^2]\ [4\,2^3]$
$[7\,3]\ [2^5]$	$[7\,2\,1]\ [4\,2^3]$	$[6\,4]\ [6\,1^4]$	$[6\,4]\ [2^5]$
$[6\,3\,1]\ [4\,2^3]$	$[6\,3\,1]\ [3^3\,1]$	$[6\,2^2]\ [7\,1^3]$	$[6\,2^2]\ [6\,2\,1^2]$
$[6\,2^2]\ [4\,2^3]$	$[6\,2^2]\ [3^3\,1]$	$[5^2]\ [4\,3\,1^3]$	$[5^2]\ [3\,2^3\,1]$
$[5\,4\,1]\ [3^2\,2^2]$	$[5\,3\,2]\ [4^2\,1^2]$	$[5\,3\,2]\ [3^2\,2^2]$	$[4^2\,2]\ [7\,1^3]$
$[4^2\,2]\ [6\,2\,1^2]$	$[4^2\,2]\ [4\,2^3]$	$[4^2\,2]\ [3^3\,1]$	$[4\,3^2]\ [4\,2^3]$
$[4\,3^2]\ [3^3\,1]$			

Genus 0: 6 edges

$[11\,1]\ [2^6]$	$[10\,2]\ [7\,1^5]$	$[10\,2]\ [2^6]$	$[10\,1^2]\ [4\,2^4]$
$[9\,3]\ [2^6]$	$[9\,2\,1]\ [4\,2^4]$	$[9\,1^3]\ [6\,2^3]$	$[9\,1^3]\ [4^2\,2^2]$
$[8\,4]\ [7\,1^5]$	$[8\,4]\ [2^6]$	$[8\,3\,1]\ [4\,2^4]$	$[8\,2^2]\ [8\,1^4]$
$[8\,2^2]\ [7\,2\,1^3]$	$[8\,2^2]\ [4\,2^4]$	$[8\,2\,1^2]\ [6\,2^3]$	$[8\,2\,1^2]\ [4^2\,2^2]$
$[7\,5]\ [4^2\,1^4]$	$[7\,5]\ [2^6]$	$[7\,4\,1]\ [4\,2^4]$	$[7\,3\,2]\ [4\,2^4]$
$[7\,3\,1^2]\ [6\,2^3]$	$[7\,3\,1^2]\ [4^2\,2^2]$	$[7\,3\,1^2]\ [3^4]$	$[7\,2^2\,1]\ [6\,2^3]$
$[7\,2^2\,1]\ [4^2\,2^2]$	$[7\,2^2\,1]\ [3^4]$	$[6^2]\ [7\,1^5]$	$[6^2]\ [3^3\,1^3]$
$[6^2]\ [3\,2^4\,1]$	$[6\,5\,1]\ [3^2\,2^3]$	$[6\,4\,2]\ [8\,1^4]$	$[6\,4\,2]\ [7\,2\,1^3]$
$[6\,4\,2]\ [3^2\,2^3]$	$[6\,4\,1^2]\ [3^4]$	$[6\,3^2]\ [5\,2^3\,1]$	$[6\,3^2]\ [4\,3^2\,1^2]$
$[6\,3^2]\ [3^2\,2^3]$	$[6\,2^3]\ [5\,3\,2^2]$	$[6\,2^3]\ [4^2\,3\,1]$	$[6\,2^3]\ [3^4]$
$[5^2\,2]\ [5\,4\,1^3]$	$[5^2\,2]\ [4\,2^4]$	$[5^2\,2]\ [3^3\,2\,1]$	$[5^2\,1^2]\ [5\,3\,2^2]$
$[5\,4\,3]\ [4\,2^4]$	$[5\,4\,2\,1]\ [3^4]$	$[5\,3^2\,1]\ [3^4]$	$[5\,3\,2^2]\ [4^2\,2^2]$
$[5\,3\,2^2]\ [3^4]$	$[4^3]\ [8\,1^4]$	$[4^3]\ [7\,2\,1^3]$	$[4^3]\ [5\,2^3\,1]$
$[4^3]\ [4^2\,2\,1^2]$	$[4^3]\ [4\,2^4]$	$[4^3]\ [3^3\,2\,1]$	$[4^2\,3\,1]\ [4^2\,2^2]$
$[4^2\,3\,1]\ [3^4]$	$[4\,3^2\,2]\ [3^4]$		

Genus 0: 7 edges

$[13\ 1]\ [2^7]$
$[11\ 3]\ [2^7]$
$[10\ 4]\ [8\ 1^6]$
$[10\ 2^2]\ [8\ 2\ 1^4]$
$[9\ 5]\ [2^7]$
$[9\ 3\ 1^2]\ [4^2\ 2^3]$
$[8\ 6]\ [2^7]$
$[8\ 4\ 2]\ [4\ 2^5]$
$[8\ 3^2]\ [4\ 2^5]$
$[8\ 2^3]\ [9\ 2\ 1^3]$
$[8\ 2^3]\ [4^3\ 1^2]$
$[7^2]\ [3\ 2^5\ 1]$
$[7\ 5\ 2]\ [3^2\ 2^4]$
$[6^2\ 2]\ [8\ 2\ 1^4]$
$[6\ 4^2]\ [9\ 1^5]$
$[6\ 4\ 2^2]\ [9\ 2\ 1^3]$
$[6\ 4\ 2^2]\ [4^3\ 1^2]$
$[6\ 3^2\ 2]\ [4^3\ 1^2]$
$[5^2\ 3\ 1]\ [6\ 2^4]$
$[5^2\ 2^2]\ [5\ 3\ 2^3]$
$[5\ 4^2\ 1]\ [3^4\ 2]$
$[5\ 3^3]\ [3^4\ 2]$
$[4^3\ 2]\ [8\ 2^2\ 1^2]$
$[4^3\ 2]\ [3^4\ 2]$

$[12\ 2]\ [8\ 1^6]$
$[11\ 2\ 1]\ [4\ 2^5]$
$[10\ 4]\ [2^7]$
$[10\ 2^2]\ [4\ 2^5]$
$[9\ 4\ 1]\ [4\ 2^5]$
$[9\ 2^2\ 1]\ [6\ 2^4]$
$[8\ 5\ 1]\ [4\ 2^5]$
$[8\ 4\ 1^2]\ [6\ 2^4]$
$[8\ 3\ 2\ 1]\ [6\ 2^4]$
$[8\ 2^3]\ [8\ 3\ 1^3]$
$[8\ 2^3]\ [4^2\ 2^3]$
$[7\ 6\ 1]\ [3^2\ 2^4]$
$[7\ 4\ 3]\ [3^4\ 1^2]$
$[6^2\ 2]\ [4\ 2^5]$
$[6\ 4^2]\ [8\ 2\ 1^4]$
$[6\ 4\ 2^2]\ [8\ 3\ 1^3]$
$[6\ 4\ 2^2]\ [4^2\ 2^3]$
$[6\ 3^2\ 2]\ [4^2\ 2^3]$
$[5^2\ 3\ 1]\ [4^2\ 2^3]$
$[5^2\ 2^2]\ [4\ 3^3\ 1]$
$[5\ 3^3]\ [6\ 2^4]$
$[4^3\ 2]\ [10\ 1^4]$
$[4^3\ 2]\ [6\ 2^4]$
$[4^2\ 3^2]\ [5\ 3\ 2^3]$

$[12\ 2]\ [2^7]$
$[11\ 1^3]\ [6\ 2^4]$
$[10\ 3\ 1]\ [4\ 2^5]$
$[10\ 2\ 1^2]\ [6\ 2^4]$
$[9\ 3\ 2]\ [4\ 2^5]$
$[9\ 2^2\ 1]\ [4^2\ 2^3]$
$[8\ 4\ 2]\ [9\ 1^5]$
$[8\ 4\ 1^2]\ [4^2\ 2^3]$
$[8\ 3\ 2\ 1]\ [4^2\ 2^3]$
$[8\ 2^3]\ [8\ 2^2\ 1^2]$
$[8\ 2^3]\ [3^4\ 2]$
$[7\ 5\ 2]\ [5^2\ 1^4]$
$[7\ 4\ 3]\ [3^2\ 2^4]$
$[6^2\ 1^2]\ [5\ 3\ 2^3]$
$[6\ 4^2]\ [4\ 2^5]$
$[6\ 4\ 2^2]\ [8\ 2^2\ 1^2]$
$[6\ 4\ 2^2]\ [3^4\ 2]$
$[6\ 3^2\ 2]\ [3^4\ 2]$
$[5^2\ 3\ 1]\ [3^4\ 2]$
$[5\ 4^2\ 1]\ [6\ 2^4]$
$[5\ 3^3]\ [4^3\ 1^2]$
$[4^3\ 2]\ [9\ 2\ 1^3]$
$[4^3\ 2]\ [4^3\ 1^2]$
$[4^2\ 3^2]\ [4\ 3^3\ 1]$

$[12\ 1^2]\ [4\ 2^5]$
$[11\ 1^3]\ [4^2\ 2^3]$
$[10\ 2^2]\ [9\ 1^5]$
$[10\ 2\ 1^2]\ [4^2\ 2^3]$
$[9\ 3\ 1^2]\ [6\ 2^4]$
$[8\ 6]\ [8\ 1^6]$
$[8\ 4\ 2]\ [8\ 2\ 1^4]$
$[8\ 4\ 1^2]\ [3^4\ 2]$
$[8\ 2^3]\ [10\ 1^4]$
$[8\ 2^3]\ [6\ 2^4]$
$[7^2]\ [5\ 4\ 1^5]$
$[7\ 5\ 2]\ [3^4\ 1^2]$
$[6^2\ 2]\ [9\ 1^5]$
$[6\ 5\ 3]\ [4\ 2^5]$
$[6\ 4\ 2^2]\ [10\ 1^4]$
$[6\ 4\ 2^2]\ [6\ 2^4]$
$[6\ 3^2\ 2]\ [6\ 2^4]$
$[5^2\ 4]\ [4\ 2^5]$
$[5^2\ 2^2]\ [6\ 5\ 1^3]$
$[5\ 4^2\ 1]\ [4^2\ 2^3]$
$[5\ 3^3]\ [4^2\ 2^3]$
$[4^3\ 2]\ [8\ 3\ 1^3]$
$[4^3\ 2]\ [4^2\ 2^3]$

Genus 0: 8 edges

$[15\ 1]\ [2^8]$	$[14\ 2]\ [9\ 1^7]$	$[14\ 2]\ [2^8]$	$[14\ 1^2]\ [4\ 2^6]$
$[13\ 3]\ [2^8]$	$[13\ 2\ 1]\ [4\ 2^6]$	$[13\ 1^3]\ [6\ 2^5]$	$[13\ 1^3]\ [4^2\ 2^4]$
$[12\ 4]\ [9\ 1^7]$	$[12\ 4]\ [2^8]$	$[12\ 3\ 1]\ [4\ 2^6]$	$[12\ 2^2]\ [10\ 1^6]$
$[12\ 2^2]\ [9\ 2\ 1^5]$	$[12\ 2^2]\ [4\ 2^6]$	$[12\ 2\ 1^2]\ [6\ 2^5]$	$[12\ 2\ 1^2]\ [4^2\ 2^4]$
$[12\ 1^4]\ [8\ 2^4]$	$[12\ 1^4]\ [6\ 4\ 2^3]$	$[12\ 1^4]\ [4^3\ 2^2]$	$[11\ 5]\ [2^8]$
$[11\ 4\ 1]\ [4\ 2^6]$	$[11\ 3\ 2]\ [4\ 2^6]$	$[11\ 3\ 1^2]\ [6\ 2^5]$	$[11\ 3\ 1^2]\ [4^2\ 2^4]$
$[11\ 2^2\ 1]\ [6\ 2^5]$	$[11\ 2^2\ 1]\ [4^2\ 2^4]$	$[11\ 2\ 1^3]\ [8\ 2^4]$	$[11\ 2\ 1^3]\ [6\ 4\ 2^3]$
$[11\ 2\ 1^3]\ [4^3\ 2^2]$	$[10\ 6]\ [9\ 1^7]$	$[10\ 6]\ [2^8]$	$[10\ 5\ 1]\ [4\ 2^6]$
$[10\ 4\ 2]\ [10\ 1^6]$	$[10\ 4\ 2]\ [9\ 2\ 1^5]$	$[10\ 4\ 2]\ [4\ 2^6]$	$[10\ 4\ 1^2]\ [6\ 2^5]$
$[10\ 4\ 1^2]\ [4^2\ 2^4]$	$[10\ 3^2]\ [4\ 2^6]$	$[10\ 3\ 2\ 1]\ [6\ 2^5]$	$[10\ 3\ 2\ 1]\ [4^2\ 2^4]$
$[10\ 3\ 1^3]\ [8\ 2^4]$	$[10\ 3\ 1^3]\ [6\ 4\ 2^3]$	$[10\ 3\ 1^3]\ [4^3\ 2^2]$	$[10\ 2^3]\ [11\ 1^5]$
$[10\ 2^3]\ [10\ 2\ 1^4]$	$[10\ 2^3]\ [9\ 3\ 1^4]$	$[10\ 2^3]\ [9\ 2^2\ 1^3]$	$[10\ 2^3]\ [6\ 2^5]$
$[10\ 2^3]\ [4^2\ 2^4]$	$[10\ 2^3]\ [3^5\ 1]$	$[10\ 2^2\ 1^2]\ [8\ 2^4]$	$[10\ 2^2\ 1^2]\ [6\ 4\ 2^3]$
$[10\ 2^2\ 1^2]\ [4^3\ 2^2]$	$[9\ 7]\ [5^2\ 1^6]$	$[9\ 7]\ [2^8]$	$[9\ 6\ 1]\ [4\ 2^6]$
$[9\ 5\ 2]\ [4\ 2^6]$	$[9\ 5\ 1^2]\ [6\ 2^5]$	$[9\ 5\ 1^2]\ [4^2\ 2^4]$	$[9\ 4\ 3]\ [4\ 2^6]$
$[9\ 4\ 2\ 1]\ [6\ 2^5]$	$[9\ 4\ 2\ 1]\ [4^2\ 2^4]$	$[9\ 4\ 1^3]\ [8\ 2^4]$	$[9\ 4\ 1^3]\ [6\ 4\ 2^3]$
$[9\ 4\ 1^3]\ [4^3\ 2^2]$	$[9\ 3^2\ 1]\ [6\ 2^5]$	$[9\ 3^2\ 1]\ [4^2\ 2^4]$	$[9\ 3\ 2^2]\ [6\ 2^5]$
$[9\ 3\ 2^2]\ [4^2\ 2^4]$	$[9\ 3\ 2\ 1^2]\ [8\ 2^4]$	$[9\ 3\ 2\ 1^2]\ [6\ 4\ 2^3]$	$[9\ 3\ 2\ 1^2]\ [4^3\ 2^2]$
$[9\ 2^3\ 1]\ [8\ 2^4]$	$[9\ 2^3\ 1]\ [6\ 4\ 2^3]$	$[9\ 2^3\ 1]\ [4^3\ 2^2]$	$[8^2]\ [9\ 1^7]$
$[8^2]\ [3\ 2^6\ 1]$	$[8\ 7\ 1]\ [3^2\ 2^5]$	$[8\ 6\ 2]\ [10\ 1^6]$	$[8\ 6\ 2]\ [9\ 2\ 1^5]$
$[8\ 6\ 2]\ [3^2\ 2^5]$	$[8\ 5\ 3]\ [3^2\ 2^5]$	$[8\ 4^2]\ [10\ 1^6]$	$[8\ 4^2]\ [9\ 2\ 1^5]$
$[8\ 4^2]\ [5\ 2^5\ 1]$	$[8\ 4^2]\ [3^2\ 2^5]$	$[8\ 4\ 3\ 1]\ [3^5\ 1]$	$[8\ 4\ 2^2]\ [11\ 1^5]$
$[8\ 4\ 2^2]\ [10\ 2\ 1^4]$	$[8\ 4\ 2^2]\ [9\ 3\ 1^4]$	$[8\ 4\ 2^2]\ [9\ 2^2\ 1^3]$	$[8\ 4\ 2^2]\ [3^5\ 1]$
$[8\ 3^2\ 2]\ [3^5\ 1]$	$[8\ 2^4]\ [7\ 3\ 2^3]$	$[8\ 2^4]\ [6\ 3^3\ 1]$	$[8\ 2^4]\ [5^2\ 4\ 1^2]$
$[8\ 2^4]\ [5^2\ 2^3]$	$[8\ 2^4]\ [4\ 3^4]$	$[7^2\ 2]\ [6\ 5\ 1^5]$	$[7^2\ 2]\ [4\ 2^6]$
$[7^2\ 1^2]\ [5\ 3\ 2^4]$	$[7^2\ 1^2]\ [3^5\ 1]$	$[7\ 6\ 3]\ [4\ 2^6]$	$[7\ 6\ 1^3]\ [5^2\ 2^3]$
$[7\ 5\ 4]\ [4\ 2^6]$	$[7\ 5\ 2^2]\ [6^2\ 1^4]$	$[7\ 5\ 2^2]\ [6\ 2^5]$	$[7\ 5\ 2^2]\ [4^2\ 2^4]$
$[7\ 5\ 2^2]\ [3^4\ 2^2]$	$[7\ 4^2\ 1]\ [3^5\ 1]$	$[7\ 4\ 3\ 2]\ [6\ 2^5]$	$[7\ 4\ 3\ 2]\ [4^2\ 2^4]$
$[7\ 3^3]\ [6\ 2^5]$	$[7\ 3^3]\ [4^2\ 2^4]$	$[7\ 3^3]\ [3^5\ 1]$	$[7\ 3^3]\ [3^4\ 2^2]$
$[7\ 3\ 2^3]\ [6\ 4\ 2^3]$	$[7\ 3\ 2^3]\ [4^3\ 2^2]$	$[6^2\ 4]\ [10\ 1^6]$	$[6^2\ 4]\ [9\ 2\ 1^5]$
$[6^2\ 4]\ [4\ 2^6]$	$[6^2\ 3\ 1]\ [6\ 2^5]$	$[6^2\ 3\ 1]\ [4^2\ 2^4]$	$[6^2\ 3\ 1]\ [3^4\ 2^2]$
$[6^2\ 2^2]\ [11\ 1^5]$	$[6^2\ 2^2]\ [10\ 2\ 1^4]$	$[6^2\ 2^2]\ [9\ 3\ 1^4]$	$[6^2\ 2^2]\ [9\ 2^2\ 1^3]$
$[6^2\ 2^2]\ [5\ 3\ 2^4]$	$[6^2\ 2^2]\ [3^5\ 1]$	$[6\ 5^2]\ [4\ 2^6]$	$[6\ 5\ 4\ 1]\ [6\ 2^5]$
$[6\ 5\ 4\ 1]\ [4^2\ 2^4]$	$[6\ 5\ 3\ 2]\ [3^5\ 1]$	$[6\ 4^2\ 2]\ [11\ 1^5]$	$[6\ 4^2\ 2]\ [10\ 2\ 1^4]$
$[6\ 4^2\ 2]\ [9\ 3\ 1^4]$	$[6\ 4^2\ 2]\ [9\ 2^2\ 1^3]$	$[6\ 4^2\ 2]\ [3^5\ 1]$	$[6\ 4\ 3^2]\ [6\ 2^5]$
$[6\ 4\ 3^2]\ [4^2\ 2^4]$	$[6\ 4\ 2^3]\ [6\ 3^3\ 1]$	$[6\ 4\ 2^3]\ [5^2\ 4\ 1^2]$	$[6\ 4\ 2^3]\ [5^2\ 2^3]$
$[6\ 4\ 2^3]\ [4\ 3^4]$	$[6\ 3^3\ 1]\ [5\ 3^3\ 2]$	$[6\ 3^3\ 1]\ [4^3\ 3\ 1]$	$[6\ 3^3\ 1]\ [4^3\ 2^2]$
$[5^3\ 1]\ [6\ 2^5]$	$[5^3\ 1]\ [4^2\ 2^4]$	$[5^3\ 1]\ [3^5\ 1]$	$[5^3\ 1]\ [3^4\ 2^2]$

$[5^2\,4\,2]$ $[6\,2^5]$ $[5^2\,4\,2]$ $[4^2\,2^4]$ $[5^2\,4\,1^2]$ $[4^3\,2^2]$ $[5^2\,3^2]$ $[5\,3\,2^4]$
$[5^2\,3^2]$ $[3^5\,1]$ $[5^2\,2^3]$ $[5\,3^3\,2]$ $[5^2\,2^3]$ $[4^3\,3\,1]$ $[5^2\,2^3]$ $[4^3\,2^2]$
$[5\,4^2\,3]$ $[3^5\,1]$ $[5\,3^3\,2]$ $[4\,3^4]$ $[4^4]$ $[11\,1^5]$ $[4^4]$ $[10\,2\,1^4]$
$[4^4]$ $[9\,3\,1^4]$ $[4^4]$ $[9\,2^2\,1^3]$ $[4^4]$ $[7\,2^4\,1]$ $[4^4]$ $[6\,2^5]$
$[4^4]$ $[5\,4^2\,1^3]$ $[4^4]$ $[5\,4\,2^3\,1]$ $[4^4]$ $[5\,3^3\,1^2]$ $[4^4]$ $[5\,3\,2^4]$
$[4^4]$ $[4\,3^3\,2\,1]$ $[4^4]$ $[4\,3^2\,2^3]$ $[4^4]$ $[3^5\,1]$ $[4^3\,3\,1]$ $[4\,3^4]$
$[4^3\,2^2]$ $[4\,3^4]$

Genus 1: 4 edges

$[5\,3]$ $[4^2]$

Genus 1: 5 edges

$[5^2]$ $[5\,3\,2]$

Genus 1: 6 edges

$[7\,5]$ $[3^4]$ $[6^2]$ $[5\,3\,2^2]$ $[5\,4\,3]$ $[4^3]$

Genus 1: 7 edges

$[7^2]$ $[5\,3\,2^3]$

Genus 1: 8 edges

$[8^2]$ $[5\,3\,2^4]$ $[8^2]$ $[3^5\,1]$ $[7\,5\,2^2]$ $[4^4]$ $[7\,3^3]$ $[4^4]$
$[6^2\,3\,1]$ $[4^4]$ $[5^3\,1]$ $[4^4]$ $[5\,4^2\,3]$ $[4^4]$

Genus greater than 1: At most 8 edges

There are no nonrealizable partitions.

11.2 For nonorientable surfaces

Genus $\tilde{1}$: 3 edges

$[3^2]$ $[3^2]$

Genus $\tilde{1}$: 4 edges

$[4^2]$ $[3^2\,2]$

Genus $\tilde{1}$: 5 edges

$[5^2]$ $[3^2\,2^2]$

Genus $\tilde{1}$: 6 edges

$[7\,4\,1]$ $[3^4]$ $[6^2]$ $[3^2\,2^3]$ $[5^2\,2]$ $[3^4]$ $[5\,4\,3]$ $[3^4]$

Genus $\tilde{1}$: 7 edges

$[7^2]$ $[3^2\,2^4]$

Genus $\tilde{1}$: 8 edges

$[8^2]$ $[3^2\,2^5]$ $[8\,4^2]$ $[3^5\,1]$ $[4^4]$ $[6\,3^3\,1]$ $[4^4]$ $[5^2\,2^3]$
$[4^4]$ $[4\,3^4]$

Genus greater than $\tilde{1}$: At most 8 edges

There are no nonrealizable partitions.

Chapter 12

Map polynomials

12.1 b-polynomials

The following tables give the b-polynomials $h(\nu, \phi, \eta; b)$ and $m(\nu, \phi, n; b)$, defined in Section 4.4, for rooted hypermaps and maps on at most 5 edges.

The tables are listed by type and number of edges. They are indexed by the vertex partition ν, the (hyper)face partition ϕ and, in the case of hypermaps, the hyperedge partition η, where the partitions, ν, are listed in increasing order of number of parts and then in reverse lexicographic order. Tables are terminated with a horizontal line if they are complete, and are left unterminated if they continue in the next column or on the next page.

Section 12.1.1 lists the tables for hypermaps beginning with type 0. Section 12.1.2 lists the tables for maps beginning with type 1/2. The polynomials for type 0 in this case are constants, and these are listed in Section 9.1 as numbers of rooted maps in orientable surfaces of given genus.

12.1.1 Hypermaps

Type 0: 1 edge

ν	ϕ	η	$h(\nu, \phi, \eta; b)$
[2]	[1]	[1]	1

Type 0: 2 edges

ν	ϕ	η	$h(\nu, \phi, \eta; b)$
$[4]$	$[2]$	$[1^2]$	1
$[4]$	$[1^2]$	$[2]$	1
$[2^2]$	$[2]$	$[2]$	1

Type 0: 3 edges

ν	ϕ	η	$h(\nu, \phi, \eta; b)$
$[6]$	$[3]$	$[1^3]$	1
$[6]$	$[2\,1]$	$[2\,1]$	3
$[6]$	$[1^3]$	$[3]$	1
$[4\,2]$	$[3]$	$[2\,1]$	3
$[4\,2]$	$[2\,1]$	$[3]$	3
$[2^3]$	$[3]$	$[3]$	1

Type 0: 4 edges

ν	ϕ	η	$h(\nu, \phi, \eta; b)$
$[8]$	$[4]$	$[1^4]$	1
$[8]$	$[3\,1]$	$[2\,1^2]$	4
$[8]$	$[2^2]$	$[2\,1^2]$	2
$[8]$	$[2\,1^2]$	$[3\,1]$	4
$[8]$	$[2\,1^2]$	$[2^2]$	2
$[8]$	$[1^4]$	$[4]$	1
$[6\,2]$	$[4]$	$[2\,1^2]$	4
$[6\,2]$	$[3\,1]$	$[3\,1]$	4
$[6\,2]$	$[3\,1]$	$[2^2]$	4
$[6\,2]$	$[2^2]$	$[3\,1]$	4
$[6\,2]$	$[2\,1^2]$	$[4]$	4
$[4^2]$	$[4]$	$[2\,1^2]$	2
$[4^2]$	$[3\,1]$	$[3\,1]$	4
$[4^2]$	$[2^2]$	$[2^2]$	1
$[4^2]$	$[2\,1^2]$	$[4]$	2
$[4\,2^2]$	$[4]$	$[3\,1]$	4
$[4\,2^2]$	$[4]$	$[2^2]$	2
$[4\,2^2]$	$[3\,1]$	$[4]$	4
$[4\,2^2]$	$[2^2]$	$[4]$	2
$[2^4]$	$[4]$	$[4]$	1

Type 0: 5 edges

ν	ϕ	η	$h(\nu,\phi,\eta;b)$
[10]	[5]	$[1^5]$	1
[10]	[4 1]	$[2\,1^3]$	5
[10]	[3 2]	$[2\,1^3]$	5
[10]	$[3\,1^2]$	$[3\,1^2]$	5
[10]	$[3\,1^2]$	$[2^2\,1]$	5
[10]	$[2^2\,1]$	$[3\,1^2]$	5
[10]	$[2^2\,1]$	$[2^2\,1]$	5
[10]	$[2\,1^3]$	[4 1]	5
[10]	$[2\,1^3]$	[3 2]	5
[10]	$[1^5]$	[5]	1
[8 2]	[5]	$[2\,1^3]$	5
[8 2]	[4 1]	$[3\,1^2]$	5
[8 2]	[4 1]	$[2^2\,1]$	10
[8 2]	[3 2]	$[3\,1^2]$	10
[8 2]	[3 2]	$[2^2\,1]$	5
[8 2]	$[3\,1^2]$	[4 1]	5
[8 2]	$[3\,1^2]$	[3 2]	10
[8 2]	$[2^2\,1]$	[4 1]	10
[8 2]	$[2^2\,1]$	[3 2]	5
[8 2]	$[2\,1^3]$	[5]	5
[6 4]	[5]	$[2\,1^3]$	5
[6 4]	[4 1]	$[3\,1^2]$	10
[6 4]	[4 1]	$[2^2\,1]$	5
[6 4]	[3 2]	$[3\,1^2]$	5
[6 4]	[3 2]	$[2^2\,1]$	5
[6 4]	$[3\,1^2]$	[4 1]	10
[6 4]	$[3\,1^2]$	[3 2]	5
[6 4]	$[2^2\,1]$	[4 1]	5
[6 4]	$[2^2\,1]$	[3 2]	5
[6 4]	$[2\,1^3]$	[5]	5
$[6\,2^2]$	[5]	$[3\,1^2]$	5
$[6\,2^2]$	[5]	$[2^2\,1]$	5
$[6\,2^2]$	[4 1]	[4 1]	5
$[6\,2^2]$	[4 1]	[3 2]	10
$[6\,2^2]$	[3 2]	[4 1]	10
$[6\,2^2]$	[3 2]	[3 2]	5
$[6\,2^2]$	$[3\,1^2]$	[5]	5
$[6\,2^2]$	$[2^2\,1]$	[5]	5
$[4^2\,2]$	[5]	$[3\,1^2]$	5
$[4^2\,2]$	[5]	$[2^2\,1]$	5
$[4^2\,2]$	[4 1]	[4 1]	10
$[4^2\,2]$	[4 1]	[3 2]	5
$[4^2\,2]$	[3 2]	[4 1]	5
$[4^2\,2]$	[3 2]	[3 2]	5
$[4^2\,2]$	$[3\,1^2]$	[5]	5
$[4^2\,2]$	$[2^2\,1]$	[5]	5
$[4\,2^3]$	[5]	[4 1]	5
$[4\,2^3]$	[5]	[3 2]	5
$[4\,2^3]$	[4 1]	[5]	5
$[4\,2^3]$	[3 2]	[5]	5
$[2^5]$	[5]	[5]	1

Type 1/2: 2 edges

ν	ϕ	η	$h(\nu, \phi, \eta; b)$
$[4]$	$[2]$	$[2]$	b

Type 1/2: 3 edges

ν	ϕ	η	$h(\nu, \phi, \eta; b)$
$[6]$	$[3]$	$[2\,1]$	$3b$
$[6]$	$[2\,1]$	$[3]$	$3b$
$[4\,2]$	$[3]$	$[3]$	$3b$

Type 1/2: 4 edges

ν	ϕ	η	$h(\nu, \phi, \eta; b)$
$[8]$	$[4]$	$[2\,1^2]$	$6b$
$[8]$	$[3\,1]$	$[3\,1]$	$8b$
$[8]$	$[3\,1]$	$[2^2]$	$4b$
$[8]$	$[2^2]$	$[3\,1]$	$4b$
$[8]$	$[2^2]$	$[2^2]$	b
$[8]$	$[2\,1^2]$	$[4]$	$6b$
$[6\,2]$	$[4]$	$[3\,1]$	$8b$
$[6\,2]$	$[4]$	$[2^2]$	$4b$
$[6\,2]$	$[3\,1]$	$[4]$	$8b$
$[6\,2]$	$[2^2]$	$[4]$	$4b$
$[4^2]$	$[4]$	$[3\,1]$	$4b$
$[4^2]$	$[4]$	$[2^2]$	b
$[4^2]$	$[3\,1]$	$[4]$	$4b$
$[4^2]$	$[2^2]$	$[4]$	b
$[4\,2^2]$	$[4]$	$[4]$	$6b$

Type 1/2: 5 edges

ν	ϕ	η	$h(\nu,\phi,\eta;b)$
[10]	[5]	[2 1^3]	$10b$
[10]	[4 1]	[3 1^2]	$15b$
[10]	[4 1]	[2^2 1]	$15b$
[10]	[3 2]	[3 1^2]	$15b$
[10]	[3 2]	[2^2 1]	$10b$
[10]	[3 1^2]	[4 1]	$15b$
[10]	[3 1^2]	[3 2]	$15b$
[10]	[2^2 1]	[4 1]	$15b$
[10]	[2^2 1]	[3 2]	$10b$
[10]	[2 1^3]	[5]	$10b$
[8 2]	[5]	[3 1^2]	$15b$
[8 2]	[5]	[2^2 1]	$15b$
[8 2]	[4 1]	[4 1]	$20b$
[8 2]	[4 1]	[3 2]	$25b$
[8 2]	[3 2]	[4 1]	$25b$
[8 2]	[3 2]	[3 2]	$15b$
[8 2]	[3 1^2]	[5]	$15b$
[8 2]	[2^2 1]	[5]	$15b$
[6 4]	[5]	[3 1^2]	$15b$
[6 4]	[5]	[2^2 1]	$10b$
[6 4]	[4 1]	[4 1]	$25b$
[6 4]	[4 1]	[3 2]	$15b$
[6 4]	[3 2]	[4 1]	$15b$
[6 4]	[3 2]	[3 2]	$10b$
[6 4]	[3 1^2]	[5]	$15b$
[6 4]	[2^2 1]	[5]	$10b$
[6 2^2]	[5]	[4 1]	$15b$
[6 2^2]	[5]	[3 2]	$15b$
[6 2^2]	[4 1]	[5]	$15b$
[6 2^2]	[3 2]	[5]	$15b$
[4^2 2]	[5]	[4 1]	$15b$
[4^2 2]	[5]	[3 2]	$10b$
[4^2 2]	[4 1]	[5]	$15b$
[4^2 2]	[3 2]	[5]	$10b$
[4 2^3]	[5]	[5]	$10b$

Type 1: 3 edges

ν	ϕ	η	$h(\nu,\phi,\eta;b)$
[6]	[3]	[3]	$1+b+2b^2$

Type 1: 4 edges

ν	ϕ	η	$h(\nu,\phi,\eta;b)$
[8]	[4]	[3 1]	$4+4b+8b^2$
[8]	[4]	[2^2]	$1+b+3b^2$
[8]	[3 1]	[4]	$4+4b+8b^2$
[8]	[2^2]	[4]	$1+b+3b^2$
[6 2]	[4]	[4]	$4+4b+8b^2$
[4^2]	[4]	[4]	$1+b+3b^2$

Type 1: 5 edges

ν	ϕ	η	$h(\nu,\phi,\eta;b)$
[10]	[5]	[3 1^2]	$10+10b+20b^2$
[10]	[5]	[2^2 1]	$5+5b+15b^2$
[10]	[4 1]	[4 1]	$15+15b+30b^2$
[10]	[4 1]	[3 2]	$10+10b+25b^2$
[10]	[3 2]	[4 1]	$10+10b+25b^2$
[10]	[3 2]	[3 2]	$5+5b+15b^2$
[10]	[3 1^2]	[5]	$10+10b+20b^2$
[10]	[2^2 1]	[5]	$5+5b+15b^2$
[8 2]	[5]	[4 1]	$15+15b+30b^2$
[8 2]	[5]	[3 2]	$10+10b+25b^2$
[8 2]	[4 1]	[5]	$15+15b+30b^2$
[8 2]	[3 2]	[5]	$10+10b+25b^2$
[6 4]	[5]	[4 1]	$10+10b+25b^2$
[6 4]	[5]	[3 2]	$5+5b+15b^2$
[6 4]	[4 1]	[5]	$10+10b+25b^2$
[6 4]	[3 2]	[5]	$5+5b+15b^2$
[6 2^2]	[5]	[5]	$10+10b+20b^2$
[4^2 2]	[5]	[5]	$5+5b+15b^2$

Type 3/2: 4 edges

ν	ϕ	η	$h(\nu,\phi,\eta;b)$
[8]	[4]	[4]	$7b+7b^2+6b^3$

Type 3/2: 5 edges

ν	ϕ	η	$h(\nu,\phi,\eta;b)$
[10]	[5]	[4 1]	$35b+35b^2+30b^3$
[10]	[5]	[3 2]	$20b+20b^2+20b^3$
[10]	[4 1]	[5]	$35b+35b^2+30b^3$
[10]	[3 2]	[5]	$20b+20b^2+20b^3$
[8 2]	[5]	[5]	$35b+35b^2+30b^3$
[6 4]	[5]	[5]	$20b+20b^2+20b^3$

Type 2: 5 edges

ν	ϕ	η	$h(\nu,\phi,\eta;b)$
[10]	[5]	[5]	$8+16b+54b^2+46b^3+24b^4$

12.1.2 Maps

Type 1/2: 1 edge

ν	ϕ	$m(\nu,\phi,1;b)$
[2]	[2]	b

Type 1/2: 2 edges

ν	ϕ	$m(\nu,\phi,2;b)$
[4]	[3 1]	$4b$
[4]	[2^2]	b
[3 1]	[4]	$4b$
[2^2]	[4]	b

Type 1/2: 3 edges

ν	ϕ	$m(\nu, \phi, 3; b)$
[6]	[4 1^2]	$9b$
[6]	[3 2 1]	$12b$
[6]	[2^3]	b
[5 1]	[5 1]	$18b$
[5 1]	[4 2]	$6b$
[5 1]	[3^2]	$6b$
[4 2]	[5 1]	$6b$
[4 2]	[4 2]	$6b$
[4 2]	[3^2]	$3b$
[3^2]	[5 1]	$6b$
[3^2]	[4 2]	$3b$
[4 1^2]	[6]	$9b$
[3 2 1]	[6]	$12b$
[2^3]	[6]	b

Type 1/2: 4 edges

ν	ϕ	$m(\nu,\phi,4;b)$
[8]	[5 1^3]	$16b$
[8]	[4 2 1^2]	$36b$
[8]	[3^2 1^2]	$16b$
[8]	[3 2^2 1]	$24b$
[8]	[2^4]	b
[7 1]	[6 1^2]	$48b$
[7 1]	[5 2 1]	$48b$
[7 1]	[4 3 1]	$56b$
[7 1]	[4 2^2]	$8b$
[7 1]	[3^2 2]	$16b$
[6 2]	[6 1^2]	$12b$
[6 2]	[5 2 1]	$32b$
[6 2]	[4 3 1]	$24b$
[6 2]	[4 2^2]	$8b$
[6 2]	[3^2 2]	$12b$
[5 3]	[6 1^2]	$32b$
[5 3]	[5 2 1]	$32b$
[5 3]	[4 3 1]	$16b$
[5 3]	[4 2^2]	$8b$
[5 3]	[3^2 2]	$8b$
[4^2]	[6 1^2]	$8b$
[4^2]	[5 2 1]	$8b$
[4^2]	[4 3 1]	$16b$
[4^2]	[4 2^2]	$6b$
[6 1^2]	[7 1]	$48b$
[6 1^2]	[6 2]	$12b$
[6 1^2]	[5 3]	$32b$
[6 1^2]	[4^2]	$8b$
[5 2 1]	[7 1]	$48b$
[5 2 1]	[6 2]	$32b$
[5 2 1]	[5 3]	$32b$
[5 2 1]	[4^2]	$8b$
[4 3 1]	[7 1]	$56b$
[4 3 1]	[6 2]	$24b$
[4 3 1]	[5 3]	$16b$
[4 3 1]	[4^2]	$16b$

ν	ϕ	$m(\nu,\phi,4;b)$
[4 2^2]	[7 1]	$8b$
[4 2^2]	[6 2]	$8b$
[4 2^2]	[5 3]	$8b$
[4 2^2]	[4^2]	$6b$
[3^2 2]	[7 1]	$16b$
[3^2 2]	[6 2]	$12b$
[3^2 2]	[5 3]	$8b$
[5 1^3]	[8]	$16b$
[4 2 1^2]	[8]	$36b$
[3^2 1^2]	[8]	$16b$
[3 2^2 1]	[8]	$24b$
[2^4]	[8]	b

Type 1/2: 5 edges

ν	ϕ	$m(\nu,\phi,5;b)$
[10]	$[6\,1^4]$	$25b$
[10]	$[5\,2\,1^3]$	$80b$
[10]	$[4\,3\,1^3]$	$70b$
[10]	$[4\,2^2\,1^2]$	$90b$
[10]	$[3^2\,2\,1^2]$	$80b$
[10]	$[3\,2^3\,1]$	$40b$
[10]	$[2^5]$	b
[9 1]	$[7\,1^3]$	$100b$
[9 1]	$[6\,2\,1^2]$	$180b$
[9 1]	$[5\,3\,1^2]$	$180b$
[9 1]	$[5\,2^2\,1]$	$90b$
[9 1]	$[4^2\,1^2]$	$90b$
[9 1]	$[4\,3\,2\,1]$	$210b$
[9 1]	$[4\,2^3]$	$10b$
[9 1]	$[3^3\,1]$	$40b$
[9 1]	$[3^2\,2^2]$	$30b$
[8 2]	$[7\,1^3]$	$20b$
[8 2]	$[6\,2\,1^2]$	$90b$
[8 2]	$[5\,3\,1^2]$	$70b$
[8 2]	$[5\,2^2\,1]$	$70b$
[8 2]	$[4^2\,1^2]$	$30b$
[8 2]	$[4\,3\,2\,1]$	$130b$
[8 2]	$[4\,2^3]$	$10b$
[8 2]	$[3^3\,1]$	$20b$
[8 2]	$[3^2\,2^2]$	$25b$
[7 3]	$[7\,1^3]$	$70b$
[7 3]	$[6\,2\,1^2]$	$110b$
[7 3]	$[5\,3\,1^2]$	$70b$
[7 3]	$[5\,2^2\,1]$	$70b$
[7 3]	$[4^2\,1^2]$	$40b$
[7 3]	$[4\,3\,2\,1]$	$90b$
[7 3]	$[4\,2^3]$	$10b$
[7 3]	$[3^3\,1]$	$10b$
[7 3]	$[3^2\,2^2]$	$20b$
[6 4]	$[7\,1^3]$	$30b$
[6 4]	$[6\,2\,1^2]$	$60b$

ν	ϕ	$m(\nu,\phi,5;b)$
[6 4]	$[5\,3\,1^2]$	$80b$
[6 4]	$[5\,2^2\,1]$	$50b$
[6 4]	$[4^2\,1^2]$	$45b$
[6 4]	$[4\,3\,2\,1]$	$80b$
[6 4]	$[4\,2^3]$	$10b$
[6 4]	$[3^3\,1]$	$10b$
[6 4]	$[3^2\,2^2]$	$15b$
$[5^2]$	$[7\,1^3]$	$30b$
$[5^2]$	$[6\,2\,1^2]$	$60b$
$[5^2]$	$[5\,3\,1^2]$	$30b$
$[5^2]$	$[5\,2^2\,1]$	$20b$
$[5^2]$	$[4^2\,1^2]$	$5b$
$[5^2]$	$[4\,3\,2\,1]$	$50b$
$[5^2]$	$[4\,2^3]$	$10b$
$[5^2]$	$[3^3\,1]$	$10b$
$[8\,1^2]$	$[8\,1^2]$	$150b$
$[8\,1^2]$	$[7\,2\,1]$	$120b$
$[8\,1^2]$	$[6\,3\,1]$	$180b$
$[8\,1^2]$	$[6\,2^2]$	$15b$
$[8\,1^2]$	$[5\,4\,1]$	$150b$
$[8\,1^2]$	$[5\,3\,2]$	$80b$
$[8\,1^2]$	$[4^2\,2]$	$20b$
$[8\,1^2]$	$[4\,3^2]$	$55b$
[7 2 1]	$[8\,1^2]$	$120b$
[7 2 1]	$[7\,2\,1]$	$210b$
[7 2 1]	$[6\,3\,1]$	$160b$
[7 2 1]	$[6\,2^2]$	$40b$
[7 2 1]	$[5\,4\,1]$	$140b$
[7 2 1]	$[5\,3\,2]$	$130b$
[7 2 1]	$[4^2\,2]$	$30b$
[7 2 1]	$[4\,3^2]$	$50b$
[6 3 1]	$[8\,1^2]$	$180b$
[6 3 1]	$[7\,2\,1]$	$160b$
[6 3 1]	$[6\,3\,1]$	$120b$
[6 3 1]	$[6\,2^2]$	$40b$
[6 3 1]	$[5\,4\,1]$	$170b$

ν	ϕ	$m(\nu, \phi, 5; b)$
[6 3 1]	[5 3 2]	$90b$
[6 3 1]	[4^2 2]	$30b$
[6 3 1]	[4 3^2]	$30b$
[6 2^2]	[8 1^2]	$15b$
[6 2^2]	[7 2 1]	$40b$
[6 2^2]	[6 3 1]	$40b$
[6 2^2]	[6 2^2]	$15b$
[6 2^2]	[5 4 1]	$40b$
[6 2^2]	[5 3 2]	$40b$
[6 2^2]	[4^2 2]	$15b$
[6 2^2]	[4 3^2]	$15b$
[5 4 1]	[8 1^2]	$150b$
[5 4 1]	[7 2 1]	$140b$
[5 4 1]	[6 3 1]	$170b$
[5 4 1]	[6 2^2]	$40b$
[5 4 1]	[5 4 1]	$110b$
[5 4 1]	[5 3 2]	$40b$
[5 4 1]	[4^2 2]	$50b$
[5 4 1]	[4 3^2]	$40b$
[5 3 2]	[8 1^2]	$80b$
[5 3 2]	[7 2 1]	$130b$
[5 3 2]	[6 3 1]	$90b$
[5 3 2]	[6 2^2]	$40b$
[5 3 2]	[5 4 1]	$40b$
[5 3 2]	[5 3 2]	$40b$
[5 3 2]	[4^2 2]	$30b$
[5 3 2]	[4 3^2]	$30b$
[4^2 2]	[8 1^2]	$20b$
[4^2 2]	[7 2 1]	$30b$
[4^2 2]	[6 3 1]	$30b$
[4^2 2]	[6 2^2]	$15b$
[4^2 2]	[5 4 1]	$50b$
[4^2 2]	[5 3 2]	$30b$
[4^2 2]	[4^2 2]	$10b$
[4^2 2]	[4 3^2]	$5b$
[4 3^2]	[8 1^2]	$55b$

ν	ϕ	$m(\nu, \phi, 5; b)$
[4 3^2]	[7 2 1]	$50b$
[4 3^2]	[6 3 1]	$30b$
[4 3^2]	[6 2^2]	$15b$
[4 3^2]	[5 4 1]	$40b$
[4 3^2]	[5 3 2]	$30b$
[4 3^2]	[4^2 2]	$5b$
[4 3^2]	[4 3^2]	$5b$
[7 1^3]	[9 1]	$100b$
[7 1^3]	[8 2]	$20b$
[7 1^3]	[7 3]	$70b$
[7 1^3]	[6 4]	$30b$
[7 1^3]	[5^2]	$30b$
[6 2 1^2]	[9 1]	$180b$
[6 2 1^2]	[8 2]	$90b$
[6 2 1^2]	[7 3]	$110b$
[6 2 1^2]	[6 4]	$60b$
[6 2 1^2]	[5^2]	$60b$
[5 3 1^2]	[9 1]	$180b$
[5 3 1^2]	[8 2]	$70b$
[5 3 1^2]	[7 3]	$70b$
[5 3 1^2]	[6 4]	$80b$
[5 3 1^2]	[5^2]	$30b$
[5 2^2 1]	[9 1]	$90b$
[5 2^2 1]	[8 2]	$70b$
[5 2^2 1]	[7 3]	$70b$
[5 2^2 1]	[6 4]	$50b$
[5 2^2 1]	[5^2]	$20b$
[4^2 1^2]	[9 1]	$90b$
[4^2 1^2]	[8 2]	$30b$
[4^2 1^2]	[7 3]	$40b$
[4^2 1^2]	[6 4]	$45b$
[4^2 1^2]	[5^2]	$5b$
[4 3 2 1]	[9 1]	$210b$
[4 3 2 1]	[8 2]	$130b$
[4 3 2 1]	[7 3]	$90b$
[4 3 2 1]	[6 4]	$80b$

ν	ϕ	$m(\nu,\phi,5;b)$
[4 3 2 1]	[5^2]	$50b$
[4 2^3]	[9 1]	$10b$
[4 2^3]	[8 2]	$10b$
[4 2^3]	[7 3]	$10b$
[4 2^3]	[6 4]	$10b$
[4 2^3]	[5^2]	$10b$
[3^3 1]	[9 1]	$40b$
[3^3 1]	[8 2]	$20b$
[3^3 1]	[7 3]	$10b$
[3^3 1]	[6 4]	$10b$
[3^3 1]	[5^2]	$10b$
[3^2 2^2]	[9 1]	$30b$
[3^2 2^2]	[8 2]	$25b$
[3^2 2^2]	[7 3]	$20b$
[3^2 2^2]	[6 4]	$15b$
[6 1^4]	[10]	$25b$
[5 2 1^3]	[10]	$80b$
[4 3 1^3]	[10]	$70b$
[4 2^2 1^2]	[10]	$90b$
[3^2 2 1^2]	[10]	$80b$
[3 2^3 1]	[10]	$40b$
[2^5]	[10]	b

Type 1: 2 edges

ν	ϕ	$m(\nu,\phi,2;b)$
[4]	[4]	$1+b+3b^2$

Type 1: 3 edges

ν	ϕ	$m(\nu,\phi,3;b)$
[6]	[5 1]	$6+6b+18b^2$
[6]	[4 2]	$3+3b+9b^2$
[6]	[3^2]	$1+b+5b^2$
[5 1]	[6]	$6+6b+18b^2$
[4 2]	[6]	$3+3b+9b^2$
[3^2]	[6]	$1+b+5b^2$

Type 1: 4 edges

ν	ϕ	$m(\nu, \phi, 4; b)$
$[8]$	$[6\,1^2]$	$20 + 20b + 60b^2$
$[8]$	$[5\,2\,1]$	$24 + 24b + 72b^2$
$[8]$	$[4\,3\,1]$	$16 + 16b + 64b^2$
$[8]$	$[4\,2^2]$	$6 + 6b + 18b^2$
$[8]$	$[3^2\,2]$	$4 + 4b + 20b^2$
$[7\,1]$	$[7\,1]$	$40 + 40b + 120b^2$
$[7\,1]$	$[6\,2]$	$16 + 16b + 48b^2$
$[7\,1]$	$[5\,3]$	$16 + 16b + 64b^2$
$[7\,1]$	$[4^2]$	$8 + 8b + 24b^2$
$[6\,2]$	$[7\,1]$	$16 + 16b + 48b^2$
$[6\,2]$	$[6\,2]$	$12 + 12b + 36b^2$
$[6\,2]$	$[5\,3]$	$8 + 8b + 32b^2$
$[6\,2]$	$[4^2]$	$4 + 4b + 12b^2$
$[5\,3]$	$[7\,1]$	$16 + 16b + 64b^2$
$[5\,3]$	$[6\,2]$	$8 + 8b + 32b^2$
$[5\,3]$	$[5\,3]$	$8 + 8b + 24b^2$
$[5\,3]$	$[4^2]$	$8b^2$
$[4^2]$	$[7\,1]$	$8 + 8b + 24b^2$
$[4^2]$	$[6\,2]$	$4 + 4b + 12b^2$
$[4^2]$	$[5\,3]$	$8b^2$
$[4^2]$	$[4^2]$	$3 + 3b + 9b^2$
$[6\,1^2]$	$[8]$	$20 + 20b + 60b^2$
$[5\,2\,1]$	$[8]$	$24 + 24b + 72b^2$
$[4\,3\,1]$	$[8]$	$16 + 16b + 64b^2$
$[4\,2^2]$	$[8]$	$6 + 6b + 18b^2$
$[3^2\,2]$	$[8]$	$4 + 4b + 20b^2$

Type 1: 5 edges

ν	ϕ	$m(\nu,\phi,5;b)$
[10]	[7 1^3]	$50+50b+150b^2$
[10]	[6 2 1^2]	$100+100b+300b^2$
[10]	[5 3 1^2]	$70+70b+250b^2$
[10]	[5 2^2 1]	$60+60b+180b^2$
[10]	[4^2 1^2]	$30+30b+120b^2$
[10]	[4 3 2 1]	$80+80b+320b^2$
[10]	[4 2^3]	$10+10b+30b^2$
[10]	[3^3 1]	$10+10b+50b^2$
[10]	[3^2 2^2]	$10+10b+50b^2$
[9 1]	[8 1^2]	$150+150b+450b^2$
[9 1]	[7 2 1]	$150+150b+450b^2$
[9 1]	[6 3 1]	$130+130b+470b^2$
[9 1]	[6 2^2]	$30+30b+90b^2$
[9 1]	[5 4 1]	$120+120b+420b^2$
[9 1]	[5 3 2]	$60+60b+240b^2$
[9 1]	[4^2 2]	$30+30b+90b^2$
[9 1]	[4 3^2]	$30+30b+130b^2$
[8 2]	[8 1^2]	$50+50b+150b^2$
[8 2]	[7 2 1]	$90+90b+270b^2$
[8 2]	[6 3 1]	$60+60b+210b^2$
[8 2]	[6 2^2]	$25+25b+75b^2$
[8 2]	[5 4 1]	$50+50b+180b^2$
[8 2]	[5 3 2]	$40+40b+160b^2$
[8 2]	[4^2 2]	$20+20b+60b^2$
[8 2]	[4 3^2]	$15+15b+65b^2$
[7 3]	[8 1^2]	$70+70b+250b^2$
[7 3]	[7 2 1]	$70+70b+260b^2$
[7 3]	[6 3 1]	$50+50b+180b^2$
[7 3]	[6 2^2]	$20+20b+70b^2$
[7 3]	[5 4 1]	$40+40b+180b^2$
[7 3]	[5 3 2]	$30+30b+120b^2$
[7 3]	[4^2 2]	$10+10b+40b^2$
[7 3]	[4 3^2]	$10+10b+40b^2$
[6 4]	[8 1^2]	$55+55b+165b^2$
[6 4]	[7 2 1]	$60+60b+180b^2$
[6 4]	[6 3 1]	$30+30b+160b^2$

ν	ϕ	$m(\nu,\phi,5;b)$
[6 4]	[6 2^2]	$15+15b+55b^2$
[6 4]	[5 4 1]	$50+50b+180b^2$
[6 4]	[5 3 2]	$30+30b+90b^2$
[6 4]	[4^2 2]	$5+5b+45b^2$
[6 4]	[4 3^2]	$5+5b+35b^2$
[5^2]	[8 1^2]	$25+25b+105b^2$
[5^2]	[7 2 1]	$30+30b+120b^2$
[5^2]	[6 3 1]	$30+30b+100b^2$
[5^2]	[6 2^2]	$10+10b+30b^2$
[5^2]	[5 4 1]	$10+10b+50b^2$
[5^2]	[5 3 2]	$30b^2$
[5^2]	[4^2 2]	$10+10b+30b^2$
[5^2]	[4 3^2]	$5+5b+25b^2$
[8 1^2]	[9 1]	$150+150b+450b^2$
[8 1^2]	[8 2]	$50+50b+150b^2$
[8 1^2]	[7 3]	$70+70b+250b^2$
[8 1^2]	[6 4]	$55+55b+165b^2$
[8 1^2]	[5^2]	$25+25b+105b^2$
[7 2 1]	[9 1]	$150+150b+450b^2$
[7 2 1]	[8 2]	$90+90b+270b^2$
[7 2 1]	[7 3]	$70+70b+260b^2$
[7 2 1]	[6 4]	$60+60b+180b^2$
[7 2 1]	[5^2]	$30+30b+120b^2$
[6 3 1]	[9 1]	$130+130b+470b^2$
[6 3 1]	[8 2]	$60+60b+210b^2$
[6 3 1]	[7 3]	$50+50b+180b^2$
[6 3 1]	[6 4]	$30+30b+160b^2$
[6 3 1]	[5^2]	$30+30b+100b^2$
[6 2^2]	[9 1]	$30+30b+90b^2$
[6 2^2]	[8 2]	$25+25b+75b^2$
[6 2^2]	[7 3]	$20+20b+70b^2$
[6 2^2]	[6 4]	$15+15b+55b^2$
[6 2^2]	[5^2]	$10+10b+30b^2$
[5 4 1]	[9 1]	$120+120b+420b^2$
[5 4 1]	[8 2]	$50+50b+180b^2$
[5 4 1]	[7 3]	$40+40b+180b^2$

ν	ϕ	$m(\nu, \phi, 5; b)$
$[5\,4\,1]$	$[6\,4]$	$50 + 50b + 180b^2$
$[5\,4\,1]$	$[5^2]$	$10 + 10b + 50b^2$
$[5\,3\,2]$	$[9\,1]$	$60 + 60b + 240b^2$
$[5\,3\,2]$	$[8\,2]$	$40 + 40b + 160b^2$
$[5\,3\,2]$	$[7\,3]$	$30 + 30b + 120b^2$
$[5\,3\,2]$	$[6\,4]$	$30 + 30b + 90b^2$
$[5\,3\,2]$	$[5^2]$	$30b^2$
$[4^2\,2]$	$[9\,1]$	$30 + 30b + 90b^2$
$[4^2\,2]$	$[8\,2]$	$20 + 20b + 60b^2$
$[4^2\,2]$	$[7\,3]$	$10 + 10b + 40b^2$
$[4^2\,2]$	$[6\,4]$	$5 + 5b + 45b^2$
$[4^2\,2]$	$[5^2]$	$10 + 10b + 30b^2$
$[4\,3^2]$	$[9\,1]$	$30 + 30b + 130b^2$
$[4\,3^2]$	$[8\,2]$	$15 + 15b + 65b^2$
$[4\,3^2]$	$[7\,3]$	$10 + 10b + 40b^2$
$[4\,3^2]$	$[6\,4]$	$5 + 5b + 35b^2$
$[4\,3^2]$	$[5^2]$	$5 + 5b + 25b^2$
$[7\,1^3]$	$[10]$	$50 + 50b + 150b^2$
$[6\,2\,1^2]$	$[10]$	$100 + 100b + 300b^2$
$[5\,3\,1^2]$	$[10]$	$70 + 70b + 250b^2$
$[5\,2^2\,1]$	$[10]$	$60 + 60b + 180b^2$
$[4^2\,1^2]$	$[10]$	$30 + 30b + 120b^2$
$[4\,3\,2\,1]$	$[10]$	$80 + 80b + 320b^2$
$[4\,2^3]$	$[10]$	$10 + 10b + 30b^2$
$[3^3\,1]$	$[10]$	$10 + 10b + 50b^2$
$[3^2\,2^2]$	$[10]$	$10 + 10b + 50b^2$

Type 3/2: 3 edges

ν	ϕ	$m(\nu, \phi, 3; b)$
$[6]$	$[6]$	$13b + 13b^2 + 15b^3$

Type 3/2: 4 edges

ν	ϕ	$m(\nu, \phi, 4; b)$
[8]	[7 1]	$104b + 104b^2 + 120b^3$
[8]	[6 2]	$52b + 52b^2 + 60b^3$
[8]	[5 3]	$40b + 40b^2 + 56b^3$
[8]	[4^2]	$19b + 19b^2 + 24b^3$
[7 1]	[8]	$104b + 104b^2 + 120b^3$
[6 2]	[8]	$52b + 52b^2 + 60b^3$
[5 3]	[8]	$40b + 40b^2 + 56b^3$
[4^2]	[8]	$19b + 19b^2 + 24b^3$

Type 3/2: 5 edges

ν	ϕ	$m(\nu,\phi,5;b)$
[10]	[8 1²]	$455b + 455b^2 + 525b^3$
[10]	[7 2 1]	$520b + 520b^2 + 600b^3$
[10]	[6 3 1]	$380b + 380b^2 + 500b^3$
[10]	[6 2²]	$130b + 130b^2 + 150b^3$
[10]	[5 4 1]	$340b + 340b^2 + 450b^3$
[10]	[5 3 2]	$200b + 200b^2 + 280b^3$
[10]	[4² 2]	$95b + 95b^2 + 120b^3$
[10]	[4 3²]	$80b + 80b^2 + 125b^3$
[9 1]	[9 1]	$910b + 910b^2 + 1050b^3$
[9 1]	[8 2]	$390b + 390b^2 + 450b^3$
[9 1]	[7 3]	$380b + 380b^2 + 500b^3$
[9 1]	[6 4]	$320b + 320b^2 + 390b^3$
[9 1]	[5²]	$150b + 150b^2 + 210b^3$
[8 2]	[9 1]	$390b + 390b^2 + 450b^3$
[8 2]	[8 2]	$260b + 260b^2 + 300b^3$
[8 2]	[7 3]	$190b + 190b^2 + 250b^3$
[8 2]	[6 4]	$160b + 160b^2 + 195b^3$
[8 2]	[5²]	$75b + 75b^2 + 105b^3$
[7 3]	[9 1]	$380b + 380b^2 + 500b^3$
[7 3]	[8 2]	$190b + 190b^2 + 250b^3$
[7 3]	[7 3]	$140b + 140b^2 + 190b^3$
[7 3]	[6 4]	$100b + 100b^2 + 150b^3$
[7 3]	[5²]	$60b + 60b^2 + 90b^3$
[6 4]	[9 1]	$320b + 320b^2 + 390b^3$
[6 4]	[8 2]	$160b + 160b^2 + 195b^3$
[6 4]	[7 3]	$100b + 100b^2 + 150b^3$
[6 4]	[6 4]	$100b + 100b^2 + 155b^3$
[6 4]	[5²]	$60b + 60b^2 + 75b^3$
[5²]	[9 1]	$150b + 150b^2 + 210b^3$
[5²]	[8 2]	$75b + 75b^2 + 105b^3$
[5²]	[7 3]	$60b + 60b^2 + 90b^3$
[5²]	[6 4]	$60b + 60b^2 + 75b^3$
[5²]	[5²]	$10b + 10b^2 + 20b^3$
[8 1²]	[10]	$455b + 455b^2 + 525b^3$
[7 2 1]	[10]	$520b + 520b^2 + 600b^3$
[6 3 1]	[10]	$380b + 380b^2 + 500b^3$
[6 2²]	[10]	$130b + 130b^2 + 150b^3$
[5 4 1]	[10]	$340b + 340b^2 + 450b^3$
[5 3 2]	[10]	$200b + 200b^2 + 280b^3$
[4² 2]	[10]	$95b + 95b^2 + 120b^3$
[4 3²]	[10]	$80b + 80b^2 + 125b^3$

Type 2: 4 edges

ν	ϕ	$m(\nu,\phi,4;b)$
[8]	[8]	$21+42b+181b^2+160b^3+105b^4$

Type 2: 5 edges

ν	ϕ	$m(\nu,\phi,5;b)$
[10]	[9 1]	$210+420b+1810b^2+1600b^3+1050b^4$
[10]	[8 2]	$105+210b+905b^2+800b^3+525b^4$
[10]	[7 3]	$70+140b+690b^2+620b^3+450b^4$
[10]	[6 4]	$65+130b+600b^2+535b^3+375b^4$
[10]	[5^2]	$33+66b+289b^2+256b^3+189b^4$
[9 1]	[10]	$210+420b+1810b^2+1600b^3+1050b^4$
[8 2]	[10]	$105+210b+905b^2+800b^3+525b^4$
[7 3]	[10]	$70+140b+690b^2+620b^3+450b^4$
[6 4]	[10]	$65+130b+600b^2+535b^3+375b^4$
[5^2]	[10]	$33+66b+289b^2+256b^3+189b^4$

Type 5/2: 5 edges

ν	ϕ	$m(\nu,\phi,5;b)$
[10]	[10]	$753b+1506b^2+2889b^3+2136b^4+945b^5$

12.2 Genus distributions

Listed below are the 142 graphs that arise as associated graphs of the maps that are drawn in the Atlas. They are indexed by number of edges and number of vertices. The genus distribution appears at the top of each drawing. The bottom left-hand corner gives the vertex partition and the bottom right-hand corner gives the index number of the graph. The actual embeddings for each graph are given in Chapter 8.

1 edge

1 vertex

1

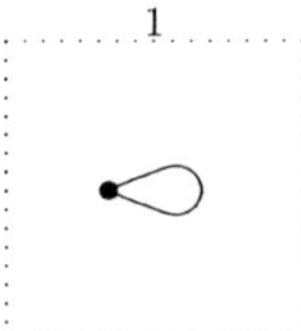

$[2]$ $\mathfrak{g}_1$

1 edge

2 vertices

1

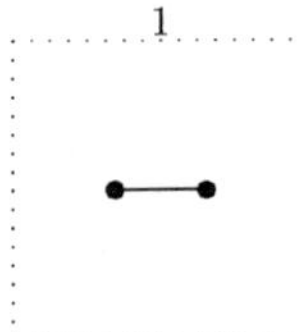

$[1^2]$ $\mathfrak{g}_2$

2 edges

1 vertex

$1+u$

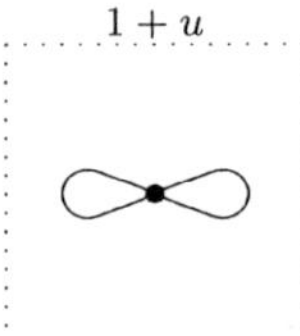

$[4]$ $\mathfrak{g}_3$

2 edges

2 vertices

1

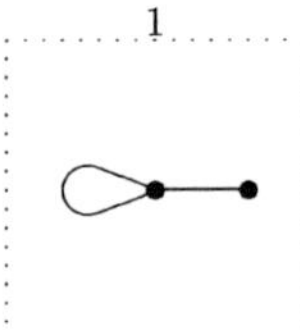

$[31]$ $\mathfrak{g}_4$

1

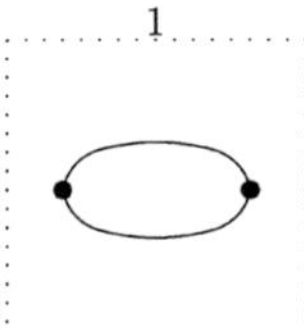

$[2^2]$ $\mathfrak{g}_5$

2 edges

3 vertices

1

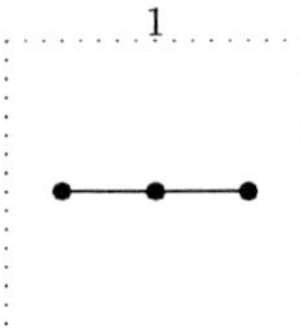

$[21^2]$ $\mathfrak{g}_6$

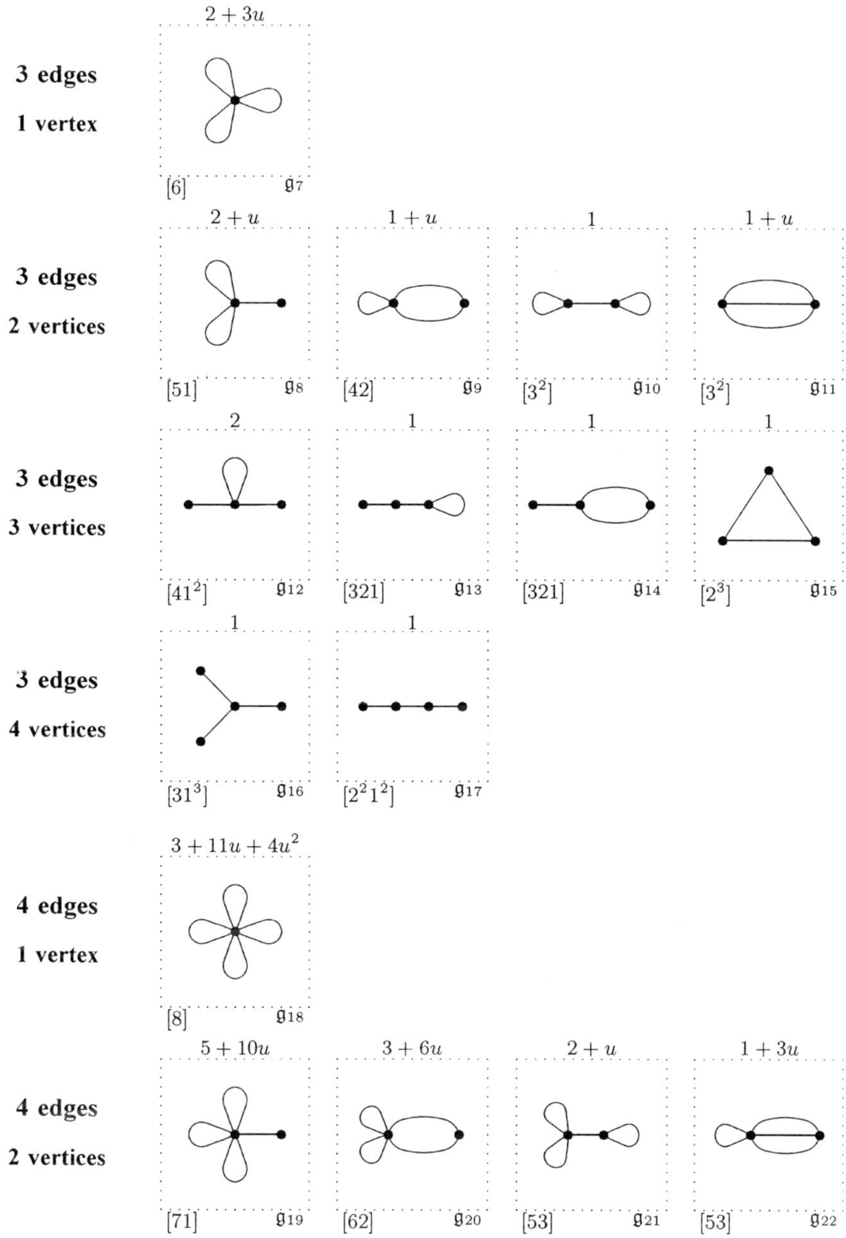
3 edges
1 vertex
2 + 3u
[6] g7
3 edges
2 vertices
2 + u
[51] g8
1 + u
[42] g9
1
[3^2] g10
1 + u
[3^2] g11
3 edges
3 vertices
2
[41^2] g12
1
[321] g13
1
[321] g14
1
[2^3] g15
3 edges
4 vertices
1
[31^3] g16
1
[2^2 1^2] g17
4 edges
1 vertex
3 + 11u + 4u^2
[8] g18
4 edges
2 vertices
5 + 10u
[71] g19
3 + 6u
[62] g20
2 + u
[53] g21
1 + 3u
[53] g22

$2+2u$

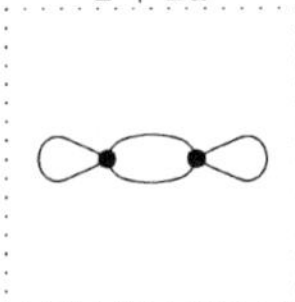

$[4^2]$ $\mathfrak{g}_{23}$

$1+2u$

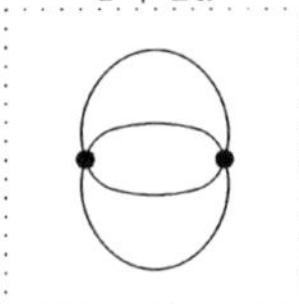

$[4^2]$ $\mathfrak{g}_{24}$

4 edges

3 vertices

$6+3u$

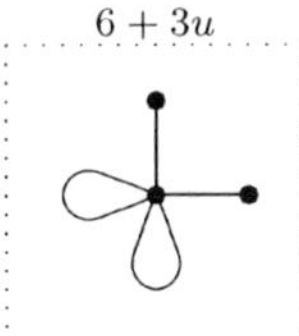

$[61^2]$ $\mathfrak{g}_{25}$

$2+u$

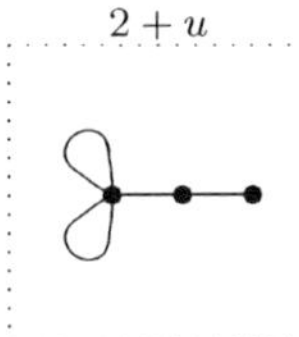

$[521]$ $\mathfrak{g}_{26}$

$4+2u$

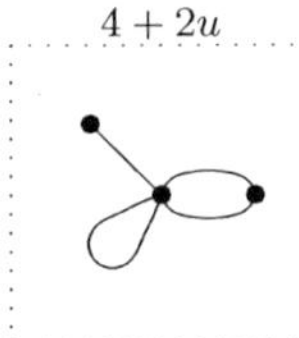

$[521]$ $\mathfrak{g}_{27}$

3

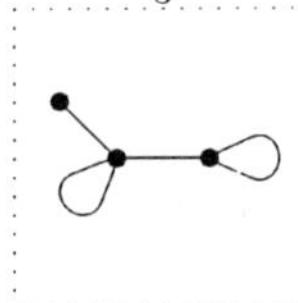

$[431]$ $\mathfrak{g}_{28}$

$2+u$

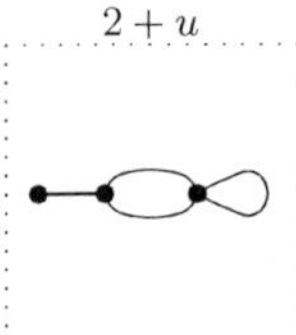

$[431]$ $\mathfrak{g}_{29}$

$1+u$

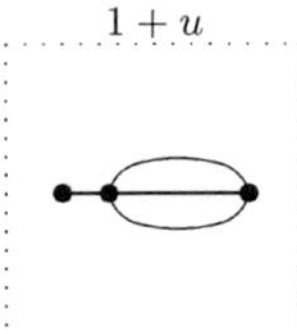

$[431]$ $\mathfrak{g}_{30}$

$1+u$

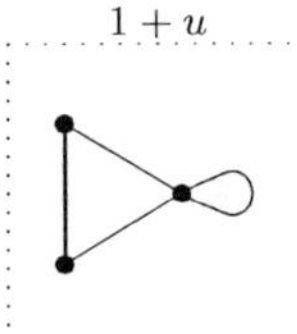

$[42^2]$ $\mathfrak{g}_{31}$

$1+u$

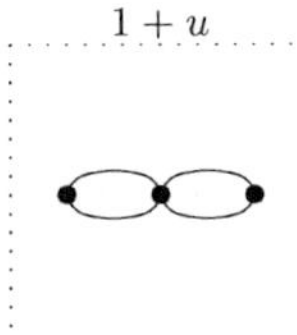

$[42^2]$ $\mathfrak{g}_{32}$

1

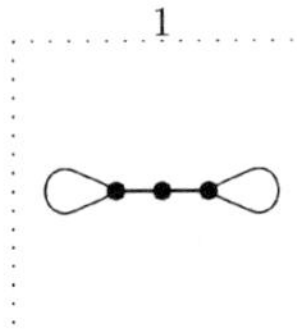

$[3^22]$ $\mathfrak{g}_{33}$

1

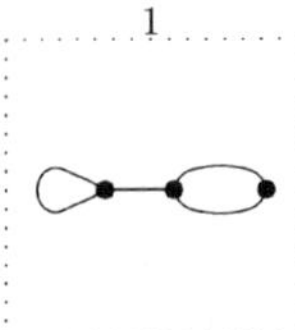

$[3^22]$ $\mathfrak{g}_{34}$

$1+u$

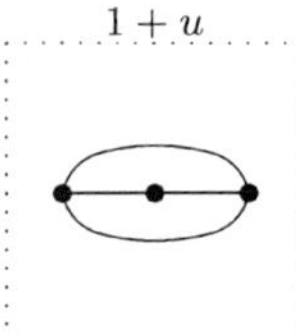

$[3^22]$ $\mathfrak{g}_{35}$

4 edges

4 vertices

2

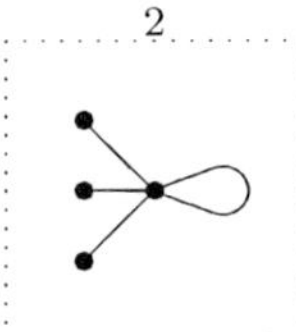

$[51^3]$ $\mathfrak{g}_{36}$

3

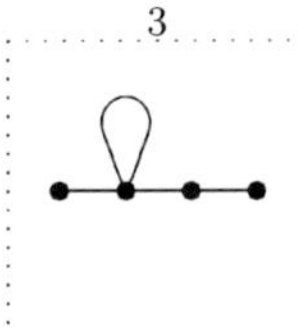

$[421^2]$ $\mathfrak{g}_{37}$

2

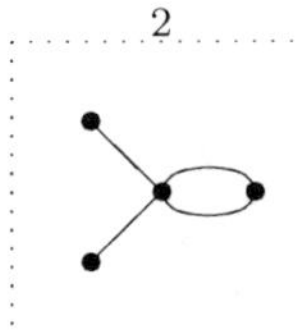

$[421^2]$ $\mathfrak{g}_{38}$

1

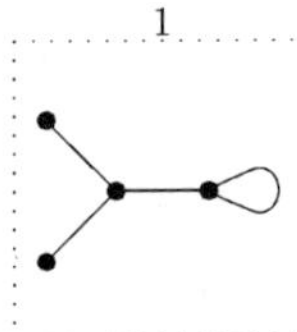

$[3^21^2]$ $\mathfrak{g}_{39}$

2

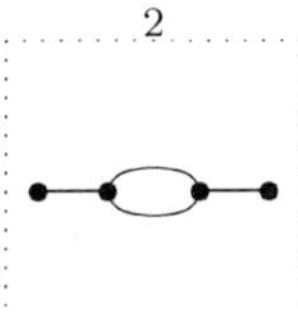

$[3^21^2]$ $\mathfrak{g}_{40}$

1

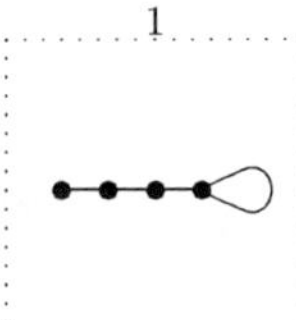

$[32^21]$ $\mathfrak{g}_{41}$

1

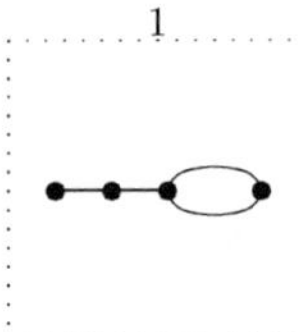

$[32^21]$ $\mathfrak{g}_{42}$

1

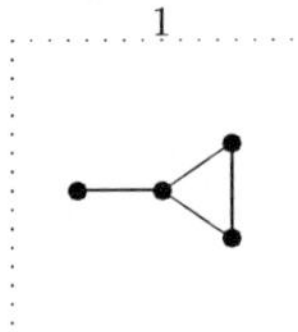

$[32^21]$ $\mathfrak{g}_{43}$

1

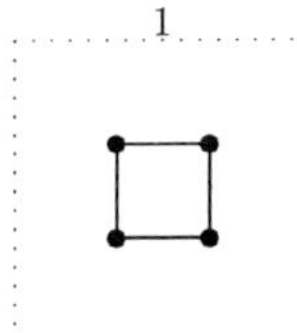

$[2^4]$ $\mathfrak{g}_{44}$

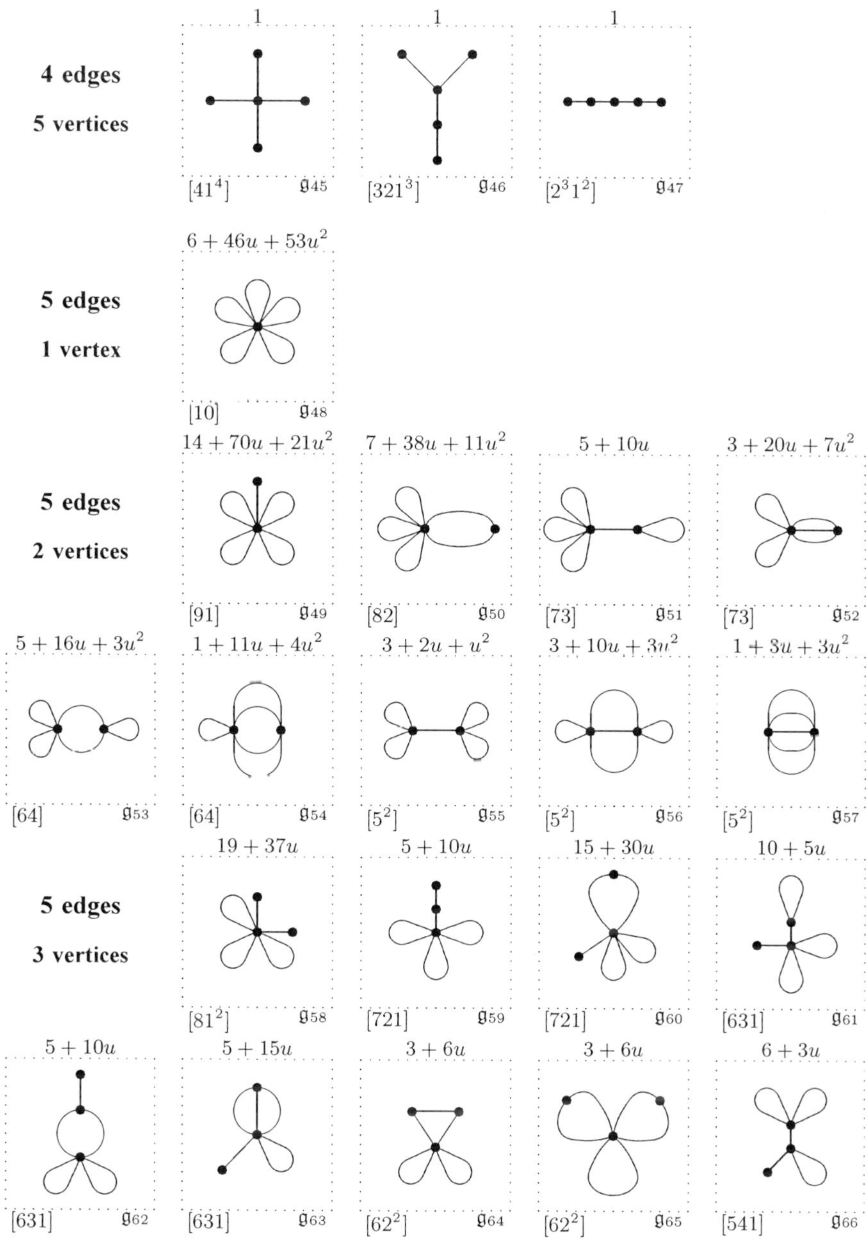
4 edges
5 vertices
1
[41⁴] 𝔤45
1
[321³] 𝔤46
1
[2³1²] 𝔤47
5 edges
1 vertex
6 + 46u + 53u²
[10] 𝔤48
5 edges
2 vertices
14 + 70u + 21u²
[91] 𝔤49
7 + 38u + 11u²
[82] 𝔤50
5 + 10u
[73] 𝔤51
3 + 20u + 7u²
[73] 𝔤52
5 + 16u + 3u²
[64] 𝔤53
1 + 11u + 4u²
[64] 𝔤54
3 + 2u + u²
[5²] 𝔤55
3 + 10u + 3u²
[5²] 𝔤56
1 + 3u + 3u²
[5²] 𝔤57
5 edges
3 vertices
19 + 37u
[81²] 𝔤58
5 + 10u
[721] 𝔤59
15 + 30u
[721] 𝔤60
10 + 5u
[631] 𝔤61
5 + 10u
[631] 𝔤62
5 + 15u
[631] 𝔤63
3 + 6u
[62²] 𝔤64
3 + 6u
[62²] 𝔤65
6 + 3u
[541] 𝔤66

$8+10u$
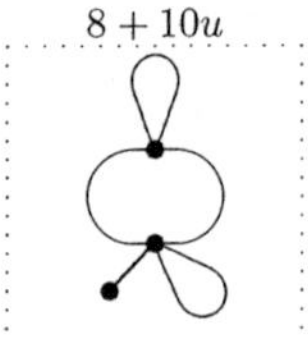
[541] $\mathfrak{g}_{67}$

$3+9u$
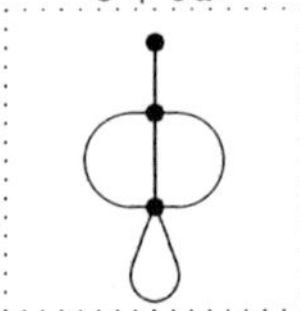
[541] $\mathfrak{g}_{68}$

$1+5u$
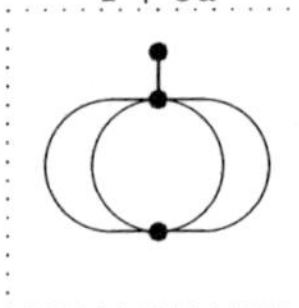
[541] $\mathfrak{g}_{69}$

$2+u$
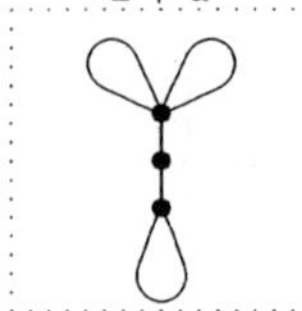
[532] $\mathfrak{g}_{70}$

$2+u$
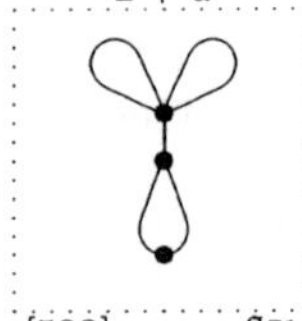
[532] $\mathfrak{g}_{71}$

$4+2u$
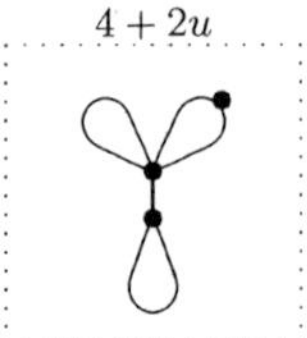
[532] $\mathfrak{g}_{72}$

$3+9u$
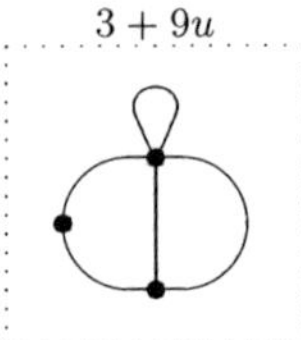
[532] $\mathfrak{g}_{73}$

$1+3u$
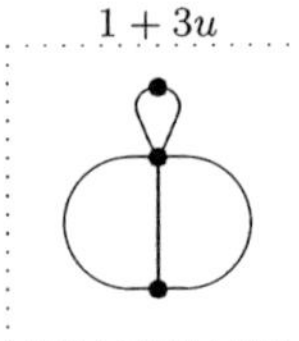
[532] $\mathfrak{g}_{74}$

$3+3u$
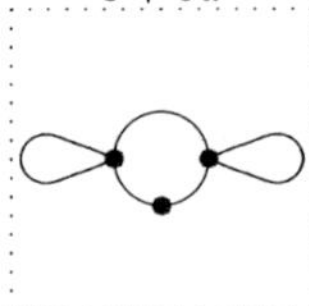
$[4^22]$ $\mathfrak{g}_{75}$

$2+3u$
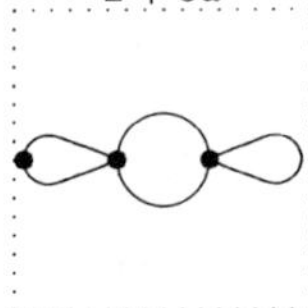
$[4^22]$ $\mathfrak{g}_{76}$

$1+4u$
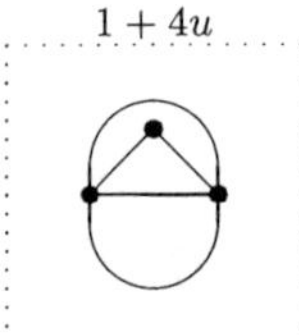
$[4^22]$ $\mathfrak{g}_{77}$

2
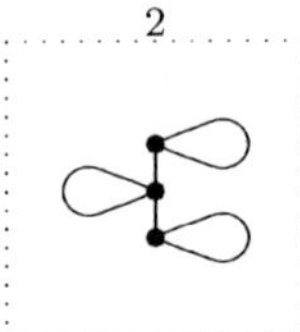
$[43^2]$ $\mathfrak{g}_{78}$

$2+u$
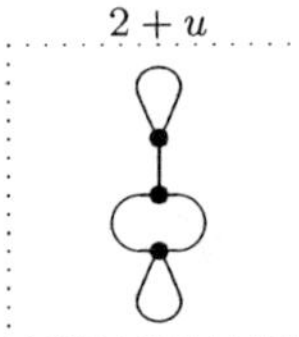
$[43^2]$ $\mathfrak{g}_{79}$

$1+u$
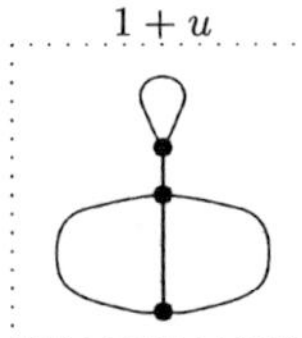
$[43^2]$ $\mathfrak{g}_{80}$

$1+3u$
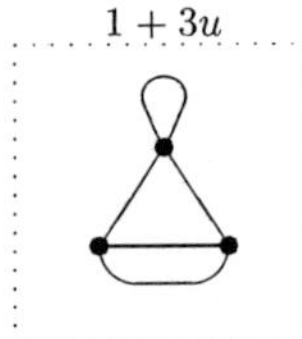
$[43^2]$ $\mathfrak{g}_{81}$

$1+3u$
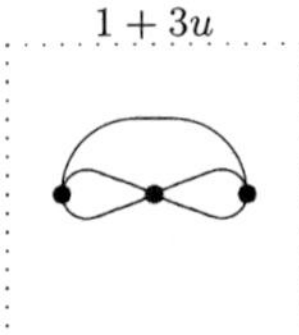
$[43^2]$ $\mathfrak{g}_{82}$

5 edges

4 vertices

$10+5u$
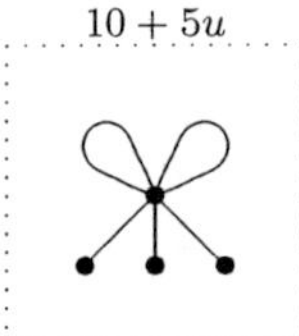
$[71^3]$ $\mathfrak{g}_{83}$

$10+5u$
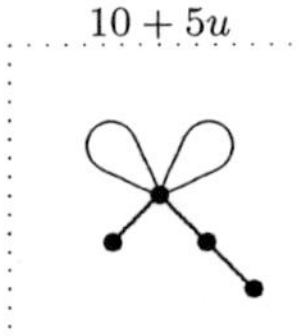
$[621^2]$ $\mathfrak{g}_{84}$

$10+6u$
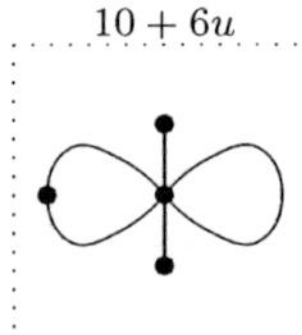
$[621^2]$ $\mathfrak{g}_{85}$

6
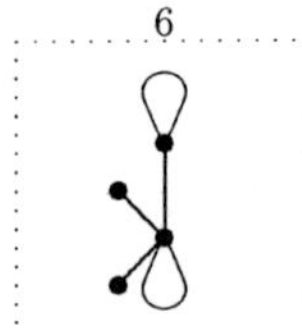
$[531^2]$ $\mathfrak{g}_{86}$

$2+u$
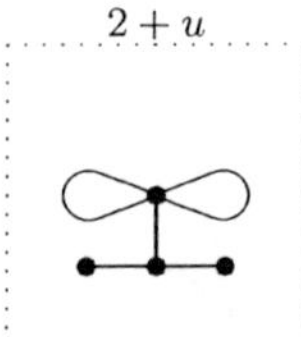
$[531^2]$ $\mathfrak{g}_{87}$

$8+4u$
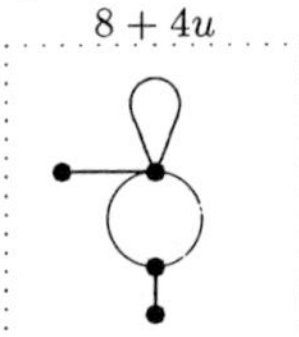
$[531^2]$ $\mathfrak{g}_{88}$

$2+2u$
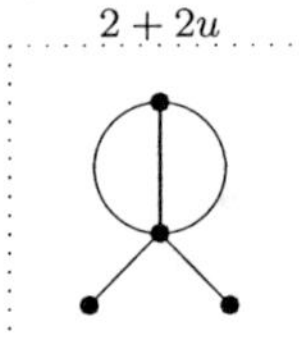
$[531^2]$ $\mathfrak{g}_{89}$

$2+u$
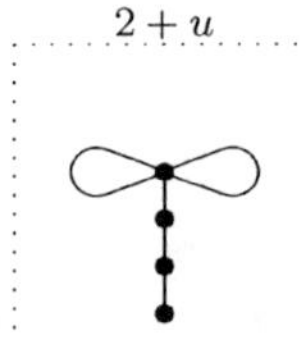
$[52^21]$ $\mathfrak{g}_{90}$

$4+2u$
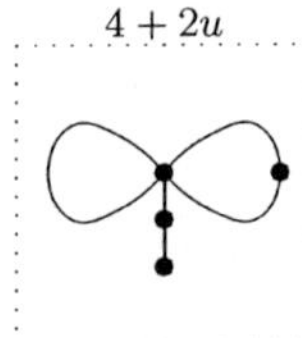
$[52^21]$ $\mathfrak{g}_{91}$

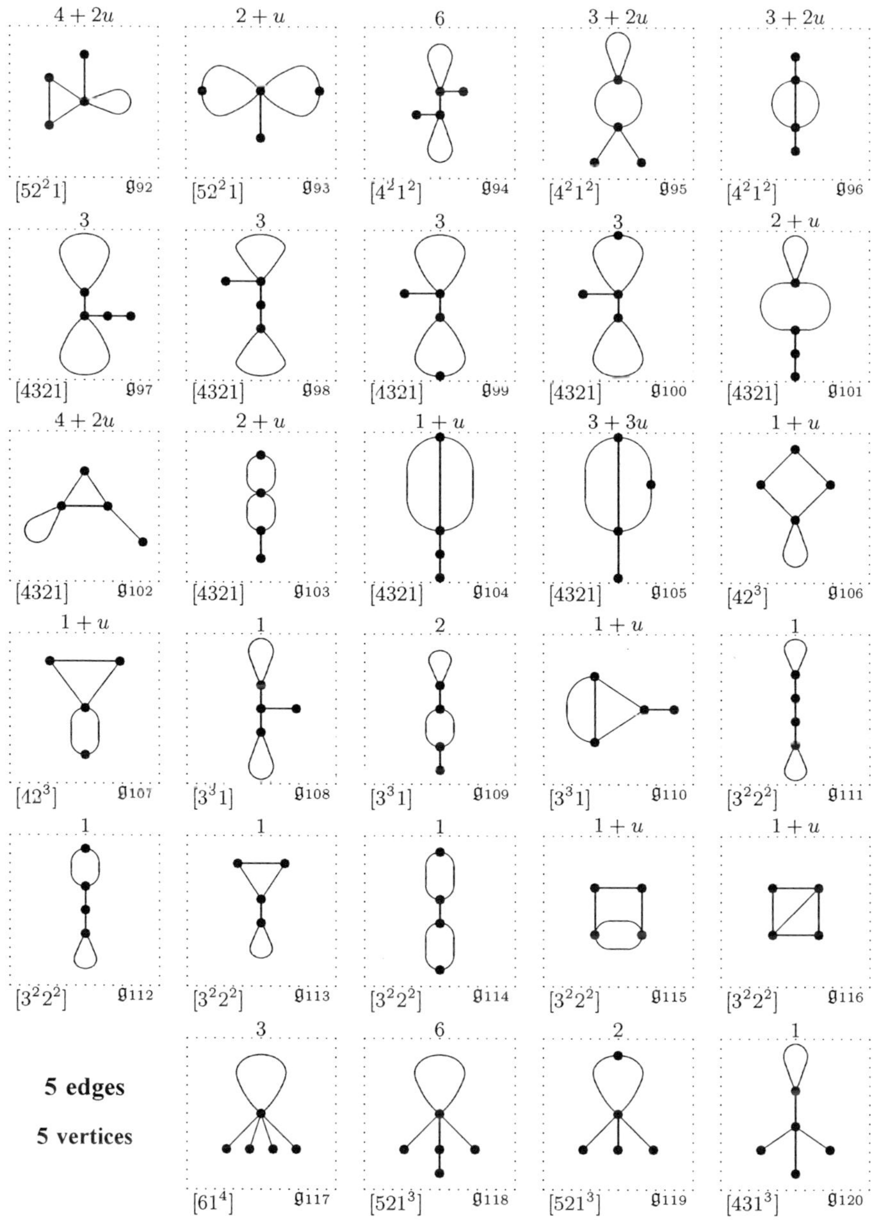
4 + 2u
[52²1] g92
2 + u
[52²1] g93
6
[4²1²] g94
3 + 2u
[4²1²] g95
3 + 2u
[4²1²] g96
3
[4321] g97
3
[4321] g98
3
[4321] g99
3
[4321] g100
2 + u
[4321] g101
4 + 2u
[4321] g102
2 + u
[4321] g103
1 + u
[4321] g104
3 + 3u
[4321] g105
1 + u
[42³] g106
1 + u
[42³] g107
1
[3³1] g108
2
[3³1] g109
1 + u
[3³1] g110
1
[3²2²] g111
1
[3²2²] g112
1
[3²2²] g113
1
[3²2²] g114
1 + u
[3²2²] g115
1 + u
[3²2²] g116
5 edges
5 vertices
3
[61⁴] g117
6
[521³] g118
2
[521³] g119
1
[431³] g120

3
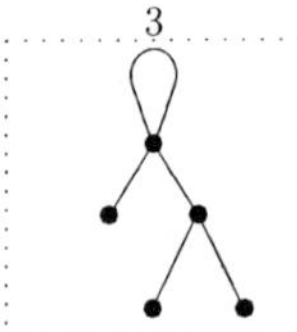
$[431^3]$ $\mathfrak{g}_{121}$

3
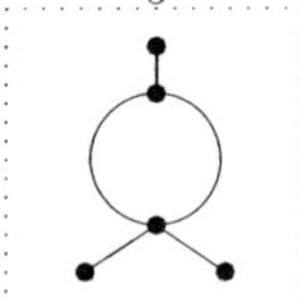
$[431^3]$ $\mathfrak{g}_{122}$

3
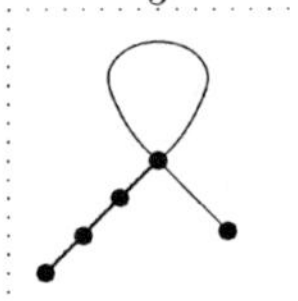
$[42^21^2]$ $\mathfrak{g}_{123}$

2
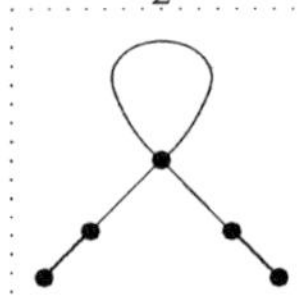
$[42^21^2]$ $\mathfrak{g}_{124}$

3
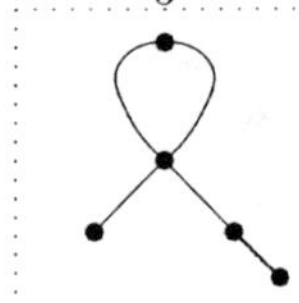
$[42^21^2]$ $\mathfrak{g}_{125}$

2
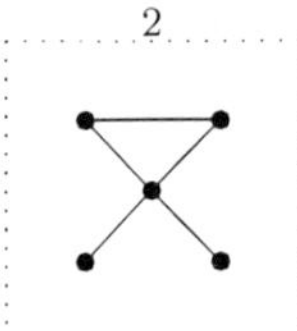
$[42^21^2]$ $\mathfrak{g}_{126}$

2
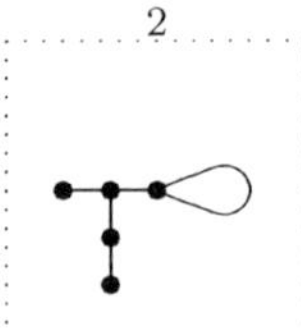
$[3^221^2]$ $\mathfrak{g}_{127}$

1
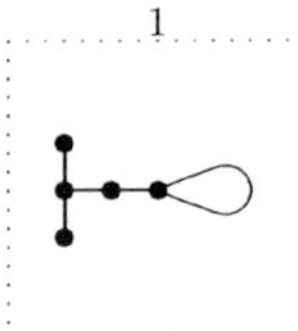
$[3^221^2]$ $\mathfrak{g}_{128}$

2
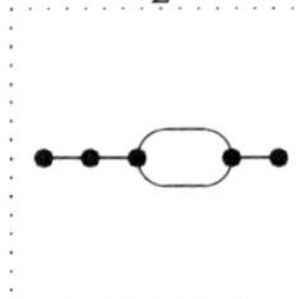
$[3^221^2]$ $\mathfrak{g}_{129}$

1
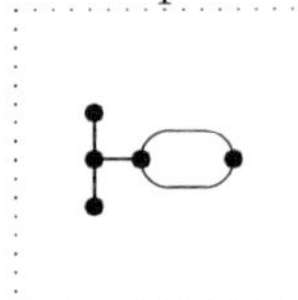
$[3^221^2]$ $\mathfrak{g}_{130}$

3
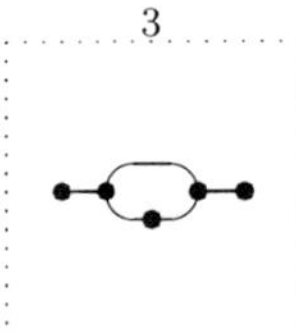
$[3^221^2]$ $\mathfrak{g}_{131}$

1
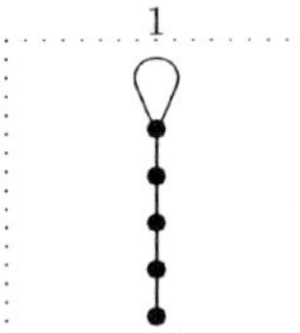
$[32^31]$ $\mathfrak{g}_{132}$

1
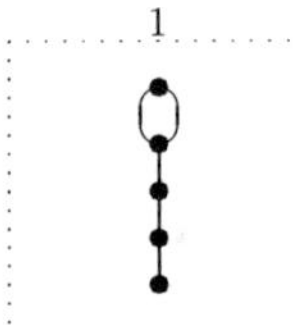
$[32^31]$ $\mathfrak{g}_{133}$

1
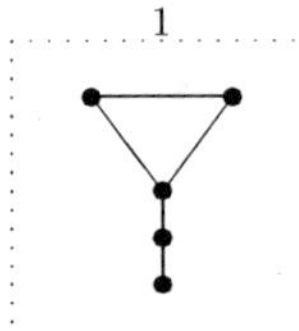
$[32^31]$ $\mathfrak{g}_{134}$

1
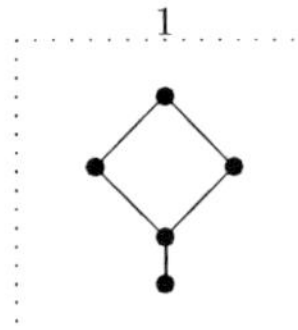
$[32^31]$ $\mathfrak{g}_{135}$

1
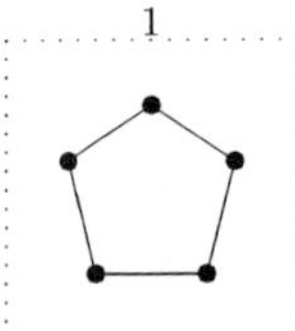
$[2^5]$ $\mathfrak{g}_{136}$

5 edges

6 vertices

1
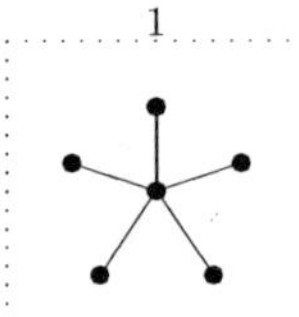
$[51^5]$ $\mathfrak{g}_{137}$

1
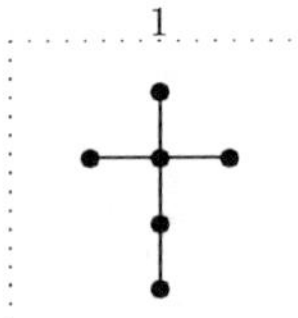
$[421^4]$ $\mathfrak{g}_{138}$

1
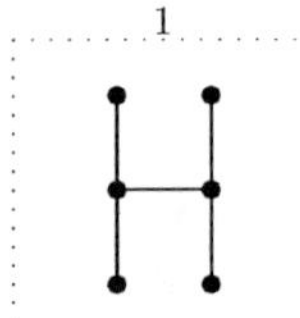
$[3^21^4]$ $\mathfrak{g}_{139}$

1
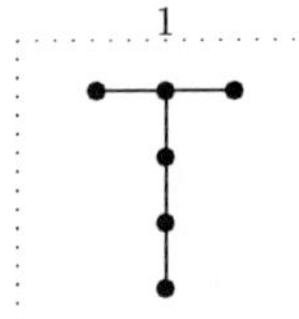
$[32^21^3]$ $\mathfrak{g}_{140}$

1
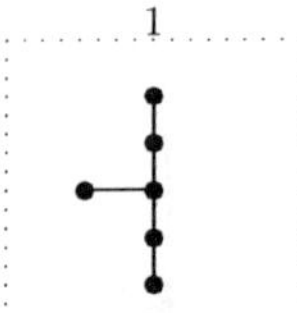
$[32^21^3]$ $\mathfrak{g}_{141}$

1
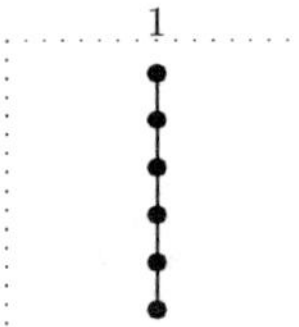
$[2^41^2]$ $\mathfrak{g}_{142}$

Bibliography

[1] N.M. Adrianov and G.B. Shabat, Unicellular cartography and Galois orbits of plane trees, in: *Geometric Galois Action*, Vol. 2, L. Schneps and P. Lochak eds., "The Inverse Galois Problem, Moduli Spaces and Mapping Class Groups," London Math. Soc. Lecture Notes Series, **243**, Cambridge University Press, Cambridge, 1997, 12–24.

[2] V.I. Arnold, Topological classification of trigonometric polynomials and combinatorial properties of graphs with an equal number of vertices and edges, *Functional Analysis and Its Applications*, **30** (1996), 1–14.

[3] G.V. Belyi, On Galois extensions of a maximal cyclotomic field, *USSR Izvestija*, **14** (2) (1980), 247–256.

[4] D. Bessis, C. Itzykson and J.-B. Zuber, Quantum field theory tchniques in graphical enumeration, *Adv. Appl. Math.*, **1** (1980), 109–157.

[5] D. Brown, Differential Equations and Depth First Search in the Enumeration of Maps in Surfaces, Ph.D. Thesis, University of Waterloo, 1999.

[6] L. Caporaso and J. Harris, Counting plane curves of any genus, *Invent. Math.*, **131** (1998), 345–392.

[7] *The Moduli Space of Curves*, Progress in Mathematics, **129**, R. Dijkgraf, C. Faber and G. van der Geer eds., Birkhaüser, Boston, 1995.

[8] L. Ernstrom and G. Kennedy, Recursive formulas for characteristic numbers of rational plane curves, *J. Algebraic Geom.*, **7** (1998), 141–181.

[9] I.P. Goulden and D.M. Jackson, Connection coefficients, matchings, maps and combinatorial conjectures for Jack symmetric functions, *Trans. Amer. Math. Soc.*, **348** (1996), 873–892.

[10] I.P. Goulden, J.L. Harer and D.M. Jackson, A parametrization of the virtual Euler characteristic for the moduli spaces of real and complex algebraic curves, *Trans. Amer. Math. Soc.*, (to appear).

[11] I.P. Goulden and D.M. Jackson, *Combinatorial Enumeration*, Wiley Interscience, New York, 1983.

[12] P.J. Hanlon, R.P. Stanley and J.S. Stembridge, Some combinatorial apects of the spectra of normally distributed random matrices, *Contemp. Math.*, vol. 138, Amer. Math. Soc., Providence, RI, 1992, 151–174.

[13] G. 't Hooft, A planar diagram theory for strong interactions, *Nuclear Phys.*, **B72** (1974), 461–473.

[14] A. Hurwitz, Ueber Riemann'sche Flächen mit gegebenen Verzweigungspunkten, *Matematische Annalen*, **39** (1891), 1–60.

[15] J.L. Gross and T.W. Tucker, *Topological Graph Theory*, Wiley Interscience, New York, 1987.

[16] J. Harer and D. Zagier, The Euler characteristic for the moduli space of curves, *Invent. Math.*, **85** (1986), 457–485.

[17] C. Itzykson and J.-M. Drouffe, *Statistical Field Theory*, Vol. 2, Cambridge Univ. Press, 1990.

[18] D.M. Jackson, M.J. Perry and T.I. Visentin, Factorisations of partition functions of random Hermitian matrix models, *Commun. Math. Phys.*, **179** (1996), 25–60.

[19] D.M. Jackson and T.I. Visentin, A combinatorial relationship between Eulerian maps and hypermaps in orientable surfaces, *J. Combinatorial Theory, Ser. A*, **87** (1999), 120–150.

[20] A. Khovanskii and S. Zdravkovska, Branched covers of S^2 and braid groups, *J. Knot Theory and Its Ramifications*, **5** (1996), 55–75.

[21] F. Knop and S. Sahi, A recursion and a combinatorial formula for Jack symmetric polynomials, *Invent. Math.*, **128** (1997), 9–22.

[22] F. Lesage, V. Pasquier and D. Serban, Dynamical correlation functions in the Calogero-Sutherland model, *Nuclear Phys.*, **B435 [FS]** (1995), 585–603.

[23] E. Looijenga, The complement of the bifurcation variety of a simple singularity, *Invent. Math.*, **23** (1973), 105–116.

[24] I.G. Macdonald, *Symmetric Functions and Hall Polynomials*, 2nd edition, Clarendon Press, Oxford, 1995.

[25] M.L. Mehta, *Random Matrices*, 2nd edition, Academic Press, New York, 1991.

[26] *Statistical Mechanics of Random Surfaces*, D. Nelson, T. Piran and S. Weinberg eds., World Scientific, Teaneck, 1989.

[27] *The Grothendieck Theory of Dessins d'Enfants*, L. Schepps ed., London Math. Soc. Lecture Note Series, **200**, Cambridge University Press, Cambridge, 1994.

[28] G.B. Shabat and V.A. Voevodsky, Drawing curves over number fields, in: *The Grothendieck Festschrift*, Vol. 3, P. Cartier et al. eds., Progress in Math. **88**, Birkhäuser, Boston, 1990, 199–227.

[29] G.B. Shabat and A. Zvonkin, Plane trees and algebraic numbers, *Contemp. Math.*, **178** (1994), 233–275.

[30] R.P. Stanley, Some combinatorial properties of Jack symmetric functions, *Advances in Math.*, 77 (1989), 76–115.

[31] J.R. Stembridge, A Maple package for symmetric functions - Version 2, Department of Mathematics, University of Michigan, Ann Arbor, July 1993.

[32] W.T. Tutte, *Graph Theory*, Encyclopaedia of Math. and Its Applications **21**, Addison-Wesley, London, 1984.

[33] W.T. Tutte, A census of planar triangulations, *Canad. J. Math.*, **14** (1962), 21–38.

[34] W.T. Tutte, A census of Hamiltonian polygons, *Canad. J. Math.*, **14** (1962), 402–417.

[35] W.T. Tutte, A census of slicings, *Canad. J. Math.*, **14** (1962), 708–722.

[36] W.T. Tutte, A census of planar maps, *Canad. J. Math.*, **15** (1963), 249–271.

[37] T.R.S. Walsh, Hypermaps versus bipartite maps, *J. Combinatorial Theory, Ser. B* **18** (1975), 155–163.

Notation

2θ partition with doubled parts, 48

$\langle\ ,\ \rangle_a$ inner product on Λ, 65

$\prec$ reverse lexicographic order, 65

$\vdash$ "is a partition of," 14

$|\theta|$ sum of the parts of θ, 14

β map invariant for b-Conjecture, 64

Δ decorating operator, 55

$\Delta_{p,q}$ decorating operator, 60

η hyperedge partition, 48

$\tilde{\theta}$ conjugate of a partition, 49

Λ ring of symmetric functions, 47

λ^2 partition with duplicated parts, 41

ν vertex partition, 14, 48

ρ edge permutation, 30

ρ edge-end permutation, 39

σ edge-side permutation, 39

τ vertex permutation, 30, 39

φ face permutation, 30, 39

ϕ face partition, 14

ϕ hyperface partition, 48

χ^θ character of irreducible representation of $\mathfrak{S}_n$, 48

χ^θ_λ evaluation of χ^θ on $\mathcal{C}_\lambda$, 48

Ω bijection for the Quadrangulation Conjecture, 59

$A(u,x,y,z)$ genus series for all maps in orientable surfaces, 54

a Jack parameter, 66

b indeterminate associated with β, 64

$\mathcal{C}_\lambda$ conjugacy class of $\mathfrak{S}$, 41

C_λ formal sum of elements in $\mathcal{C}_\lambda$, 50

$d_{\mathfrak{g}}(g)$ number of maps (not rooted) in an orientable surface of genus g, with $\mathfrak{g}$ as their associated graph, 15

$E(u,x,\mathbf{y},z)$ genus series, dually Eulerian maps, 58

$\mathcal{E}$ set of dually Eulerian maps, 58

$\mathcal{E}_{g,2\mathrm{j}}$ set of dually Eulerian maps of genus g, 60

$\mathcal{F}_\theta$ Ferrers graph, 48

f^θ degree of irreducible representation of $\mathfrak{S}_n$, 52

g genus, 15

$\mathfrak{g}_i$ associated graph with index number i, 15

$H(\mathbf{x},\mathbf{y},\mathbf{z},0)$ hypermap genus series, orientable surfaces, 48

$H(\mathbf{x},\mathbf{y},\mathbf{z},1)$ hypermap genus series, locally orientable surfaces, 48

$H(\mathbf{x},\mathbf{y},\mathbf{z},b)$ hypermap genus series with respect to β, 67

$H^\star_\theta(x)$ principal specialization of zonal polynomial, 52

$H_\theta(x)$ principal specialization of Schur function, 52

Index